AAPS Advances in the Pharmaceutical Sciences Series

Volume 29

AF783089

Series Editor

Yvonne Perrie, Strathclyde Institute of Pharmacy, University of Strathclyde, Bearsden, Dunbartonshire, UK

The AAPS Advances in the Pharmaceutical Sciences Series, published in partnership with the American Association of Pharmaceutical Scientists, is designed to deliver volumes authored by opinion leaders and authorities from around the globe, addressing innovations in drug research and development, and best practice for scientists and industry professionals in the pharma and biotech industries.

More information about this series at http://www.springer.com/series/8825

Henk G. Merkus · Gabriel M. H. Meesters
Wim Oostra
Editors

Particles and Nanoparticles in Pharmaceutical Products

Design, Manufacturing, Behavior and Performance

aaps®

Springer

Editors
Henk G. Merkus
Pijnacker, Zuid-Holland
The Netherlands

Wim Oostra
Hilversum, Noord-Holland
The Netherlands

Gabriel M. H. Meesters
Delft, Zuid-Holland
The Netherlands

ISSN 2210-7371 ISSN 2210-738X (electronic)
AAPS Advances in the Pharmaceutical Sciences Series
ISBN 978-3-030-06809-7 ISBN 978-3-319-94174-5 (eBook)
https://doi.org/10.1007/978-3-319-94174-5

This Springer imprint is published by the registered company Springer Nature Switzerland AG
The registered company address is: Gewerbestrasse 11, 6330 Cham, Switzerland

Preface

Pharmaceutical products occur in a wide variety of formulations, viz. powders, tablets, capsules, liposomes, extended-release formulations, crèmes, ointments, nanomedicines, coatings and theranostic particles. They are in the frontrunner group of particulate products in terms of value, both in relation to human well-being and money, despite the fact that their tonnage is relatively small in comparison to, e.g. cement and paint. Powders take a dominant position in the pharmaceutical base products, and tablets are the main final product. Formulations contain excipients in addition to the Active Pharmaceutical Ingredient (API). Both API and excipients mainly consist of particles. Their nature—particle size distribution, particle shape and morphology—is essential for adequate quality during processing (e.g. powder flowability and tablet strength) as well as for adequate functionality of the final product (e.g. efficacy, content uniformity, dissolution rate and dispersibility in air for dry powder inhalers). Pharmaceutical products based on nanoparticles are relatively new and undergo intensive R&D in view of their potential for local treatment of, e.g. cancer.

Of course, the chemical structure of the API is of prime importance for best efficacy without unwanted side effects. Therefore, much research effort is spent to find new and better APIs. Here, molecular design by computer programs shows good promise. This is, however, outside the scope of this book, which deals with the influence of particulate nature on product quality, both during processing and in the final product.

Conventionally, pharmaceutical products are manufactured following extensive research, development and testing of both intermediate products and the final product. If the final product has been proven successful, it is approved while including a description of all base materials and processing steps. This results not only in a complex and long process of testing and acquiring permission for application but also in the necessity for new tests and permission in case any process step or base material is changed. Therefore, the FDA has launched the idea of applying Quality by Design (QbD) in combination with Process Analytical Technology (PAT) to pharmaceutical products and their manufacture early in the twenty-first century to minimize the development time and costs. It means that next to clinical aspects the decision for approval, wherever possible, is based on knowledge of the material properties and adequate control of the manufacturing process.

Pharmaceutical products are very complex due to their particulate nature. Typically, any base material contains particles of different sizes, i.e. have a size distribution. Moreover, the particles are typically not spherical but have a deviating shape, e.g. angular, elongated or flakey. Then, their measured size distribution (a distribution of equivalent sphere diameters) depends on the principle of measurement and different techniques lead to different results. And finally, the particles may have a different morphology; they may be amorphous or have one or more crystal structures, which may depend on the method of preparation. Although knowledge and measurement instrumentation have greatly improved in the last decennia, this complexity makes the design and development of pharmaceutical products still a kind of art.

This book covers all the particulate aspects that are relevant for manufacture, behaviour and performance of the various types of pharmaceutical products. Also, attention is given to the potentially harmful effects of particles in medicines. Moreover, all relevant measurement techniques and their potential for in-line measurement are shortly described. Contributors to this book are acknowledged experts in the world of pharmaceutical products coming from both industry and academia. We are indebted for their contribution in which they share their experience. Their names and affiliations have been listed separately. We hope and trust that the information provided will promote optimization in design, development and quality of pharmaceutical products.

Pijnacker, The Netherlands — Henk G. Merkus
Delft, The Netherlands — Gabriel M. H. Meesters
Hilversum, The Netherlands — Wim Oostra

Contents

Editors and Contributors

About the Editors

Henk G. Merkus graduated in 1960 in physical organic chemistry at the University of Amsterdam. He worked several years at the Royal Dutch Shell Laboratories in Amsterdam on research in the field of detergents and industrial chemicals, followed by development work on thermal wax cracking for production of C_2–C_{14} olefins and on acid-catalyzed synthesis of carboxylic acids from C_3–C_6 olefins. Then, he made the change to analytical chemistry, involving both measurements and method development with a large variety of techniques and methods, first at Shell's process development department in Amsterdam and later in the chemical engineering department of Delft University of Technology. Gradually, his analytical horizon widened: first surface area and porosity measurements were added to chemical analysis, later followed by particle size analysis. In those areas he participated in many university, national and international courses and national and international education and research projects. He is the author of about 45 journal articles. He retired from Delft University in 2000 but remained active in the field of particle characterization. He is author of the book 'Particle Size Measurements—Fundamentals, Practice, Quality' (2009; Springer) and many journal articles, as well as co-author and co-editor of the books 'Particulate Products—Tailoring Properties for Optimal Performance' (2014; Springer) and 'Production, Handling and Characterization of Particulate Materials' (2016, Springer). Moreover, he participates since 1987 in standardization activities regarding particle size measurement, both in the Dutch standards organization NEN and the international ISO organization (TC24).

Gabriel M. H. Meesters has a B.Sc. and M.Sc. in Chemical Engineering with a major in Bio Process Technology from the Delft University of Technology. He has a Ph.D. in Particle Technology also from the Delft University of Technology. He worked at biotechnology companies like Gist-Brocades in The Netherlands, as well as for Genencor International and currently at DSM in research and development in

The Netherlands. In all these functions, he was working on formulation and product development. Since 1996 he holds a part-time position at the Delft University of Technology, as Assistant Professor at the Faculty of Applied Sciences, first in the Particle Technology group, later the Nanostructured Materials Group and currently in the Product and Process Engineering group. He supervised over 15 Ph.D. students and more than 50 M.Sc. students. He published over 70 refereed papers, holds around 15 patents and patent applications and is co-author and co-editor of the books 'Particulate Products—Tailoring Properties for Optimal Performance' (2014; Springer) and 'Production, Handling and Characterization of Particulate Materials' (2016, Springer). He (co-) organized several international conferences in the field of particle technology and was president of the World Congress on Particle Technology in 2010.

Wim Oostra obtained his Ph.D. in chemical engineering from Delft University of Technology. He works in Pharma since 1998. Originally, he started his career in formulation development, later moving to process development and upscaling. He currently works for Abbott's Established Products Division where he is a Sr. Technical Manager in the Manufacturing Science and Technology department. While working previously for Organon, Schering Plough and MSD, Wim was a member of the corresponding Quality by Design (QbD) and Process Analytical Technology (PAT) teams, and was actively involved in two QbD filings of new products, including the online control of Blend uniformity by Near-Infrared spectroscopy (NIR). Wim was a member of the International Society for Pharmaceutical Engineering (ISPE) Product Quality Lifecycle Implementation® (PQLI) initiative and is co-chair of the European Federation for Pharmaceutical Sciences (EuFEPS) PAT and QbD network.

Contributors

Carl Allenspach Bristol-Myers Squibb, New Brunswick, NJ, USA

Carl A. Anderson Center for Pharmaceutical Technology, Duquesne University, Pittsburgh, USA

Christopher L. Burcham Eli Lilly and Company, Lilly Corporate Center, Indianapolis, IN, USA

Carla M. Caramella Department of Drug Sciences, University of Pavia, Pavia, Italy

Carrie A. Coutant Eli Lilly and Company, Lilly Corporate Center, Indianapolis, IN, USA

Maria Cristina Bonferoni Department of Drug Sciences, University of Pavia, Pavia, Italy

Ana Filipa Cruz FFUC—Faculty of Pharmacy, Pólo das Ciências da Saúde, CNC—Center for Neuroscience and Cell Biology, University of Coimbra, Coimbra, Portugal

Anneke M. Dijkhuis-Bouwman Meppel, The Netherlands

Franca Ferrari Department of Drug Sciences, University of Pavia, Pavia, Italy

Nuno A. Fonseca CNC—Center for Neuroscience and Cell Biology, University of Coimbra, Coimbra, Portugal; TREAT U, SA—Parque Industrial de Taveiro, Coimbra, Portugal

Antonietta M. Gatti President of Health, Law and Science Association, Geneva, Switzerland

Ana C. Gregório CNC—Center for Neuroscience and Cell Biology, University of Coimbra, Coimbra, Portugal; TREAT U, SA—Parque Industrial de Taveiro, Coimbra, Portugal

David Harris Merck & Co. Inc., Kenilworth, USA

Anthony J. Hickey RTI International, Research Triangle Park, NC, USA

Jon Hilden Eli Lilly and Company, Lilly Corporate Center, Indianapolis, IN, USA

Sheila Khodadadi Delft University of Technology, Delft, Netherlands

Martin Kluenker Institute for Inorganic Chemistry and Analytical Chemistry, Johannes Gutenberg-University, Mainz, Germany

Sven Kurch Institute for Inorganic Chemistry and Analytical Chemistry, Johannes Gutenberg-University, Mainz, Germany

Jim Miesle Eli Lilly and Company, Lilly Corporate Center, Indianapolis, IN, USA

Stefano Montanari Nanodiagnostics, San Vito, MO, Italy

João Nuno Moreira FFUC—Faculty of Pharmacy, Pólo das Ciências da Saúde, CNC—Center for Neuroscience and Cell Biology, University of Coimbra, Coimbra, Portugal

Vera Moura CNC—Center for Neuroscience and Cell Biology, University of Coimbra, Coimbra, Portugal; TREAT U, SA—Parque Industrial de Taveiro, Coimbra, Portugal

Silvia Rossi Department of Drug Sciences, University of Pavia, Pavia, Italy

Giuseppina Sandri Department of Drug Sciences, University of Pavia, Pavia, Italy

Sérgio Simões FFUC—Faculty of Pharmacy, Pólo das Ciências da Saúde, CNC—Center for Neuroscience and Cell Biology, University of Coimbra, Coimbra, Portugal

Stephen D. Stamatis Eli Lilly and Company, Lilly Corporate Center, Indianapolis, IN, USA

Muhammad Nawaz Tahir Institute for Inorganic Chemistry and Analytical Chemistry, Johannes Gutenberg-University, Mainz, Germany

Peter Timmins Bristol-Myers Squibb, Moreton, Merseyside, UK; Department of Pharmacy, University of Huddersfield, Huddersfield, UK

Wolfgang Tremel Institute for Inorganic Chemistry and Analytical Chemistry, Johannes Gutenberg-University, Mainz, Germany

Natasha L. Velez Center for Pharmaceutical Technology, Duquesne University, Pittsburgh, USA

Chapter 1
Introduction

Henk G. Merkus

Abstract Most pharmaceutical products contain particles, as active ingredient (API) and/or as excipient. The nature of these particles—e.g. particle size distribution (PSD), particle shape, morphology and powder flowability—is generally essential for product quality, during processing of powders and liquids to tablets, capsules, suspensions, emulsions and ointments as well as for the quality of the final product in terms of content uniformity, efficacy and stability. In view of their complex and particulate nature, pharmaceutical products and their manufacturing process require careful design and control. The quality of the final products should meet various strict requirements for e.g. content uniformity, tablet strength and stability and viscosity of dispersions, as well as for patient safety. Most often, quality aspects relate to the particle size distribution, particle shape, specific surface area and particulate concentration of the base materials. Since determination of these aspects is relatively easy, optimum relevant characteristic parameters are laid down in the form of specifications. In addition, some performance aspects of powders and dispersions, such as flow and viscosity behavior, are usually specified. This chapter describes the properties and characteristic features of particles, powders and dispersions and their relevance to pharmaceutical products. Also, it gives an overview of the main measurement methods that are relevant for pharmaceutical products. At the end of this chapter, the contents of this book are summarized.

1.1 Objective of This Book

Most pharmaceutical products contain particles, both as active ingredient (API) and as excipient in formulations in the form of e.g. binder, filler, disintegrant, stabilizer, lubricant or colorant. The particles may be solid or liquid, occurring as powders in capsules or in tablets or dispersed in an immiscible liquid in the form of suspensions

H. G. Merkus (✉)
Retired Associate Professor, Delft University of Technology, Park Berkenoord 30,
2641 CZ Pijnacker, The Netherlands
e-mail: henkmerkus@hetnet.nl

H. G. Merkus et al. (eds.), *Particles and Nanoparticles in Pharmaceutical Products*,
AAPS Advances in the Pharmaceutical Sciences Series 29,
https://doi.org/10.1007/978-3-319-94174-5_1

or emulsions or in air as sprays. Second to the chemical nature of the active ingredient, the nature of all of these particles—e.g. particle size distribution (PSD), particle shape, morphology and powder flowability—is essential for product quality, both during processing and for the final quality.

The objective of this book is to highlight the role of particulate nature in the wide variety of pharmaceutical products, with focus on the very large range of particle sizes involved, as well as to give guidance in the choices of particle and powder characteristics and their measurement for setting product specifications, which have an optimum relationship with product quality. A further objective is to promote the approach of Quality by Design (QbD) and Process Analytical Technology (PAT) for pharmaceutical products and processes. Expert authors have contributed to the chapters on different types of pharmaceutical products, for which we are very grateful to them.

In the next sections of this introductory chapter, particle and powder characteristics, measurement methods and their relevance for product manufacture, performance and quality are described.

1.2 Pharmaceutical Products

Pharmaceutical products occur in many forms, viz. powders, capsules, tablets, solutions, emulsions, suspensions, ointments and sprays. Their composition may be very complex. The Active Pharmaceutical Ingredient (API) is the heart of the product. It should have optimum pharmaceutical activity for curing a specified disease in given surroundings with least negative side effects. It usually is a solid material, often crystalline, having a complex organic chemical structure. Among other things, adequate bioavailability requires that the substance is sufficiently soluble in the surrounding liquid. Following synthesis and purification, the active ingredient is typically combined with other components, each with its specific function, in a pharmaceutical formulation. These formulations may contain:

- binders, to facilitate tabletting and granulation
- lubricants or glidants, to aid flow of the powder
- fillers, to give tablets the desired size, API concentration and strength
- disintegrants, to facilitate break-up and dispersion of tablets into smaller particles for quick dissolution when in contact with water
- polymer, to form a matrix for extended release of the drug
- particle coatings, to promote liberation of the API in the desired region of the gastro-intestinal system, to promote extended release and/or to mask an unwanted taste
- liquid, to form suspensions or emulsions
- surfactants, to facilitate wetting of powders
- dispersants, to stabilize particle dispersions in liquids
- antimicrobial preservatives

- antioxidants
- colorants, to give an attractive and distinctive color to tablets
- flavoring agents, to mask bad tastes and increase patient acceptance
- viscosity modifiers, to adjust the viscosity of ointments.

In addition to the pharmaceutical products for curing diseases, there are the particulate theranostics. These are particles that are useful for diagnosis of the presence of a specific, often localized disease and its therapeutic treatise (see Chap. 3).

The majority of pharmaceutical products consists of tablets and capsules. Typically, the production process of these products starts with dry powders that are stored in hoppers and must be blended to the desired, constant composition, prior to die filling for compaction into tablets or introduction into capsules. Adequate, steady flow behavior of the mixture at the exit of the blender is essential to reach a constant quality for tablets or capsules. Most often, either of three blending processes is applied:

1. Dry blending. The API and the excipients are mixed as powders and fed into capsules or into separate dies where they are compressed to tablets. The blending process can be executed in batch or continuous equipment [106].
2. Dry granulation. Here, the various powders are mixed in the dry state and immediately and continuously squeezed to form a thin ‘ribbon’, which is then cut into pieces [24]. Compacts, having sufficient strength, require adequate binders and lubricants. They can be produced in roller compactors.
3. Wet granulation. Here, some liquid is sprayed over the powders while they are mixed, which causes agglomeration of particles. In this process, particles having a hydrophobic surface require addition of a suitable surfactant to promote adequate wetting. There are two basic designs for granulators. Mixer granulators can be applied when small forces are necessary for mixing and granulation. They typically result in near-spherical mixed particles. When strong shear forces are required, then impellers in the granulator cause mixing as well as partial breakage of the lumps formed and, thus, re-mixing and densification of the final agglomerated product [24]. Following drying, the usually porous granules can be used as such or be milled to some extent and then pressed into more compact tablets.

Dry processes are favorable as they avoid the use of a liquid and related extra processing steps. However, they are very sensitive to adequate flow behavior of the particles and to segregation, and, thus, require careful design and control to reach a homogeneous composition of the product. This sensitivity to particle flow behavior is much less in wet granulation due to the increased cohesivity of wetted particles and the intermediate break-ups of lumps, followed by re-agglomeration, which improves the mixing process. On the other hand, at least an additional drying step is required to reach the final product.

Typically, particle size applied for manufacture of tablets is in the range of 10–300 μm. But particle sizes in pharmaceutical products can be much smaller. For example, medical inhalers require an aerodynamic size of about 1–5 μm for particles to reach the lungs, (micro-) emulsions contain liquid and/or solid particles in a size range of about 10 nm–100 μm (the exact range depending upon way of preparation and type of application) and nanoparticles have a size range of about 1–100 nm. These latter particles are subject of much research directed towards diagnostic purposes (e.g. for cancer cells), antibacterial applications, or as carriers for delivery of substrates (e.g. drug molecules) to specific targets. The advantages of nanoparticles are their large surface area for encapsulation to enhance stability, their potential for good access to specific cells and tissues, and their possibility for being triggered from the outside or for acting as contrast agent. This offers possibilities to target drug delivery to diseased cells and tissues, and to regulate drug release, while minimizing side effects and improving therapeutic efficacy.

The complex nature of pharmaceutical products requires extensive research and development of all composing materials, their processing and process control, designed to reach optimum, constant product quality. This complex nature and the long process of R&D, testing and acquiring permission for application ask for a good design and planning of activities. Various design methods exist. In Chap. 2, the method of Quality by Design (QbD; for products and processes) in a Process Analytical Technological (PAT; for process monitoring and control) framework is discussed. This approach has been launched in the first decennium of the 21st century by the US Food and Drug Administration (FDA) for the development and manufacture of pharmaceutical products and adopted by its European counterpart EMA and other national regulatory agencies. It promotes that the desired good quality of pharmaceutical products in combination with an improved efficiency of its manufacturing process be reached by design on the basis of (available) knowledge and understanding of the product, its processing steps and their control, rather than by trial and error and analysis of the final product alone. In an earlier book, I described the basic steps in a general method with similar goals named Quality Function Deployment (QFD) or House of Quality (HoQ), which also can be applied for both products and processes [55, 59].

Both methods aim at optimum product quality at minimum costs and time. Their guidelines agree on the importance of defining explicit targets for the desired product (a ***Quality Target Product Profile***), collecting all relevant information, selecting the ***critical quality attributes*** of final products, components and manufacturing process steps and of adequate reporting. However, also differences exist. For example, the QFD/HoQ approach also emphasizes the necessity of making a realistic planning scheme for all activities, including research and development and decision moments, whereas the QbD/PAT approach is more specific on the quality attributes for the drug products, their components and their manufacturing process steps as well as on process monitoring and control issues for application by regulatory agencies.

1.3 Relevance of Particle and Powder Characteristics for Product Quality

Product quality can be defined as '*The totality of features and characteristics of a product that bear on its ability to satisfy stated or implied expectations, needs and requirements in exchange for monetary considerations*'. A more simple definition is '*Conformance to product specifications*'. This simple definition implies, of course, that all product performance properties are covered by the specifications.

The characteristic features of the final product typically do not only depend upon the characteristics of the active material but also upon those of its components and the processing steps. Important quality attributes for pharmaceutical products and components are:

- pharmaceutical efficacy and potential side effects, in relation to:
  - content uniformity in dose quantities
  - product stability in dispersions
  - dissolution properties of solid particles, without or with a coating
  - dispersability of the product in air
  - rheological properties of emulsions, suspensions and ointments
  - safety for patients.

Additional characteristics are often to be addressed in relation to product processing steps, for example

- flowability of powders
- segregation behavior of the particles in powders during transport and storage
- tendency of particles to form agglomerates
- ability to compact particles into tablets having sufficient strength
- dusting behavior during transport and processing, which may lead to health hazards
- safety, toxicity and explosion sensitivity when processing
- ability to sufficiently break up agglomerates or aggregates
- vulnerability to breakage and attrition of particles and tablets
- potential to optimize API particle shape during production
- drying rate of solid particles coming from liquid dispersions
- influence of moisture in these behavioral aspects
- absence of unwanted pollutants.

In relation to performance characteristics, component and product specifications must be set, process equipment chosen and process conditions defined in order to make the products conform to the specified quality. First, requirements are set for the performance properties that are listed above. Many of them are subsequently transformed into specifications for particles, powders and dispersions, as well as into process conditions. These specifications require availability of adequate analytical methods and techniques. Typically, a variety of specifications is necessary since different performance aspects relate to different characteristics of particles, powders and dispersions. Note that, sometimes, different performance characteristics ask for contradictory particle characteristics. Then, an optimum has to be found. As stated in Sect. 1.2, often additional, non-particulate components are necessary to reach optimum product quality, such as particle coatings, surfactants, dispersants and viscosity modifiers.

1.4 Particle, Powder and Liquid Dispersion Characteristics [54–56, 58]

Relevant performance characteristics of products and their processing may be related to chemical composition as well as to characteristics of the individual particles, of the powders and of the liquid dispersions. An overview of identifiers follows below. Measurement methods and techniques are described in Sect. 1.5.

1.4.1 Composition

Adequate product composition is of prime importance for pharmaceutical products. This holds primarily for the active ingredient (API) where each dose should contain the targeted amount within narrow limits. It requires that the composing components are well mixed and segregation in the manufacturing process is suppressed. But also the concentration of unwanted impurities and contaminants below a stated level should be guaranteed.

The concentration of API in a dose is often determined through infrared (IR, in the form of Fourier transform, FTIR, or of attenuated total reflection, ATR-IR), or Raman spectroscopy, or by dissolving samples followed by liquid chromatography (HPLC), ultraviolet (UV) spectrometry or mass spectrometry (MS).

A fairly new method for control and monitoring of pharmaceutical processes and for analysis of tablets uses near-infrared (NIR) analysis through application of chemometric data analysis and multivariate calibration. A major advantage of this technique is that the analysis no longer requires dissolution but can be done directly on powders and tablets. In addition to API concentration, typically also that of other components can be measured. Another new technique for determination of e.g. the content and dispersion of water and various APIs, the morphology of crystals as well as thickness of the coating layer of tablets is terahertz (THz) spectroscopy. An in-line application for this latter process has been reported by May et al. [51].

Presence, identity and concentration of metallic elements and anions in powders and solutions can be measured by energy-dispersive spectroscopy (EDS), atomic absorption spectroscopy (AAS) or atomic emission spectroscopy (AES).

See Sect. 1.5.3 for further explanation of the analysis techniques.

1.4.2 Particle Size and Particle Size Distribution (PSD)

Particle size is an important parameter for the properties of the final product as well as in the unit operations during its production and handling. For example, it has a strong influence on solubility, dissolution rate, flowability of powders and the intrinsic stability of suspensions. It has the advantage that many different techniques and instruments are available for easy measurement. Moreover, milling techniques are available for size reduction, and specific methods for particle preparation in the nanometer size range.

Homogeneous spheres are the ideal particles for characterization since their size can be described by a single parameter, the diameter (or, in some areas of scientific work, the radius). Surface area and volume of a sphere can be easily calculated from its diameter D through πD^2 and $1/6\pi D^3$, respectively; specific surface area is $6/D\rho$ (where ρ is particle density). Moreover, all particle sizing techniques yield the same size result, regardless of the measurement principle. In the world of particulate materials, however, spherical particles are very rare in comparison to other shapes.

Particles having a regular but non-spherical shape occur more often, especially at sizes below 1 μm, but are still fairly rare for larger sizes; e.g. crystals may have a regular shape, like cubic sugar crystals. Here, full description of size requires at least the size of the ribs and the angles between them, or any other typical feature. Also, calculation of surface area and volume is more difficult.

Particles having an irregular shape are the normal case for particulate materials, especially in the above-micrometre size range. Now, adequate characterization of individual particles requires many descriptors, such as Feret diameter, length, breadth, thickness and angularity. Unfortunately, many descriptors are usually dependent on particle size as well as on particle orientation during measurement. For full description of collections of non-spherical particles having different sizes and shapes this would lead to an excessively large number of descriptors, which would make setting relationships with product properties impossible.

In order to avoid such excessive numbers and to find relation to measurement techniques, the concept of equivalent spheres has been introduced. Here, an arbitrary particle and its equivalent sphere have the same property in relation to a given measurement principle. Again, the diameter of the equivalent sphere characterizes the size[1] of the particle. The nice consequence of this concept is that an arbitrary particle can still be characterized by a single parameter, the equivalent sphere diameter. Note, however, that such diameters often differ when different measurement principles are applied. This holds especially when length, breadth and thickness of the particle are very different and different particle cross sections are offered in instruments for measurement. It must be clear that also conversion of these diameters to surface area and volume may yield different results, and vice versa. Because the characterization as equivalent spheres includes both different sizes and different shapes in a distribution, derivation of particle volume from

[1]I have chosen symbol D to represent particle size (equivalent diameter).

equivalent size becomes especially problematic for e.g. acicular, fibrous and flaky particles, for which length, breadth and thickness are largely different. Then, a decision has to be made for best possibilities to adequately describe the particle characteristics with respect to performance.

Some examples of equivalent sphere diameters are:

- Equivalent projected area diameter, i.e. diameter of a circle having the same area as the particle's projection
- Equivalent surface area diameter, i.e. diameter of a sphere having the same surface area as the particle
- Equivalent volume diameter, i.e. diameter of a sphere having the same volume as the particle
- Equivalent sieve diameter, i.e. diameter of a particle that just passes through the apertures of a sieving medium
- Stokes' diameter or equivalent settling diameter, i.e. diameter of a sphere having the same settling rate as the particle under conditions of Stokes' law (low Reynolds number)
- Equivalent laser diffraction diameter, i.e. diameter of a sphere having the same scattering pattern as the particle
- Aerodynamic diameter, i.e. diameter of a sphere with density 1000 kg/m^3 having the same aerodynamic property as the particle (especially used for aerosols).

The potential differences between the different equivalent sphere diameters necessitates that their measurement technique is stated in the measurement results.

Typically, the particles in particulate materials do not have the same size but show a distribution of different sizes. These distributions are usually presented in the form of numbers or volumes of particles in given size classes, in relation to the measurement method. In graphical form the histogram curves are usually smoothened to continuous, differential or cumulative, curves [32, 54]. An example is given in Fig. 1.1.

The figure clearly illustrates that presence of small amounts of e.g. large particles in a distribution, such as agglomerates, is more easily visible in the differential presentation where they may appear as shoulders on the main peak, or even as a separate peak. On the other hand, PSD characteristics relating to the amounts of material smaller or greater than a stated size and quantile values such as the D_{10}, D_{50} and D_{90} are more easily derived from the cumulative curve.

A point of attention is that different size measurement techniques yield the amount of particles in the size classes in a different manner. For example, particles are recorded by number in e.g. microscopy, whereas e.g. laser diffraction reports them by volume. This also may lead, depending upon PSD width, to considerable differences as is shown in Fig. 1.2.

For conversion of one distribution into the other, the volume of the number of particles in each size class can be calculated by assuming spheres, when errors due to particle shape are acceptable. It will be clear that small volumes of particles in the

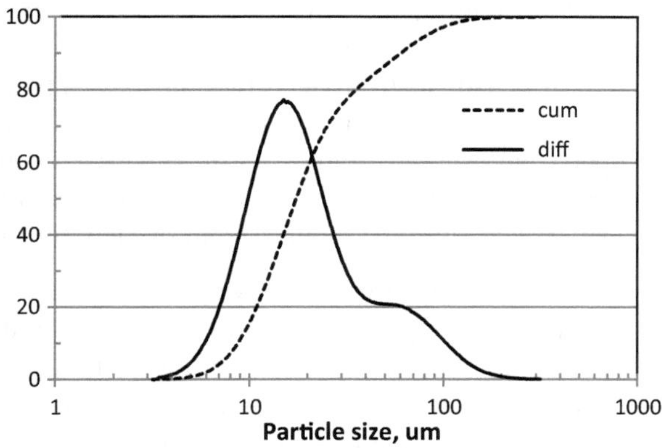

Fig. 1.1 Example of a size distribution as differential and cumulative curve

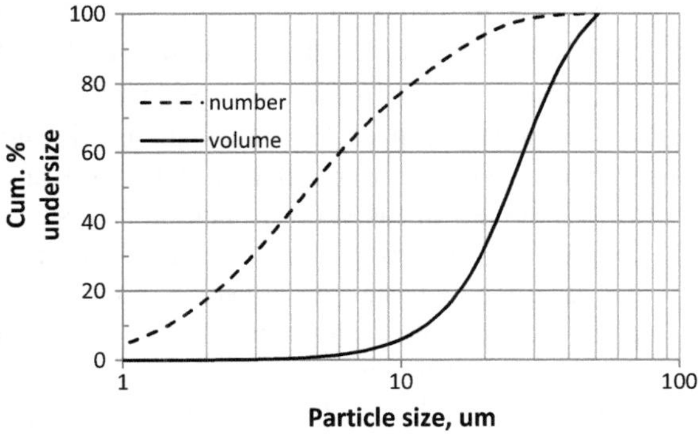

Fig. 1.2 Differences between a number-based and a volume-based size distribution

low size end of a distribution relate to a large number and are more easily visible in number-based distributions, whereas small numbers of large particles, corresponding to a large volume, are more easily seen in volume-based distributions. In relation, the uncertainty in measured results at both ends of the size distribution after transformation from number to volume or vice versa may be easily underestimated. Since the relative errors in the measured percentages remain the same in these transformations, this holds especially if the amount of sample in terms of total number or mass of particles is not considered. We take Fig. 1.2 as the example, while assuming perfect spheres. In case a microscopic measurement would include few, e.g. 500 particles in total, about 5% n/n above 20 μm would relate to 25 particles This corresponds to a relative standard deviation (RSD) of 20%, which would also mean 20% RSD in about 70% v/v of the upper side of the distribution, i.e. a RSD of about 14% v/v. The same can be said when a volume- (or mass-) based technique would only measure the 5% v/v of particles below 8 μm with a standard deviation of 1% v/v, or 20% RSD. When this 5% v/v corresponds to about 70% n/n, this 20% relative would mean a RSD of about 14% n/n for the lower size part. The differences between number- and volume-based distributions must especially be considered if an instrument is only capable of measuring part of the total size range of the particles.

Industrial particulate products typically have a fairly wide size distribution that is composed of amounts of material of many sizes. Therefore, it is usually required to compress the data into some kind of summary for an easier translation to quality parameters. For such a summarized representation, following PSD parameters seem logical choices from the product point of view for the cases of equant particles:

- Mean size of the distribution, weighted according to number, area, volume, etc., depending on application and theoretical background. The mean values have the advantage that the contribution of all particles to the performance is taken into account. For example, the arithmetic, number-weighted mean size $(\langle D_{1,0} \rangle)$ is important e.g. in number-based health effects. The surface area-weighted Sauter mean diameter $(\langle D_{3,2} \rangle)$ is important in the rate of dissolution and evaporation as well as in explosion behavior. The mean volume diameter $(\langle D_{3,0} \rangle)$ is important for e.g. dose effects. Content uniformity has been found to relate to the volume-weighted mean volume diameter $(\langle D_{6,3} \rangle)$ (see Chap. 7).
- Size of the largest particle if this value is relevant to product quality. Often, the fraction of particles larger than a stated size or the quantile value D_{90} is used instead, in view of easier determination.[2] This parameter is important when large particles have a strong impact on dose uniformity or when they are harmful, such as in e.g. pastes or polishing powders, which should have a smooth appearance and should not cause scratches. As shown above, here the $D_{90;3}$ is a good choice for precision.
- Fraction of particles smaller than a stated size (e.g. 45 μm) or the quantile value D_{10} if these smaller particles are relevant to product quality.[2] This parameter may be important for e.g. filtration, powder dusting or powder flow properties. As shown above, here the $D_{10;0}$ is a good choice for precision.
- The width of the size distribution, in addition to one or some PSD parameters chosen, expressed as ratio D_{90}/D_{10}, D_{84}/D_{16} or the width parameter of a modeled distribution (see Footnote 2), such as the log-normal or Rosin-Rammler distribution. This parameter may be important for use in process control when narrow size distributions are desired.
- Stated volume fractions in given size classes of the distribution, e.g. to minimize voids in compacts such as tablets.

The choice of PSD parameters for representation of the distribution is far from arbitrary. Figure 1.3 and Table 1.1 illustrate, as an example, similarities and differences between three types of volume-based size distributions and some of their characteristic parameters. Distribution #1 is a monomodal, lognormal distribution

[2]In all cases an indication in the form of a second subscript has to be given for the type of weighting of the value, viz. by number (0), area (2) or volume (3), etc. and for the measurement technique.

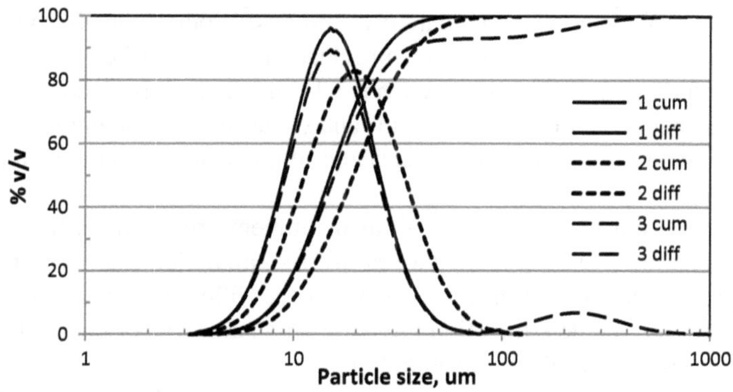

Fig. 1.3 Three cumulative and differential volumetric size distributions: #1 = distributions of $D_{50;3}$ = 15 μm and $s_g = 1.6$; #2 = mixture of $D_{50;3}$ = 15 μm and $s_g = 1.6$ (50%) and $D_{50;3}$ = 25 μm and $s_g = 1.6$ (50%); #3 = mixture of $D_{50;3}$ = 15 μm and $s_g = 1.6$ (93%) and $D_{50;3}$ = 225 μm and $s_g = 1.6$ (7%)

around a geometric mean size of 15 μm. Distribution #2 is a 50/50 combination of #1 with a close and similar neighbor (geometric mean size 25 μm); it looks still monomodal. Distribution #3 is a combination of #1 with a small amount (7% v/v) of much larger particles (geometric mean size 225 μm) and is clearly bimodal.

The changes in the distribution of #2 in comparison with #1 are visible in all PSD parameters, although differences are more pronounced for the parameters that relate to the higher size end (where the change occurred). For these monomodal distributions, mean values and statistical parameters show similar trends. However, for bimodal mixture #3, containing a small amount of large particles, the mean values $\langle D_{6,3}\rangle$, $\langle D_{4,3}\rangle$, $\langle D_{3,2}\rangle$ and $\langle D_{3,0}\rangle$ give a much better indication than the quantile PSD parameters $D_{50;3}$ and $D_{90;3}$. The $D_{95;3}$ also indicates the difference between #1 and #3 well; the reason is that it lies in the PSD region of the 7% large particles. Note, however, that the precision of measured $D_{95;3}$ values is much worse than that of mean sizes since it involves only few particles.

In the application of PSD parameters, two types must be discriminated, viz. application for instrument testing and representation for properties of product performance and process control.

Certified standard reference materials for ***instrument testing*** are usually mono-modal (see also Chap. 2). Their values have been certified by governmental institutes or commercial manufacturers with a stated accuracy. Most often, precision, accuracy and stability of instruments are tested through repeated measurements of D_{10}, D_{50} and D_{90}, which can be derived directly from the cumulative distributions. Testing results can be set for monitoring and control over long periods of time in quality control charts, such as the as the Shewhart chart (cf. Sect. 2.5) [54]. These reference materials are often quite different from own products and, thus, the methods for sampling and dispersion may be quite different. Instead of certified standard reference materials, a standard own in-house product can also be taken as a reference material within a company, provided that it is sufficiently stable, and reference values and their tolerances have been characterized adequately. The advantage is that the influence of the full analytical method (sampling, dispersion and dilution in addition to the instrument) and of the analyst can be included in the quality testing.

If also resolution and sensitivity of an instrument are at stake, then reference materials containing two or more modes are required. For testing of resolution, typically peak width of and distances between the differential peaks are checked, and for sensitivity, the determination limits for small amounts of material as well as sensitivity to changes in their concentration.

For relationships of PSD parameters with ***product performance*** and ***process control*** it is important to know whether the size distribution of the product is monomodal or contains more modes. Monomodal distributions are the normal case in industrial particulate products. Sometimes, the distributions are reported as parameters of a model distribution. For example, those of normal (Gaussian) distributions for narrow size distributions, which relates to small deviations during preparation. For wider distributions, the parameters of lognormal, Rosin-Rammler or other model distributions are sometimes applied. The advantage of these model

Table 1.1 Some characteristic parameters of the size distributions presented in Fig. 1.3

Parameter	#1 $D_{50;3} = 15$; $s_g = 1.6$	#2 50% $D_{50;3} = 15$; $s_g = 1.6$ + 50% $D_{50;3} = 25$; $s_g = 1.6$	#3 93% $D_{50;3} = 15$; $s_g = 1.6$ + 7% $D_{50;3} = 225$; $s_g = 1.6$
$D_{10;0}$ (μm)	4.6	5.4	4.6
$D_{50;0}$ (μm)	7.9	10.2	8.2
$D_{90;0}$ (μm)	14.1	20.1	18.6
$D_{95;0}$ (μm)	16.7	24.2	91.8
$D_{10;3}$ (μm)	8.2	9.7	8.4
$D_{50;3}$ (μm)	15.0	19.3	15.7
$D_{90;3}$ (μm)	27.4	38.5	35.7
$D_{95;3}$ (μm)	32.5	46.5	172.3
$\langle D_{1,0} \rangle$ (μm)	8.8	11.7	17.5
$\langle D_{3,0} \rangle$ (μm)	10.9	15.0	67.3
$\langle D_{3,2} \rangle$ (μm)	13.5	19.1	191.1
$\langle D_{4,3} \rangle$ (μm)	16.7	23.3	249.8
$\langle D_{6,3} \rangle$ (μm)	20.7	26.9	310.0
$D_{90;3}/D_{10;3}$	3.3	4.0	4.3
>50 μm (%v/v)	0.5	3.6	7.4
>80 μm (%v/v)	0.0	0.3	6.2

distributions is that they can be characterized by two parameters, a size location parameter and a distribution width parameter [54]. A drawback is that the model most often corresponds with only a part of the actual distribution. Still, if a proper model is chosen in relation to the size region of interest, its parameters can be attractive for process control.

Reasons for extra modes may be intentional e.g. by additions in a given size class to improve product quality, or unintentional by abrasion causing the presence of fines, or by agglomeration of particles. Sometimes the different modes are even clearly visible in the analytical results or microscopic images. Alderliesten recently described a program for fitting distributions to combinations of lognormal distributions [2].

For selection of appropriate PSD parameters in relation to product performance, it is important to know in which performance aspects all particles play a role and in which aspects smaller or larger particles have a more dominant influence. In the general case of product specification, it deserves preference to use some weighted mean size that can be reasoned from theory or has been found empirically to relate most significantly to a performance property, and can be measured with sufficient precision. For monomodal size distributions also an appropriate quantile parameter might be used, as can be seen in the above discussion on the three PSDs in Fig. 1.3 and Table 1.1. However, caution is required since quantile parameters have a statistical nature and do not have an intrinsic relationship with product performance properties. Moreover, the median size D_{50} alone gives no indication of the distribution width, whereas size changes in the lower and upper 10% of the PSD will not show up in the D_{10} and D_{90}, respectively, as is shown in the difference between $D_{90;3}$ and $D_{95;3}$ for sample #3.

Parameters applied for process control require above all good precision, stability and sensitivity for process changes. Here, the quantile values D_{10}, D_{50} and D_{90} and the ratio D_{90}/D_{10} usually suffice.

For best understanding of the behavior of products containing polymodal size distributions, it can be beneficial for understanding to fit the distribution to a combination of lognormal distributions and use the corresponding parameters.

In general, the choice of PSD parameters for representation of product performance should be based on theoretical considerations, literature or sound investigation as well as on best measurability to reach optimum correlation. Their measurement should be done through using an optimized, written measurement method, including sampling and dispersion—i.e. a standard operating procedure (SOP)—in instruments having optimum quality for the defined parameters. An overview of techniques is presented in Sect. 1.5.4.4 and [54]. Note that different product properties may require different PSD parameters, which may have a contradictory correlation with particle size. On the other hand, the number of parameters taken for product specification should be minimized in view of measurement costs.

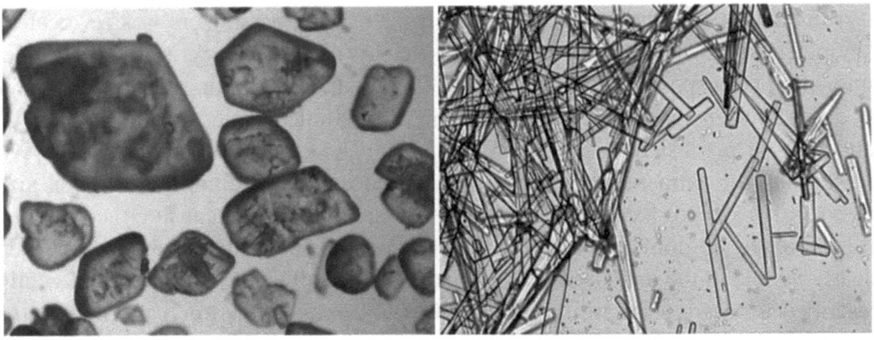

Fig. 1.4 Images of Succinic acid particles (left) and Almorexant. HCl needles (right) [105]; copyright Springer, reproduced with permission

1.4.3 Particle Shape

Although particle size is usually given first attention in relation to the properties of particulate materials, particle shape can have a significant or even dominant effect on product performance. This is especially the case for needles and flakes, having a small aspect ratio, i.e. large length to breadth ratio, which makes them vulnerable to breakage and gives them very poor flow properties, in contrast to rounded equant particles. Two examples of particle shapes are given in Fig. 1.4.

Conceptually, particle shape is the pattern of all points on the boundary of a particle [54]. Thus, it includes every aspect of external morphology of the particle. Three scales can be discriminated, viz.:

- macroscale, related to the general 3-dimensional form of particles, for example the ratio of their main three dimensions
- mesoscale, which relates to the general aspects of roundness and angularity of the particle's contour
- microscale, which involves surface roughness and porosity.

Often, only the dominant particle shape in a collection of particles is described in a qualitative manner in addition to the PSD [80]. This is due to the fact that the description of both size and shape distributions yields too many parameters for application at the present time. An example of such qualitative shape descriptors is given in Fig. 1.5.

Automated microscopy techniques (light, SEM, TEM) in combination with image analysis offer good possibilities for quantitative measurement (see Sect. 1.5.4.5).

By looking at the vast differences between some of the above shapes and spheres it will be clear that application of the equivalent sphere concept in some cases can be of little value to a proper description of the particle. This is the more true when a preferred particle projection is measured by the sizing technique. Then, a choice is

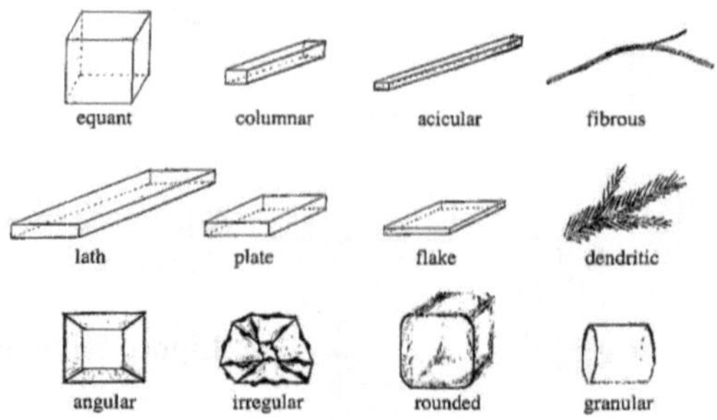

Fig. 1.5 Qualitative descriptors for particle shape [54]; copyright Springer, reproduced with permission

to be made for using e.g. (distributions of) particle length or breadth or aspect ratio to represent the particles instead of or in combination with equivalent sphere diameter.

1.4.4 Particle Morphology

The morphology of API particles is also important with regard to product quality, both during processing and of the final product.

Simple solid particles can be amorphous or crystalline. Moreover, many materials can crystallize in different crystal structures (polymorphs) depending upon the type and conditions of the crystallization process. The degree of crystallinity and the crystal structure of a solid material are important for two reasons: (a) solubility and dissolution rate may be different, (b) different crystal structures often lead to different particle shapes and different strengths with respect to breakage and attrition.

For complex particles, however, there is more. Liposomes containing a particulate pharmaceutical product exemplify such complex product (see Fig. 1.6). Complex liposomes may contain:

- an aqueous core that can contain an API
- one or more (phospho-) lipid bilayers containing hydrophilic heads and hydrophobic tails, which restrain the transport of pharmaceutical product from the core to the outside and may also contain API for direct dissolution and specific colorants or elements for imaging
- a protective layer of a polyethylene-glycol derivative, to which specific targeting agents may have been added.

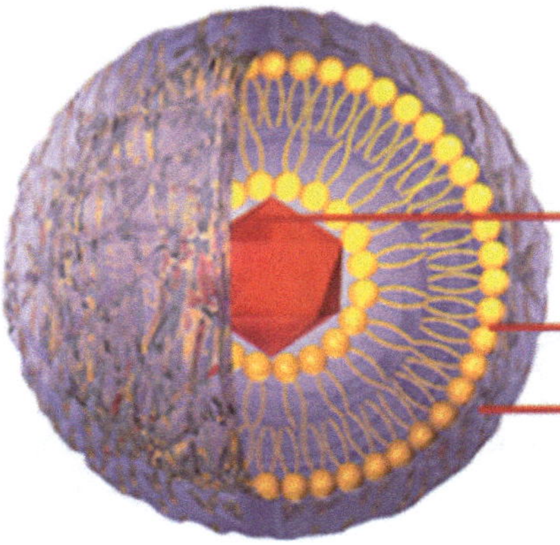

Fig. 1.6 Schematic composition of a pharmaceutical liposome

Their size is typically in the order of 100 nm, leaving a maximum size of about 80 nm for the core. It should be noted that crystalline pharmaceuticals for which the length exceeds the size of the core may change the spherical form of the liposome, or even disrupt it. This may have severe consequences its release rate. Cryogenic transmission electron microscopy (cryo-TEM), where the samples are quickly and deeply frozen before analysis, seems to be the only technique for characterization to a resolution of about 1 nm. Its rapid development in the last decades has enabled detailed characterization of these complex nanostructures.

Crystallinity and crystal type are usually determined by means of X-ray diffraction and differential scanning calorimetry (DSC), crystal shape by microscopy (optical, SEM or TEM). Infrared, (Terahertz-) Raman or Terahertz spectroscopy can be applied to identify the structure. See further Sects. 1.5.4.2 and 1.5.4.5.

1.4.5 Particle Density and Porosity

The density of particles is important as it relates volume to mass. Pharmaceutical particles are usually non-porous; mesoporous (nano-) particles are the exception. Then, particle density means skeleton density. For porous particles, the 'effective' density is smaller than the true skeleton density if closed pores are present, since their volume is incorporated in the measured density. This is also the case for the 'apparent' density coming from measurements where some part of the open pores has not been entered by the liquid used. Porosity of particles is important when sorption of components from the surrounding medium (gas or liquid) plays a role as it greatly increases particulate surface area. It can also influence PSD analysis, e.g. in laser diffraction, where scattering at the pore walls contributes to the total scattering pattern, which leads to the artificial presence of small particles.

The porosity of a pharmaceutical tablet usually mainly consists of voids in between the particles, but also porosity of the particles themselves, if present, contributes. Porosity of tablets can be important since it may facilitate, for example, access of water for easy disintegration.

Pore sizes in particles are classified in three categories, viz.:

- *macropores*, with pore diameters larger than 50 nm
- *mesopores*, with pore diameters in the range 2–50 nm
- *micropores*, with pore diameters smaller than 2 nm.

Voids in between particles in a powder or in agglomerates depend upon particle size and are generally larger than 50 nm. Thus, they may overlap with macropores.

1.4.6 Surface Area

The (external) surface area of particles is important in relation to the dissolution rate of solids and evaporation properties of droplets as well as for stabilization of suspended particles and for explosion behavior of air dispersed dry particles or small droplets. Note that dust and mist explosions require very little ignition energy if particle size drops below about 50 μm and may have very severe consequences of casualties and destruction of equipment [47, 49].

When particles or tablets contain accessible pores, then the total surface area often is greatly magnified. Surface area can be directly measured (see Sect. 1.5.4.6) or be derived from the particle size distribution. Note that the latter calculation only yields the external particle surface area, since the area of pores is not taken into regard. Moreover, it only gives a rough estimate for non-spherical particles, since equivalent sphere diameters are used in the calculation.

The internal surface area is relevant for cases where adsorption properties are required.

1.4.7 Wettability

The wetting properties of particulate surfaces—of both particles and compacts—determines to what extent a surface is wetted by a liquid. This property is important since adequate wetting of powders is required in wet granulation and wetting of the final product precedes its dissolution. Wetting occurs when the surface energy of the solid (S) exposed to the atmosphere (V) is greater than that of the liquid (L), that is if the surface (interface) tension $\gamma_{SV} > \gamma_{SL} > \gamma_{LV}$. The driving force for wetting is $\gamma_{SV} - \gamma_{SL}$. The resisting force is the energy required for increasing the surface area of the liquid drop. The Young-Dupré equation presents the balance of these forces [58]:

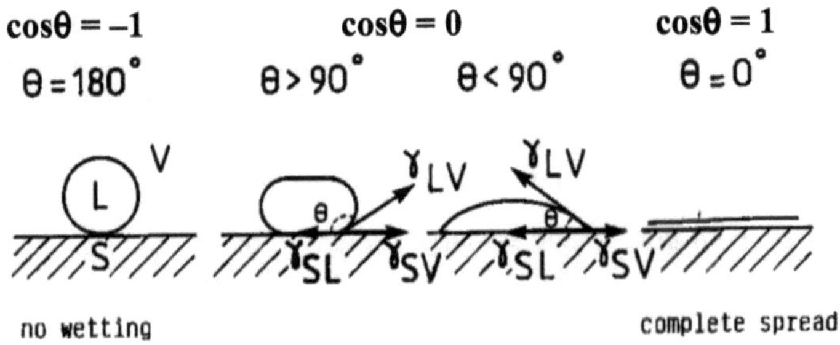

Fig. 1.7 Spreading of liquid on a solid surface [58]; copyright Springer, reproduced with permission

$$\gamma_{SV} - \gamma_{SL} = \gamma_{LV} \cdot \cos\theta \tag{1.1}$$

As indicated in Fig. 1.7, no wetting occurs when the contact angle θ is 180° ($\cos\theta = -1$). Neither is the liquid capable to penetrate into inter- or intra-particle pores. Wetting starts to occur when the driving force approaches zero, that is near a contact angle $\theta = 90°$ and $\cos\theta = 0$. At still smaller contact angles wetting is reasonable, at $\theta = 0°$ and $\cos\theta = 1$, wetting is complete. Only at small contact angles, liquid will be able to penetrate into the voids in between the particles of a compact and into the pores of particles.

The range of surface tensions of liquids and solids is about 20–70 mN/m (20–70 dyne/cm). Polar solids and liquids, such as water, exhibit high values, non-polar materials low values. It will be clear that a liquid can only wet a solid surface if its surface tension is smaller than that of the solid. Thus, water cannot wet surfaces of low or medium polarity.

Surfactants (surface active agents), containing a long non-polar chain and a polar head, can provide a solution to such wetting challenges, since they reduce the surface tension of water. Thus, they facilitate wetting of solid surfaces as well as penetration of water into the voids and pores of dry powders and compacts.

Surfactants are also applied for stabilization of emulsion droplets. Here, they form a stabilizing layer in between a hydrophobic surface and a hydrophilic surrounding, or vice versa.

1.4.8 Zeta Potential

The zeta (ζ) potential of particles dispersed in an ionic liquid such as water is the electric charge that exists at the surface of shear of the particles; it includes counter ions and some molecules of the dispersion liquid [34, 55]. Its value is an essential

Table 1.2 Zeta potential and stability of colloidal dispersions

Zeta-potential absolute value (mV)	Stability behavior
<10	Rapid coagulation/flocculation
10–30	Instable
30–60	Moderate/good stability
>60	Excellent stability

parameter for the stability of suspensions and emulsions, in relation to absence or presence of electrostatic repulsion between the particles (see Table 1.2). The pH of the dispersion liquid may have a dominant influence on the zeta potential as it influences the degree of ionization of oxidic groups at the particle surface. In addition, the zeta potential is influenced by the type and concentration of ionized electrolyte molecules in the solution.

1.4.9 Powder Packing, Density and Porosity

Powder packing, density and void fraction describe the bulk properties of a powder in rest. They have a close relationship and relate to the mass of powder in a given volume. The degree of packing is usually expressed as effective powder density, product porosity or void fraction. A higher degree of packing means a greater powder density and less voids in between the particles, and vice versa [55, 60].

The degree of packing relates to the flow properties of a powder. It depends upon the number and area of contacts between particles, the attractive and frictional forces between the particles, the size distribution and the history and magnitude of forces of gravity and vibration exerted during packing. Generally, particles smaller than about 20 μm have stronger contacts, show poorer flowability, and have smaller degree of packing than particles greater than about 50 μm.

Maximum density for random packing of monosized particles is about 64%, but in practice it is typically smaller, and void fractions are greater than 40%. Particle size distributions may result in smaller void fractions, down to a minimum of about 15%, since the smaller particles can fill the voids in between the larger ones.

In general, two descriptors for bulk density of powders are used, viz. loose, poured density and tapped density [8, 10, 69, 71]. The loose, poured density is the bulk density obtained after free pouring of the powder into a vessel. The tapped density is obtained after tapping/vibration of the contents to a minimum volume. Thus, tapped density has the greatest value of the two and least voids. Still, the present voids cause the tapped density value being significantly smaller than the particle density of a material.

Note that high compressive strengths may lead to higher powder densities than the tapped density (see Sects. 1.4.10 and 1.5.5).

1.4.10 Powder Flowability

Flowability, cohesivity and fluidization ability are important characteristics of dry powders, for example in emptying of hoppers, blending, dosing or drying. Simply said, powder flowability is the ability of a powder to flow, and cohesivity is the resistance to flow, similar to the viscosity of Newtonian liquids. However, this comparison is too simplistic since powder flowability depends upon attractive van der Waals forces, attractive or repulsive electrical forces between particles, restrictive mechanical interactions due to elongated or rough particle shape as well as degree of aeration or packing and the degree of consolidation, which typically relates to the powder's history. These factors make the flow properties highly specific for a material and its history. Moreover, the size, shape and wall friction of hoppers used for storage have an effect on the exiting powder flow from the hopper [97, 98].

Good, i.e. free-flowing behavior of a powder requires a particle size of at least 20–50 μm and a smooth, equant shape. In contrast, smaller particles and/or particles having unequal dimensions or rough surfaces have poor flow properties. Note that the addition of small amounts of small particles (e.g. SiO_2 or Mg-stearate) may improve poor flow behavior of cohesive powders. Note further that moist powders usually exhibit poor flow properties due to formation of liquid bridges between particles.

Simple tests, such as the Carr Compressibility index CI and the Hausner ratio H, are often used to indicate both compressibility in relation to powder packing and powder flowability [6, 55, 60, 83]:

$$CI = 100\frac{V_f - V_t}{V_f} = 100 \times \left(1 - \frac{\rho_f}{\rho_t}\right) = 100 \times (1 - 1/H) \quad (1.2)$$

and:

$$H = \frac{V_f}{V_t} = \frac{\rho_t}{\rho_f} \quad (1.3)$$

where:

CI Carr Compressibility Index (%)
H Hausner Ratio
V_f Freely settled volume per unit mass of powder
V_t Tapped volume per unit mass of powder
ρ_f Freely settled bulk density of powder
ρ_t Tapped bulk density of powder.

An impression of the flowability of a powder can also be obtained from the value of the angle of repose after free deposition of powder into a heap on a flat surface. Other simple ways to characterize flow properties are the rate of flow through orifices and the frequency of avalanches of powder in a slowly rotating drum.

Table 1.3 Classification of flow properties of powders

Flow property	Carr index (%)	Hausner ratio	Angle of repose (°)
Excellent	≤ 10	1.00–1.11	25–30
Good	11–15	1.12–1.18	31–35
Fair; no aid needed	16–20	1.19–1.25	36–40
Passable; borderline	21–25	1.26–1.34	41–45
Poor; must agitate, vibrate	26–31	1.35–1.45	46–55
Very poor	32–37	1.46–1.59	56–65
Very, very poor	≥ 38	≥ 1.60	>66

Some values for classification of the flow properties of powders are given in Table 1.3. For example, Carr Index < 15, Hausner Ratio < 1.18 and Angle of Repose < 35° are considered as indications for good flowability, against Carr Index > 25, Hausner Ratio > 1.34 and Angle of Repose > 45° for poor flowability.

Taylor et al. [102] have applied principal component analysis to reveal a flowability index for pharmaceutical powders based on measurements of critical orifice, compressibility and angle of repose. They conclude that a combined index provides for better characterization of powder flow than the individual tests. This can be explained by the fact that the different tests challenge different aspects of the complex flow behavior of powders.

Extensive research has not yet lead to adequate understanding and basic relationships of flowability with size and shape characteristics of particles and their distributions. The above characteristics appear to be only part of a full characterization.

Basic information on cohesivity/flowability of powders in relation to consolidation can be obtained from measurements in shear cells (see Sect. 1.5.5). The measured data points are plotted in normal stress—shear stress diagrams, in which they result in so-called Mohr stress circles (see Fig. 1.8) [1, 97].

From these plots, e.g. a flow function ff_c of a product can be derived as the ratio of consolidation stress σ_1 to compressive strength or unconfined yield strength σ_c at different points of consolidation (see Fig. 1.9, which also gives the limits of different flow regions):

$$ff_c = \sigma_1 / \sigma_c \tag{1.4}$$

The value of this flow function for a product at any point gives a good indication of its flow properties in relation to the degree of consolidation. Often, the flowability of bulk solids is dependent on consolidation stress. As shown in Fig. 1.9, many bulk solids exhibit poor flowability at low consolidation stress, while flowability improves at greater consolidation stresses [97, 98].

Note that the unconfined yield strength of some bulk solids increases when they are stored for a longer time at rest under a compressive strength, e.g. under their

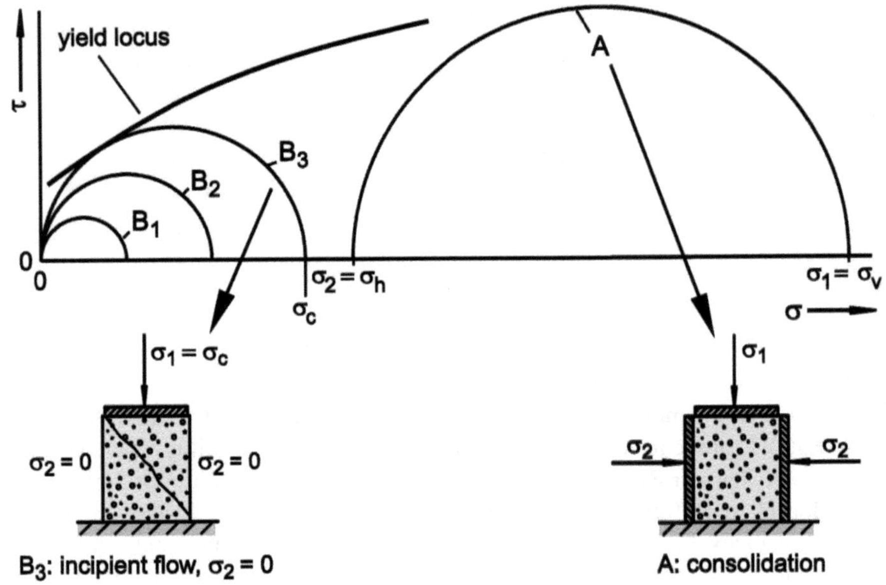

Fig. 1.8 Measurement of unconfined yield strength in a $\sigma - \tau$-diagram [97, 98]; copyright Springer; reproduced with permission

own weight in a hopper. This effect is named time consolidation or caking; it also leads to poorer flow.

Similar shear tests can be performed to determine the wall friction between a bulk solid and the surface of a solid material such as the wall of a hopper. This information is important for the design of hoppers with respect to flow and strength, as well as for other equipment where a bulk solid flows along a solid surface [97].

'Spider' diagrams can be used to give an integrated view on different aspects of flowability for bulk solids. They show the influence on powder flowability of e.g. measured material characteristics such as bulk density (ρ), shear strength (τ) and Hausner ratio (*HR*) and equipment related characteristics such as wall friction (φ), hopper wall angle for mass flow (β), hopper outlet size (D_{cr}). The diagrams are constructed, through using available information for a variety of products and equipment, by drawing three concentric circles, where values at the inner circle are related to easy flow, at the middle circle to modest flow and at the outer circle to poor flow (for examples, see Fig. 1.10) [52].

Such spider diagrams allow an easy and fast impression of the limiting factors for e.g. powder flow from a hopper. For example, the arbitrary material at the right shows that special attention in hopper design must be paid to shear strength (τ) and hopper outlet size (D_{cr}).

The measured values of the material can then be used for adequate design of the angle of the hopper and the width of its outlet slot in order to promote mass flow and to prevent funnel flow and arch formation, or to make the decision to improve powder characteristics, for example, through increasing particle size by granulation [46].

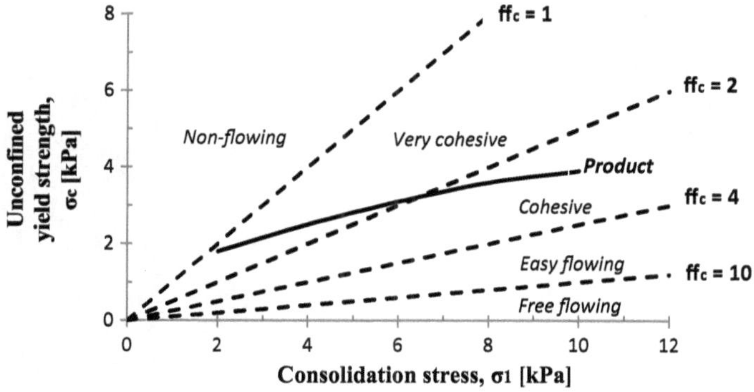

Fig. 1.9 Flow function of a product in a $\sigma_1 - \sigma c$ diagram

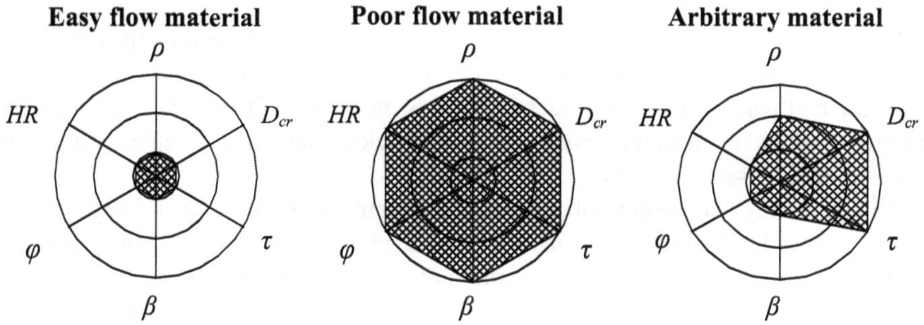

Fig. 1.10 Spider diagrams of materials having good, poor and arbitrary flow behavior. Adapted from [52]

1.4.11 Segregation

Dry powder mixtures form at best a random mixture. This means that the probability of finding a particle of any type is equal to the proportion of that particle type in the bulk and the same at all locations. But the exact proportion varies according to statistical laws. A 'fundamental error' can be calculated that depends upon the number of particles taken into account [54].

However, deviation from random mixing often occurs in mixtures. One reason is unsteady flow of the different components into the mixture. Thus, funnel flow from hoppers may contribute to segregation in the product. Often, however, the different elements (particle sizes) of a powder also undergo segregation during transport and deposition in hoppers or heaps through impulse differences, percolation or elutriation. Segregation is promoted by wide size distributions, different particle densities and good powder flow properties (lack of cohesivity). These properties may lead to

Fig. 1.11 Segregation in a heap, formed by pouring a free-flowing mixture of particles (note that some segregation is apparent in the hopper as well)

different flow paths of the particles within the particulate material, thus causing different size distributions and component concentrations at different locations in a particulate stream or in the product bulk. It is easily visualized when particles of different size have a different color (smaller particles often have a lighter color than larger particles; see Fig. 1.11).

Unfortunately, the degree of segregation (degree of de-mixing or 'segregation error') cannot be predicted but only assessed experimentally by analyzing samples from or at different locations. Conventionally this is done by taking samples of sufficient amounts (governed by the fundamental error) at different locations in a product bulk or from a transport line and analyzing them off-line (see Chap. 6). This has the drawback that the sampling technique itself may contribute significantly to the results found as well as to local changes of the bed contents and, for a blending process, that the blender has to be stopped for sampling. However, measurement can also be accomplished directly for mixtures of different components, without physically removing samples, by in-line or in-process analysis using e.g. near-infrared, Raman or terahertz spectroscopy.

The degree of segregation or the degree of de-mixing is usually expressed as variance, standard deviation or relative standard deviation (coefficient of variation) of a relevant parameter like API concentration or $\langle D_{6,3} \rangle$, which can be calculated from multiple measurement results [106]. From these measurements, a confidence interval for this parameter value can be determined in relation to the number of samples measured, and compared to the desired level of confidence.

Representative sampling of segregated product batches may pose serious challenges in case of free-flowing or cohesive pharmaceutical powders and their mixtures with particulate excipients, where small tolerances for the API concentration are required in small doses. Such cases require adequate care. If the content uniformity is unsatisfactory, then blending conditions should be improved, an alternative blender installed, some pre-mixing step added to the process, and/or API particle size decreased. See also Chaps. 6 and 7 and Sect. 1.5.1.

1.4.12 Rheology of Emulsions and Suspensions [55, 60]

The rheology of emulsions and suspensions is of particular interest in relation to processability, stability and applicability of such products.

Dilute liquid dispersions show similar Newtonian flow as liquids. Here, the required force (shear stress) to maintain flow is proportional to the shear rate, with the viscosity as the proportionality constant. In other words, the viscosity is a measure of the resistance of a liquid to flow. The viscosity of dilute dispersions is directly proportional to the concentration of the dispersed particulate phase, up till concentrations of about 5% v/v. The proportionality constant depends upon particle shape and particle-particle interactions, but not on particle size.

In the concentration range of about 5–30% v/v interactive forces between the particles cause that the linear relationship between viscosity and concentration changes into a non-linear one. In addition to other components present that influence particle-particle interactions, now also particle size becomes important. For example, decreasing particle size cause higher viscosity at the same volume concentration [18, 19, 55]. This is due to the fact that smaller size means a greater number of particles per unit volume, which, thus, come in closer proximity and have more interaction.

At particle concentrations above 30% v/v, especially if they come close to maximum particle packing, the viscosity behavior often becomes non-Newtonian. Then, the viscosity becomes dependent upon shear rate and exposure time. Possibilities are:

- Pseudoplastic or shear thinning behavior occurs most often (at small shear rates). The apparent viscosity decreases with increasing shear rate. Yoghurt and ketchup are typical examples. Typically, some kind of network structure has been formed that breaks down under shear. This behavior is advantageous to processability, stability and applicability of a product, but may cause difficulties in stirring processes when the existing network is only broken down over a short distance.
- Dilatant or shear thickening behavior, where the apparent viscosity increases with increasing shear rate. Here, increased shear causes less freedom for movement of the particles. It occurs at high shear rates for many products.
- Bingham plastic behavior shows a more complex relationship between shear stress and shear rate. Typically, a certain minimum stress—the yield stress—is required to overcome particulate structures, after which pseudoplastic or dilatant behavior starts.

1.4.13 Stability of Emulsions and Suspensions [55, 60]

Commercial emulsions and suspensions are generally required to be stable during storage over a long period of time. This means that no changes may occur due to sedimentation, creaming, agglomeration, coagulation or coalescence (or that simple shaking will bring the mixture back to its original state). This requires that the particles have sizes that are smaller than about 10 μm and are subject to some kind of stabilization by dispersants and/or thickening agents. Stabilization may be induced by Brownian motion (particle diffusion) if the particles are very small, e.g. in the nanometer range. But most often it results from repulsive forces caused by electric charges and/or steric hindrance induced by adsorbed species at the surface of the particles. The zeta potential is the governing parameter for electric stabilization (see Sect. 1.4.7). The size of adsorbed species is important for steric effects. At high particle concentrations, formation of network structures between the different particles and their adsorbed species may enhance stability (see Sect. 1.4.11).

1.5 Measurement Methods and Techniques [52, 54, 55]

Any reliable measurement method for a particulate material requires analyzing test samples of sufficient quantity that are representative within stated limits for the product bulk. Often, the samples are physically separated from the bulk or transport line and analyzed off-line or at-line. However, when analyzing in-process, in-line or on-line, the samples are not physically removed but the size of the measured sample is determined by e.g. optical boundaries of the measurement technique. For off-line measurement of PSD or particle shape, the separated samples are dispersed and/or diluted to yield a collection of individual, non-touching particles at adequate concentration, if not the state of agglomeration is the goal for measurement. Surface area and crystal structure are measured in the dry state. Measurements are conducted directly in the test sample taken for characterization of flowability of powders and of rheological behavior and stability of liquid dispersions.

The measurement technique and its conditions, as described in the analytical method, form the heart of the product characterization. It should provide for adequate precision, resolution, sensitivity and robustness for the critical product attributes to discriminate between good and poor products. For best results of product control standardized, written procedures as well as instruments and trained analysts of proven quality are required, all of which are tested on a regular basis.

A general remark should be made here for the case of the Quality by Design approach. Here, it is necessary to investigate all sources of potential error in the composition of the final product. Since errors in different process steps are combined as variance (standard deviation squared), major attention should be given to improve those steps in the total process of production and analysis that pay a major contribution to the overall error in the product quality aspects.

1.5.1 Sampling [39, 57, 70]

Small samples coming from large batches of powders often show large variations in composition, both for single products having a size distribution and for mixtures of different products. Several reasons may be responsible for these variations: (a) an insufficient number of particles in the sample, (b) segregation of powders in pipelines and powder beds, (c) mode of sampling, i.e. from a powder batch or stream, (d) inability of a sampler to collect a representative sample for the location where sampling is performed, (e) fluctuations in production processes and loading sequences and (f) insufficient blending of components (see also Chaps. 6 and 7).

Measurement errors due to an insufficient number of particles taken into account are called *fundamental errors*. They have a statistical nature and can be calculated from the number of particles involved in the measurement of a parameter through using the statistics of binomial distributions. Larger particle numbers mean smaller errors and better precision.

Heterogeneous distribution of product particles in a heap or hopper causes so-called *segregation errors*. Segregation of particles is usually the main cause for errors, especially in case of more or less free-flowing powders (see also Sect. 1.4.11). It relates to differences in particle size, shape and density, and is promoted by vibrations. However, also fluctuations during processing of the product may contribute; they can be minimized through careful design of the equipment. Segregation errors can only be calculated by analysis of a number of different samples from the product batch. The size of these errors can be decreased by analyzing greater numbers of primary samples, or combining them as increments into a composite sample, which is then split into a test sample.

In general, maintaining general 'golden' rules for collecting test samples from bulk particulate products that take regard of potential segregation of particles can minimize errors. More detailed information on sampling fundamentals, methods and techniques is given in Chap. 6 and Refs. [52, 54].

The conventional way for pharmaceutical powder sampling is collection of samples from heaps with spear or thief probes. It may cause significant errors [64]. Problems encountered are bed disturbances by the samplers, inability of the samplers to reach all locations in a batch and uneven flow or even resistance to flow into the samplers for cohesive mixtures. Another drawback is that the blending process has to be stopped for sampling.

PAT related, non-invasive measurements will not disturb the powder bed [12, 23, 70, 78]. For example, modern spectroscopic analyzers using e.g. near-infrared (NIR) spectroscopy with chemometric data analysis and multivariate calibration techniques allow such non-invasive in-process analysis of component heterogeneity in powders and tablets. For powder blending operations, such analyses have the advantage that they can be executed during operation. If a location in the blender can be chosen where the composition is reasonably representative for the total blender content, this location can be applied for measurements over time to follow progress. If such location cannot be found, multiple measurement locations are necessary for adequate representation of the blending progress.

Similar to powders, testing of tablets and capsules also involves sampling from the production line according to an accepted procedure that ensures randomness and independence of production process equipment and its fluctuations. As with the particles, a sufficient number of tablets/capsules should be analyzed for statistical significance of test results.

Note Gross particulate samples for chemical analysis are often milled before further splitting. The (much) smaller particle size together with the resulting powder cohesivity enables collection of even very small quantities of test samples to be representative with respect to chemical composition. It will be clear that milling is not an option when particle size, shape or surface area is to be investigated.

1.5.2 Dispersion/Dilution [40, 58]

Adequate dispersion/dilution of particulate samples requires that the measured state of the particles agrees with the goals for measurement. Measurements of size and shape are typically executed on primary particles. Thus, dispersion and/or dilution is required in order to reach both complete de-agglomeration and an acceptable concentration for measurement. Sometimes, however, the state of dispersion and the degree of agglomeration or flocculation are of interest; then it is required that this state doesn't change during dilution. *In general it can be said that the conditions of powder dispersion for analysis should correspond with those during powder processing if the results are to be used to judge product quality.*

Dry powders can be fully dispersed either in air or in a suitable liquid. Dry dispersion is simpler as it does not require a liquid. Also, it allows larger amounts of sample for analysis than typically possible in wet dispersions. Free-flowing powders can be dispersed in air by gravity. Dry powders that are cohesive require some mechanical energy for dispersion to produce individual, primary particles. The increasing adhesion strength between particles at decreasing particle size requires an increasing energy for de-agglomeration, while avoiding breakage of particles. Typically, this makes dry dispersion impossible for particles smaller than about 1 μm, because they are usually strongly agglomerated, and for larger fragile particles, such as acicular crystals.

Full dispersion of dry powders in a liquid requires that the particles are adequately wetted, that no dissolution or swelling occurs, that all particle clusters are broken up to primary particles and that the dispersion remains stable during measurement. Typically, this asks for a surfactant and/or dispersant in addition to the liquid, as well as for application of some type of dispersion energy (stirring or ultrasound). The advantage of wet dispersions is that the dispersion conditions can be varied easily and that dispersion quality can be checked by microscopy. A drawback may be that the issue of fundamental error is easily overlooked when small dispersion units are applied in a particle sizing instrument. Note that nanoparticles form very stable clusters in the dry state. To avoid this clustering, these particles must be prepared in such a way that the primary particles formed are immediately stabilized and not allowed to cluster. Otherwise, full conversion of nanoparticle clusters into individual, primary particles is virtually impossible.

For pharmaceutical powders, finding a suitable liquid and dispersant is often a major challenge since they are prone to swelling or dissolution in many liquids, which would make size determination impossible. Dilution of suspensions, emulsions and ointments requires that the particles remain in their original state of dispersion. This point asks for special care since these products are usually stabilized by specific dispersants or surfactants. Also, stability of dispersions is very sensitive to the concentration of dispersants and dissolved ions.

More details on dispersion are given in Sect. 2.4 and [58].

Solid particles are sometimes diluted in a solid matrix. This is typically done when using NIR or THz spectroscopy for analysis. Similar to diluting liquids, the diluent particles should be transparent at the wavelengths of interest for the analysis.

Chemical analysis of samples by e.g. HPLC requires their dissolution (see below).

1.5.3 Measurement Techniques for API Concentration and Coating

The composition of pharmaceutical products varies with the type of product and its application. It may be very complex. In view of this, analytical techniques should have sufficient sensitivity and resolution for the aimed substances. A variety of techniques and methods for quantitative analysis of API concentration is available; for example, liquid chromatography, UV-, IR-, Raman and mass spectrometry are often used in the field of pharmaceutical products. Application of near-infrared absorption is relatively new for analysis of pharmaceutical formulations and in-process monitoring of manufacturing operations; in combination with chemometric methods it has become an attractive method. Terahertz spectroscopy is also a new technique; its application to pharmaceutical products is increasing.

As indicated above, a wide range of wavelengths is applied for spectroscopic analysis. The broad field of electromagnetic radiation is shown in Fig. 1.12.

This figure clearly indicates the large differences in frequencies and wavelengths when going from γ and X-rays via ultraviolet, visible, infrared and terahertz radiation to micro-, radio- and TV-waves.

Liquid chromatography is most often used in the form of its high performance variant HPLC. Here, test samples are dissolved in a suitable liquid and a sample taken from the solution is introduced into a mobile liquid that flows through a column containing a suitable stationary phase for separation of the components through a process of continuous ad- and desorption. A detector at the end of the

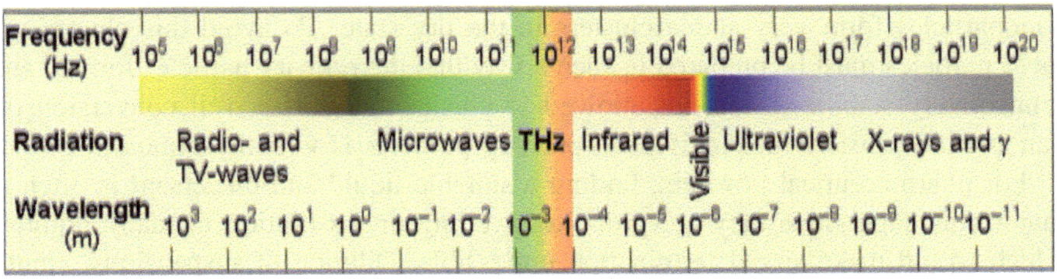

Fig. 1.12 Electromagnetic spectrum

column senses component elution. Retention time in the column is typically used for component identification, peak area relates to component concentration. HPLC offers very good separation between components. The concentration can be calculated through comparison with solutions of pure reference materials, or by standard addition. In view of the fact that different products may ask for different stationary phases and elution liquids, that different detectors can be used and that this book is intended to deal with the particulate aspects, no further explanation is presented here [75].

UV spectrometry is an old technique for quantitative determination of organic components, which are aromatic or have conjugated double bonds and show specific absorption in the UV region. Measurements are done after dissolution in a suitable solvent. The absorbance at a suitable wavelength or combination of wavelengths is used for calculation of the component concentration. UV is also applied for detectors in HPLC [67, 75].

UV analysis is also often used for direct determination of the dissolution rate of solid products in liquids.

Mass spectrometry (MS) is another technique for component identification as well as for measuring composition. Here, the samples are vaporized in some way and then ionized, after which the charged species (coming from the complete molecules and/or from their fragments) are separated in a magnetic field according to their mass-charge ratio. Sometimes, HPLC is combined with MS in order to yield best separation and identification of mixtures of components [76].

Infrared (IR) ***spectroscopy*** is typically divided into three regions, viz. near-IR (see later), mid-IR (wavelength 2.5–25 μm, wave number 4000–400 cm^{-1}) and far-IR (wavelength 25–1000 μm, wave number 400–10 cm^{-1}). The latter is nowadays usually named terahertz spectroscopy (see later). Mid-infrared analysis typically concerns resonance of fundamental vibrations of covalent bonds by interaction with infrared radiation, whereby a change of dipole moment occurs. The spectra are usually complex since all bonds in a molecule may contribute at a specific wavelength, which depends on the atoms involved, the type and strength of the bond and its dipole moment. Qualitative analysis for chemical identification of components is related to the specific wavelengths in the spectrum where light is absorbed. Quantitative analysis is executed by measuring the absorbance at one or more of these wavelengths and relating this by means of the component-specific extinction coefficients to component concentration. A library is available containing the specific wavelengths of absorption for many covalent bonds. The absorbance coefficients at different wavelengths for a specific component are usually determined experimentally. Conventionally, measurements were dispersive, through using a monochromator and determining the absorption at subsequent single wavelengths. Nowadays, Fourier transform measurements are more popular, where the whole (transformed) spectrum is measured at once by means of an interferometer and then translated into the spectrum. A specific way for measurement offers the attenuated total reflection technique (ATR-IR), which is most often applied for powders. Here, the material of interest (e.g. a powder) is brought in close proximity with a crystal, which causes that the absorption spectrum comes from multiple reflections at the material and, thus, has an improved signal-to-noise ratio. In this technique, the penetration depth in powders is typically limited to ≤ 5 µm, depending on material. Sometimes, chemometric methods are applied for analysis where a combination of several absorption wavelengths is applied for better component identification. *Note* water has a strong absorption spectrum, which often disturbs the analysis of other components [72].

Raman spectroscopy is similar to infrared in that it gives component specific spectra that relate to molecular vibrations—and gives chemical information on the components—in the so-called fingerprint region. The difference is that no light is absorbed resulting in absorption spectra, but that interaction of (laser) light with the bonds causes molecular excitation, which in turn results in light emission at characteristic wavelengths for the vibrations of the covalent bonds in molecules. For organic molecules, the wave number range is 500–2000 cm^{-1}. The intensity of Raman spectra is much weaker than that of IR spectra [84].

Recently, the conventional Raman spectroscopy has been extended into the low frequency spectral region (low wave number, 5–200 cm^{-1}) and beyond into the anti-Stokes region through application of terahertz radiation (see section on terahertz spectroscopy). This THz-Raman spectroscopy allows determination of structural information of solids, since these vibration spectra are affected by the collective motion of atoms that surround the corresponding covalent bond. The THz-Raman spectra have much higher intensity and better signal-to-noise ratio than the conventional Raman spectra. Moreover, the symmetrical nature of the anti-Stokes and the Stokes spectra improves overall reliability and provides inherent self calibration (see Figs. 1.13 and 1.14 for carbamazepine examples) [16, 17]. Similar to IR and THz, typically chemometric methods using absorption at multiple wavelengths are applied for qualitative and quantitative analysis.

Near-infrared (NIR) analysis is a technique that gains increasing interest in pharmaceutical manufacture as it allows in-process and in-line analysis of particulate mixtures. It operates in the wavelength range of about 0.8–2.5 µm (wave number about 12,500–4000 cm^{-1}). Typically, the diffuse reflectance of powders and tablets is measured, which shows absorption of NIR radiation in specific wavelength regions due to overtones or combination bands resulting from vibration of hydrogen containing bonds. Here, the wavelengths of absorption are specific for the components whereas the degree of absorption relates to their concentration. The

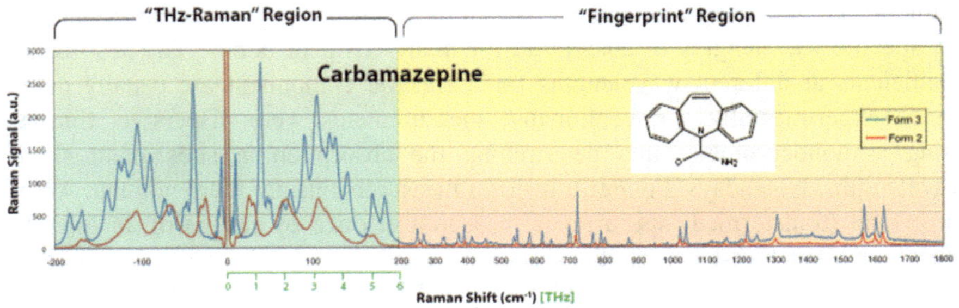

Fig. 1.13 Raman spectra of two forms of anhydrous carbamazepine [16]. Reproduced with permission of Ondax, Inc

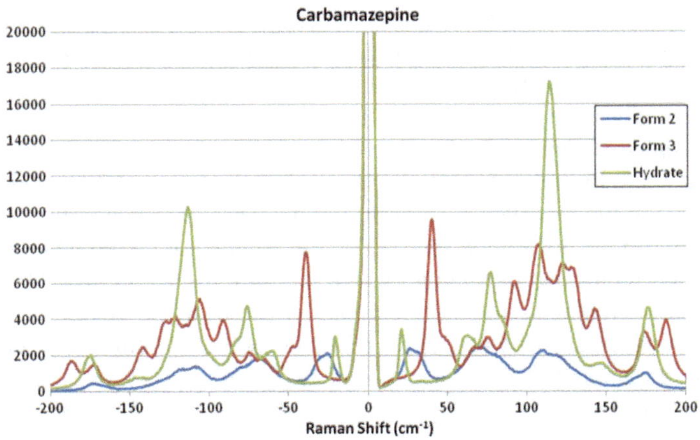

Fig. 1.14 Expanded THz-Raman spectra of two forms of anhydrous carbamazepine and a hydrate [17]. Reproduced with permission of Ondax, Inc

technique is fast, non-destructive, simple in its execution and requires no sample preparation. On the other hand, the spectra obtained are often quite complex since the absorption bands of different components, including moisture, are overlapping. Thus, quantitative analysis typically requires special mathematical procedures and multivariate data analysis (chemometrics) as well as sufficient reference data to extract the most relevant data from the spectra and relate them in a quantitative way to the components of interest. Often, the second derivative of absorbance with respect to wavelength is applied in analysis, which converts absorbance maxima into minima with positive side lobs, to improve spectral resolution and to eliminate offsets and slope changes in the baseline due to particle size differences. Subsequently, (linear combinations of) the most selective wavelengths for the components are identified, e.g. by Principal Component Analysis (PCA) and/or other statistical tools, to minimize the influence of interferences and noise. Then, a calibration line for conversion of absorbance to concentration is prepared by regression (e.g. by a least squares method or an artificial neural network) while using a representative set of reference standards that cover the full range of components and their concentrations. Finally, the calibration is validated while using a second set of standards. Note that penetration depth in a particulate sample is at maximum few millimeters; it decreases with decreasing particle size and increasing packing density (compaction). In relation, also absorbance decreases in the same direction. Note further that some particle size information may be obtained from the baseline changes in the spectrum, which relate to scattering of the NIR radiation by the particles. Analysis over time at a single location during blending operations, or simultaneous analysis at different locations in powder streams, tablets or processing equipment, through using fiber optics bundles, sapphire windows and a chemical image analysis system, enables assessment of blend quality and content uniformity at the selected locations and within tablets [12–15, 20, 23, 63, 65, 78, 95]. For more information see also Chap. 6.

Terahertz spectroscopy can be considered as an extension of infrared spectroscopy into the far-infrared region up till the microwave region [66]. The typical spectral range is about 0.3–3 THz, corresponding to wavelengths of 1000–100 μm and wave numbers of 10–100 cm^{-1}. In this range, vibration frequencies of covalent bonds show characteristic spectroscopic signatures in dependence upon the chemical environment of the molecules (collective motion of more atoms than present in the corresponding covalent bond). Absorption occurs at wavelengths of the incoming radiation which cause resonance with the vibrations. The fact that the chemical environment of the of covalent bonds is eminent in the THz absorption frequencies contrasts to the vibration frequencies in NIR, which do not or only slightly depend upon the environment of the molecules. The THz technique is very versatile as well as complex due to the many intermolecular interactions; it is mostly executed in the form of time-domain spectrometry. It can reveal morphology and internal structures of complex molecules, yield fingerprints for mixtures of different polymorphic and amorphous forms, detect phase transitions upon temperature changes, perform chemical imaging/chemical mapping in relation to differences in composition and refractive index (important for determination of e.g. spatial distribution of API in and thickness distribution of coating layers of pharmaceutical tablets, for which an in-process application has started) as well as determine water content in materials and tissues (e.g. detection of cancer tissues) [99, 108]. Further advantages of the THz technique are that it is fast, contact free, non-invasive and non-destructive. Also, the energy of THz radiation is too low to cause ionization of atoms and the very low power levels applied (often <1 μW) make the technique safe for probing biological samples without thermal effects. Moreover, sample preparation is fairly easy and instruments are robust. Finally, refractive index values can be derived from the measurements [101, 108]. The refractive index can be converted to material density by proper calibration and further to its porosity while using the true density of the material. For transmission measurement, solid samples are shaped to thin, uniform slabs with parallel faces (typically 100–300 μm thick), whereas powders are compressed to pellets, mostly diluted with a low-absorbing binder (often polyethene or Teflon) to accommodate both the degree of transmission and pellet strength. Chemical imaging of pharmaceutical tablets is typically directly executed in reflectance without any sample preparation. Full tablets can be examined since the excipients usually exhibit little or no absorption.

As with NIR, multivariate data analysis (chemometrics) as well as sufficient reference data are often used to extract the most relevant data from the spectra and relate them in a quantitative way to the components of interest. THz spectroscopy has made great progress in the last two decades by the development of tunable and ultrafast (femtosecond) pulsed lasers, adequate optics and detectors as well as in the modeling of molecular vibrations in a solid environment. In relation to its recent history, still much research on fundamentals and applications is being directed towards improving instruments and expanding the field of applications in relation to the high potential of this technique.

1.5.4 Measurement Techniques for Particle Characteristics

1.5.4.1 True Skeleton Density

The true skeleton density of solid non-porous materials is usually determined by conventional ***pycnometry*** using a liquid. For particles in pharmaceutical powders it is usually determined by the volumetric technique of ***gas displacement*** in order to avoid effects by dissolution or swelling of particles by a liquid. For both non-porous and porous materials, helium is often used as the fluid for displacement because it can easily enter into all open pores of particles and voids [44, 71, 103]. If particles contain closed pores, which have no access to the external surface, the volume of these pores is included in the effective density measurement of particles. An apparent density may be measured when using liquids for porous materials, the pores of which are not easily accessible to the liquid due to limited wetting.

1.5.4.2 Crystallinity and Crystal Structure

Crystallinity and crystal structure are usually determined by X-ray diffraction, differential scanning calorimetry and differential thermal analysis.

X-ray diffraction (XRD) uses X rays of a specific wavelength (in the range of 0.1–10 nm) to measure the angular diffraction pattern produced by a crystal lattice consisting of a regular, three-dimensional array of atoms or molecules. Crystalline materials show a diffraction pattern that contains sharp peaks resulting from constructive interference of diffracted rays. The positions of these peaks (diffraction angles) are specific for the component's crystal structure. The spacing of the lattice planes is calculated from the diffraction angle and the wavelength of the X-rays through application of Bragg's law. Databases are available for comparison and identification. In contrast to crystalline materials, amorphous products do not show a diffraction pattern with sharp peaks as they have no regular arrangement of atoms or molecules. Measurements are often done on crystalline powders (powder XRD or PXRD) [94]. (see also Chap. 9).

Differential scanning calorimetry (DSC) is a technique where the differences in heat flux between a material and a reference to keep them at the same temperature are measured as a function of temperature during warming-up or cooling-down at a given rate. When increasing the temperature and starting with a solid material, its transformation from one phase into another can be indicated by two types of change: (a) an endothermic peak at a given temperature or temperature range (e.g. glass or melting temperature of a solid) indicating that the phase transfer consumes energy, or an exothermic peak when it releases energy (e.g. crystallization), and (b) a change in het flux related to a changed heat capacity of subsequent phases. By this technique conversions from amorphous (glassy) state via rubbery and crystalline state to liquid state can be observed as well as stability of the amorphous state and free energy contents of the various phases estimated [77, 89] (see also Chap. 9). Alternatively, the same DSC technique can be used for crystallization studies; then, the temperature is decreased at given rates. The resulting crystalline phases, which may depend upon the cooling rate, can then be identified by means of e.g. X-ray diffraction.

Differential thermal analysis (DTA) is similar to DSC. The only difference is that the heat flows to sample and reference are the same instead of the heat flux being necessary to keep their temperature constant. The resulting temperature differences between sample and reference are measured to indicate phase changes and their energy contents.

1.5.4.3 Elemental Analysis

Elemental analysis of (nano-) particles is usually executed by energy dispersive spectroscopy in combination with electron microscopy. Metallic components in solution or powders can be determined by atomic absorption or emission spectroscopy.

Energy dispersive spectroscopy (EDS), also named energy dispersive X-ray spectroscopy (EDXS or EDX), is a technique where focused high-energy electron beams or X-rays remove an electron from an inner orbital of an atom. It causes a vacancy in this orbital that is filled by an electron from an outer orbital, which results in emission of X-rays at a characteristic frequencies related to the energy difference between the two orbitals in the emitting atom, while the X-ray intensity relates to the concentration of the element. This emission is measured in an energy-dispersive spectrometer. Typically, measurements are executed at solid surfaces identified in electron microscopy. See Chap. 13 for examples.

In ***atomic absorption spectroscopy*** (AAS) samples are atomized at about 2500 ° C through using a pneumatic nebulizer plus air-acetylene or nitrous oxide-acetylene flame for liquids or an electrically heated graphite tube atomizer for solids. The electrons of the atoms are then transferred to higher orbitals (excited state) by absorption of radiation at given wavelengths that correspond to the energy differences between normal and excited states. Here, the wavelength is characteristic for the element, whereas the absorbance can be converted to concentration through using reference standards.

Atomic emission spectroscopy (AES) is similar to AAS in that samples are atomized and excited at high temperature in a flame, or by an inductively coupled plasma (ICP-AES). Here, however, the emitted light is measured when the electrons fall back from their excited state into their normal orbital. Again, the wavelength is characteristic for the element, whereas the intensity relates to concentration.

1.5.4.4 Particle Size Distribution (PSD)

Commercial **PSD measurement** techniques and instruments are available in a wide variety. Here only a selection of techniques, which appear to be or may become popular for pharmaceutical products, is presented in a summarized way. More detailed information is given in [54].

Microscopic techniques allow the direct inspection of individual particles. Thus, they can provide information not only on particle size but also on particle shape. This information is typically qualitative or semi-quantitative. ***Digital image analysis*** most often provides for quantitative results. High resolution information of particle size and shape requires high magnification, i.e. a sufficient number of pixels per particle. The results are number-based since particles are measured on an individual basis. Quantitative information on particle shape is especially necessary if it has a dominating influence in product performance.

Three types of techniques can be discriminated, viz. light microscopy (overall size range about 0.3–500 μm), scanning electron microscopy (SEM; overall size range about 10 nm–500 μm) and transmission electron microscopy (TEM; overall size range about 1 nm–5 μm). Emulsions, micro-emulsions and liposomes are typically measured in frozen form (at liquid nitrogen temperature) by cryogenic TEM (cryo-TEM). Size classes can be set by the user, in relation to the magnification applied. Low particle concentrations are required since particle overlap leads to biased results. Thus, many image frames have to be measured for adequate precision of PSD parameters, which asks for automation. Moreover, optimum precision of PSD parameters and particle shape depends upon magnification (number of pixels per particle image) and is at best about 1% relative. On the other hand, qualitative or semi-quantitative results can be reached by measuring only one or a few frames with particles. As an example, Fig. 1.15 shows TEM pictures of three formulated doxorubicin products.

Figure 1.15 clearly illustrates the morphological and structural changes that may occur during drug crystal growth in the vesicular aqueous space. The crystal morphology is important, since a needle-shaped crystal may grow to an extent causing vesicular deformation or even vesicular rupture by penetration through the bilayer, thus affecting vesicular stability and drug release rate.

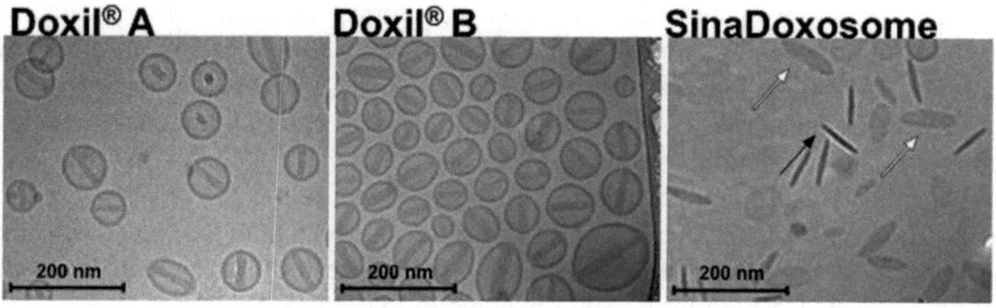

Fig. 1.15 Cryo-TEM images of regulatory-approved liposomal doxorubicin formulations (2 mg/ml). The products are two different batches of Doxil® (batch B is also known as Caelyx®, the European trade name for Doxil®) and SinaDoxosome (manufactured and marketed in Iran). Modified from [107]. Courtesy Prof. S. Moein Moghimi

Detailed information is given in [38, 54, 79, 85]. An additional advantage of microscopy is that typical images of particles can be stored. Note that light microscopy may also be helpful in checking the quality of dispersions as it can indicate presence of agglomerates and aggregates.

Laser diffraction (LD) measures the angular light scattering patterns of collections of dispersed particles in air or a liquid and converts them, aided by an appropriate scattering model and deconvolution program, to a particle size distribution. Since the scattering model is based on volumes of spherical particles in size classes, the PSD is volume-based. The overall size range is about 0.1–10,000 μm. Typically, about 30 size classes are presented by instruments in a single measurement in part of this size range, in relation to the number of detectors and the degree of smoothing used for deconvolution. This means a step ratio for size classes of about 1.2, i.e. medium size resolution. For transparent particles smaller than about 20–50 μm, depending upon their relative refractive index, the exact Mie theory has to be applied in the scattering model, which takes all types of scattering and absorption into account and, thus, requires information on the refractive index of particles and dispersion medium. For larger or opaque particles the approximating Fraunhofer theory can be applied in the model, which only considers diffraction at the particle contours and does not require such refractive index information. The repeatability of the technique is about 0.5–3% v/v. Note that alignment in the measurement zone of particles having large differences between length, breadth and/or width may cause a biased PSD, since only a limited number of cross sections is offered for scattering. This makes the LD technique less suitable for such products. Detailed information is given in [36, 54, 74].

Dynamic light scattering (DLS), also named ***photon correlation spectroscopy*** (PCS) or ***quasi-elastic light scattering*** (QELS), senses the variation of the scattered light intensity by an ensemble of dispersed particles with time at some defined angle. The rate of change of the intensity is related to the diffusion coefficient of the particles, which in turn inversely relates to the hydrodynamic diameter through the Stokes-Einstein equation. The time-dependent intensity spectrum is then converted into an intensity-weighted PSD by means of a digital correlator or by spectrum analysis. Since the data are regarded as ill-conditioned, usually the PSD information is limited to only a few parameters, e.g. an intensity-weighted mean size and a polydispersity index related to the width of a lognormal size distribution (cumulants procedure). The overall size range of the technique is about 5–1000 nm. The repeatability of the mean size is about 2% relative, that of the polydispersity index about 20% relative. Size resolution is fairly low. Note that the hydrodynamic diameter is slightly larger than that of the particle size itself, since it includes an adsorbed layer. Detailed information is given in [37, 45, 54].

Particle tracking analysis (PTA), or ***nanoparticle tracking analysis*** (NTA), is similar to dynamic light scattering in that it senses the diffusion of dispersed particles in a liquid. A major difference is that it follows movement of individual particles illuminated by a focused laser beam. Then, the rate of diffusion movement is again converted to the hydrodynamic particle size. Thus, a number-based size distribution is obtained. The lower size limit depends upon the degree of scattering of the particles, i.e. laser strength and the refractive index of particles and medium, and is about 15–75 nm. Measurement precision strongly depends upon the number of particles measured in relation to the width of the distribution. Size resolution is fairly low. Detailed information is given in [11, 42].

Ultrasound extinction is comparable to laser diffraction in that it applies electromagnetic radiation. However, the attenuation of ultrasound and/or the retardation of ultrasound velocity is measured at a series of frequencies instead of the angular light scattering pattern at a single wavelength. Again, a model matrix containing attenuation patterns per unit volume of particles in defined size classes having stated properties is applied to convert the spectrum into a PSD. Often, a modeled, e.g. lognormal, PSD is calculated. The overall size range is about 0.01–3000 μm. The repeatability of PSD parameters is about 3% relative. An advantage of the technique is that it allows higher particulate concentrations than laser diffraction. Size resolution is fairly low. Detailed information is given in [43, 54].

Electrical sensing zone or ***electrozone sensing*** is a technique invented by W. Coulter in 1953 [22]. It measures the change of resistance when particles dispersed in a conductive liquid (electrolyte) pass one by one through a narrow aperture across which typically a constant voltage is maintained. The particle passage causes an increase of the resistance within the aperture and is seen as a voltage pulse, the height of which is proportional to the particle volume. Particle size is derived after calibration with standard reference particles. The overall size range is about 0.5–1000 μm; per aperure it covers a range of about a factor 30. High-resolution, number-based size distributions can be obtained for narrow size distributions. Detailed information is given in [25, 35, 54].

Recently application of microchannels in microfluidic cartridges or tunable pores in elastomeric membranes has decreased the lower size limit to the nanometer range (to about 30 nm) [4, 21, 26, 28, 48, 53, 100, 104]. This new technique is typically referred to as ***(tunable) resistive pulse sensing***. Following calibration, this allows high-resolution size and concentration analysis for nanoparticles and micro-emulsions. With the membranes also the zeta-potential of particles can be determined.

Sieving is an old, very simple and cheap technique for size measurement. It measures the mass of material that retains on or passes through the apertures of a sieve or series of sieves having specified mesh sizes. Typically, it is applied for particles greater than 32 μm. The step ratio between apertures sizes of subsequent sieves is at best 1.12, but mostly 1.4, meaning medium-low size resolution. The precision and accuracy of the results depends upon the quality and the calibration of the sieving medium, the size of the test sample and the sieving procedure followed. The optimum repeatability of the mass of sieve fractions is about 0.5% relative to sieve load. An advantage of sieving is that it allows fairly large sample sizes as well as results in size fractions that can be used for further analysis. Since typically square sieve apertures are applied, sieving is less suited for particles having quite different length, breadth and/or thickness, since the major two dimensions determine passage through the apertures. Moreover, fragile particles such as acicular crystals may break by the vibratory forces during sieving. For more information see [30, 31, 54, 68, 82, 86, 87].

Impaction is a direct technique for determination of aerodynamic size of aerosol particles. Here, air dispersed particles are forced to flow through a nozzle of known diameter, at a known flow rate, and directed towards an impaction plate positioned across the flow. So-called cascade impactors use a stack of up to 7 nozzle-impactor combinations. Particles with large enough inertia (large enough aerodynamic diameter) will cross the air streamlines, impact onto the plate and are assumed to stay there. Particles with less inertia (smaller diameter) follow the air streamlines and may enter into the next stage of the cascade impactor. The Stokes' number, which relates the impactor design parameters and flow conditions to particle size, governs whether a particle will impact or not. The overall size range of impactors is about 0.01–50 μm (aerodynamic diameter), the actual measuring range is limited to about a factor 25 [29, 54, 73]. The step ratio for size in the stacks is about 1.6, meaning medium-low size resolution. Dissolution of the impacted material followed by HPLC analysis or weighing of the impactor plate(s) usually gives the amount of impacted particles. Impactors can be used for both solid and liquid particles; for the latter, often the word impinger is used instead of impactor. Measurement repeatability can be about 5%. It strongly depends upon the experience of the operator.

Focused Beam Reflectance Method (FBRM) or ***Time of Transition Technique*** (TOT)

Here, a rotating or vibrating, focused laser senses or 'scans' the light reflection by particles that traverse the measurement zone. The rate of rise of the measured signal is used to select signals that come from particles in the focal plane [54]. The measured time of reflection together with the velocity of the laser beam is used to evaluate the chord length at the specific position of the crossing laser beam. The typical size range is about 0.5–3000 μm. A difficulty of the technique is that it may provide a range of chord lengths for the same particle, which range also depends on particle shape. Often, the chord lengths as such are used for interpretation during process monitoring, or for process control. Chord lengths can be converted into a PSD if an adequate model has been developed for instrument response to particle size and shape in their orientation(s) while in process. When such a model exists, it can also be used for process control. An advantage of the in-process instrument version is that it can be applied at high concentrations, pressures and temperatures (up to about 40% v/v, 100 bar/5 MPa and 150 °C). This enables in-line measurement in e.g. crystallization processes.

1.5.4.5 Particle Shape

For measurement of **particle shape**, most often microscopy (light, SEM or TEM) is applied. Image analysis is often used additionally for obtaining (semi-) quantitative results. Typically, only the dominant macroscale and mesoscale features are assessed in a qualitative way through comparison with standard shapes (cf. Fig. 1.5), or in a quantitative way, e.g. by expression as aspect ratio, elongation, flakiness, roundness and angularity [54, 61, 80].

1.5.4.6 Surface Area, Porosity and Pore Size Distribution

Measurement of **surface area** of particles in dry powders is usually executed with the gas adsorption method. **Porosity and pore size distribution** can be measured by both the gas adsorption method and the mercury intrusion technique [33, 41, 55, 81, 88].

Gas adsorption method

In this method the amount of gas molecules adsorbed at the particle surface is determined at constant temperature and at various relative gas pressures as a so-called adsorption isotherm. Nitrogen and argon are most often used as adsorptive at liquid nitrogen temperature (about 77 K or −196 °C). The volumetric measurement method is most popular. To an 'outgassed' amount of test sample known volumes of adsorptive are added stepwise and the resulting equilibrium pressure and the amount of adsorbate determined until all pores are filled. Subsequently, the desorption isotherm may be determined by gradual decrease of the pressure. Adequate outgassing, i.e. removal of all adsorbed components, of the sample prior to analysis is essential for obtaining good results.

For determination of **surface area**, the first part of the adsorption isotherm is used, up till a relative pressure P/P_0 of about 0.3. Below this pressure, the adsorptive molecules predominantly adsorb in a monolayer. The BET equation (Brunauer-Emmett-Teller) is then used to calculate the number of gas molecules adsorbed in the monolayer and, from this, the specific surface area. The repeatability of this method is about 1% relative or $\pm 0.1\ m^2/g$. The method can only be applied if the adsorption isotherm shows an inflection point when going over from monolayer to multilayer adsorption. A standard ISO method is available for this technique [33].

For determination of **porosity and pore size distribution** both the adsorption and the desorption isotherm can be used. The choice depends among others on the type and shape of the pore system, for which the type of hysteresis between the two isotherms gives some information. Most often, cylindrical pores are assumed and the Kelvin equation used for calculation of pore size. The range of pore diameters in this method is about 0.5–200 nm. The repeatability is about 3% relative. A standard ISO method is available for this technique [41].

Mercury intrusion technique

In this technique the amount of mercury that penetrates into pores at increasing pressures is determined at room temperature. Here too, the measurement starts after adequate 'outgassing' of a known amount of sample. Then mercury is added to the sample and the amount of mercury intruded into the pores measured while stepwise increasing the pressure. The Washburn equation, which shows an inverse relationship between pressure and pore size, is used to calculate the pore size distribution. A pressure range of about 0.001–400 MPa is covered in commercial instruments, which allows pore diameter measurements of about 800 μm to 4 nm. Note that voids are included in the results. The repeatability is about 3% relative. A standard ISO method is available for this technique [41, 81].

Note Mercury is poisonous. Therefore, its use is restricted in most countries.

1.5.4.7 Wettability

The **wettability** of flat particle surfaces of suitable size can be easily determined by the ***sessile drop technique*** through measurement of the contact angle θ (cf. Fig. 1.6). The same technique can be applied for powder compacts, but surface curvature and roughness and presence of voids may cause deviating results.

Alternatively, the ***penetration rate*** of the liquid from a droplet into the solid can be measured for compacts [3, 50]. It should be noted that high compression strengths, as often applied in this technique for compaction of powders, may alter the physical properties of the surface and, thus, the measurement results [3].

Several techniques exist for measurement of wetting properties of powders; reference [3] reviews the techniques and compares their advantages, limitations and disadvantages. An often used technique is the ***Washburn capillary rise method***, where the liquid penetration rate into a powder bed packed in a column (typically in a 5–10 mm diameter glass tube, standing in a pool of liquid) is measured and compared with the uptake of a completely wetting reference liquid. Following a short initial period, the squared mass uptake of liquid in the void volume of the column packing proceeds linearly with time until gravity slows the uptake down. The cos θ is then calculated from this linear relation between the liquid uptake and its density, viscosity and surface tension, while a necessary geometric factor—containing the effective equivalent diameter of capillary channels in the powder bed, the cross-sectional area of the packed tube and the bed porosity—is derived from the measurement with the completely wetting reference liquid. Repeatable column packing leading to similar channels and bed porosity as well as application of a suitable support for the packing are essential for reaching good results in this method. Narrow size distributions of powders are often favorable for repeatable packing. Particle size is typically at least about 50 μm.

Recently, a new technique for measuring ***drop penetration*** was published [50]. Here, the penetration dynamics of a sessile drop deposited on a slightly compressed powder bed are measured and compared with a drop of a completely wetting reference liquid. The cos θ is then calculated from the ratio of penetration times, viscosities, surface tensions and squared droplet diameters of test liquid and reference liquid.

1.5.4.8 Zeta Potential

The **zeta potential** is determined via measurement of the electrophoretic mobility of the particles in a liquid. This mobility can be determined in an electric field under a microscope, in a dynamic light scattering instrument, or in electro-acoustic instruments by measurement of the colloid vibration potential (CVP), the colloid vibration current (CVI) or the electrokinetic sonic amplitude (ESA). Some more information is given in [55].

1.5.5 Measurement Methods for Powder Characteristics

1.5.5.1 Bulk Density

The **bulk density** of a powder relates not only directly to the mass per unit volume but also to its flow properties and packing history. The *poured density* relates to the loosest powder packing. It is usually determined by pouring a given mass of powder from a standard height into a graduated cylinder and measuring the volume. Alternatively, it can be determined by filling a cup-like vessel of known volume to overflowing, then level with a straight edge and measure the mass. The *tapped density* uses the cylinder with powder again, in which the powder volume is measured after tapping until the powder volume does not change anymore. As a variant to the latter method, known weights are placed on top of the powder in a vessel and the remaining volume measured. This allows measurement of bulk density as a function of compressive stress [8, 10, 69].

The **density of tablets** can be calculated through dividing their mass by their volume (dimensions). Also, terahertz spectroscopy offers possibilities for determination of density and porosity of tablets (see Sect. 1.5.3).

The **void fraction** in a powder or in a tablet can be derived from the difference between the measured apparent density and the true skeleton density. For tablets, also the mercury intrusion technique can be applied for measurement (see above).

1.5.5.2 Flow Properties

A simple estimate of the **flow properties** of a powder can be obtained from the ratio of poured and tapped density in the form of the *Carr compressibility index* or the *Hausner ratio* (see Sect. 1.4.9).

Measurement of *the angle of repose*, the *number of avalanches* per unit time in a slowly rotating drum and the *flow through orifices* give additional information on powder flow properties [6].

Basic information on cohesivity and flowability of powders following consolidation by compressive stresses can be obtained from measurements in *shear cells*, such as the testers designed by Jenike, Peschl or Schulze (see Sect. 1.4.9) [5–9, 97, 98]. In these testers compressive stresses by external weights are applied for consolidation. In practice this consolidation usually results from the powders own mass during storage and by vibration during powder handling. Following consolidation of the powder in the tester, the forces necessary for lateral or rotational displacement are measured. Increased consolidation typically causes increased powder density and decreased powder flowability.

1.5.6 Characterization Techniques for Liquid Dispersions

For liquid dispersions—emulsions, suspensions and ointments—the **rheological behavior** is one of the aspects determining product quality (see Sect. 1.4.11). It is a function of particulate concentration as well as depends upon dissolved components that influence viscosity.

For dispersions of low and medium particulate concentrations, similar techniques can be applied as for the dispersion liquids, e.g. measurement of flow rate through capillaries. At high particulate concentrations and for ointments, where the rheological behavior is non-Newtonian due to the presence or formation of particulate structures, the relation between shear stress and shear rate has to be measured. Note that this relation is sometimes also time dependent. For measurement, usually rotational testers with parallel plates or falling needle viscosimeters are applied [55, 62, 90–93, 96].

Dispersion stability is another quality aspect. It is usually checked visually or by determining the degree of light absorption at different times and heights in dispersions, which may change due to sedimentation or creaming of particles. For design of stable dispersions usually a large enough ζ-potential is chosen (see Sect. 1.4.7).

1.5.7 *Characterization Techniques for Sprays*

Oral sprays (inhalers) that are meant to reach the lungs through inhalation should preferably contain particles with an aerodynamic size range of 1–5 μm [27–29]. Smaller particles are prone to be exhaled, larger particles are deposited in mouth, nose, throat or bronchial tubes and do not reach the lung surface. Nasal sprays typically have an aerodynamic size in the range 20–200 μm. The particles may be solid or liquid. The design of and conditions in the sprayer/inhaler together with the rheological properties of powders (flowability) and liquids (viscosity) determine the size distribution of the sprays.

Conventionally, the aerodynamic size distribution is measured directly by inertial impactors in special equipment, as described in the pharmacopeia [73].

Nowadays, the PSD of sprays is often measured by means of laser diffraction, which provides for a wider measurement range and better size resolution than impactors. The sizes are then converted into aerodynamic sizes through using the known density of the particulate material. Measurements in different regions of the spray may indicate potential differences of the PSD in different locations in the spray.

1.6 Set up of the Book

The main objective of this book is to highlight the role of particulate nature in the wide variety of pharmaceutical products and to give guidance in the choices of particle and powder characteristics and their measurement for setting product specifications, which have an optimum relationship with product quality, both during product processing and in the quality of the final product. This chapter gives an introduction to the general aspects of particulate nature and the multitude of measurement methods that are relevant to pharmaceutical products. Chapter 2 deals with main regulatory requirements and guidelines for approval of pharmaceutical products. It also discusses the method of Quality by Design and Process Analytical Technology for solid oral dosage forms. Moreover, the chapter considers the quality aspects of the main PSD analysis techniques for pharmaceutical products and the importance of quality management. Theranostic possibilities at the cellular level by nano-particles are discussed in Chap. 3. It describes the importance of functionalized nano-particles for targeted drug delivery, diagnostics and multimodal imaging. Adequate functionalization is here the major challenge and several possibilities are discussed. Also, it pays attention to potential of nano-particle toxicity. Chapter 4 reviews methods for preparation of (nano-) liposomes and for drug loading, together with the critical changes necessary for up-scaling to industrial manufacture. Also, characterization techniques are reviewed in their relation to successful application in the clinical setting of oncology. Chapter 5 deals with dry powder inhalers as alternatives to pressurized metered dose inhalers for administration of aerosols to the lungs. Both preparatory methods for powders where the particle properties lead to adequate dispersibility and delivery devices are described. Chapter 6 describes powder blending and sampling for analysis. It also compares old and new methods for characterization of the degree of segregation/mixing of powders. Chapter 7 gives guidance to control of PSD parameters to achieve adequate uniformity of solid dosage units for ensuring patient safety. Moreover, it discusses ways to estimate and mitigate over-potent doses in tablets through application of risk analysis with respect to the presence of small large particles (agglomerates). Chapter 8 overviews the characteristics that determine dissolution rate of particulate materials, which is essential for bioavailability. The effects of particle size, surface nature and crystal type are reviewed, with special attention to preparation and stabilization of nano-particles. In Chap. 9 amorphous state technologies are described for quicker release of APIs, with special attention for preparation of amorphous APIs and characterization of API morphology. Chapter 10 gives an overview of hydrophilic matrix tablets for extended drug release. It discusses the mechanism for release and the influence of the properties of the particulate components in this mechanism. In Chap. 11 the critical quality attributes of coating liquids, their preparation and the coating process of pharmaceutical tablets are described, which coatings are applied for a multitude of reasons, from advanced drug delivery to cosmetic purposes and easy recognition. Chapter 12 overviews the particulate attributes relevant for the functionality of the formulation of semi-solid pharmaceutical products—creams, ointments, gels, pastes and lotions—and the principal techniques for characterization of these attributes. The final Chap. 13 describes several case studies on the nano-pathological hazards and risks that particles, coming from the medicine itself or present as an unwanted contaminant, can exert on patients.

1.7 Definitions, Abbreviations and Symbols

Adsorbent	Solid that adsorbs the adsorptive
Adsorbate	Enriched adsorptive at the surface of an adsorbent
Adsorptive	Gas or vapor to be adsorbed to the surface of a solid particle
Aerodynamic diameter	Diameter of a sphere with density 1000 kg/m^3 having the same aerodynamic property as the particle
Apparent particle density	Mass of material per unit volume, including closed and (some) open pores (depending upon the accessibility of the liquid to the open pores)
Aspect ratio	Ratio of minimum to maximum Feret diameter of a particle (this is standardized by ISO [32] because it yields a value between 0 and 1; in some fields of application the ratio of length to breadth is still used)

Breadth	Minimum Feret diameter of a particle projection when the particle is resting in its position of maximum stability
Cohesivity	Stickiness of a powder, caused by strong attractive forces between individual particles relative to gravitational forces as well as by mechanical particle interactions that limit their movement
Effective particle density	Mass of material per unit volume, excluding open pores but including closed pores if present
Elongation	One minus the ratio of (geodesic) breadth to length of a particle image (typically used for long particles, which can reach a value close to 1 spheres have a value = 0)
Equant	Particle shape showing similar length, breadth and thickness (typical dimensional deviations are smaller than 40%)
Equivalent size	Diameter of an equivalent sphere (or circle)
Equivalent sphere	Sphere that shows the same properties as the observed particle in relation to a given measurement principle
Feret diameter	Distance between two parallel tangents on opposite sides of the image of a randomly oriented particle
Hydrodynamic diameter	Particle size calculated from the diffusion coefficient of the particle in a liquid according to the Stokes-Einstein equation it includes a layer of adsorbed molecules to the particles
Length	Feret diameter perpendicular to the breadth of a particle (often the maximum Feret diameter is meant)
Polymorphism	Ability of a substance to exist in two or more crystalline forms that have different arrangements of the molecules within the crystal lattice. Polymorphs have different physical properties, such as solubility, dissolution rate and melting point
Pore	Cavity or channel in an object, such as a particle
Segregation	De-mixing of particles in a particulate mixture (e.g. powder) due to differences in size, shape and/or density of the particles
Skeleton density	Mass of a non-porous material per unit volume
Specific surface area	Surface area of particles per unit mass
Step ratio	Ratio of upper to lower size of a size class
Test sample	Representative sample of adequate size that is used entirely for testing
True particle density	Mass per unit volume of non-porous materials (skeleton density)
Void	Space in between particles, usually in a powder

AAS	Atomic absorption spectroscopy
AES	Atomic emission spectroscopy
API	Active pharmaceutical ingredient
ASTM	American Society for Testing and Materials, USA
ATR	Attenuated total reflection
DLS	Dynamic light scattering
DSC	Differential scanning calorimetry
DTA	Differential thermal analysis
EDS	Energy-dispersive spectroscopy
EDX	Energy-dispersive X-ray spectroscopy
EDXS	Energy-dispersive X-ray spectroscopy
EMA	European medicines agency
FDA	Food and drug administration, USA
HoQ	House of quality
HPLC	High performance (pressure) liquid chromatography
ICP	Inductively coupled plasma
IR	Infrared
ISO	International organization for standardization
MS	Mass spectrometry
NIR	Near-infrared (spectroscopy)
NTA	Nanoparticle tracking analysis
PAT	Process analytical technology
PCA	Principal component analysis
PCS	Photon correlation spectroscopy
Ph.Eur.	European pharmacopoeia
PSD	Particle size distribution
PTA	Particle tracking analysis
PXRD	Powder X-ray diffraction
QbD	Quality by design
QFD	Quality function deployment
SEM	Scanning electron microscopy
SOP	Standard operating procedure
TDS	Time-domain spectroscopy
TEM	Transmission electron microscopy
THz	Terahertz (spectroscopy)
USPharm	USA pharmacopoeia
UV	Ultraviolet (spectroscopy)
VIS	Visible light (spectroscopy)
XRD	X-ray diffraction

CI	Carr compressibility index
D	Particle size; equivalent sphere diameter
$\langle D_{1,0} \rangle$	Number-weighted mean size
$\langle D_{3,0} \rangle$	Mean volume diameter

$\langle D_{3,2} \rangle$	Surface area-weighted mean size; Sauter mean diameter
$\langle D_{4,3} \rangle$	Volume-weighted mean size
D_{10}	Particle size at 10% point of cumulative, undersize distribution; a second subscript indicates the type of distribution (0 for number based, 2 for area based and 3 for volume based)
D_{50}	Median particle size (at 50% point of cumulative distribution); a second subscript indicates the type of distribution (0 for number based, 2 for area based and 3 for volume based)
D_{90}	Particle size at 90% point of cumulative, undersize distribution; a second subscript indicates the type of distribution (0 for number based, 2 for area based and 3 for volume based)
ff_c	Flow function
H	Hausner ratio
n	Number of particles
P	Pressure
P_0	Saturation pressure, where all pores are filled with adsorptive
s_g	Geometric standard deviation
subscript	A subscript 0 to a PSD parameter relates to number, 1 to length (size), 2 to area and 3 to volume
ε	Void fraction in a powder bed (powder bed porosity)
γ	Surface (interface) tension
ρ	Density
σ_1	Consolidation stress
σ_c	Compressive strength or unconfined yield strength.

References

1. Alavi S, Kingsley Ambrose RP (2016) Particulate flow and agglomeration in food extrusion, Chap. 5 in [56]
2. Alderliesten M (2016) Mean particle diameters, part VIII; computer program to decompose mixtures of (truncated) lognormal particle size distributions using differential evolution to generate starting values for nonlinear least squares. Part Part Syst Charact 33:675–697
3. Alghunaim A, Kirdponpattara S, Newby BZ (2016) Techniques for determining contact angle and wettability of powders. Powder Technol 287:201–215
4. Anderson W, Kozak D, Coleman VA, Jämting ÅK, Trau M (2013) A comparative study of submicron particle sizing platforms: accuracy, precision and resolution analysis of polydisperrse particle size distributions. J. Colloid Interface Sci 405:322–330
5. ASTM D6128, Standard test method for shear testing of bulk solids using the Jenike shear cell. ASTM International
6. ASTM D6393, Standard test method for bulk solids characterization by Carr indices. ASTM International
7. ASTM D6682, Standard test method for measuring the shear stresses of powders using the Peschl rotational split level shear tester. ASTM International
8. ASTM D6683, Standard test method for measuring bulk density values of powders and other bulk solids as a function of compressive strength. ASTM International

9. ASTM D6773, Standard test method for bulk solids using the Schulze ring shear tester. ASTM International
10. ASTM D7481, Standard test method for determining loose and tapped bulk densities of powders using a graduated cylinder. ASTM International
11. ASTM E2834, Standard guide for measurement of particle size distributions of nanomaterials in suspension by nanoparticle tracking analysis. ASTM International
12. Besseling R, Damen M, Tran T, Nguyen T, van den Dries K, Oostra W, Gerich A (2015) An efficient, maintenance free and approved method for spectroscopic control and monitoring of blend uniformity: the moving F-test. J Pharm Biomed Anal 114:471–481
13. Blanco M, Alcalá M, González JM, Torres E (2006) A process analytical technology approach based on NIR spectroscopy: tablet hardness, content uniformity and dissolution test measurements of intact tablets. J Pharm Sci 95(10):2137–2144
14. Blanco M, Villarroya I (2002) NIR spectroscopy: a rapid-response analytical tool. Trends Anal Chem 21(4):240–250
15. Burns DA, Ciurczak EW (eds) (2008) Handbook of near-infrared spectroscopy, 3rd edn. CRC Press
16. Carriere J, Heyler R (2016) Low frequency/THz-Raman: real-time monitoring of structural forms in pharmaceutical applications. Application note. Eur Pharm Rev, 19 Dec 2016
17. Carriere J, Heyler R, THz-Raman: advancing efficiency and reliability in pharmaceutical analysis and quality control. Application Note Ondax
18. Chen S, Øye G, Sjöblom J (2005) Rheological properties of aqueous silica particle suspensions. J Disp Sci Technol 26:495–501
19. Chen S, Øye G, Sjöblom J (2005) Rheological properties of silica particle suspensions in mineral oil. J Disp Sci Technol 26:791–798
20. Ciurczak EW, Igne B (2015) Pharmaceutical and medical applications of near-infrared spectroscopy, 2nd edn. CRC Press
21. Cleland AN, Fraikin J-L, Meinhold P, Mouzon F (2016) Nanoparticle characterization—one size does not fit all: nanoparticle size analysis for nanomedicine applications. Drug Dev Deliv 16(2):20–26
22. Coulter W (1953) Means for counting particles in a fluid. US Patent 2,656,508
23. El-Hagrasy AS, Drennen JK III (2006) A process analytical approach to near-infrared process control of pharmaceutical powder blending. Part III. Quantitative NIR calibration for predicting blend homogeneity and characterization of powder mixing kinetics. J Pharm Sci 95(2):422–434
24. Emady H, Hapgood K, Smith R (2016) Granulation and tabletting, Chap. 4 in [56]
25. Figueiredo MM (2000), Electrozone sensing in particle size analysis. In: Meyers RA (eds) Encyclopedia of analytical chemistry, Wiley, pp 5358–5372
26. Fraikin J-L, Teesalie T, McKenney CM, Ruoslahti E, Cleland AN (2011) A high-throughput label-free nanoparticle analyzer. Nat Nanotechnol 6:308–313
27. Gemci T, Chigier N (2016) Atomization, spraying and nebulization, Chap. 7 in [56]
28. Henriquez RR, Ito T, Sun L, Crooks RM (2004) The resurgence of coulter counting for analyzing nanoscale objects. Analyst 129(6):478–482
29. Hickey AJ, Xu Z (2014) Dry powder inhalers, Chap. 10 in [55]
30. ISO 2591, Test sieving. International Organization for Standardization
31. ISO 3310, Technical requirements for test sieves. Part 1: test sieves of metal wire cloth. Part 2: test sieves of perforated metal plate. Part 3: test sieves of electroformed sheets. International Organization for Standardization
32. ISO 9276, Representation of results of particle size analysis. Part 1: graphical representation. Part 2: means and moments. Part 3: adjustment to a reference model. Part 4: characterization of a classification process. Part 5: calculation of a lognormal probability distribution. Part 6: particle shape and morphology. International Organization for Standardization
33. ISO 9277, Surface area by gas adsorption. International Organization for Standardization

34. ISO 13099, Zeta potential of colloidal systems. Part 1: by electro acoustic and electro kinetic phenomena. Part 2: by optical methods. Part 3: by acoustic methods. International Organization for Standardization
35. ISO 13319, Determination of particle size distributions—electrical sensing zone method. International Standards Organization
36. ISO 13320, Laser diffraction. International Organization for Standardization
37. ISO 13321, Photon correlation spectroscopy. International Organization for Standardization
38. ISO 13322, Image analysis methods. International Organization for Standardization. Part 1: static methods. Part 2: dynamic methods
39. ISO 14488, Sampling and sample splitting. International Organization for Standardization
40. ISO 14887, Dispersion procedures for powders in liquids. International Organization for Standardization
41. ISO 15901, Pore size distribution and porosity. Part 1: mercury porosimetry. Part 2: macro- and mesopores by gas sorption. Part 3: micropores by gas sorption. International Organization for Standardization
42. ISO 19430, Particle tracking analysis. International Organization for Standardization
43. ISO 20998, Ultrasound attenuation spectroscopy. Part 1: concepts and procedures. Part 2: guidelines for linear theory. Part: 3: guidelines for non-linear theory. International Organization for Standardization
44. ISO 21154, Determination of skeleton density by gas pycnometry. International Organization for Standardization
45. ISO 22412, Dynamic light scattering. International Organization for Standardization
46. Ji J, Fitzpatrick J, Cronin K, Fenelon MA, Miao S (2017) The effects of fluidised bed and high shear mixer granulation processes on water adsorption and flow properties of milk protein isolate powder. J Food Eng 192:19–27
47. Klinzing GE (2016) Solids transport and handling, Chap. 14 in [56]
48. Kozak D, Anderson W, Vogel R, Chen S, Antaw F, Trau M (2012) Simultaneous size and zeta-potential measurement of individual nanoparticles in dispersion using size-tunable pore sensors. ACS Nano 6:6990–6997
49. Lemkowitz SM, Pasman HJ (2014) Assessment and control of fire and explosion hazards and risks of particulates, Chap. 4 in [55]
50. Liu Z, Wang Y, Muzzio FJ, Callegari G, Drazer G (2017) Capillary drop penetration method to characterize the liquid wetting of powders. Langmuir 33:56–65
51. May RK, Evans MJ, Zhong S, Warr I, Gladden LF, Shen Y, Zeitler JA (2011) Terahertz in-line sensor for direct coating thickness measurement of industrial tablets during film coating in real time. J Pharm Sci 100:1535–1544
52. McGlinchey D (2016) Sampling and characterization of bulk particulate materials and products, Chap. 15 in [56]
53. Mehn D, Caputo F, Rösslein M, Cabrolai L, Saint-Antonin F, Courant T, Wick P, Gilliland D (2017) Larger or more? Nanoparticle characterization methods for recognition of dimers. RSC Adv 7:27747–27754
54. Merkus HG (2009) Particle size measurements, fundamentals, practice, quality. Springer
55. Merkus HG, Meesters GMH (eds) (2014) Particulate products, tailoring properties for optimal performance. Springer
56. Merkus HG, Meesters GMH (eds) (2016) Production, handling and characterization of particulate materials. Springer
57. Merkus HG (2009) Sampling of particulate material, Chap. 4 in [54]
58. Merkus HG (2009) Dispersion of powders in air and liquids, Chap. 5 in [54]
59. Merkus HG (2014) Introduction, Chap. 1 in [55]
60. Merkus HG (2014) Basic information for design of particulate products, Chap. 2 in [55]
61. Merkus HG (2014) Measurement of particle size, shape, porosity and zeta-potential, Chap. 3 in [55]
62. Merkus HG (2014), Chocolate, Chap. 8 in [55]
63. Metrohm, Monograph NIR Spectroscopy (2013). Metrohm, Herisau, Switzerland

64. Muzzio FJ, Goodridge CL, Alexander A, Arratia P, Yang H, Sudah O, Mergen G (2003) Sampling and characterization of pharmaceutical powders and granular blends. Int J Pharm 250:51–64
65. Palou A, Cruz J, Blanco M, Tomas J, de los Rios J, Alcala M (2012) Determination of drug, excipients and coating distribution in pharmaceutical tablets using near-infrared—chemical imaging (NIR-CI). J Pharm Anal 2(2):90–97
66. Peiponen K-E, Zeitler JA, Kuwata-Gonokami M (eds) (2013) Terahertz spectroscopy and imaging. Springer series in optical sciences, vol 171. Springer
67. Pharmacopeia, Absorption spectrophotometry UV and VIS, USPharm section ⟨1092⟩, Ph. Eur. Sect. 2.2.25
68. Pharmacopeia, Analytical sieving, USPharm section ⟨786⟩, Ph.Eur. Sect. 2.9.38
69. Pharmacopeia, Bulk density, USPharm section ⟨616⟩, Ph.Eur. Sect. 2.9.34
70. Pharmacopeia, Bulk powder sampling, USPharm section ⟨1097⟩
71. Pharmacopeia, Density of solids by gas pycnometry, USPharm section ⟨699⟩, Ph.Eur. Sect. 2.9.23
72. Pharmacopeia, Infrared spectroscopy, USPharm section ⟨197⟩ and ⟨854⟩, Ph.Eur. Sect. 2.2.24
73. Pharmacopeia, Inhalation and Nasal drug products, performance quality test—impactors, USPharm section ⟨601⟩, Ph.Eur. Sect. 2.9.19
74. Pharmacopeia, Laser diffraction, USPharm section ⟨429⟩, Ph.Eur. Sect. 2.9.31
75. Pharmacopeia, Liquid chromatography, USPharm section ⟨621⟩, Ph.Eur. Sect. 2.2.29
76. Pharmacopeia, Mass spectrometry, USPharm section ⟨736⟩, Ph.Eur. Sect. 2.2.43
77. Pharmacopeia, Microcalorimetry (crystallinity), USPharm section ⟨696⟩, Ph.Eur. Sect. 2.2.61
78. Pharmacopeia, Near-infrared spectroscopy, USPharm section ⟨1119⟩, Ph.Eur. Sect. 2.2.40
79. Pharmacopeia, Optical microscopy, USPharm section ⟨776⟩, Ph.Eur. Sect. 2.9.37
80. Pharmacopeia, Particle shape, USPharm section ⟨776⟩, Ph.Eur. Sect. 2.9.37
81. Pharmacopeia, Porosity and pore size distribution by mercury intrusion, USPharm section ⟨267⟩, Ph.Eur. Sect. 2.9.32
82. Pharmacopeia, Powder fineness, USPharm section ⟨811⟩, Ph.Eur. Sect. 2.9.35
83. Pharmacopeia, Powder flowability, USPharm section ⟨1174⟩, Ph.Eur. Sect. 2.9.36
84. Pharmacopeia, Raman spectroscopy, USPharm section ⟨197⟩, Ph.Eur. Sect. 2.2.48
85. Pharmacopeia, Scanning electron microscopy, USPharm section ⟨1181⟩
86. Pharmacopeia, Sieves, USPharm section ⟨786⟩, Ph.Eur. Sect. 2.1.4
87. Pharmacopeia, Sieve test, USPharm section ⟨786⟩, Ph.Eur. Sect. 2.9.12
88. Pharmacopeia, Specific surface area by gas adsorption, USPharm section ⟨846⟩, Ph.Eur. Sect. 2.9.26
89. Pharmacopeia, Thermal analysis, USPharm section ⟨891⟩, Ph.Eur. Sect. 2.2.34
90. Pharmacopeia, Viscosity, USPharm section ⟨911–913⟩, Ph.Eur. Sect. 2.2.8
91. Pharmacopeia, Viscosity by capillary viscometer method, USPharm section ⟨911⟩, Ph.Eur. Sect. 2.2.9
92. Pharmacopeia, Viscosity by falling ball viscometer, USPharm section ⟨913⟩, Ph.Eur. Sect. 2.2.49
93. Pharmacopeia, Viscosity by rotating viscometer method, USPharm section ⟨912⟩, Ph.Eur. Sect. 2.2.10
94. Pharmacopeia, X-ray powder diffraction, USPharm section ⟨941⟩, Ph.Eur. Sect. 2.9.33
95. Reich G (2005) NIR spectroscopy and imaging: basic principles and pharmaceutical applications. Adv Drug Delivery Rev 57:1109–1143
96. Scholten E (2014) Ice cream, Chap. 9 in [55]
97. Schulze D (2016) Storage and discharge of bulk solids, Chap. 13 in [56]
98. Schulze D (2008) Powders and bulk solids—behavior, characterization, storage and flow. Springer, Berlin
99. Shen Y-C (2011) Terahertz pulsed spectroscopy and imaging for pharmaceutical applications. Int J Pharm 417:48–60

100. Sivakumaran M, Platt M (2016) Tunable risistive pulse sensing: potential applications in nanomedicine. Nanomedicine 11(16):2197–2210
101. Sullivan M, King E, Kato E, Heaps D, McKay R, Zhou XH (2013) Tablet analysis, using terahertz spectroscopy to map tablet cores and coatings. Tablets & Capsules, CSC Publishing, Oct 2013
102. Taylor MK, Ginsburg J, Hickley AJ, Gheyas F (2000) Composite method to quantify powder flow as a screening method in early tablet or capsule formulation development; AAPS Pharm Sci Tech 1(3) Article 18
103. UOP 851, Density of powders and solids by helium displacement. ASTM International
104. Vogel R, Willmott G, Kozak D, Roberts GS, Anderson W, Groenewegen L, Glossop B, Barnett A, Turner A, Trau M (2011) Quantitative sizing of nano/microparticles with a tunable elastomeric pore sensor. Anal Chem 83(9):3499–3506
105. Vonk P (2016) Industrial aspects of crystallization, Chap. 2 in [56]
106. Weinekötter R (2016) Mixing of solid materials, Chap. 9 in [56]
107. Wibroe P, Ahmadvand D, Oghabian MA, Yaghmur A, Moghimi SM (2016) An integrated assessment of morphology, size, and complement activation of PEGylated liposomal doxorubicim products Doxil®, Caelyx®, DOXOrubicin, and SinaDoxosome. J Control Release 221:1–8
108. Zeitler JA, Shen Y-C (2013) Industrial applications of terahertz imaging, Chap. 18 in [66]

Henk G. Merkus graduated in 1960 in physical organic chemistry at the University of Amsterdam. He worked several years at the Royal Dutch Shell Laboratories in Amsterdam on research in the field of detergents and industrial chemicals, followed by development work on thermal wax cracking for production of C_2–C_{14} olefins and on acid-catalyzed synthesis of carboxylic acids from C_3–C_6 olefins. Then, he made the change to analytical chemistry, involving both measurements and method development with a large variety of techniques and methods, first at Shell's process development department in Amsterdam and later in the chemical engineering department of Delft University of Technology. Gradually, his analytical horizon widened: first surface area and porosity measurements were added to chemical analysis, later followed by particle size analysis. In those areas he participated in many university, national and international courses and national and international education and research projects. He is the author of about 45 journal articles. He retired from Delft University in 2000 but remained active in the field of particle characterization. He is author of the book 'Particle Size Measurements—Fundamentals, Practice, Quality' (2009; Springer) and many journal articles, as well as co-author and co-editor of the books 'Particulate Products—Tailoring Properties for Optimal Performance' (2014; Springer) and 'Production, Handling and Characterization of Particulate Materials' (2016, Springer). Moreover, he participates since 1987 in standardization activities regarding particle size measurement, both in the Dutch standards organization NEN and the international ISO organization (TC24).

Chapter 2
Guide to Pharmaceutical Product Quality

Henk G. Merkus

Abstract Most drug constituents have a particulate nature, the properties of which may be critical for both the manufacturing process and the final drug quality. This chapter discusses means to reach and maintain optimum quality of pharmaceutical products. A scientific basis, as promoted by the approach of Quality by Design (QbD), for identification, measurement and control of the critical quality attributes of both pharmaceutical products and processes, with special attention for particle size, shape and morphology, may provide good opportunities for decreasing their development time. In addition, the Process Analytical Technology (PAT) framework, which promotes application of modern analytical instruments and chemometric tools for process monitoring and control, offers good opportunities to shorten and optimize manufacturing processes, including changes from batch to continuous processes. Thus, good and stable product quality may result at reduced manufacturing costs if pharmaceutical products and manufacturing processes are based on modern knowledge and instrumentation. This chapter further summarizes the general ICH guidelines for quality control of pharmaceutical products. Moreover, it discusses the quality aspects of some methods and techniques for particle size analysis and particle shape, acceptance criteria for particle size distributions, as well as qualification of instruments and methods, and management of their quality.

2.1 Introduction

Pharmaceutical products occur in many forms, viz. powders, capsules, tablets, solutions, emulsions, suspensions, ointments and sprays. Solid dosage forms are the main product type. Besides the active pharmaceutical ingredient (API), products typically contain excipients for optimum control of the desired product quality.

H. G. Merkus (✉)
Delft University of Technology, Park Berkenoord 30, 2641 Pijnacker,
The Netherlands
e-mail: henkmerkus@hetnet.nl

H. G. Merkus et al. (eds.), *Particles and Nanoparticles in Pharmaceutical Products*,
AAPS Advances in the Pharmaceutical Sciences Series 29,
https://doi.org/10.1007/978-3-319-94174-5_2

A major part of both API and excipients has a particulate nature. For users of pharmaceutical products adequate quality and safety must be ensured. This means for the quality attributes:

- Optimum efficacy for the intended purpose, with no or least negative side effects
- Content uniformity in dose quantities
- Optimum bioavailablity in direct or extended release of tablets
- Adequate product stability in liquid dispersions
- Adequate dispersability of powders in air for inhalers
- Optimum rheological properties for emulsions, suspensions and ointments
- Maximum safety for patients
- Absence of hazardous amounts of impurities and contaminants.

These properties also depend upon characteristics that relate to the processing steps of the products, e.g.:

- Flowability or cohesivity of powders
- Segregation behavior of dry particles during transport and storage
- Blend uniformity
- Tendency of small particles to form agglomerates
- Compactability of powders to tablets
- Vulnerability of large particles to breakage and attrition
- Non- or low dusting of dry powders for safe processing.

The particle size distribution is highly relevant for most quality attributes. For it may impact solubility and dissolution rate, bioavailablity, dispersability in air and possibilities to reach specific parts of the human respiratory system, powder flowability, tendency of particles to form agglomerates, particle breakability and powder dusting behavior. Often, also particle shape is important. Thus, sufficient characterization of particle size and shape is one of the crucial aspects in drug product development and its quality control. One of the prerequisites is that the critical quality aspects of the particle size distribution and particle shape have been assessed and laid down in specifications.

Adequate specifications for drug substances and products are essential for approval by the regulatory authorities. A specification is defined as a list of tests, references to analytical procedures and appropriate acceptance criteria, which are numerical values, ranges, or other criteria for the tests described. It establishes the set of criteria to which the material should conform in order to be considered acceptable for its intended use. 'Conformance to specifications' means that the drug product meets the listed criteria when tested according to the listed analytical procedures. Specifications are proposed and justified by the manufacturer. They are chosen to confirm the quality of the drug product, with focus on the characteristics ensuring the safety and efficacy of the product rather than full characterization [16].

2.2 Quality by Design and Process Analytical Technology

In early times, APIs originated from extraction from natural sources. Nowadays, they are mostly manufactured by chemical or biotechnological synthesis, followed by purification. Drug products are based on experience and/or knowledge on the relationship between chemical structure, formulation and bio-pharmaceutical effect. In the development phase, they are produced in small batches for testing purposes in formulations with excipients. The production process of a final solid product involves several unit operations, such as synthesis and purification of the API, milling, screening, blending of API with selected excipients to yield an appropriate formulation, compaction, and storage and handling of intermediate and final products. For liquid dispersions—suspensions and emulsions—one or more liquids come in the place of compaction. Sufficient API purity at small scale is most often reached through a batch crystallization process. In order to obtain an optimal final product from formulation and processing, chemical and physical tests are executed on API, excipients and final product to show e.g. the chemical structure, morphology, purity, solubility, powder flowability and particle size distribution as well as potential variations of product performance. Moreover, clinical tests should provide proof on therapeutic efficacy and safety of the drug product (potential negative side effects). When all tests are positive, the manufacturing processes are scaled-up. Scaling-up of conventional batch processes, however, provides serious challenges. This holds, for example, for mixing processes, where the size of equipment cannot be simply increased but good engineering requires that dimensionless numbers of process variables as well as mathematical modeling must be considered in the scaling up (cf. [50, 61]). Also for conventional batch crystallization, which is typically executed through cooling down of a solution to (super-) saturation and seeding with small crystals, local changes of process conditions with equipment size have to be accounted for. Over the last decades, process understanding of crystallization has strongly increased and is still further increasing. This has enabled changes of batch processes by using temperature programming, by making the process semi-continuous through introduction of external loops for milling and dissolution of fines, or by applying a small-scale continuous process, most often through application of anti-solvents, as well as by using (adaptive) process models for control (see e.g. [31, 64, 66]). Batch processes are usually less versatile than (semi-) continuous processes since the latter allow more freedom for addition of different components to the mixture at different times, and control of residence time is easier. Moreover, fully continuous processes may offer possibilities for process integration and intensification to reduce plant size and energy usage and to improve constant product quality at decreased overall costs, e.g. by integration of the API production with a formulation step with excipients [32].

Following scale-up, further tests are executed, the results of which are the basis for (preliminary) specifications and application for approval by the regulatory agencies.

Often, product quality is analyzed off-line in an analytical laboratory, whereas process conditions such as e.g. pH and temperature are measured in-process. The off-line type of analysis typically results in a considerable time delay in between product manufacture and availability of analysis results. This delay can be decreased significantly by analyzing at-line, on-line, in-line or in-process (see Table 2.1, for these PSD analysis types) [40]. This gives opportunities for better/faster process monitoring and control.

It must be realized, however, that an analytical technique can only be applied for process control if its intrinsic measurement time—varying from less than a second to half an hour—is short in comparison to the typical time constant of the process, which may vary from seconds to over an hour, and the measurement location has been chosen properly.

Over the years it has been recognized that quality cannot be tested into products, but that it should be built in by design. Also, the knowledge of both products and processes increased significantly and some continuous processes for pharmaceutical products were copied from other industrial sectors. In view of this, the US Food and Drug Administration (FDA) has launched early in the 21st century the Quality by Design (QbD) approach in combination with the Process Analytical Technology (PAT) framework as an alternative to the traditional way for process design and product approval, typically consisting of defining set points and operating ranges for process parameters while demonstrating process reproducibility to meet established acceptance criteria through tests [8, 63]. Few years later this approach was adopted by the European Medicines Agency (EMA), the International Conference on Harmonisation (ICH) and other national regulatory agencies [5, 18–21]. This harmonization aims in general at removal of redundancy and duplication in the development and review processes of drug products.

The QbD approach promotes that the desired good quality of pharmaceutical products and an improved efficiency of their manufacturing processes be reached by

Table 2.1 Types of PSD analysis

Type of analysis	Analysis location	Degree of automation	Typical delay time
Off-line	Analysis of a separated sample from the production equipment in a laboratory at some distance from the production site	Mostly manual	1–4 h
At-line	Analysis of a separated sample from and near the production equipment	Partly automated	$\leq$ 10 min
On-line	Analysis in a limited part of a side-line of the production equipment, where conditions may be changed (e.g. concentration)	Fully automated	$\leq$ 1 min
In-line	Analysis in a limited part of the main line of the production equipment (conditions cannot be changed)	Fully automated	$\leq$ 1 min
In-process	Analysis in a limited part of the production equipment at prevailing conditions	Fully automated	$\leq$ 1 min

design based on available scientific knowledge and understanding of the product and of the steps in their manufacturing process, and by adequate control of the process steps, rather than by trial and error. This is regarded possible since much progress has been made in chemical engineering of both processes and products, medical, biotechnological and pharmaceutical sciences and several new analytical methods and techniques for process monitoring and control have become available (see Sect. 1.5). In QbD, part of the approval of a product may be shifted from a full description of all ingredients, process steps and process conditions to a 'design space', which describes the relationships between the process inputs—material attributes and process parameters—and the critical quality attributes of the drug product, e.g. in mathematical equations or as components of a multivariate model. These relationships may be based on science and/or on experimental data [18, 20, 21, 30]. Together with optimum selection of process equipment, they help to identify the variables and their ranges within which consistent product quality can be achieved. Adaptations of product or process within this design space are not considered as a change and do not require further regulatory review. Movement out of the design space is considered to be a change, and would normally necessitate a regulatory post-approval process. Description of this design space including specifications for both ingredients and final product is now required for approval by the regulatory agencies to guarantee adequate product quality and safety.

QbD is a systematic approach to pharmaceutical development that begins with definition of product objectives. It emphasizes the importance of product and process understanding as well as of process control, all based on sound science, knowledge management and quality risk management. In many respects QbD is similar to the approach of Quality Function Deployment (QFD) and House of Quality (HoQ), used in other industrial sectors [45].

PAT comprises modern systems for science-based design, analysis, monitoring and control of all steps of manufacturing processes based on timely measurements of critical quality and performance attributes of raw and in-process materials as well as identification of the critical process variables that affect product quality. The measurement systems include appropriate statistical and information technology tools. The PAT tools can be categorized as [65]:

- Multivariate data acquisition and analysis tools.
- Modern process analysis tools.
- Endpoint monitoring and process control tools.
- Knowledge management tools.

Adoption of QbD and PAT offers beneficial opportunities for pharmaceutical manufacturing, such as:

- To improve process understanding and identification of causes for process deviations.
- To develop risk-mitigation strategies and reduce process failures.
- To ensure high and stable quality of products.
- To improve manufacturing efficiency by reducing cycle time.

- To base regulatory scrutiny on process knowledge and scientifically based risk management.
- To reduce manufacturing costs.

Altogether, the QbD/PAT approach promotes definition of a multidimensional design space in relation to a defined Quality Target Product Profile, in view of efficacy and patient safety and based on chemical and engineering knowledge, that contains:

- Molecular design of the active pharmaceutical ingredient and design of its synthesis and purification route
- Design of optimum API particulate properties, viz. size, shape and morphology, and assessment of the limits for adequate product quality
- Design of optimum product formulation in relation to the defined Quality Target Product Profile, and assessment of the limits for adequate product quality
- Design of optimum functionalization of nanoparticles, and assessment of the limits for adequate product quality
- Design of optimum route and conditions for product manufacturing, and assessment of the limits for adequate product quality
- Design of optimum process control instrumentation.

Preferably, the science-driven QbD/PAT approach for pharmaceutical development and manufacture should more specifically include the following issues [18–22]:

1. Define the Quality Target Product Profile (QTPP) of API and drug product. It consists of a prospective summary of the quality characteristics of the product with regard to safety and efficacy, that ideally will be achieved to ensure the desired quality for a stated disease. Thus, it includes e.g., intended use in clinical setting, route of administration, dosage form, delivery systems, dosage strength(s), bioavailability and stability.
2. Identify critical quality attributes (CQAs) of the drug product, including impurities, considering all aspects of QTTP, so that these quality attributes can be studied and controlled.
3. Design, test and specify a formulation (composition) for the drug product and identify critical material attributes (CMAs) of all composing materials.
4. Identify potential process steps and equipment for manufacture of the drug product, their critical process parameters (CPPs) and process robustness.
5. Identify optimum techniques and methods for analysis of API, excipients and final product as well as for process monitoring and control.
6. Execute clinical, chemical, physical and safety studies to find effects of formulation, processing equipment and conditions on product quality. Include studies on presence and source of impurities and contaminants.
7. Relate the CQAs to the CMAs of API, excipients, etc. and to the CPPs. Include addressing particulate material attributes that have high potential to affect or be affected by formulation and processing, such as agglomeration, cohesivity,

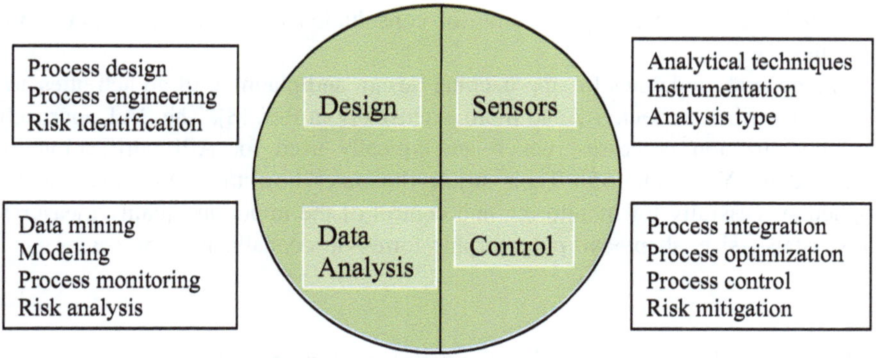

Fig. 2.1 Four elements of the QbD/PAT approach. Modified from [50]

compactability, friability and segregation of particles and powders. If necessary, further investigate the functional relationships that link CQAs to CMAs and CPPs and define them in the design space.

Note: Relating CMAs of the components to CQAs of the final product requires that the component attributes do not change during further processing, or that these changes are included in the relationships.

8. Perform a risk assessment of the formulated product and its manufacturing process, and select the optimum drug product formulation for the desired quality and the most appropriate process steps, equipment and conditions in view of expected or known simplicity, robustness and costs. Specify the results in the design space.
9. Set initial specifications for the CMAs and CQAs of the drug substances and the drug product, and design and implement a control strategy for reaching optimum process conditions and control limits to ensure product quality.
10. Describe the quality of the drug product, the control strategy, the risk-benefit analysis and the related design space in sufficient detail for formal approval.
11. Design a management plan to improve, during the lifecycle of the product, knowledge and understanding about the product and its manufacturing process and to evaluate the risks of side effects, and, if necessary, reconsider the initial specifications for the CQAs of the drug product as well as for CMAs, CPPs, control strategy and dosage.
12. Document all findings and the conclusions drawn.

The QbD/PAT approach combines four elements as illustrated in Fig. 2.1, viz. design, sensors, data analysis and control.

Of course, these issues can only be addressed by experts who have a broad knowledge and experience with regard to similar products, potential process steps, equipment and costs. To this end, usually expert teams are formed by a company that combine this knowledge and experience for discussion of potential equipment in relation to complexity and costs. Literature is, of course, helpful and specific research is typically executed to fill in gaps in the knowledge. Moreover, the FDA and EMA have formed expert teams for consultation to support new QbD/PAT activities.

The general guidelines for approaching design and planning of all activities are shortly described in Sect. 1.2 and more extensively in Ref. [8]. The PAT approach for crystallization processes, which are typically used for API purification, is reviewed by Yu et al. [65]. Their article illustrates how the PAT tools can be applied successfully for monitoring and control of the important quality measures for crystals—size, shape and polymorphic form—in crystallization processes.

2.3 Guidelines for Quality Control of Drug Substances

The International Conference on Harmonisation (ICH), a tripartite cooperation between the regulatory bodies of the European Union, Japan and USA, has approved a series of harmonized guidelines for quality control and consistency of new drug substances [11–22]. Guideline Q1 (stability testing) [11], Q4 (pharmacopoeia harmonisation) [14] and Q5 (viral safety of biotechnological products) [15] are less relevant in the context of this book.

Guideline Q6 [16] provides guidance on the setting and justification of acceptance criteria and the selection of test procedures for new drug substances of synthetic chemical, biotechnological and biological origin, and new drug products produced from them. This is done by requiring specifications, defining the criteria for a drug substance or drug product in order to be considered acceptable for its intended use. "Conformance to specifications" means that the drug substance and/or drug product, when tested according to the listed analytical procedures, will meet the stated acceptance criteria. Specifications are critical quality standards that are proposed and justified by the manufacturer and approved by regulatory authorities as conditions of approval of the drug product.

Decision tree #3 of this guideline defines when acceptance criteria for the PSD must be defined for e.g. solid dosage forms (see Fig. 2.2). Note that similar schemes can be used for other dosage forms such as suspensions, emulsions or sprays.

It will be clear from Fig. 2.2 that PSD criteria are only required if the size distribution has a significant influence on the performance quality of the solid drug product or its processing. In view of the fact that PSD parameters often play an essential role in product processing in addition to product quality, the related critical parameters should be addressed and specified. This role seems to be an underestimated topic in pharmaceutical development.

Guidelines Q9 provide guidance on the principles and examples of tools involved in quality risk management [19]. With respect to development and manufacture of drug substances these guidelines are further clarified in guidelines Q8, Q10, Q11 and Q12 [18, 20–22]. As in other industrial areas, risk is defined as the combination of the probability of occurrence of harm (damage to health) and the severity of that harm. Potential causes for harm are named hazards. An effective quality management approach requires:

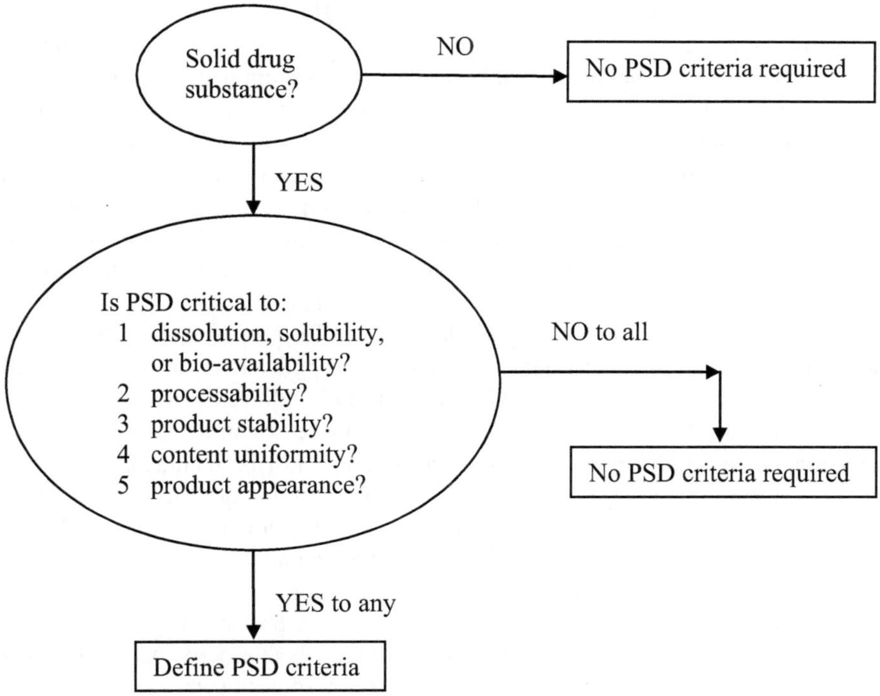

Fig. 2.2 Decision tree #3 for setting PSD acceptance criteria for e.g. solid dosage forms. Adapted from [16]

- Systematic and pro-active identification of hazards of the product itself, its constituents and its manufacturing process
- Analysis of the probability of occurrence and the severity of harm
- Reduction of risk by process optimization, effective monitoring and control
- Evaluation of risk through comparison with given criteria and a benefit-risk analysis
- Understanding that the quality of the drug product attributes should remain consistent with those used in the clinical trials
- Periodic monitoring and review of events during the lifecycle of the product to improve knowledge of product composition and its manufacturing process
- Effective communication of the risk involved in application of the product.

Adequate risk management, based on science and practical considerations, can improve the decision making in events when a quality problem arises.

Good manufacturing practice (GMP) is, of course, always required for the manufacturing process in order to guarantee that the specifications will be met during regular product manufacture. For this purpose, the manufacturer must establish a quality unit that is independent of production to fulfill both quality assurance and quality control responsibilities [17].

Guidelines Q10 complements Q7, Q8, Q9 and Q11 in that it sets requirements for knowledge management in addition to risk management and good manufacturing practice [20]. It defines a quality management system that contains a quality policy, quality planning, a quality monitoring system with periodic quality reviews and resource management with commitment by the company management. It serves to gather and improve knowledge on both product and manufacturing process during the lifecycle of the product—from the development stage, via technology transfer, through commercial manufacturing up till product discontinuation—to facilitate continual improvement of both product and manufacturing process through reduction of process variability and innovations.

Guidelines Q3 are intended to provide guidance for setting and justification of acceptance criteria (specifications) for registration applications on the content of impurities in new drug substances [13]. An impurity is defined as any component of the new drug product that is neither the drug substance nor an excipient. Although contaminants are not mentioned as impurities, they must be included. The specifications regard both organic impurities—coming from e.g. starting materials, by-products, intermediates, degradation products and residual solvents—and inorganic impurities, such as heavy metals or substances coming from reagents, ligands or catalysts. The typical minimum level for identification is set at 0.1%; lower levels are required for compounds that are toxic or have a pharmacological effect. In contrast to inorganic impurities coming from chemicals that participate in the API synthesis, contaminants may come from unexpected sources (see e.g. Chap. 13).

It should be noted that the sizes of particles in solid dosage forms (about 10–300 μm) are large in comparison to e.g. micro-emulsions, liposomes and (functionalized) nano-particles, which are submicrometre and have dimensions that go down to 1 nm. Also they often have a very complex structure. These (nano-) products may have unique properties that are quite different from their larger counterparts in terms of e.g. rate of dissolution, solubility, bioavailability and access to specific cells. For example, such products can be important for targeted drug delivery, therapeutics, diagnostics and specific imaging. However, also variability of these products and their potential toxicity deserves concern. The complexity of most of these products causes that their biological and physicochemical properties must be evaluated on an individual basis in relation to their specific composition and application. This is in accordance with the general FDA and EMA approach to assess each product on its own merits, through thorough characterization, utilizing established principles of benefit-risk analysis, with the same legal requirements for all products [6, 9, 10]. This framework for safety assessment is considered sufficiently robust and flexible to be appropriate for inclusion of nano-materials.

The analytical procedures for measurement of all relevant quality aspects of pharmaceutical products should conform to either US or EU Pharmacopoeia or other appropriate procedures, which are validated using appropriate reference standards. Guideline Q2 gives guidance to validation of the analytical procedures used for pharmaceutical products in terms of accuracy, precision, specificity, detection limit, quantitation limit, linearity, range and robustness [12]. However, neither this Guideline nor most Pharmacopoeia methods give guidance to validation of PSD measurement methods and instruments. Only the Pharmacopoeia laser diffraction procedure specifies the required quality [52]. Therefore, I will describe quality criteria and management in the following sections of this chapter.

2.4 Analysis of PSD and Particle Shape and Acceptance Criteria

The relevance of particle and powder characteristics to the performance and quality of drug products has already been described in Chap. 1 of this book. Also, the decision tree in Fig. 2.2 clearly identifies that particle size specifications are required if the particle size of the drug substance is critical to drug product performance, i.e., dissolution rate, solubility, bioavailability, content uniformity, stability, or to product appearance, or to drug product manufacturability (i.e. processability). The effect of PSD on processability may easily be overlooked when PSD parameters do not directly relate to drug product quality. However, the particle size of powders and pharmaceutical product quality can be significantly affected in most unit operations for manufacturing solid dosage forms [60]. For example:

- mixing and blending are meant to result in homogenous blends, but particles having different sizes, shapes and/or density typically show segregation if their size is greater than about 30 μm, which may have severe consequences for content uniformity if not taken into regard in the equipment design of mixers and hoppers
- particles larger than about 30 μm, especially crystals having a small breadth to length ratio, may be prone to breakage and attrition during processing, which results in presence of smaller particles
- small dry particles (especially sub-micrometre particles) typically form agglomerates having a substantially larger size, which may be difficult or impossible to break up again
- screening and sieving relate to separation of particles having different sizes; however, the separation is never perfect and brittle particles may break during sieving
- during tabletting, uniformity of content and weight strongly depend upon absence of segregation in the feed (typically counteracted by prior granulation of ingredients) and smooth particulate flow into the equipment
- milling and grinding relate to particle size reduction, but the product may show a too wide size distribution if no screening is applied.

Therefore, in addition to drug performance aspects also good understanding of the impact of particle characteristics of pharmaceutical powders and excipients on the manufacturing process, and vice versa, is critical for each drug application. If particle characteristics have significant impact, they can be regarded as critical material attributes (CMAs). Then, particle size specifications are required for each of the particulate constituents, which includes not only the drug substance but also the (major) excipients. Thus, their CMAs should be controlled to ensure manufacturing consistency, or their changes during processing must be taken into regard. In addition, specification of particle shape is required when e.g. the ratio of particle dimensions deviates strongly from unity, and a relation to segregation and particle breakability can be expected. Such adverse effects of particulate nature should be minimized by intelligent control of CMAs and adequate equipment design. On the other hand, if one or more processing steps have a significant influence of these CMAs, these effects should be taken into account as well.

Here, I will deal with quality aspects of particle sizing instruments, method validation and acceptance criteria for the three most popular techniques for size (and shape) analysis of particles, viz. sieving, laser diffraction and microscopy.

The general question to be answered is, of course, does the measurement instrument meet the requirements set by the product and the process.

Of course, all requirements should be defined properly. The most popular techniques for chemical analysis and measurement of particle size distributions are described in Chap. 1 and Refs. [23, 24, 26–29, 51–54, 56, 57]; Ref. [3] gives an overview of worldwide published standards relating to particle and spray characterization. The ISO methods also give guidance to qualification of measurement techniques and methods [23–29].

First, I will give definitions of terms that are relevant in the world of particle sizing for the quality of measurements, instruments and measurement procedures, as well as their qualification and validation, including some explanation [4, 34, 58]:

Accuracy is the closeness of agreement between a test result and an accepted reference value. It includes both random and systematic errors. Tests can be executed with a reference material for testing an instrument or an analysis procedure, or with an actual product to verify its quality, while using a standard operation procedure (SOP). Acceptance or rejection of the test result is based on preset acceptance values for the instrument, the operation procedure and the product (see later Sect. 2.5 on Shewhart chart). For (equivalent) size distributions of non-spherical particles, the same type of measurement instrument must be used for test values and reference values.

Bias is the difference between the mean value of a measurement (check), coming from a set of test results, and an accepted reference value. It represents the systematic error of the test measurement. Given the fact that most particles are non-spherical and particle size is the diameter of an equivalent sphere governed by the principle of measurement, bias is less important in particle sizing than precision. Still, it may be important if the accepted reference value comes for the same product, the same type of instrument and the same measurement method, as in proficiency tests and instrument qualification.

Precision is the closeness of agreement between multiple measurement results of a given property by the same operation procedure. Precision only relates to random errors (see also repeatability, instrument repeatability, method repeatability, reproducibility and intermediate precision). It is usually expressed as absolute standard deviation (s), or as relative standard deviation (RSD) or coefficient of variation (*CV*), both given as percentage with respect to the corresponding mean value.

Repeatability is the closeness of agreement between multiple measurement results of a given property, executed by the same operator with the same operation procedure in the same instrument within a short period of time, while using different aliquots of a sample as test sample. Thus, this kind of precision includes the effects of (sub-) sampling and sample dispersion and can be used to find optimum conditions to be implemented in a standard operation procedure (SOP).

Instrument repeatability is repeatability, while using the same dispersed sample aliquot in the same instrument under identical conditions. This kind of repeatability relates most closely to instrument quality and can be used to test the influence of instrument conditions for a given product and find optimum values for implementation in a SOP.

Method repeatability is repeatability, while using different aliquots of a sample with the same operation procedure in the same instrument, executed by different operators. Based on an optimized operating procedure, this kind of repeatability tests the quality of the operator to perform this analysis. The robustness of the method can be assessed through variation of the operational parameters of the procedure.

Reproducibility is the closeness of agreement between multiple measurement results of a given property in different aliquots of a sample, prepared and executed by different operators in similar instruments in different surroundings by the same operation procedure. Through using a SOP, it becomes apparent whether the surroundings have an influence. If so, the effect should be assessed in a robustness test.

Intermediate precision is the closeness of agreement between multiple measurement results of a given property in different aliquots of a sample by the same operation procedure. In contrast to reproducibility, here the same operator may test different instruments or different operators may test the same instrument. The latter test can give also information on the quality of the analyst and the robustness of the method. Different aliquots of a sample and the same operation procedure are used for the tests.

Resolution is the ability of a measuring device to distinguish meaningfully between different components of a mixture, or in case of PSD measurement between adjacent values of size.

Sensitivity is the change of (instrument) response with a change in absolute analyte concentration, or in case of PSD measurement in the amount or relative concentration of material in a specific size class of the distribution.

Quantitation limit is the lowest concentration or amount of analyte in a sample, or in case of PSD measurement in the relative concentration of material in a specific size class of the distribution, that can be determined with suitable precision and accuracy.

Robustness of a measurement procedure is its capability to remain unaffected by small variations in operational parameters. Typically, it involves a deliberate examination of the operational parameters that may influence the test result. The results of this examination yield the optimum values for the parameters as well as the robustness of the procedure in relation to (accidental) variations and, thus, gives an indication of the method's reliability during normal use.

Operation procedure describes all essential points of an analysis, including sampling, sample division, sample dispersion and analysis. Typically, a standard operation procedure (SOP) has been written for testing of each product or instrument of interest, which describes all operational details.

Method means a short description of the essential elements of an operation procedure without operational details.

Instrument qualification is providing documented evidence with a reference material that an instrument is operating in agreement with its specifications. Following installation of a new instrument, an ***installation qualification*** should provide evidence and documentation that the instrument has been installed properly in a suitable environment. The testing typically involves accuracy and precision while using a suitable reference material in combination with an appropriate SOP. It is executed by a trained engineer of the instrument manufacturer. At regular time intervals, a trained analyst of the user company can repeat this test to demonstrate continued evidence of satisfactory instrument performance, especially if doubt has arisen about its functioning (e.g. during validation or after significant repair or maintenance operation). Reference [7, 55] gives general guidelines for qualification of analytical instruments, whereas [58] gives guidance for particle sizing instruments.

Performance qualification is providing documented evidence with a reference material that a designed operation procedure is acceptable for all elements of its scope with respect to intended use (to be specified in product specifications). The testing typically involves accuracy, precision, quantitation limit and robustness of specified product quality attributes following development of an appropriate SOP for a specific own product. It is executed by a trained analyst of the product manufacturer.

Analyst qualification is providing documented evidence that an analyst is capable of measuring specified product quality attributes, while considering all types of potential measurement influences, and designing a suitable operating procedure for their measurement in a defined instrument. The testing typically involves optimum accuracy, precision, quantitation limit and robustness of the method applied. Both in-house reference materials and standard reference materials can be used for this qualification.

Calibration is the experimental determination, at defined conditions, of the relationship between the test result(s) or signal(s) of a measurement instrument and the corresponding value(s) of a certified reference material used as calibrant (e.g. for microscopic measurement of particle size or size of sieve apertures).

Method validation is providing documented evidence with a reference material that an operation procedure for measurement of a product quality attribute in a qualified instrument will consistently and at high degree of assurance operates in agreement with predetermined criteria. Here, typically a standard own product of the instrument user company is used for the testing, which involves preparing a test sample, sample dispersion and measurement. It is tested by different analysts of the product manufacturer, who have a different level of education.

Verification is judgment by an authorized person that data provided and conclusions drawn from them are correct.

Note: Similar to analytical instruments, also process equipment has to be qualified as for '*Installation Qualification*' (IQ, verifying proper installation), '*Operational Qualification*' (OQ, verifying performance of the equipment for the intended operation or range of operations) and '*Process Validation*' (PV, verifying repeatability and consistency of producing a drug product).

Now, we come to the qualifying and disqualifying aspects of the whole procedure for measurement of particle size distributions.

Particle size analysis involves preparation of a representative test sample from a bulk product, sample dispersion and measurement. As stated in Sect. 1.5.1 and Chap. 6, it is emphasized that inadequate **sampling** (sampling of a powder batch or stream and its subdivision to reach a test sample) is often the main source of measurement errors in case of broader size distributions [1, 2, 28, 33, 35, 47]. This is especially true if the bulk product is segregated.

In the practice of dry powders, following measures should be taken to reach optimum test samples:

- Minimize sampling errors due to segregation of bulk powder; use a fair number of primary samples/increments at adequate amount in relation to the required precision.
- Use the golden rules stated in Chap. 6.
- Apply rotating rifflers for (sub-) sampling of powders whenever possible.
- Define and use at least the minimum amount of test sample for measurement required by the fundamental error of the PSD characteristics.
- Use written, detailed standard operating procedures for preparation of test samples having defined and proven accuracies for the relevant PSD characteristics.
- Let well-trained operators perform manual operations.

In view of the fact that similar issues are involved in the feeding of e.g. tabletting equipment, and that measured PSD parameters are meant to be used to predict product performance during manufacture, the sampling steps between batch and tablets need similar consideration. In relation to potential segregation, API and excipient powders are usually premixed and granulated before further usage.

Sampling of emulsions and suspensions typically is more simple. Often good mixing by shaking or stirring before sampling is sufficient to reach a representative sample. The requirement for a minimum number of particles related to the accuracy of characteristics PSD parameters remains valid.

Given the challenges in obtaining representative test samples, it may be concluded that the full sampling procedure must be well designed, tested and laid down in a SOP, which also describes the related uncertainty of the PSD characteristics. Of course, these SOPs must be followed in full detail.

In relation to the above, ICH Guideline Q5 gives selection criteria for sampling of biotechnological/biological samples to be measured [15].

The next step in a measurement procedure for particle size distributions and particle shape of dry powders is **dispersion**, meaning for particle size measurement that the particles are presented as single units which do not overlap during measurement. This dispersion may proceed in air or in liquid; application of liquids requires that they neither swell nor dissolve the particles. Free-flowing dry particles, having a size greater than about 30 μm and equant shape, typically are present as non-clustered single particles. Then, dispersion can proceed simply from flow by gravity. Dry particles, having a size smaller than about 20 μm, typically form clusters. For particles greater than about 0.5 μm the attractive forces between the particles in these clusters are usually medium strong. Here, dispersion in air is still possible although stronger forces are required at decreasing particle size to reach primary particles. At wish, also dispersion in a suitable liquid can be applied, which involves wetting, dispersion (stirring) and stabilization (see Sects. 1.4.7, 1.4.8 and 1.4.13) [36]. When dry dispersion does not lead to full de-agglomeration, then dispersion of dry powders in a suitable liquid still may successfully lead to de-agglomerated primary particles. Here, dispersion forces coming from stirring or ultrasound are usually sufficient to break-up particle clusters. On the other hand, adequate care is required that the dispersion forces are indeed sufficiently strong for full de-agglomeration, but still weak enough not to cause particle breakage. However, for very small particles attractive forces are typically too strong for full de-agglomeration. This is the case for nanometer particles, and for particles that originate from a drying process, where the drying has caused the presence of some solid material in between neighboring particles (so-called solid bridging). When the bonding between the particles is too strong, usually only partial break-up of clusters can be achieved during liquid dispersion.

Dispersion of emulsions and suspensions for PSD measurement typically means dilution. Here, it should be ensured that the dilution liquid contains adequate stabilizers to maintain dispersion stability.

Note: If one wants to relate the parameters of the particle size distribution resulting from particle size measurement to their behavior in the manufacturing process, the conditions for de-agglomeration of clustered particles for the purpose of PSD measurement should agree with those during the manufacturing process. If these conditions have not been tuned appropriately, the measured PSD parameters cannot be used to predict product performance during manufacture.

The last step in PSD characterization is **measurement**. Sieving, laser diffraction and microscopy are the most common techniques for pharmaceutical materials. They differ in various ways (see Table 2.2).

In addition to the technical differences between the three techniques sketched in Table 2.2, there are a few more points that need attention:

Sieving. For test sieving to obtain a particle size distribution, woven-wire test sieves are most often applied. The apertures of these sieves show a fairly wide size distribution, to which tolerances set limitations [24, 38]. For some examples, see Table 2.3. Tolerances for electroformed sieves are much smaller, i.e. 2 μm; the thin

Table 2.2 Main differences between sieving, laser diffraction and microscopy

	Dry sieving	Laser diffraction	Microscopy		
			Optical	SEM	TEM
Size range, μm	32–10^4	0.1–10^4	0.3–500	0.01–500	0.001–5
Size type	Sieve	Scatter	Length, area, + shape[b]		
Quantity	Mass	Volume	Number		
Sample size, g	10–100	0.1–10	Depend on particle size and number		
Anal. time, min.	20	1–10	1–30		
Meas. quality[a]	+	++	++		
In-/on-line	No	Yes	Yes	No	No

[a]The quality indication is the author's impression of quality composed of mainly precision, resolution, robustness and sensitivity
[b]For shape measurement, generally higher magnification is required than for size measurement, since at least two size aspects are taken into account

surfaces of these sieves are, however, more vulnerable than that of woven-wire sieves.

It will be clear that the separation of particle sizes by sieving is never a sharp one due to the allowed differences in aperture size and the fact that each particle can only approach a limited number of apertures. Also, it must be clear that the size separation is dependent upon the test sieve applied, the load of material and the sieving conditions (i.e. the possibilities for each of the particles to approach and pass one or more apertures). Moreover, the shape of particles plays a role, since only their smallest cross section may be able to pass through an aperture. The general quality of sieves can be tested by microscopy and image analysis, or through using appropriate reference materials to find some kind of 'effective' aperture size at given conditions. Periodic qualification is recommended, since abrasive products may cause aperture wear and cleaning may cause defects in the surface of the sieve. Sieving performance of a product can be investigated by microscopic inspection of the sieve fractions.

Laser diffraction analysis. Modern instruments for this type of analysis are well designed and very stable during operation. Typically, accessory equipment is available for dry and wet analysis. A first point of attention during operation is that the particles to be measured can contact the window of the measurement cell or a lens at which they can adhere, which causes faulty signals as a background in subsequent measurements. Thus, background should be checked regularly while using a clean dispersion medium, and, if necessary, cell windows and lenses should be cleaned. A second point of attention is the optical model to be applied in the deconvolution from detector signals to volume-based particle size distribution. If most particles are smaller than about 30 μm and transparent, the Mie model must be used, which involves refraction of light in addition to diffraction. Thus, the refractive index of particles and suspension medium are important and these refractive index values must be reported. For larger particles, the approximating

Table 2.3 Aperture tolerances of woven-wire test sieves (adapted from [24])

	Tolerances	
Nominal aperture size (μm)	Max. aperture size (μm)	Average aperture size (μm)
500	589	482–518
250	308	240.1–259.9
125	163	119.2–130.8
63	89	59.3–66.7

Note In addition to the above tolerances, also a limit is set to the standard deviation of the aperture size distribution

Fraunhofer model for diffraction can be applied, which does not require refractive index information. Application of an inappropriate model or inappropriate refractive index values can cause significant errors at the lower side of the PSD. A third point of attention is the fact that elongated particles typically align to some extent during flow in the measurement zone and present a scattering profile to the detector, which is different in different directions. Thus, if different flow profiles exist in the measurement zone, or the detectors of an instrument are at another position (which may be the case in instruments of different manufacturers), then the recorded scattering pattern is different. This means that laser diffraction instruments of different manufacturers may yield different PSD results for the same product. A final point of attention concerns the dispersion conditions for clustered materials during dry analysis. Here, conditions have to be tuned to reach full de-agglomeration without breakage of particles. Especially for brittle particles such optimization can be a real challenge since there may be no distinct point of change from de-agglomeration to breakage. The general performance of laser diffraction instruments can be tested by well trained analysts through using appropriate reference materials (see also later in this section) [26, 39].

Microscopy. The particle size measured by static microscopy is usually related to the area of the largest cross-section of particles (most stable position in static microscopy). Good measurement requires low particle concentrations to avoid particle overlap, which means measurement of many frames for sufficiently large numbers of particles. In dynamic microscopy, potential alignment of particles has to be considered. The number-based PSD results of microscopy mean that a very large number of particles have to be measured for reaching a good repeatability. This holds especially for the upper end of the PSD, where a small number of particles represent a significant volume percentage. The general performance of microscopes can be tested through using appropriate reference materials or standards [27, 37].

In view of the above, optimum measurement results can only be obtained if following conditions are fulfilled:

- Standard Operating Procedures (SOPs) have been written that describe collection and preparation of a representative test sample from a batch of material, adequate powder dispersion and measurement instrument settings in full details.

- The SOPs have been tested to yield values for precision, resolution and quantitation limit for the critical PSD parameters of the material involved, which allow adequate characterization of the material.
- Regular qualification of procedures and measurement instruments is prescribed.
- All operations are executed conform the SOPs by well-trained analysts.
- Any deviation from normal practice is recorded.

With the exception of the laser diffraction method [52], methods for PSD analysis of pharmaceutical materials in the US and EU Pharmacopoeia do not give guidance themselves to qualification of and acceptance criteria for measurement techniques and methods, but they refer to ISO methods [23–29]. For such qualification of particle sizing instruments, three types of reference materials can be applied, viz. standard certified reference materials, instrument qualification standards and in-house product reference materials [41]. The standard reference materials in their primary form have been certified by an official certification institute. In their secondary form their certification is made by a commercial manufacturer through comparison with an official primary standard. These standards are used for general qualification of instruments (especially traceability and accuracy). The instrument qualification standards have been produced by single instrument manufacturers to qualify their instruments before and after delivery to users. Typically, values for D_{10}, D_{50} and D_{90} have been determined in many instruments (of the same type and manufacture) and bottles, and acceptance values set. Thus, they are applied for specific qualification of the instruments of that company, i.e. to ensure that they are operating in agreement with the instrument specifications.

Note: Typically, official certification institutes give an uncertainty indication for each certified property value of a measurand at a 95% level of confidence (expanded uncertainty, U_{crm}; standard uncertainty, u_{crm}, = $0.5 \cdot U_{crm}$). Other providers of reference materials often present an upper and lower limit for acceptance of an instrument or measurement procedure, which relates to the 95% limits of the population of measured data.

The ISO standards give acceptance values for PSD analysis instruments and methods (see Table 2.4) [23, 24, 26, 27].

The drawback of the standard certified reference materials is that their composition and properties usually are quite different from the pharmaceutical products. The in-house reference materials are typical own or related products, the critical parameters of which have been assessed by competent analysts in well-functioning instruments. Their advantage is that their performance can be quite similar to that of the actual product itself, so that sampling, dispersion and analyst quality can be included in the testing. Here too, adequate, written analysis procedures for the material should be prepared for the full measurement method, including sampling and dispersion. Regular validation of the SOPs with these in-house products is advised, in case of deviations supported by instrument qualification with suitable primary or secondary reference materials.

Table 2.4 Acceptance issues for laser diffraction instruments, microscopes and sieves

Laser diffraction[a]	– Regular instrument qualification/validation recommended – Instrument repeatability $D_{10,3} < \pm 2\%$, $D_{50,3} < \pm 1.5\%$, $D_{90,3} < \pm 3\%$ – Method repeatability $D_{10,3} < \pm 3\%$, $D_{50,3} < \pm 2.5\%$, $D_{90,3} < \pm 4\%$
Microscopy/image analysis	– Instrument qualification required to check homogeneity of the full viewed field (potential lens aberration) – Calibration required with standard certified reference material to convert pixel size into (micro-) meters – Repeatability ± 1 pixel; thus, it strongly depends upon magnification (number of pixels per particle image) – Repeatability of PSD parameters strongly depends upon the number of particles measured
Test sieves	– Most often woven-wire sieves are used – Regular calibration of aperture size required (typically by microscopy/image analysis, or by using reference material) – Tolerances depend upon aperture size (cf. Table 2.3)
Test sieving	– Accuracy sieve fractions better than 0.1% of mass of charge

[a]According to the new draft ISO 13320 (WD 13320 Revision V9, May 2018); the values for acceptable accuracy and precision are still subject to discussion until a definitive version of ISO 13320 has been approved

The acceptance criteria for the PSD of products must reflect the critical quality attributes of the product, which may not be the case for the statistical D_{10}, D_{50} and D_{90} values of the distribution. Thus, optimum PSD parameters for this purpose must be chosen, often in addition to the above quantile values. (cf Sect. 1.4.3). Moreover, the acceptance criteria should not only include the potential variation of the product but also that of the analysis method. Thus, optimum product specifications require analysis methods that have a significantly smaller variability than the allowed variation of the product.

Note: For instrument testing with reference materials having wide PSDs, the overall measurement precision is usually expressed as a relative standard deviation (RSD) of the D_{10}, D_{50} and D_{90} quantile values, which relate to 10, 50 and 90% of the cumulative, undersize size distribution, respectively, or of the quantities of material at more or less related size values (e.g. for sieve aperture sizes). For materials having narrow PSDs typically the modal size value and its RSD are taken. As indicated above, specific materials may require other PSD parameters for optimum characterization.

Data of a NIST certified primary reference material consisting of glass beads (SRM 1004b) are shown first (Fig. 2.3) [48]. Certification has been done by both scanning electron microscopy (SEM; the number-based data have been transformed into mass-based data) and sieve analysis while using woven-wire test sieves.

The figure shows a mass-based size distribution (equal to volume-based distribution, since the density of all particles is the same) ranging from about 40 to 150 μm. The expanded uncertainty of the certified values, corresponding to 2 RSD, coming from the SEM data is about 1–2% relative of the mass percentage. It must be noted that the good quality of the SEM data is reached by the fact that the size

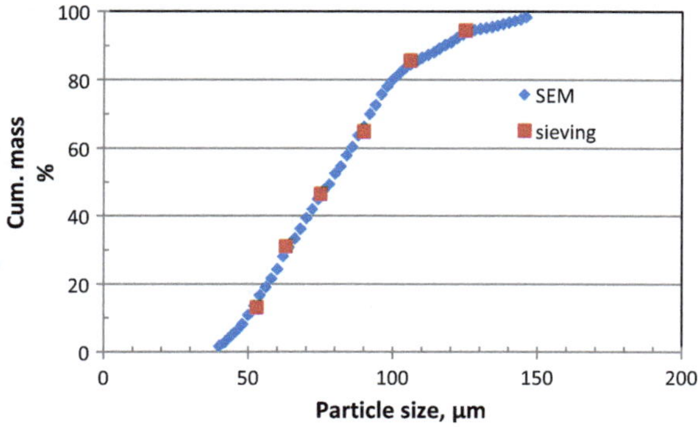

Fig. 2.3 Cumulative PSDs of NIST reference material SRM 1004b [48]

measurement was done for about 4000 particles divided over seven sieve fractions of the material at six different magnifications. This allowed optimum magnification for all particles in addition to adequate counting statistics and size resolution.

For the sieve data the expanded uncertainty of the certified values is somewhat poorer. This may be due to the variability of the aperture sizes (cf Table 2.3) as well as to the sieving procedure.

Figure 2.3 clearly shows that the microscopic technique allows many more calibration points than the sieving technique since more size classes can be defined. The reference material is intended primarily for use in evaluating and calibrating particle size measurement instruments covering the 40–150 μm range. It is commonly applied for and illustrates determination of the effective size of apertures of test sieves.

Different laser diffraction instrument manufacturers use different materials for qualification of their instruments. For example, Malvern uses glass beads (size range about 18–109 μm), Sympatec uses silicon carbide powders (size ranges about 2–10, 15–50 and 250–750 μm).

Unfortunately, no certified data for laser diffraction instruments are available for SRM 1004b from NIST. Therefore, also data are shown, in Fig. 2.4, for a secondary reference material—also glass beads—prepared and certified by Whitehouse Scientific [62].

This material has been certified using light microscopy and sieving in a NIST traceable manner. For this standard, the test sieving was executed by using electroformed sieves, having a much more uniform aperture size than woven wire sieves and following calibration with a NIST traceable reticule. Optical microscopy in combination with image analysis yielded somewhat poorer results; this is probably due to the single magnification applied and the relatively small number of particles measured in relation to the width of the PSD.

Since the mean values of sieving and microscopy agreed well, all data were combined, which resulted in certified values for the 1–95% values of the cumulative distribution to have a very small expanded uncertainty of about 0.1–0.4% for the certified measurands. For example, the certified $D_{10,3}$, $D_{50,3}$ and $D_{90,3}$ values are 93.7, 150.5 and 238.8 μm, respectively. The application of a multistage spinning

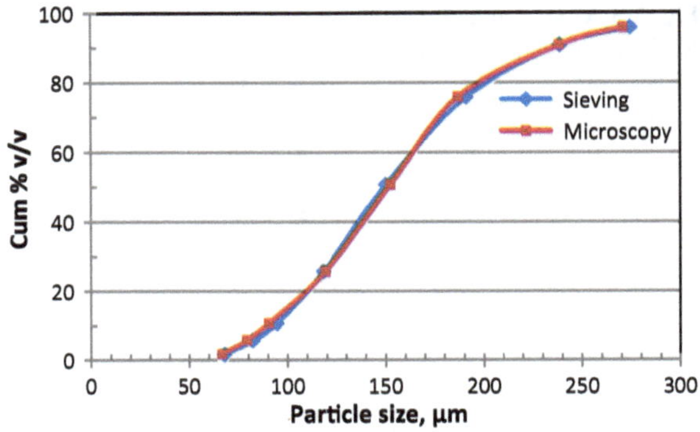

Fig. 2.4 Whitehouse polydisperse particle standard 50–350 µm [62]

riffler for sample subdivision to yield test samples from the larger batch of material promoted this good precision.

For this material, laser diffraction analyses were performed in a round-robin proficiency test under auspices of LGC Ltd [59]. Laser diffraction results for the $D_{10,3}$, $D_{50,3}$ and $D_{90,3}$ values coming from different instruments were compared with the assigned values and their uncertainties (see Fig. 2.5).

Figure 2.5 clearly shows that the results of the participating laboratories in the round-robin test vary largely. The overall mean quantile sizes $D_{10,3}$, $D_{50,3}$ and $D_{90,3}$ (expressed by the solid lines) agree well with the certified data. However, only about half of the results fall within the tolerances set at that time (2004) by ISO [26]. Many results exhibited a Z-score[1] greater than 2 and, thus, are considered questionable. Reasons for these deviations have not been investigated. They may be lack of experience of the analyst to collect representative test samples from a bottle containing free-flowing powders, application of an inadequate operation procedure for measurement and/or use of different types and age of instruments. In some cases —Z-score ≫ 3—results are unsatisfactory and seem to indicate failure of instrument or analyst, which should necessitate further investigation. Altogether, the results emphasize that presence of an adequate operation procedure as well as sufficient analyst education and regular qualification of instrument and analyst performance are necessary for obtaining and maintaining optimum results.

For analysis of nanoparticles typically electron microscopy and dynamic light scattering are applied. In relation to the difficulties in the size measurement of such very small particles and to the fact that most distributions are fairly narrow, often only the median size or the mean size of the distribution is certified. Figures 2.6 and 2.7 present the measurement results of NIST certified reference material RM 8013 [49].

This Reference Material is intended primarily to evaluate and qualify methodology and/or instrument performance related to the physical/dimensional characterization of nanoscale particles used in pre-clinical biomedical research and for

[1]Z-score = $(x - X)/y$, where x is the property value reported by the participant, X is the assigned property value and y is the standard uncertainty of the assigned property value.

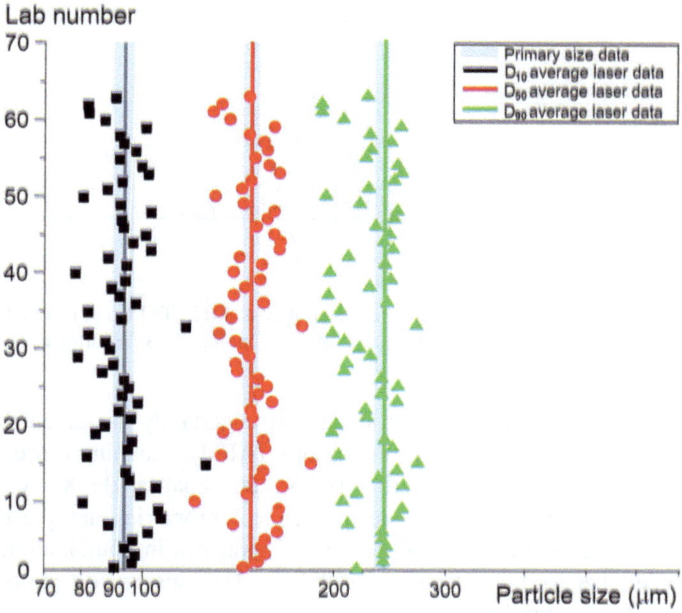

Fig. 2.5 Laser diffraction results of a round-robin with Whitehouse 50–350 μm reference material (tolerances for the $D_{10,3}$ (±5%), $D_{50,3}$ (±3%) and $D_{90,3}$ (±5%) are indicated by grey shade) [59]; reproduced by courtesy of Whitehouse Scientific

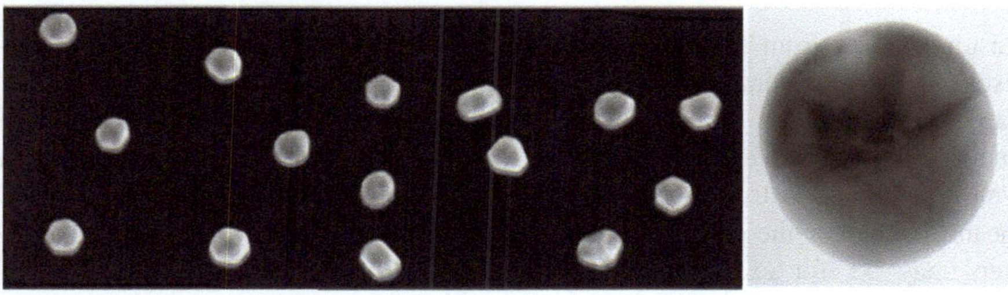

Fig. 2.6 SEM/TEM pictures of NIST certified gold nanoparticles RM 8013 at 2 magnifications [49]; images used with permission of National Institute of Standards and Technology, US Department of Commerce

inter-laboratory comparison tests. It consists of an aqueous suspension of citrate-stabilized, nearly spherical gold nanoparticles having a nominal size of 60 nm. The differential TEM analysis results—presented as number of particles per size class—show that the suspension contains mainly primary particles (distributed around a modal size of 55 nm) with a small number-percentage (about 1% n/n) of clusters of primary particles (size up till about 85 nm).

The number-based mean value for TEM was certified to be 56.0 ± 0.5 nm (value and its expanded uncertainty). Next to transmission electron microscopy

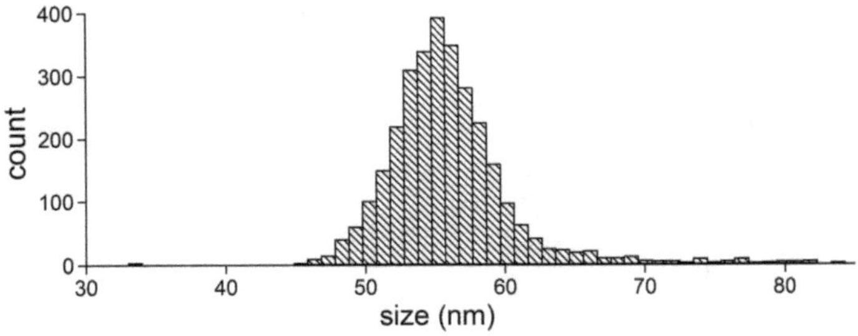

Fig. 2.7 Differential PSD of NIST reference material RM 8013 (TEM data) [49]. Figure used with permission of National Institute of Standards and Technology, US Department of Commerce

(TEM), certified values together with expanded uncertainty are given for several other techniques, viz. atomic force microscopy (AFM), scanning electron microscopy (SEM), dynamic light scattering (DLS) and small-angle X-ray scattering (SAXS). Moreover, chemical and electrochemical properties are presented; for example, the zeta-potential value of—37.6 mV indicates in combination with the small particle size good suspension stability. The suspension is completely translucent and has a bright red color.

2.5 Quality Management

Results coming from analytical measurements should show good day-to-day consistency. Measurement quality should be significantly better than product quality variations in order to allow adequate product specifications. Basic to optimum quality of any particle sizing result are a written standard operating procedure, a well-functioning instrument and a qualified analyst. Quality control charts are meant to monitor the consistency of analytical methods [25, 42]. Control parameters may be, for example, concentration of API in a dose, concentration of impurities, or PSD parameters such as the D_{10}, D_{50} and D_{90} (having a second subscript in relation to e.g. the number (0) or volume (3) basis of the distribution).

The **Shewhart chart** is an example of a control chart. It needs defining an acceptable quality level for each relevant parameter. Usually, this level is derived from determination of the mean value and the standard deviation of the parameter(s) in a given standard sample in a series of measurements, preferably spread over an appreciable time (e.g. some weeks), in view of the fact that small time intervals for setting up the control chart tend to flatter the standard deviation since long-term fluctuations are not taken into account. Then, a chart is constructed of data against time, in which the mean parameter value is marked as a horizontal line together with two other sets of parallel lines. One set relates to '**warning limits**', often defined by the 95% confidence level (±2 RSD) of measured values. According to statistical reasoning, then there is 5% probability for a good result to lie outside this

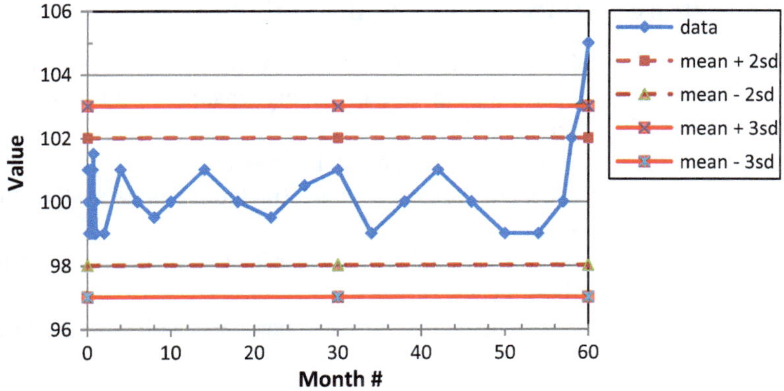

Fig. 2.8 Example of a Shewhart control chart

interval. The other set relates to '**action or alarm limits**', often defined by the 99.7% confidence level (±3 RSD), where there is only 0.3% probability that a good result lies outside this area. Other criteria for these limits may be taken as well, for example based on required performance (e.g. quite fit-for-purpose versus just good enough) instead of pure statistics, if this allows at least the same level of significance. Now, this chart is used to plot measured data in time. A simulated example of a Shewhart chart is given in Fig. 2.8 for a mean value 100 and a RSD of 1%.

The data show fluctuations up to month #58, followed by a rising trend and exceeding the warning limit at month #59 and the alarm limit at month 60. Exceeding the warning limits warns the operator that the instrument comes close to the stated limits of acceptable operation. A second determination may clarify the situation. Two or more successive points in between the warning and the alarm limits as well as a single point exceeding the alarm limit should lead to the conclusion that the instrument is out of specification and immediately needs cleaning or more intensive maintenance or repair. Also, it means disposal of all results after the last control test, since the instrument may have been out of specification in this period of time. Following simple maintenance the same Shewhart chart may be continued again. More intensive maintenance or repair may lead to the necessity of starting a new Shewhart chart.

The above testing relates to the probability (or significance) that a measured value does or does not belong to a population of values, given their mean and standard deviation. The above Shewhart chart assumes that the probability for rejecting a measured value relates to 3 standard deviations (which may not be predefined). This is usually named significance testing.

Other approaches exist as well, all of which are based on some kind of statistical evaluation. A first example is hypothesis testing, where acceptance or rejection of a measured value is set prior to measurement. A second example is an outlier test, according to e.g. the Grubbs' test or the Dixon's or Q-test. A third example is testing through the method of analysis by variance (ANOVA). This method is meant to give a decisive answer to the question whether several samples have the same or different quality [43].

2.6 Definitions, Abbreviations and Symbols

Acceptance	Decision that a product is produced, or a system is operating, in a satisfactory manner with respect to a predefined quality level
Accuracy	Closeness of agreement between a test result and an accepted reference value (including both systematic and random errors)
Agglomerate	Assemblage of primary particles with intermediate attractive forces
Aggregate	Assemblage of primary particles with strong attractive forces
Analysis method	Short description of the essential points of the analysis procedure
Analysis procedure	Written document describing all details of one or more operational steps in the characterization of a material or product property (primary sampling, sample splitting, dispersion, instrument calibration/qualification, instrument conditions and reporting)
Analysis technique	Type of instrument, including instrument settings, applied for analysis
Alarm/action limits	Points in a control chart, exceeding of which is unacceptable and should lead to immediate action (e.g. disposal of earlier results, maintenance)
Cluster	Assemblage of primary particles with undefined attractive forces
Detection limit	Lowest concentration or amount of analyte in a sample that can be determined but not necessarily quantified as an exact value
Mean size	Arithmetic mean particle size for a given population of particles, weighted according to number, volume, etc.
Median size	Particle size at the point in the cumulative size distribution, where 50% of the particles is smaller and 50% larger
Modal size	Most frequently occurring particle size in a size distribution
Polymorphism	Occurrence of different crystalline forms of the same drug substance; this may include hydration products (also known as pseudopolymorphs) and amorphous forms
Qualification	*General*: action of proving and documenting that equipment or ancillary systems are properly installed, work correctly, and actually lead to the expected results *For instruments*: proof with reference material that the instrument is operating in agreement with its specifications
Quantitation limit	Lowest concentration or amount of analyte in a sample that can be determined with suitable precision and

	accuracy; in case of size distributions it may be the lowest percentage of material in a stated size class that can be determined
Repeatability	Closeness of agreement between multiple measurement results of a given property, executed by the same operator with the same method within a short period of time
Representative sample	Sample that has the same properties as a defined batch of material that it is supposed to represent, within a defined confidence interval
Robustness	Measure of the capacity of an analytical procedure or a process to remain unaffected by small variations of its parameters or conditions
Specification	Precise statement of all requirements to be satisfied by a material, product, system, or service together with procedures for determination of the requirements and appropriate acceptance criteria
Test sample	Representative sample of adequate size that is used entirely for testing
Traceability	Property of a measurement result whereby it can be related in a quantitative way to appropriate (inter)national standards through an unbroken chain of comparisons (with stated uncertainties involved in all steps of comparison)
Validation	*General*: documented program that provides a high degree of assurance that a specific process, method, or system will consistently produce a result meeting pre-determined acceptance criteria. *For analysis procedure*: proof with reference material that the procedure is acceptable for all elements of its scope
Z-score	$(x - X)/y$, where x is the property value reported by the participant, X is the assigned property value and y is the standard uncertainty of the assigned property value.
CMA	Critical material attribute
CPP	Critical process parameter
CQA	Critical quality attribute
EMA	European Medicines Agency
FDA	Food and Drug Administration (USA)
GMP	Good manufacturing practice
HoQ	House of quality
ICH	International conference on harmonisation
IQ	Installation qualification (of an instrument or piece of equipment)
ISO	International Standards Organisation
OQ	Operation qualification (of an instrument or piece of equipment)

PAT	Process analytical technology
PQ	Performance qualification of instrument, or process qualification of equipment
PSD	Particle size distribution
QbD	Quality by design
QFD	quality function deployment
QTTP	Quality target product profile
RSD	Relative standard deviation
sd	(absolute) standard deviation
SEM	Scanning electron microscopy
SOP	Standard operating procedure
TEM	Transmission electron microscopy
D	Equivalent particle size; diameter of an equivalent sphere
D_{10}	10% value of a cumulative undersize PSD (second subscript 0 for number-based and 3 for volume-based PSD)
D_{50}	50% value of a cumulative undersize PSD (second subscript 0 for number-based and 3 for volume-based PSD)
D_{90}	90% value of a cumulative undersize PSD (second subscript 0 for number-based and 3 for volume-based PSD)
u_{crm}	Standard uncertainty of certified value
U_{crm}	Expanded uncertainty of certified value
x	Measured property value
X	Assigned property value
y	Standard uncertainty of assigned property value
Z	Z-score in proficiency test

References

1. Allen T (1997) Particle size measurement. Chapman & Hall, London
2. Allen T, Khan AA (1970) The chemical engineer, pp. 108–112
3. ASTM E1919 Standard guide for worldwide published standards relating to particle and spray characterization. American Society for Testing Materials, USA
4. Chan CC, Lam H, Zhang X-M (2010) Practical approach to method validation and essential instrument qualification. Wiley, New York
5. EMA (2017) Report from the EMA-FDA QbD pilot program; EMA/213746/2017. European Medicines Agency
6. EMA (2006) Reflection paper on nanotechnology-based medicinal products. EMA/CHMP/79769/2006. European Medicines Agency
7. FDA (2000) Guidance for Industry, analytical procedures and methods validation. www.fda.gov/CDER/Guidance/2396 dft.pdf

8. FDA (2004) Guidance for Industry, PAT—a framework for innovative pharmaceutical development, manufacturing, and quality assurance. Food and Drug Administration, Department of Health and Human Services, USA
9. FDA (2014) Guidance for industry, considering whether a FDA-regulated product involves the application of nanotechnology. Food and Drug Administration, Office of the Commissioner, USA
10. FDA (2014) Guidance for industry, safety of nanomaterials in cosmetic products. FDA, Center for Food Safety and Applied Nutrition, USA
11. ICH (1996–2006) Guideline Q1A(R2)—Q1F, stability testing
12. ICH (1996) Guideline Q2(R1), validation of analytical procedures: test and methodology
13. ICH (2006) Guideline Q3A(R2)—Q3D, impurities in new drug substances
14. ICH (2007) Guideline Q4–Q4B, Pharmacopoeia Harmonisation
15. ICH (1995–2004) Guideline Q5A(R1)–Q5E, quality/viral safety of biotechnological products
16. ICH (2000) Guideline Q6A, specifications: test procedures and acceptance criteria for new drug substances and new drug products: chemical substances
17. ICH (2000) Guideline Q7, good manufacturing practice for active pharmaceutical ingredients
18. ICH (2009) Guideline Q8(R2) pharmaceutical development
19. ICH (2005) Guideline Q9, quality risk management
20. ICH (2009) Guideline Q10, pharmaceutical quality system
21. ICH (2012) Guideline Q11, development and manufacture of drug substances
22. ICH (2014) Guideline Q12(draft), lifecycle management
23. ISO 2591. Test sieving; International Standards Organisation
24. ISO 3310. Technical requirements for test sieves. International Standards Organisation. 1—Test Sieves of Metal Wire Cloth. 2—Test Sieves of Perforated Metal Plate. 3—Test Sieves of Electroformed Sheets
25. ISO 8258-1991. Shewhart control charts. International Standards Organisation
26. ISO 13320. Laser diffraction. International Standards Organisation
27. ISO 13322. Image analysis methods. International Standards Organisation. 1—Static Methods. 2—Dynamic Methods
28. ISO 14488. Sampling and sample splitting. International Standards Organisation
29. ISO 14887. Dispersion procedures for powders in liquids. International Standards Organisation
30. Kulshreshtha AK, Singh ON, Wall GM (eds) (2010) Pharmaceutical suspensions—from formulation development to manufacturing. Springer/AAPS, Berlin
31. Lakerveld R, Benyahia B, Braatz RD, Barton PI (2013) Model-based design of a plant-wide control strategy for a continuous pharmaceutical plant. AIChE J 5910:3671–3685
32. Mascia S, Heider PL, Zhang H, Lakerveld R, Benyahia B, Barton PI, Braatz RD, Cooney CL, Evans JMB, Jamison TF, Jensen KF, Myerson AS, Trout BL (2013) End-to-end continuous manufacturing of pharmaceuticals: integrated synthesis, purification, and final dosage formulation. Angew Chem 5247:12359–12363
33. McGlinchey D (2016) Sampling and characterization of bulk particulate materials and products; Chap. 15 in [46]
34. Merkus HG (2009) Particle size measurements, fundamentals, practice, quality. Springer, Berlin
35. Merkus HG (2009) Chapter 4, Sampling of particulate material, in [34]
36. Merkus HG (2009) Chapter 5, Dispersion of powders in air and liquids, in [34]
37. Merkus HG (2009) Chapter 7 Microscopy, in [34]
38. Merkus HG (2009) Chapter 8, Sieves and sieving, in [34]
39. Merkus HG (2009) Chapter 10, Laser diffraction, in [34]
40. Merkus HG (2009) Chapter 14, In- and on-line measurement, in [34]
41. Merkus HG (2009) Chapter 16, Reference materials, in [34]
42. Merkus HG (2009) Chapter 17, Quality management and calibration, in [34]
43. Merkus HG (2009) Chapter 20, Statistical background, in [34]

44. Merkus HG, Meesters GMH (eds) (2014) Particulate products—tailoring properties for optimal performance. Springer, Berlin
45. Merkus HG (2014) Chapter 1, Introduction, in [44]
46. Merkus HG, Meesters GMH (eds) (2016) Production, handling and characterization of particulate materials. Springer, Berlin
47. Muzzio FJ, Goodridge CL, Alexander A, Arratia P, Yang H, Sudah O, Mergen G (2003) Sampling and characterization of pharmaceutical powders and granular blends. Int J Pharm 250:51–64
48. NIST (2011) Certificate of standard reference material 1004b—glass beads, Particle size distribution. National Institute of Standards and Technology, USA
49. NIST (2015) Report of investigation, reference material 8013—gold nanoparticles, nominal diameter 60 nm. National Institute of Standards and Technology, USA
50. Pathak Y, Thassu D (2010) Scale up and technology transfer of pharmaceutical suspensions; Chap. 8 in [30]
51. Pharmacopeia. Analytical sieving. USPharm section <786>, Ph.Eur. section 2.9.38
52. Pharmacopeia. Laser Diffraction, USPharm section <429>, Ph.Eur. section 2.9.31
53. Pharmacopeia. Optical Microscopy, USPharm section <776>, Ph.Eur. section 2.9.37
54. Pharmacopeia. Particle Shape, USPharm section <776>, Ph.Eur. section 2.9.37
55. Pharmacopeia. Qualification of Analytical Instruments, USPharm section <1058>, General European OMCL Network (GEON), Quality Management Document PA/PH/OMCL (08) 73 R3
56. Pharmacopeia. Sieves, USPharm section <786>, Ph.Eur. section 2.1.4
57. Pharmacopeia. Sieve Test, USPharm section <786>, Ph.Eur. section 2.9.12
58. Rawle AF (2010) Instrument qualification and performance verification for particle size instruments; in [4]
59. Rideal GR, Mitchell JP, Bunker R, Brookman B (2004) Characterisation of reference standards for low-angle laser light scattering (laser diffraction) particle size analysis. In: PSA conference
60. Sun Z, Ya N, Adams RC, Fang FS (2010) Particle size specifications for solid oral dosage forms: a regulatory perspective; particle size/oral dosage April, May, June 2010, pp 68–73
61. Weinekötter R (2016) Mixing of solid materials. Springer, Berlin; Chap. 9 in [46]
62. Whitehouse Scientific, Certificate of Analysis of Polydisperse Particle Standard 50–350 μm
63. Willis RC (2004) Process analytical technology. Today's Chemistry at Work, Feb 2004, pp 21–22, ACS (www.tcawonline.com)
64. Yang Y, Nagy ZK (2014) Model-Based systematic design and analysis approach for unseeded combined cooling and antisolvent crystallization (CCAC) systems. Cryst Growth Des 142:687–698
65. Yu LX, Lionberger RA, Raw AS, D'Costa R, Wu H, Hussain AS (2004) Applications of process analytical technology to crystallization processes. Adv Drug Deliv Revs 56:349–369
66. Zang H, Lakerveld R, Heider PL, Tao M, Su M, Testa CJ, D'Antonio AN, Barton PI, Braatz RD, Trout BL, Myerson AS, Jensen KF, Evans JMB (2014) Application of continuous crystallization in an integrated continuous pharmaceutical pilot plant. Cryst Growth Des 145:2148–2157

Henk G. Merkus graduated in 1960 in physical organic chemistry at the University of Amsterdam. He worked several years at the Royal Dutch Shell Laboratories in Amsterdam on research in the field of detergents and industrial chemicals, followed by development work on thermal wax cracking for production of C_2–C_{14} olefins and on acid-catalyzed synthesis of carboxylic acids from C_3–C_6 olefins. Then, he made the change to analytical chemistry, involving both measurements and method development with a large variety of techniques and methods, first at Shell's process development department in Amsterdam and later in the chemical engineering department of Delft University of Technology. Gradually, his analytical horizon widened: first surface area and porosity measurements were added to chemical analysis, later followed by particle size analysis. In those areas he participated in many university, national and international courses and national and international education and research projects. He is the author of about 45 journal articles. He retired from Delft University in 2000 but remained active in the field of particle characterization. He is author of the book 'Particle Size Measurements—Fundamentals, Practice, Quality' (2009; Springer) and many journal articles, as well as co-author and co-editor of the books 'Particulate Products—Tailoring Properties for Optimal Performance' (2014; Springer) and 'Production, Handling and Characterization of Particulate Materials' (2016, Springer). Moreover, he participates since 1987 in standardization activities regarding particle size measurement, both in the Dutch standards organization NEN and the international ISO organization (TC24).

Chapter 3
Bio-nano: Theranostic at Cellular Level

Martin Kluenker, Sven Kurch, Muhammad Nawaz Tahir and Wolfgang Tremel

Abstract Functionalized nanoparticles are important platforms for targeted drug delivery and multimodal imaging. Materials scientists provide tailor-made tools for medical research, diagnosis and treatment. These tools are rationally designed to have defined functions. Still, the value of these tools can only be determined by the users in medical sciences that develop assays for applying these tools. Until now, little is known about the impact of multifunctional particles that display intrinsic chemical and physical asymmetry which poses new challenges for cells associated with the amphiphilicity, dipole moments and chemical diversity/patchiness of the functionalized nanoparticles. Why is it important to study the impact of anisotropic multifunctional particles on biological cells extending the intricacy of the problem even further? Current nanotechnology projects that started during the past few years focus on the "supramolecular" weak binding of functionalized particles with the goal to form larger ensembles with new functionalities. Thus, one may anticipate new phenomena associated with the exposure of human tissue to the primary building blocks of these new materials. Despite the challenges that still have to be met, multifunctional nanoparticles provide fascinating opportunities for tailoring properties that are not possible with other types of therapeutics. As more clinical data become available, the nanoparticle strategy will improve to such an extent that more sophisticated tools actually reach the clinic. Results from current trials are fueling the enthusiasm of researchers.

3.1 Introduction

Cancer is one of the world most leading causes of death. Cancer therapy relies on surgery, radio- and chemotherapy, but can prolong a patient´s life span only by several months. Chemotherapy, besides surgery, is currently still the most

M. Kluenker · S. Kurch · M. N. Tahir · W. Tremel (✉)
Institute for Inorganic Chemistry and Analytical Chemistry,
Johannes Gutenberg-University, Duesbergweg 10–14, 55128 Mainz, Germany
e-mail: tremel@uni-mainz.de

H. G. Merkus et al. (eds.), *Particles and Nanoparticles in Pharmaceutical Products*,
AAPS Advances in the Pharmaceutical Sciences Series 29,
https://doi.org/10.1007/978-3-319-94174-5_3

promising therapy and known for cancer treatment for over 60 years [149]. Because of its severe side effects with often massive burdens for the patient, researchers have concentrated on therapy improvement by delimiting chemotherapy to the tumor region, thereby improving therapeutic results and reducing systemic side effects. Drug delivery systems (DDS) using nanoparticles have evolved as a promising strategy. Their nanoscopic structures exhibit exceptional properties and characteristics and differentiate them noticeably from their bulk counterparts. The origin of most new properties in multicomponent nanoparticles is related to their small size and special morphology. Their size significantly increases the number of surface atoms by increasing the surface-to-volume ratio, which leads to new chemical and physical properties. At the same time, the surface structure, determined by the shape and morphology of the nanoparticles, plays a pronounced role for the chemical and physical behavior. One of the first examples for the use of colloidal nanoparticles is an ancient Roman glass cup called "Lycurgus Cup" which contains gold nanoparticles causing a change of the glass color from red to green depending on the direction of illumination [117]. It was until the 19th century before Michael Faraday conducted breakthrough experiments on colloidal gold nanoparticles by reducing tetrachloroaurate with white phosphorus [148]. Since these experiments, important progress has been made in nanoscience: Father and son Ostwald studied the influence of particle size on the growth process via ripening [384], further synthetic procedures were devised such as the trisodium citrate reduction established by Turkevich [514], and a nucleation model, adapted later for nanoparticles, was proposed by LaMer and Dinegar [271, 272]. This knowledge led to a better understanding of nanomaterials and assisted to tune them towards specific applications. Nowadays the focus of researchers working in the area of nanoscience has stretched to demanding topics such as human health care, catalysis, renewable energy and nano-intelligence. Consequently, the diversity of nanostructures has reached a climax. Especially, DDSs have advanced as inorganic, organic or mesoporous carrier systems with a shell to encapsulate the drug and targeting ligands to target specific cell populations or tissue to release the cargo. The key functionalities protect the drug from premature degradation, stopping premature interaction with the biological environment, enhancing absorption of the drugs into specific tissue, controlling pharmokinetics, improving intracellular penetration and therefore extending circulation half-life, lowering the rate of aggregation and prolonging shelf life [402]. The first FDA approved carrier platform in 1995 was a liposome-based DDS Doxil® (Doxorubicin) [27]. Subsequently, different combinations of liposome and drug, like Abraxane® (albumin-bound paclitaxel) and DaunoXome (daunorubicin), were introduced successfully into the market. Preclinical studies showed high potential while phase III clinical trials showed only marginal improvements over existing therapies. Doxil® showed not even statistically a significant efficiency gain and FDA approval was largely based on its reduced cardiotoxicity. Another example is the liposome-based drug carrier Abraxane® whose success was based on the linkage between toxicity reduction and elevated doses. Since then, much effort has been undertaken to develop nanosystems as anti-cancer therapeutics.

3.2 Janus Particles, Superparticles and Core-Shell Particles

Along with (spherical) monodisperse metal, metal oxide and semiconductor nanoparticles [180, 398] and nanoalloys [152], three forms of heterodimeric nanoparticles are most intriguing: metal-metal oxide heterodimers synthesized as (i) two-domain systems with solid state interface, often also with different surface functionalization of the metal and the metal oxide domain, known as eponym Janus particles (from the Roman god of beginnings and gates Janus, who has two distinctive and opposite faces) [169], (ii) as colloidal superparticles, i.e. mesoscale particles defined as size- and shape-controlled assemblies from nanoscale building blocks [536, 538], and (iii) core-shell nanoparticles, where a core material is covered by a shell of another material often with variable shell thickness, including mesoporous core-systems for drug load and shells as gate system [52, 201]. We will discuss the role of these three morphologies in more detail starting with Janus-type particles.

3.2.1 Janus-Particles

A key factor for the synthesis of monodisperse nanocrystals is the control of thermodynamics and kinetics by separating a short burst-nucleation stage from subsequent size-focusing by a diffusion-controlled growth process. Size-tuning can then be obtained by harmonizing the monomer consumption between the nucleation and growth stage. Generally, two methodologies are used to initiate a burst-nucleation process by oversaturation of monomers in the reaction mixture. In the heat-up approach active monomers are formed gradually from precursor molecules until a critical supersaturation is reached leading to burst nucleation. The hot-injection approach creates the critical supersaturation instantaneous by injecting the precursor solution into a hot reaction mixture. This also lowers the reaction temperature and prevents additional nucleation processes after burst nucleation [266, 504, 541]. Surfactants can stabilize the highly reactive surfaces and prevent agglomeration during the size focusing growth process [565, 582]. A critical supersaturation level of the monomer is required because of the large energy barrier of homogeneous nucleation. In classical nucleation theory (CNT) the positive surface energy term increases because of the formation of new surface, exceeds the negative volume energy term of newly generated volume resulting in an overall positive Gibbs free energy term until a critical radius of the nucleus is reached [504, 541, 585]. In recent years much progress was made to carefully evaluate these parameters and control the growth of monodisperse particles [398]. However, to obtain anisotropic structures [233] with cubic [598], truncated [498] or rod-like morphologies, it is crucial to induce the symmetry break required for promoting anisotropy and to control the reaction kinetics and the facet specific adsorption of organic stabilizing ligands

[106, 565] with dynamic temperature-dependent adhesion and diffusion rates [582], twinning defects [541, 573, 575], oriented attachment [10, 317], chemical etching/galvanic displacement [488, 581], or template assisted growth [201, 309]. Moving to heterodimer structures, the interplay of energy and structural effects becomes more complex as more interaction factors occur simultaneously. One widely employed technique for the synthesis of heterodimeric nanoparticles, containing at least two different materials, is the seed mediated growth approach where preformed seed nanoparticles are used to grow one or more domains of a second material onto it [498, 565]. According to CNT heterogeneous nucleation is favored compared to homogeneous nucleation because of a lower energy barrier which results in a lower Gibbs free energy and a lower chemical potential [504, 541, 585]. The reaction takes place at a critical thermodynamic-kinetic crossover. In addition, atomic diffusion, exchange facet specific reactivity, surface wetting properties as well as interfacial stress and strain play major roles. Epitaxial growth, where the crystallographic orientation of the seed particle is preserved during addition of a second material, is important for tuning the properties of the different domains after they have merged [455, 472, 487]. As an interface between two materials is formed, several synergistic effects can be observed. In typical solid-state vapor deposition techniques a lattice mismatch of 1–3% is tolerated for epitaxial growth [58]. In solution-phase or beaker epitaxy a lattice mismatch of 7–12% can be surmounted when chemical match, temperature, solvents and surfactants are chosen carefully [243, 495]. For controlling the formation of heterodimers two factors are important: (i) The amount of seed particles and the amount of precursor for the second domain determines the size ratio between the seed and second domain. (ii) The polarity of the solvent regulates the quantity of second domains. As shown for Au@Fe_3O_4 [442] and Au@MnO (Fig. 3.1a) [443, 444, 446], the nucleation of the metal oxide domain onto Au seeds in an apolar solvent leads to a withdrawal of electron density from Au and transfer to iron oxide through the interface and subsequent electron deficiency of Au, which prevents the nucleation of more than one iron oxide domain. When a polar solvent is used instead, electron density can be provided to the Au seed by the solvent, allowing the formation of multiple nucleation sites. For heterogeneous nucleation, the change of the Gibbs free surface energy is associated with the difference between the surface energies of both materials and a positive solid-solid interface term. Heterogeneous nucleation with layer-by-layer deposition for epitaxial growth is typically related to the Frank–van der Merwe mechanism where the overall Gibbs free surface energy is positive because the surface energy of the second material is lower than that of the first material, and the interface term is small [58, 330]. The surface energy is influenced by the exposed surface and the stabilizing ligands, whereas the interface term is controlled by binding strength and lattice mismatch.

For tuning the chemical and physical properties of heterodimeric Janus particles interfacial electronic communication is essential. Direct contact of a noble metal and a semiconductor leads to the formation of a Schottky barrier [528, 606]. Electrons from the Fermi level of the noble metal are transferred to the semiconductor conduction band which causes a downwards bending of the conduction band. The Fermi levels of both materials are equilibrated. This leads to a charge

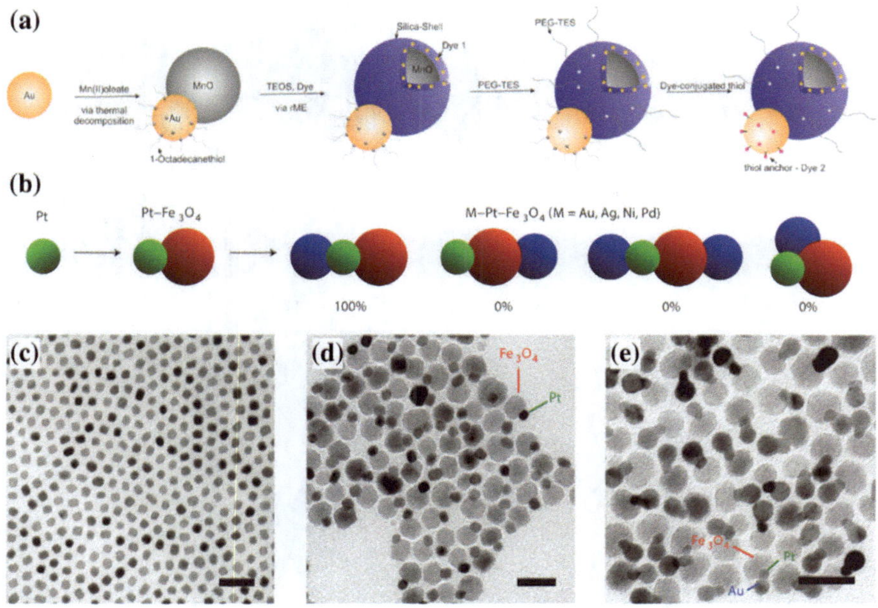

Fig. 3.1 **a** Seed-mediated synthesis of Au@MnO heterodimers, subsequent encapsulation with silica and functionalization of the SiO_2-shell as well as the Au domain. Adapted with permission from Ref. [443]. **b** Schematic showing the multistep synthesis of M–Pt–Fe_3O_4 heterotrimers, along with the most significant possible products and their observed frequencies (expressed as the percentage of observed heterotrimers, not total yield). Representative TEM images show **c** Pt nanoparticle seeds, **d** Pt–Fe_3O_4 heterodimers and **e** Au–Pt–Fe_3O_4. All scale bars are 25 nm. Adapted with permission from Ref. [51]

accumulation at the interface and blocks electron transfer from the noble metal to the semiconductor. First observed for Au@CdS, the Schottky barrier blocks the electron-hole recombination after photo excitation of the semiconductor and therefore quenches photoluminescence [460]. This is especially relevant for (photo)catalysis [224, 426, 451, 495, 593]. The electronic interplay is also reflected in altered surface plasmon resonance (SPR) and magnetic properties of Au@Fe_3O_4. The SPR is red-shifted compared to that of pure Au nanoparticles because of the change in the local dielectric environment caused by the high refractive index of iron oxide [460, 528, 587, 595]. The magnetic coercivity is higher and the remanence is lower for heterodimeric Au@Fe_3O_4 compared to Fe_3O_4 owing to a decreased number of nearest Fe neighbors at the interface. This leads to a smaller interatomic exchange coupling and larger spin canting effects [93]. As a result of the anisotropy, higher magnetic fields are needed to align the spins [460, 528, 587, 595]. The electronic communication can also be influenced by surface chemistry e.g. by functionalizing Ag in Ag@Fe_3O_4 with thiols the red shift of the SPR can be suppressed [528, 602]. A broad variety of heterodimers such as Au@TiO_2 [83, 257, 478], Pt@Fe_3O_4 [527] or Au@ZnO [257, 291] with different morphologies have

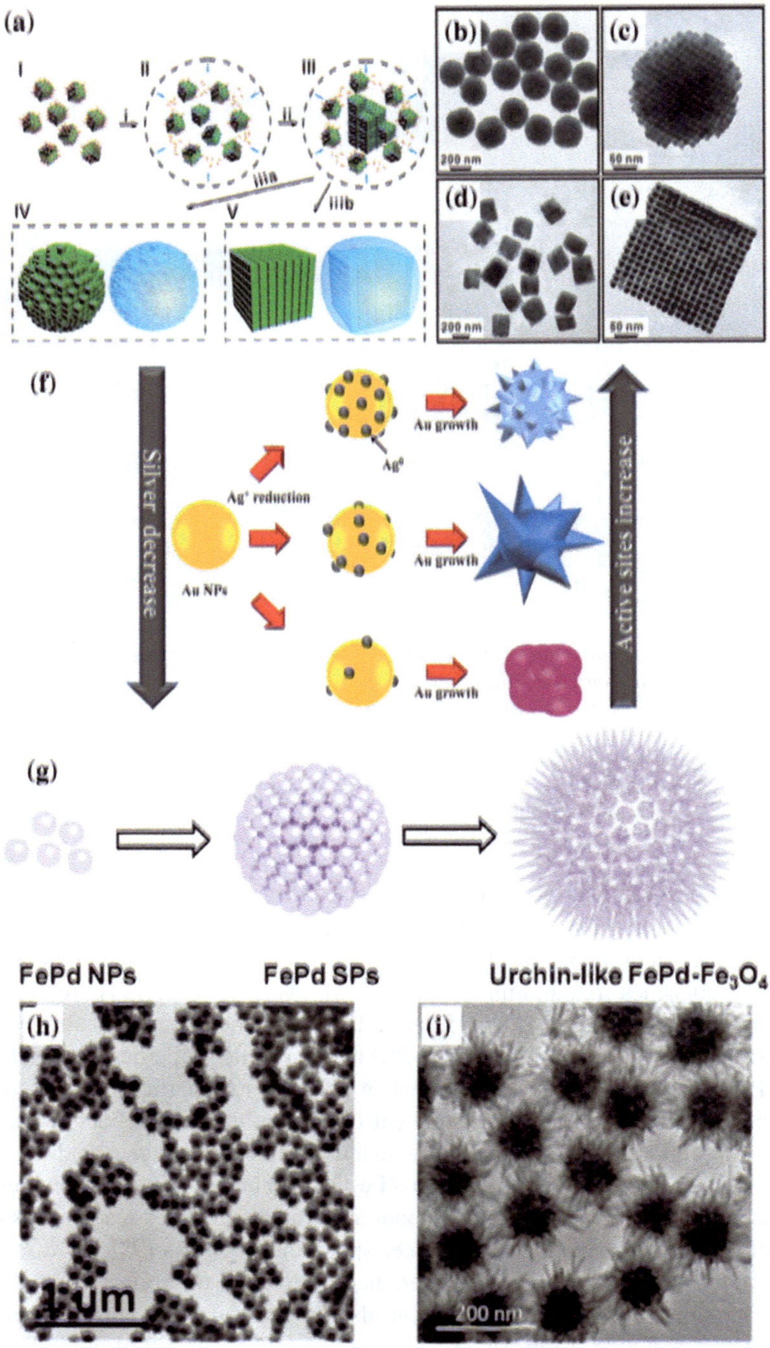
(a)
I
II
III
IIIa
IIIb
IV
V
(b)
(c)
(d)
(e)
200 nm
50 nm
200 nm
50 nm
(f)
Silver decrease
Ag⁺ reduction
Au NPs
Au growth
Ag⁰
Au growth
Au growth
Active sites increase
(g)
FePd NPs
FePd SPs
Urchin-like FePd-Fe₃O₄
(h)
1 μm
(i)
200 nm

◂**Fig. 3.2** **a** Formation of colloidal superparticles from iron oxide nanocubes. Nanocube micelles (I), superparticles embryo (II, III), spherical (IV) and cube-shaped superparticles (V). Embryo formation (i), nanocube crystallization (ii) and superparticle formation (iii). **b** TEM images of sphere-shaped superparticles, **c** a spherical superparticle viewed along the [001] zone axis, **d** cube-shaped superparticles and **e** a cubic superparticle viewed along the [001] zone axis. Adapted with permission from Ref. [536]. **f** Reduction of Ag^{+}-ions on the surface of Au nanoparticles to obtain star-shaped Au/Ag bimetallic superparticles. Decrease of $AgNO_3$ precursor leads to less formation of active Ag^0 sites resulting in the formation of fewer spikes. Adapted with permission from Ref. [89]. **g** Schematic illustration of self-aggregation of FePd NPs into FePd superparticles and the formation of urchin-like FePd–Fe_3O_4 composite superparticles. **h, i** TEM images of the urchin-like FePd–Fe_3O_4 composite superparticles with an overall Fe/Pd ratio at 74:26. Adapted with permission from Ref. [590]

been reported. Similarly, trimer systems pointing towards multifunctional lab-on-a-particle chemistry have been demonstrated [459, 460]. Schaak and coworkers combined Au, Ag, Ni or Pd with Pt@Fe_3O_4, but those structures are still challenging due to the complex reaction framework (Fig. 3.1b–e) [51].

The ability to functionalize the different domains of heterodimeric nanoparticles individually and generating Janus particles (Fig. 3.1a) makes them promising candidates for multiple applications reaching from catalysis [289, 408, 593] to biomedical applications [273, 274] in theranostics [88]. Amphiphilic Janus particles can be self-assembled for nanocapsules and – membranes [173].

3.2.2 *Superparticles*

Colloidal superparticles are size- and shape-controlled nanoparticle assemblies in the form of colloidal particles [617, 618]. The individual nanoparticles can be viewed conceptually as artificial atoms, organized in an ordered superlattice structure, in order to generate new collective properties [104, 105]. This has been valuable for various applications in optoelectronics [127], biochemistry [187], plasmonics [386], or storage devices [483]. Typically, solvophobic interactions are used to assemble the superparticles from their nanoparticle building blocks (Fig. 3.2a–e) [536, 619]. The nucleation of superparticles does not have a characteristic free-energy barrier like that of nanocrystals. The molecular or ionic precursors of nanocrystals are in true solutions and therefore have very different surface properties (no surface) from those of nuclei (solid nanophase, large surface area) that are formed during nucleation. Nanocrystal nucleation is accompanied by an increase in the volume and surface of the nuclei.

The change of the total Gibbs free energy during nucleation is a balance between the energy gain due to the new volume (lattice energy) and the energy loss invested to generate the new surface (surface energy). The nucleation energy barrier ΔG^* needed to form a spherical nucleus is important for the separation of the nucleation and growth steps in nanocrystal synthesis, which is relevant for the synthesis of monodisperse nanocrystals. The precursors for superparticle growth are nanocrystals retaining their own surface boundary and thus form a separate phase in the dispersion solution. The chemical and physical properties of the nanoparticle precursors are very similar to those of the resulting nuclei and/or superparticles. The free energy to form the surface superparticle nuclei will be small. The decrease of the surface area will lead to a decrease in surface energy due to the surface binding between nanocrystals. In addition, the release of solvent molecules will add a large entropic contribution to the Gibbs free energy. Superparticle nucleation does not exhibit a characteristic free energy barrier as observed in nanocrystal nucleation [582]. The lack of a thermodynamic nucleation energy barrier makes it difficult to achieve the separation of nucleation and growth in superparticle synthesis, which leads to concurrent nucleation and growth, thus resulting in particles with a poor size distribution.

The Gibbs–Thomson equation does not apply to superparticle growth systems. Superparticles do not exhibit size-dependent "solubility" in their dispersion media because their surface properties are very close to those of their nanocrystal building blocks. In contrast, colloidal nanocrystals can be dissolved into atomic or molecular species in growth solution, and exhibit size-dependent solubility owing to a surface effect governed by the Gibbs–Thomson equation [103, 504]. The solubility of the nanocrystals increases as their size decreases. This size-dependent property is the thermodynamic basis of the Ostwald ripening process: Smaller nanocrystals redissolve and redeposit onto larger particles.

Another point is the surface functionalization itself. The surface ligands protecting colloidal particles allow for numerous particle interactions with multiple assembly geometries [39]. The identical surface functionalization can be a disadvantage in the organization of the colloidal particles to superparticles as well [337]. Their dispersions are typically not stable against aggregation or agglomeration. The forces responsible for the assembly of superparticles are long-range van der Waals/ London forces based on induced dipole [337] along with solvophobic interactions [535, 618, 619]. Upon addition of a poor solvent for the colloidal nanoparticles the surface tension at the interface rises, and the solvent shell is minimized. As a result, colliding particles reduce their surface area and surface energy by aggregation and superparticle formation. Electrostatically charged surfaces should be avoided because of high-range electrostatic repulsion forces. Steric interactions from uncharged organic ligands can be both, attractive and repulsive. When such modified colloidal nanoparticles come into close proximity in a poor solvent, the solvent shell is reduced and solvent molecules are released. The concomitant entropy gain forms an attractive force. When the organic ligands come into contact by virtue of nanoparticle assembly, attractive van der Waals forces can arise but also steric hindrance may lead to repulsive forces [535].

Nevertheless, superparticles by nanoparticle assembly lack a direct connection of the building blocks and therefore a direct electronic communication as heteroparticles obtained by epitaxial growth. This may be a disadvantage for applications where electronic or magnetic interfacial crosstalk combined with large surface areas is necessary (e.g. biomedical theranostic [89, 213], (photo)catalysis [99], enzyme mimetics [485]). Here the concept of heterodimers with the benefits of Schottky junctions, diversive functionalization possibilities and reaction control can be extended to *multidomain systems on a single seed particle*, i.e. colloidal superparticles. Anisotropic and branched structures can be designed and synthesized using colloidal chemistry protocols [581]. However, it is worth mentioning that in order to achieve such epitaxially grown superparticles, one needs some intelligently designed synthetic protocols. As an example, when seedless growth is desired, the high symmetry of the closed-packed *fcc-type* structures must be broken. Strategies include axial growth along five-fold twin defects as observed in the early stages of noble metal nanocrystals [139, 498] or initial stage surface protection of thermodynamically stable Wulff polyhedra by chemisorption of capping agents to control the growth rate of different surface facets [563, 565]. Superparticle-like branched and anisotropic structures can also be realized through ion assistance or chemical etching. Urchin-like growth of Au nanoparticles can be induced through the deposition of Ag nuclei on spherical Au nanoparticles (Fig. 3.2f) [89]. Anisotropic Pd nanoparticles can be synthesized with the aid of copper(II)-ions [99], whereas nitrate anions can promote anisotropic growth of Pt nanoparticles [198]. The difference in the chemical potential of various noble metals and associated redox reactions can be used to reshape nanocrystals via a seed mediated growth process [549] or to generate branched Au structures by a Fe nanoparticle assisted templated growth [297]. Typical oxidative etching by addition of facet-specific etching agents like H_2O_2 or chlorine ions can also transform isotropic structures [319, 360]. Hedgehog-like particles can be designed by the growth of iron oxide nanorods on preformed seed particles with iron pentacarbonyl (Fig. 3.2g–i) [590]. Recently our group demonstrated that this synthetic approach can be extended on different seed particle materials (Ni, Pd) and different seed particle morphologies (nanotetrahedra and nanoplates) with epitaxial growth of the nanorods. The superparticles exhibit enhanced catalytic activity compared to their seed particles [253, 494].

3.2.3 Core-Shell Particles

Core-shell particles are another type of heteroparticles where the second domain is present as a shell around a core material e.g. through multiple nucleation spots in a polar solvent (vide supra). Similarly, any organic or silica coating of inorganic nanoparticles is a wide synthetic field for core-shell nanoparticles [74]. Therefore core-shell nanoparticles can be divided into many categories based on their composition. A few important types will be discussed here. A widely used type of core-shell particles includes an inorganic core combined with an inorganic shell material with silica as the most common example. Advantages of silica coatings on inorganic nanoparticles include extended suspension stability in aqueous media, various surface functionalization possibilities as well as improved biocompatibility [177, 222, 255]. Especially for plasmon active noble metals like Au, a silica shell has been shown to strongly influence the surface plasmon properties as a function of the shell thickness [316]. A typical synthesis of silica uses a modified sol-gel processing (Stöber process) based on hydrolysis and polycondensation of metal/silicon alkoxides under specific pH, temperature and alkoxide concentration conditions [473]. Particularly the pH value plays an important role for the hydrolysis of tetraalkoxy silanes under acidic as well as basic conditions. Usually a basic catalysis is used to increase the rate of condensation and to control the particle morphology by inhibiting a strong gelation to network structures [49, 421].

Another possibility to obtain monodisperse and small silica particles/coatings is the reverse microemulsion technique. By applying surfactants and water in oil a microemulsion in form of aqueous nanodroplets is generated. These droplets are used subsequently as reaction containment for the formation of silica particles/coatings. The size of the micelles determines the particle size and morphology [177, 222]. The shell thickness can be controlled by the ratio of base/tetraalkoxy silane/number of seed particles. A template chemistry serves also for the formation of mesoporous silica structures for drug loading and triggered payload release [562]. Apart from silica coatings, inorganic core materials can be coated with other inorganic materials to form core-shell structures mainly to alter the chemical and physical properties of the core material. An Au coating is frequently used to provide chemical stabilization from oxidation/corrosion and add surface functionality by introducing the opportunity to apply amine and thiol functionalization. For catalytic applications surface-enhanced Raman scattering (SERS) of Pd is low compared to that of Au or Ag due to the red-shifted SPR and strong morphology dependence [211, 214, 373]. By coating Au nanoparticles with Pd the SERS activity can be increased and fine-tuned owing to the electromagnetic communication [211]. Au coatings on silica core particles also exhibit an excellent tunability of the optical properties, suitable for optical imaging strategies and photothermal therapy [310]. Here the properties depend on the size of the core and the shell thickness. Further attention has been given to core-shell structures where both components are magnetic [594] or one is a semiconductor and second one the magnetic domain. This is particularly useful for bio-imaging [620], catalysis [471] and energy applications [215]. Other valuable systems contain one or both components of the core-shell structure from organic/polymeric nature. Often an organic shell is used to protect the core material or to create biocompatibility and surface functionality. Common strategies include the coating of inorganic core materials with hydrophilic polymers such as polyethylene glycol (PEG) [204, 280, 505, 555] or polysaccharides such as dextran [36]. Further details on nanomaterial functionalization and surface modification strategies will be given in the section surface modification (vide infra). Full organic core-shell structures mainly use polymers on the basis of their glass transition temperature to prepare biodegradable drug encapsulation carriers for biomedical applications [25].

3.3 Overview Drug Delivery Systems (DDS)

Researchers at Mobil Oil Corporation developed the silica based mesoporous material MCM-41 (Mobile Crystalline Material) [31, 262], and its application for drug delivery was first introduced in 2001 [518]. Silica nanoparticles show good biocompatibility, options for a variety of functionalizing strategies, high surface area and adjustable pore size [23, 57, 131, 343]. They are functionalized easily with alkine [376], carboxy [566], amine [470] and azido [327] functional groups. The template associated polycondensation of silica precursors leads to various porous structures and shapes, tailored specifically to the individual therapeutic agent size and their applications. Starting from non-porous silica particles [71, 72, 383] made by the Stöber approach [473] by the microemulsion technique [570, 571], the first systematic study to biorelated degradation and biodistribution was completed. Dendritic silica nanoparticles [130, 456, 532, 567, 569, 574, 579] and other unique pore shapes like cubic (Pm3n [161]), yolk-shell [114, 303], fibrous (KCC-1 [21, 164]), wrinkled [394] and cylindrical (KIT-6 [342]) were developed for drug delivery and imaging. Organosilica hybrid nanoparticles, combining the organic flexible network with various functionalizing possibilities and the rigid, inorganic silica structure with adjustable pore size, are promising drug delivery systems (DDSs) as well [107, 109, 110, 315]. Other carrier systems deployed for drug delivery or diagnostic purposes are gold, iron oxide and CdSe quantum dots (QD). These metal-based systems combine an imaging of the tumor tissue and blood vessels with the possibilities for treatment with the encapsulated drug. Theranostic carrier systems, combining therapy and diagnostic, are utilized in various disease treatment scenarios, making in vitro and in vivo research cornerstones for translation of nanomedicines into the clinic. Liposomes and polymer based nanoparticle are another class of DDSs [284, 508]. First steps to the translation of nanocarriers were carried out in 2006, when small interfering RNA (si-RNA) was encapsulated in liposomes and delivered in non-human primates [621].

3.4 Surface Modification

Surface modification is *a conditio sine qua non* for nanoparticle synthesis to be eligible for biomedical application. The nanoparticles are prepared either in aqueous solution or in organic solvents. In both cases, a suitable application of surfactants is required to prevent agglomeration and to provide colloidal stability. These surfactants also protect pure metal nanoparticles whose surfaces are highly reactive for oxidation. When the nanoparticles prepared in apolar solvents are dispersible only in organic non-polar media because of their long aliphatic chain surfactants [391, 411], a ligand exchange is mandatory for biomedical applications to achieve water solubility and physiological compatibility. Furthermore, surface functionalization can provide a control on the intrinsic chemical reactivity of the particles, which may unintentionally affect reaction pathways at the cellular level and protect from opsonization, because the nanoparticles are recognized by the immune system and subsequently covered with serum proteins and surrendered to phagocytosis [484]. The latter is often referred to as the stealth effect [50, 70, 348]. Besides silica functionalization there are a number of strategies for surface modification based on polymers available resulting in core-shell-like structures. The most important ones will now be discussed.

3.4.1 Amphiphillic Micelle Structures

This strategy does not change the initial apolar surfactant coating. Instead, a layer of amphiphilic surface active molecules like cetyltrimethylammonium bromide (CTAB) is assembled on the nanoparticles. The long hydrophobic alkyl chains of CTAB interact with the initial apolar surfactant on the surface (e.g. oleyl amine, oleic acid) through van der Waals forces. The polar ammonium moiety enables water dispersability and stabilization through electrostatic repulsion by forming a micelle-like structure [232, 249]. This strategy is often applied for stabilizing Au colloids [174], but it lacks long-term stability and robustness against shear forces [232]. Further evolution of this strategy includes amphiphilic polymer structures where the apolar part consist of phospholipids and the polar part comprised of polyethylene glycol (PEG) which can also serve as an anchor for additional cell targeting moieties (Fig. 3.3a, b) [133, 167]. Thermoresponsive polymers are also available for temperature dependent drug release mechanisms as demonstrated by Qin et al. who combined thermosensitive poly(N-isopropylacrylamide) (PNIPAAm) with a hydrophobic poly(maleic anhydride-*alt*-1-octadecene) backbone to coat Fe_3O_4 nanoparticles [410]. Amphiphilic saccharide structures [382] and amphiphilic triblock copolymer structures were used as well [167]. Nevertheless, the hydrodynamic radius is increased due to the micellar structures [589], which can affect the in vivo application. Due to the dynamic equilibrium a ligand exchange by washing or diluting is possible.

3.4.2 Multidentate Polymeric Ligands

Another strategy involves the replacement of the original apolar surfactant coating through a multifunctional polymeric ligand (Fig. 3.3c, d). Here PEG is commonly used because of its antifouling, low polydispersity, non-ionic and non-toxic character [254]. For nanoparticle attachment the PEG chains are equipped with a functional anchor group. Most suitable is a chemisorption of the anchoring group on the nanoparticle based on Pearson's HSAB principle [400], where Pearson-soft materials like Au show high affinity towards thiol groups, and Pearson-hard materials like transition metal oxides (e.g. iron oxides, manganese oxides) show

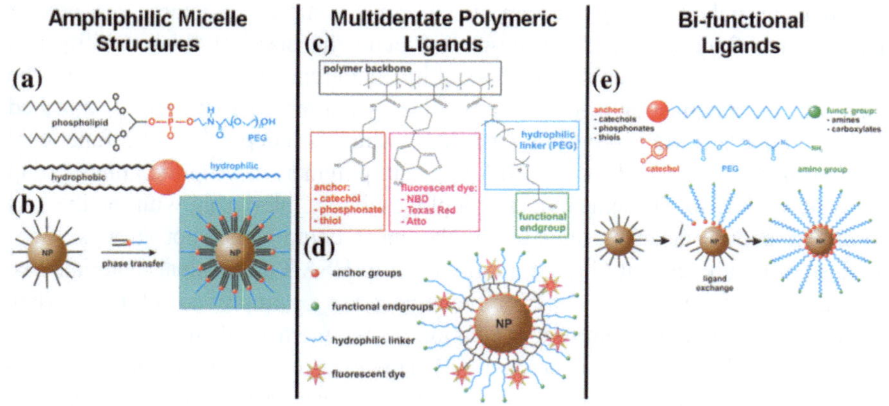

Fig. 3.3 **a** Chemical structure of a PEG-phospholipid. **b** Coating procedure: the hydrophobic lipid tails are incorporated between the alkyl chains of the capping molecules on the nanoparticle surface, whereas the hydrophilic PEG chains point to the outside creating a hydrophilic shell. **c** Chemical structure of a typical multifunctional polymeric ligand. The co-polymer contains 3-hydroxytyramine (dopamine) as anchor group, a fluorescent dye, PEG chains and free amino groups for further functionalization. **d** Illustration of a magnetic NP modified with a multifunctional polymer. **e** Surface modification of magnetic nanoparticles using bi-functional ligands. These ligands consist of an anchor group conjugated to a hydrophilic linker carrying a functional end group. A key requirement for a successful exchange of the hydrophobic capping agents is the complete dispersability of the nanoparticles and the ligand molecules in the same solvent. Adapted with permission from Ref. [445]

high affinity towards oxygen groups. A variety of polymers like poly(ethyleneimine) [564] or poly(amidoamine) [414] and anchoring groups like phosphine oxide [250], amines [372] or carboxylates [87] have been used. When these anchoring ligands are executed as chelating ligands like catechols [434] or bidentate thiols [516], the binding affinity is reinforced. Accordingly, a large variety of polymer–anchor combinations have been explored covering cross-linked polymers [228], which provide more mechanical stability, and introduction of terminal functional groups for attachment of dyes or targeting ligands through click chemistry [577]. Nevertheless, a drawback is that the nanoparticles can be linked through one polymer chain resulting in aggregates disadvantageous for biomedical application.

3.4.3 Bi-functional Ligands

Bi-functional ligands are used to replace the surface active capping agents from the particles. Those ligands contain an anchor group (catechol [415, 416], thiol [15], phosphonate [510] or carboxylate [522]), a spacer (typically alkyl chain or PEG) and a reactive head group to bind functional molecules like dyes, targeting ligands or drug molecules (Fig. 3.3e). Advantages of bi-functional ligands compared to the multidentate polymeric ligands are: (i) a more compact structure, (ii) precise adaptability of the spacer chain length to obtain the best colloidal stability and (iii) independent functionalization possibility. Different polymer moieties can be used for functionalization of the same particle: A fluorescence dye can be attached to the nanoparticle via a short polymer chain, which is independent of a functional amine group connected via a second long chain polymer, making the amine group easily accessible for subsequent drug coupling without interfering with the dye. As described above, catechol ligands are the most common anchor groups [555] because of their excellent binding affinity (also to HSAB ambivalent materials like transition metals [435]) and superior chelating effect by improved orbital overlap, reduced steric hindrance through generation of five-membered rings and stabilization through electron delocalization [464]. However, catechols are prone to oxidation and subsequent polymerization resulting in insoluble polymers [625].

3.5 Biomedical Application—Diagnostics and Therapy

Engineered multifunctional nanoparticles cover a wide range of applications in biomedicine and material science. They can easily be adjusted for specific use in diagnostics, therapy or materials science. Synthetic control combined with multiple surface functionalization strategies make them useful for multipurpose applications towards lab-on-a-particle structures.

3.5.1 Magnetic Resonance Imaging (MRI)

First theranostic approaches using magnetic iron oxide nanoparticles were made in the 1970s [453]. Since then, magnetic iron oxide nanoparticles have proven valuable for clinical use as MRI contrast agents [479]. However, first formulations of dextran coated superparamagnetic iron oxide (SPION) nanoparticles in clinical use (e.g. AMI-25) [551] showed poor monodispersity and increasable relaxation performance [232]. Until now many efforts have been made to understand the physical processes associated with magnetic relaxation and nanodimensional size as well as in vivo behavior of nanoparticles to improve their biocompatibility and performance. When protons are exposed in nuclear magnetic resonance (NMR) spectroscopy to an external homogeneous magnetic field (B_0) their nuclear spins start precessing with specific larmor frequencies along the direction of applied magnetic field (z-axis). Due to the two possible orientations of the nuclear spin of the protons and the resulting spin (dictated by the Boltzmann distribution) oriented along the direction of field, a small but detectable magnetization $M_z = M_0$ results along B_0. When a 90° radiofrequency pulse (P90) with larmor frequency is applied, the spins change their orientation with respect to the field and M_z decreases to zero. Additionally, the applied radiofrequency pulse bears a circular polarization perpendicular to B_0, which induces a rotation of the deflected magnetization vector in the *xy* plane where a detector coil is able to picks up the magnetization signal. Directly after the pulse $M_{xy} = M_0$ [183, 351], and two relaxation processes occur, which can be influenced by magnetic nanoparticles.

The longitudinal, positive or spin-lattice relaxation T_1 is associated with the return of the spins to the Boltzmann equilibrium and the concomitant restoration of M_z (Fig. 3.4a). This arises from the realignment of the spins which were oriented opposite to the field B_0. The excess energy released through this process can be transferred to the water protons present in immediate vicinity (the "lattice", enthalpic process), and a T_1 enhancement or bright field imaging can be applied. The corresponding nanomaterials and their molecular counterparts (e.g. paramagnetic f^7-Gd^{3+} complexes) are strongly paramagnetic substances which can generate a fluctuating interference of the magnetic environment and additionally take over the excess relaxation energy through water adsorbance. A typical nanomaterial for T_1 contrast enhancement are MnO nanoparticles due to their antiferromagnetism. The T_1 activity results from the size of the particles and a magnetically disordered spin-glass-like surface layer, exhibiting spin-canting [93, 355, 486]. Here the surface spins are mostly uncompensated and not antiferromagnetically ordered. Additionally, the spins are canted and cannot fully align with the imposed magnetic field. The effect becomes more distinct with smaller particle size and larger surface area [231].

The transversal, negative or spin-spin relaxation T_2 is coupled to the magnetization in the *xy* plane and occurs without energy transfer (entropic process, no effect on Boltzmann distribution, Fig. 3.4a). It is based on the dephasing of the deflected spins which precess in the *xy* plane upon excitation through the radiofrequency pulse. Directly after the pulse they are all in phase, and therefore M_{xy} is maximized. The phase shift is induced by two factors: (i) fluctuating inhomogeneity of the microscopic environment based on Brownian relaxation (rotational external orientational fluctuations at the boundary to the solvent due to diffusion processes, i.e. particle rotation) and Néel-relaxation (internal rotation of the magnetic moments in the tissue due to thermal energy, i.e. spin rotation) [462], known as T_2 and (ii) stationary local inhomogeneity of B_0, which additionally alters the Larmor frequency of several spin bundles, known as T_2' (Fig. 3.4b). Both processes ($T_2^* = T_2 + T_2'$) induce a drop of the free induction decay (FID) signal in the detector coil, but only T_2 is the desired relaxation time linked with the information about the microscopic environment. The T_2' part can be compensated by a 180° refocusing pulse (P180) in a spin-echo pulse sequence which brings the spins back to their original Larmor frequency. Thus, T_2 is accessible, which leads to less artifacts and higher resolution compared to T_2^* measurements (gradient-echo pulsing sequence). The prerequisite for a good T_2 contrast agent is a high magnetic moment which supports the inhomogeneity of the microscopic environment by building an own magnetic field interfering with the detection field and therefore leading to a darkening the MR image by fast reduction of the M_{xy} magnetization [183, 351].

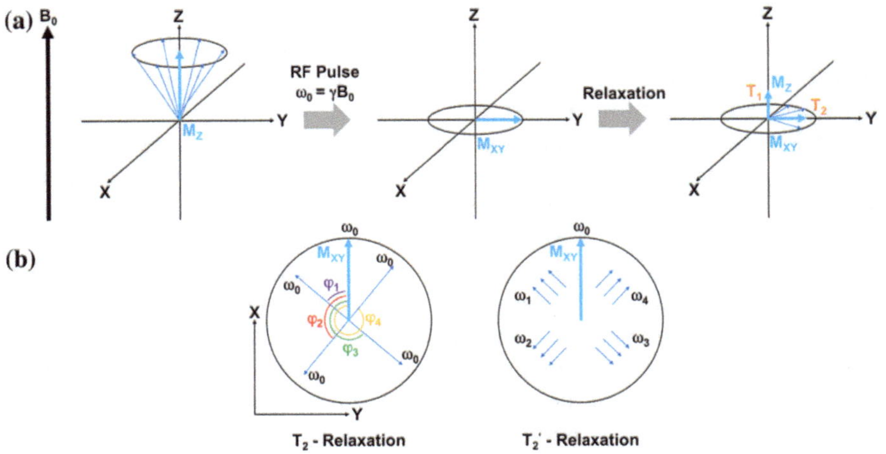

Fig. 3.4 **a** Simplified nuclear magnetic resonance experiment. After application of the 90° excitation pulse with larmor frequency two different relaxation processes take place, restoring the M_z magnetization (T_1-relaxation) and dephasation of the deflected spins reducing M_{xy} (T_2-relaxation). **b** T_2-relaxation with spontaneous phase differences and identical lamor frequency due to Brown- and Nèel-relaxation and T_2'-relaxation due to stationary local inhomogeneity of B_0 which additionally alters the Larmor frequency of several spin bundles. Idea for graphics from Ref. [511]

Accordingly, one-domain superparamagnetic nanoparticles are good candidates because of their large magnetic moment.

Tailoring magnetic nanoparticles for application in MRI diagnostics requires engineering of their magnetic characteristics. To achieve the best results one needs to adjust both, magnetic and physical properties of nanomaterials. For tuning the MRI relaxivities, magnetic properties such as saturation magnetization, coercivity and magnetocrystalline anisotropy as well as physical properties such as particle size, dispersity and surface stability must be evaluated and tuned [584]. For enhanceing the R_2 relaxivity ($R_2 = 1/T_2$) the quadratic dependency of the particle size and saturation magnetization should be considered, which leads to enhanced transversal contrast if the effective radius of the superparamagnetic core and the saturation magnetization is increased [507]. This was demonstrated by Zhao et al. who synthesized iron oxide octapods as transversal contrast agent (Fig. 3.5a, b) [611]. They combined iron oxide with a saturation magnetization (M_s) up to 71 emu/g and a star-like octapod morphology which doubled the effective radius of the superparamagnetic core compared to a spherical iron oxide nanoparticle with same geometric core volume. Additionally, the anisotropy supports a more inhomogeneous local magnetic field compared to spherical nanoparticles resulting in better proton dephasing and reduced T_2 relaxation time. Nevertheless, enhancing the saturation magnetization might be a challenge due to spin canting effects that lower the maximum saturation magnetization of the nanoparticles compared to their bulk counterparts [374] and agglomeration phenomena even in the absence of an external magnetic field due to strong internal magnetization [283]. Lee et al. demonstrated a large saturation magnetization for manganese doped magnetite nanoparticles (Fig. 3.5c) [281]. Owing to the inverse spinel structure of magnetite where a cubic face centered structure of oxide anions with Fe^{3+} in tetrahedral (T_d) and octahedral holes (O_h) and Fe^{2+} in O_h sites, a total magnetic moment of 4 μ_B is expected [339]. Under an external magnetic field the ions in the octahedral site align parallel to the magnetic field delivering 5 μ_B from Fe^{3+} (d^5) and 4 μ_B from Fe^{2+} in high spin (d^6 hs) state. The Fe^{3+} ions in the tetrahedral sites align antiparallel to the magnetic field reducing the magnetic moment by 5 μ_B, leaving the 4 μ_B from Fe^{2+} as overall magnetic moment per unit. If Fe^{2+} is replaced by Ni^{2+} (d^8), Co^{2+} (d^7 hs) or Mn^{2+} (d^5) the magnetic moment is reduced to 2 μ_B, 3 μ_B or increased to 5 μ_B. The saturation magnetization is increased as well. In a consecutive work Jang et al. investigated the effect of non-magnetic Zn^{2+} on those iron and manganese spinels (Fig. 3.5d, e) [223]. They showed that Fe^{2+} in O_h sites is exchanged wit Zn^{2+} which is located in T_d sites instead. Accordingly, Fe^{3+} from the T_d site moves to the O_h site and the antiparallel compensation of T_d and O_h sites is less, which in turn increases the overall magnetic moment.

3.5.2 Hyperthermia

A therapeutic approach based on the magnetic properties is hyperthermia treatment [159, 356]. Physically, magnetic nanoparticles can serve for energy dissipation and as heat generator. In an alternating current (AC) field heat emission via induction can be generated through repeated Néel- and Brownian-relaxation [584]. As a result of the thermal motion of the magnetic moments, the magnetization M is not in phase with the magnetic field (H) varying with time. This leads to a conversion of magnetic energy into heat [159, 282]. The heat emission, defined as specific loss power (SLP), is proportional to the saturation magnetization and reaches a maximum for a certain particle size and magnetic anisotropy [431]. It is also dependent on the heat capacity of the solvent and the mass of the magnetic material [332, 526]. Iron oxide nanocubes have been shown to exhibit higher specific absorption rates (SAR i.e. SLP) than their isotropic counterparts [526]. Additionally, dipolar effects depending on anisotropy can occur leading either to nanomaterial agglomeration or alignment in chains, which can reduce or increase the heating properties, respectively [277]. Iron oxide nanodiscs can be used to double the SAR values of oriented nanodiscs with respect to random orientation [578]. The nanodiscs align parallel to the AC field generating a maximum SAR of 4.66 kW/g. Zn substituted magnetite nanoparticles show potential for hyperthermia treatment owing to their high saturation magnetization [223].

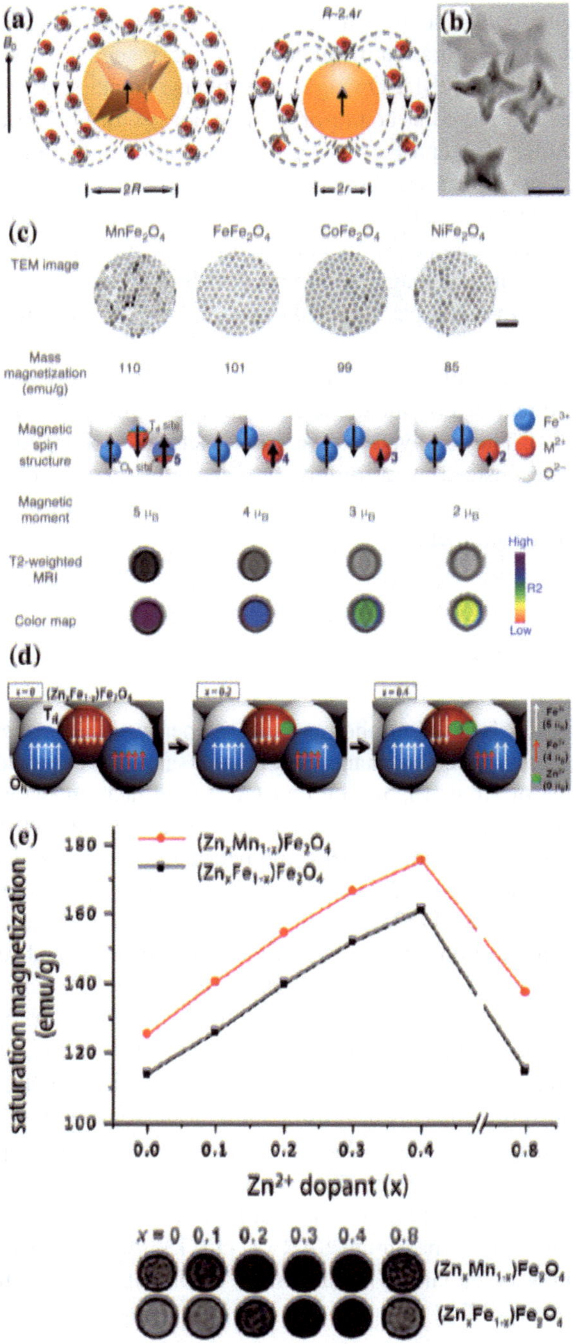
(a)
B_0
R~2.4r
2R
2r
(b)
(c)
MnFe$_2$O$_4$
FeFe$_2$O$_4$
CoFe$_2$O$_4$
NiFe$_2$O$_4$
TEM image
Mass magnetization (emu/g)
110
101
99
85
Magnetic spin structure
T_d site
O_h site
Fe^{3+}
M^{2+}
O^{2-}
Magnetic moment
5 μ_B
4 μ_B
3 μ_B
2 μ_B
T2-weighted MRI
Color map
High
R2
Low
(d)
(Zn$_x$Fe$_{1-x}$)Fe$_2$O$_4$
T_d
O_h
(e)
(Zn$_x$Mn$_{1-x}$)Fe$_2$O$_4$
(Zn$_x$Fe$_{1-x}$)Fe$_2$O$_4$
saturation magnetization (emu/g)
100
120
140
160
180
0.0
0.1
0.2
0.3
0.4
0.8
Zn^{2+} dopant (x)
x = 0 0.1 0.2 0.3 0.4 0.8
(Zn$_x$Mn$_{1-x}$)Fe$_2$O$_4$
(Zn$_x$Fe$_{1-x}$)Fe$_2$O$_4$

◂**Fig. 3.5** **a** Schematic cartoon shows the ball models of octapod and spherical iron oxide nanoparticles with the same geometric volume. The black dotted lines represent the magnetic field of the octapod and spherical iron oxide nanoparticles. The same length of black arrow means the same M_s of octapod and spherical iron oxide nanoparticles. With the same geometric core volume, the octapod nanoparticles have much larger effective volume (radius, *R*) than the spherical nanoparticles (radius, *r*). **b** TEM image of octapod with average edge lengths of 30 nm. Scale bar 20 nm. Adapted with permission from Ref. [611]. **c** Magnetism-engineered iron oxide nanoparticles and effects of their magnetic spin on MRI. TEM images show $MnFe_2O_4$, Fe_3O_4, $CoFe_2O_4$ and $NiFe_2O_4$. Scale bar 50 nm. Additional data of mass magnetization values, magnetic spin structure, magnetic spin moment and T_2-weighted spin echo MR images with color map of R_2 relaxivity. Adapted with permission from Ref. [281]. **d** Zn-doped magnetic nanoparticles for strong MRI contrast effect. Undoped (x = 0) and Zn^{2+} doped (x = 0.2, 0.4) magnetic spin alignment diagrams of inverse spinel-structured nanoparticles under magnetic field. **e** Graphs of M_s versus Zn^{2+} dopant moiety including T_2-weighted MR images of $(Zn_xM_{1-x})Fe_2O_4$. Adapted with permission from Ref. [584]

3.5.3 Computed Tomography (CT)

CT diagnostics measures the X-ray attenuation of X-ray photons penetrating tissue. The attenuation is linked to the mass attenuation coefficient (μ). Different tissues can be differentiated by differences in their mass attenuation. However, the difference in the μ values across the human soft tissue is not very substantial which makes application of contrast agents with distinct uptake in different tissue mandatory [307]. Two main processes are responsible for X-ray attenuation: Compton scattering and the photoelectric effect. Whereas a low X-ray energy results in a high exposure dose for the patient, because most X-rays are absorbed, a high X-ray energy lowers the patient′s exposure and reduces the amount of Compton scattering through elastic impact. Consequently, the photoelectric effect dominates the attenuation which also leads to a strong increase of μ at the K-shell electron binding energy (K-edge) [392, 588]. Typical X-ray spectra are taken between 57 and 69 keV with contrast agents exhibiting high atomic numbers for a good mass attenuation and K-edge values within the spectral boundaries [588]. A typical candidate for CT-contrast agents is Au because of its non-toxic and chemical inert behavior as well as morphologic and synthetic diversity combined with a high atomic mass for X-ray attenuation [180, 247, 457]. Combined with magnetic materials like Fe [15] or Gd chelates an application for dual mode imaging is possible [480]. Bi_2S_3 with an even higher relative atomic mass of Bi can serve as CT-contrast agent when the Bi^{3+} ions are sufficient stabilized by oleic acid or poly(vinylpyrrolidone) (PVP) to prevent reduction to chemically active Bi^0 [4]. Recently, an Yb based CT-contrast agent was developed with improved X-ray attenuation because the Yb K-edge energy of 61 keV is located within the typical energy range for CT spectra [306].

3.5.4 Photothermal and Photodynamic Therapy

Visible or near infrared (NIR) light in combination with plasmon active metal nanoparticles can be used as an alternative heat or radical generating treatment called phototherapy. Photothermal therapy is based on irradiation of plasmonic or photocatalytically active nanoparticles which transform the radiation into thermal energy [182]. Photodynamic therapy differs in the way the radiation energy is converted into chemically reactive species such as reactive oxygen species (ROS) via photosensitizers that subsequently initiate apoptosis of malignant tissue [212]. Especially Au [213], Au/ZnO hybrids [235] and Pd [458, 496] nanoparticles have gained interest for these applications since they exhibit a broad visible and tunable NIR SPR. The Zink group reported the use of octopus-type Janus nanoparticles for simultaneous photothermal and chemotherapy [601]. They equipped Au nanoparticles with a poly(acrylic acid) (PAA) domain to form Janus particles. Subsequently, the PAA domain was used as template to grow mesoporous silica. Additional decoration of the Au domain with spikes lead to improved SPR. Finally, the particles were functionalized with methoxy-poly(ethylene glycol)-thiol (PEG) on the Au domain and lactobionic acid for active cell targeting on the mesoporous silica domain, carrying doxorubicin hydrochloride as chemotherapeutic agent. These designed multifunctional Janus particles were then used in an in vivo mouse model for photothermal therapy and NIR responsive drug release for chemotherapy. Similar dual therapeutic systems where NIR irradiation is used both as photothermal therapy source and drug release have been also established for Pd nanosheets in combination with mesoporous silica nanoparticles (MSNs) [147]. Tang et al. presented Pd nanosheets for NIR photothermal therapy and doxorubicin drug release. [497].

3.6 Tumor Microenvironment

The solid tumor is composed of three basic components, that is >50% tumor cells, 1–10% vasculature and <35% interstitium (collagen-rich matrix) [265]. In humans, vessels range in diameter from 2.5 cm in large arteries to about 7 μm in capillaries, and the volume of blood in an adult is about 4–5 L [5]. In comparison, blood vessels in the tumor microenvironment (TMI) are more chaotically organized due to uncontrolled cell growth and constant release of pro-angiogenetic factors. Unlike the endothelial lining of normal vasculature, which has a turnover of approximately 1000 days, the endothelium in tumors can double approximately every 10 days [199]. Increased proliferation rate combined with confined space leads to an extremely high cell density, approaching or even exceeding 60% [265]. This leads to dense areas of collagen rich vasculature, inducing enormous mechanical stress onto the cells. Spatial restriction and lack of supportive stromal structures leads to compress and collapse blood vessels, limiting blood—and oxygen supply and resulting in hypoxic and necrotic tumor tissue [387]. Hypoxia, a hallmark of growing tumor tissue, is responsible for tumor properties ranging from chemo –and radioresistance to invasiveness, metastasis and resistance to cell death [544]. It is caused by limited oxygen diffusion in abovementioned chaotic tumor blood vessels. Especially critical is the increased intercapillary distance [22, 586], exceeding oxygen diffusion range (up to 200 μm [381]), and establish a relationship between tumor vascularity and hypoxia [552]. Hypoxia is known to aid tumor progression and metastasis (angiogenesis, inducing genetic instability, resistance to cell death by apoptosis and autophagy) [143, 144, 236] and confers resistance against numerous cancer therapy (chemotherapies, radiation) due to the lack of oxygen [219, 242]. Optical frequency domain imaging (OFDI) as an intravital microscopy was utilized to clearly image vascular abnormality and necrotic/apoptotic areas of the tumor microenvironment in vivo [517]. In addition to visualizing necrotic/apoptotic areas and chaotic blood vessels, anti-angiogenesis therapy via vascular epidermal growth factor receptor (VEGFR) targeting showed promising results in controlling rampant tumor growth (Fig. 3.6). In addition to hypoxia, chaotic blood vessels and necrotic/apoptotic areas, the two characteristics of interstitial fluid pressure (IFP) and reactive oxygen species (ROS), define the TMI. Increased IFP relative to the tumor periphery is largely related to the inability of the tumor stroma to maintain pressure gradients across the blood vessel walls, leading to reduced or impaired ability of DDS for deep penetration and increases the supply shortage of the tumor core. Increasing tumor sizes lead to increased IFP in both human and animal tumors [44], and therefore large tumors are more difficult to treat than small ones [2]. Baxter and Jain investigated the therapeutic effect of monoclonal antibodies in solid tumors. They observed an inhomogeneous distribution in the tumor despite increased vascular permeability. They concluded that elevated interstitial pressure combined with heterogeneous blood perfusion, extravascular binding and hampered diffusion is responsible for the poor therapeutic outcome [220]. By a mathematical description of these phenomena Jain et al. demonstrated for the first time an intratumoral elevated IFP which drops abruptly to a normal level in the surrounding tissue [43]. Besides elevated IFP limited delivery of carrier systems to cells is linked to the high oncotic pressure (colloid osmotic pressure) in tumors [474]. Further reduction in transport and diffusion of macromolecules and nanoparticles in the extracellular matrix (ECM) is based on the existing electrostatic interactions and collagen-derived resistance [418]. Tissue resistance to macromolecular transport may be promoted by mucopolysaccharide in the ECM [368]. It was suggested that maximum delivery capabilities are possible through cationic charged particle for vasculature targeting followed by change to neutral charge after exiting the bloodstream [477]. ROS have essential functions in living organisms but display highly increased values in tumors [492]. They are known through metal oxide nanoparticle-induced ROS generation, inducing oxidative stress [160]. They are also produced in the mitochondria during oxidative phosphorylation. The highly proliferating tumor microenvironment uses ROS for oxidative stress-mediated signaling, advancing energy metabolism, proliferation and cell survival [475]. Upregulation of ROS production and increased ROS level can be linked to the

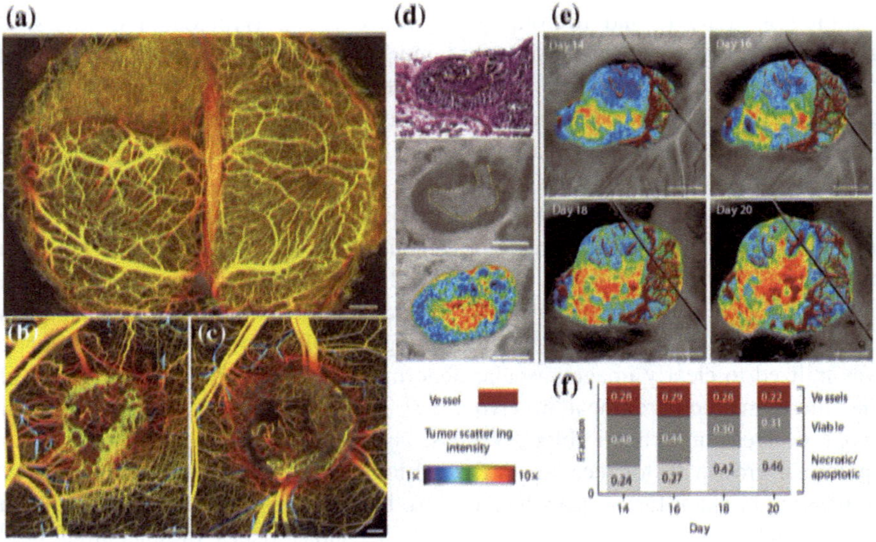

Fig. 3.6 Vessel density and hypoxic/necrotic areas in developing tumors and the impact of antiangiogenic treatment onto tumor blood vessels. **a** Depth-projected vasculature within the first 2 mm of mouse brain bearing a xenotransplanted U87 human glioblastoma multiform tumor imaged with OFDI. Depth is denoted by color: yellow (superficial) to red (deep). Scale bar, 500 μm. **b** OFDI images of representative control and **c** treated tumors 5 d after initiation of antiangiogenic VEGFR-2. The lymphatic vascular networks are also presented (blue) for both tumors. Scale bars, 500 μm. **d** Comparison of standard H and E staining (top) with OFDI (middle) reveals association of tissue necrosis with highly scattering regions. Viable and necrotic regions within the same tumor are highlighted by color gradients indicating scattering intensity (bottom). **e** Scattering properties correlated with the microvasculature during tumor progression, illustrating the expansion of necrotic or apoptotic regions in areas with minimal vascular supply. **f** Quantitative analysis of tissue viability and vascular regions in vivo, revealing an increase in the fraction of necrotic or apoptotic tissue during tumor progression. Scale bars (**d**) 500 μm; (**e**) 1 mm. Adapted with permission from Ref. [517]

stimulative response of different growth factors and cytokines (TNF-α, IFN-γ) on the TMI. The adaption of tumor tissue to these harsh environmental factors is possible through multiple pathways, ranging from increased expression of anti-oxidant molecules to function alteration in cell death factors. This ROS-mediated adaptions lead to altered drug metabolism and to drug -and chemotherapeutic resistance of the cancer tissue [509]. To counterbalance these highly energetic and chemically reactive species, detoxification of cancer cells is achieved through increased levels of antioxidant protein (glutathione, superoxide dismutase, flavonoids and different vitamins) [301]. The vicious cycle of increased ROS levels leading to ROS induced gene mutations and consequently to additional metabolic malfunctions [509] makes a ROS-mediated mechanism for selective therapeutic treatment plausible. One can distinguish between ROS generation [323, 363, 559], promoting ROS production to unsustainable levels for cell killing or

ROS elimination through inhibiting redox regulatory mechanism. This approach aims to disable redox adaption and redox-sensitive survival molecules, tipping the balance in intracellular ROS level control [129, 230].

3.7 Biodistribution and Functionalization of the DDS

FDA requirement for all injected contrast agents includes, but is not limited to, full clearance from the body in a reasonable time period [97]. Two potential ways of body excretion are relevant for the carrier systems. Renal clearance and urinary excretion covering particle and macromolecules smaller 5.5 nm, filtrating them out of the body [138, 304]. Spleen and liver, both major components of the mononuclear phagocyte system (MPS), clean the blood from particles >10 nm, raising toxicity concerns for non-degradable carrier systems due to increased accumulation in these organs [365]. Kupfer cells are responsible for the phagocytic activity of the liver and represent 80–90% of the total body macrophage population [37]. Macrophage phagocytose combined with the reduced velocity of nanomaterials passing through the liver (up to 1000 times slower than in arteries and veins) explains the increased accumulation of nanomaterials [513]. In the past, there were different approaches implemented to minimize cellular interactions, to avoid opsonization and phagocytose as well as complement activation. Opsonization is based on opsonin protein corona formation (immunoglobulins and components of the complement system) onto foreign materials in the blood. Complemented by the phagocytose, these two mechanism are responsible for blood cleaning of materials larger than the renal threshold limit [385]. PEGylation of the nanoparticle surface via adsorption or covalent binding is known to prevent opsonization and increase the blood circulation time [179]. The known stealth effect of PEG leads to numerous reports in the field of liposome- and inorganic nanoparticle-based drug delivery [254]. The surface coverage with PEG alters the composition of proteins adsorbed on the surface of the materials [53], and these proteins (mainly clustering) are responsible for preventing non-specific cellular uptake [447]. Functionalization of the silica nanoparticle surface with organosilanes and PEG showed strong a correlation between the ligand density, the chain length and charge to biorelated degradation and colloidal stability [64, 65, 264, 566]. Avoiding MPS recognition is essential for prolonging blood longevity and favoring passive targeting. Camouflaging of DDS through pre-formed protein or lipid shells can be a beneficial approach. Red blood cell and white blood cell membrane coating showed prolonged blood circulation time of polymeric nanoparticles due to camouflage of the nanoparticle surface [209, 399]. Platelet membrane-cloaked nanoparticles possess a unilamellar membrane coating functionalized with immunomodulatory and adhesion antigens. This stealth property leads to a lack of particle-induced complement activation, avoiding MPS recognition [208]. Discher et al. utilized the membrane protein CD47 ("marker of self") for intravenous injection of virus-sized particles, hampering phagocytosis and promoting persistent circulation [425].

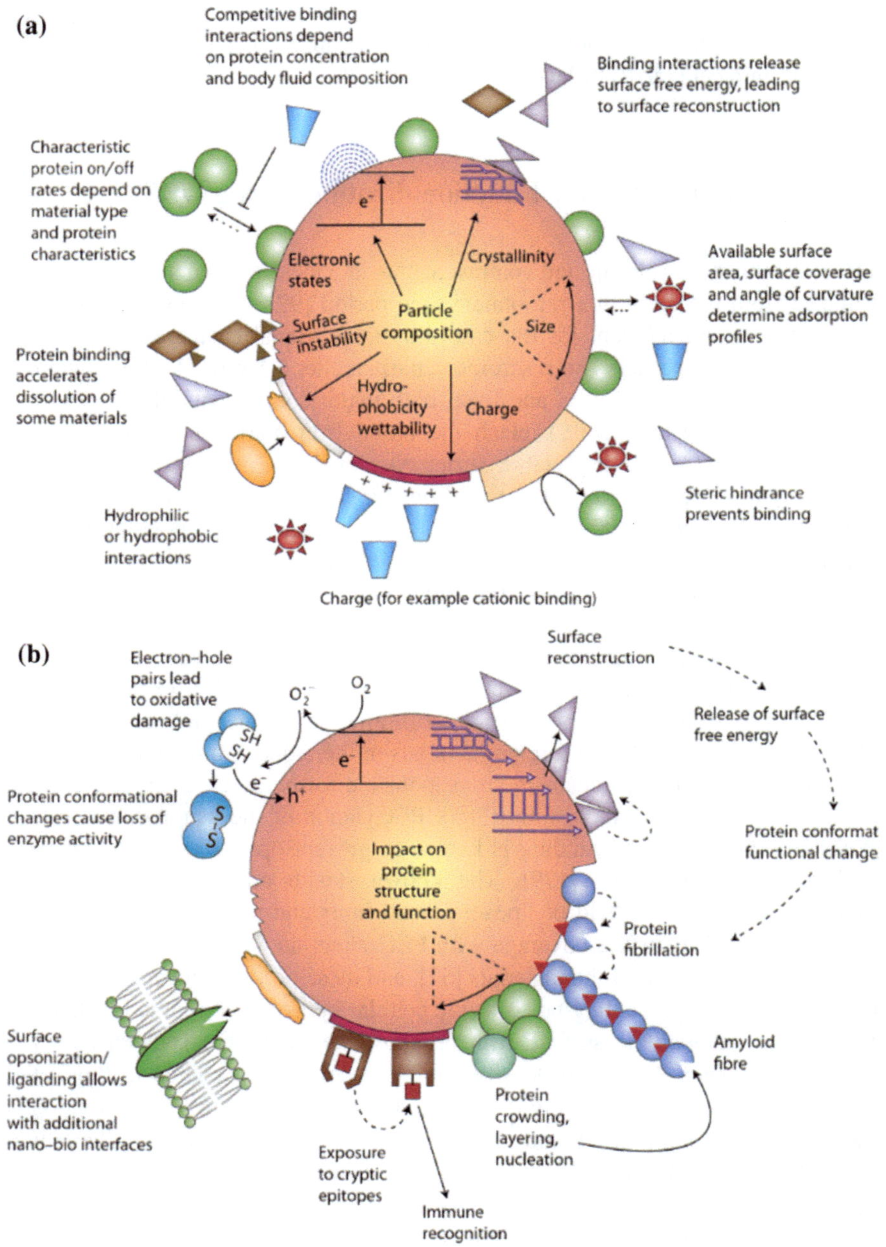
(a)
Competitive binding interactions depend on protein concentration and body fluid composition
Binding interactions release surface free energy, leading to surface reconstruction
Characteristic protein on/off rates depend on material type and protein characteristics
e⁻
Electronic states
Crystallinity
Particle composition
Surface instability
Size
Available surface area, surface coverage and angle of curvature determine adsorption profiles
Protein binding accelerates dissolution of some materials
Hydro-phobicity wettability
Charge
Hydrophilic or hydrophobic interactions
Steric hindrance prevents binding
Charge (for example cationic binding)
(b)
Electron-hole pairs lead to oxidative damage
Surface reconstruction
Release of surface free energy
Protein conformational changes cause loss of enzyme activity
Protein conformat functional change
Impact on protein structure and function
Protein fibrillation
Surface opsonization/ liganding allows interaction with additional nano-bio interfaces
Amyloid fibre
Protein crowding, layering, nucleation
Exposure to cryptic epitopes
Immune recognition

◂**Fig. 3.7** Effects of protein corona surrounding a nanoparticle. The corona constitutes a primary nano–bio interface that determines the fate of the nanoparticle and can cause deleterious effects on the interactive proteins. **a** Pre-existing or initial material characteristics contribute to the formation of the corona in a biological environment. Characteristic protein attachment/detachment rates, competitive binding interactions, steric hindrance by detergents and adsorbed polymers, and the protein profile of the body fluid lead to dynamic changes in the corona. The corona can change when particles move from one biological compartment to another. **b** Potential changes in protein structure and function as a result of interacting with the nanoparticle surface can lead to potential molecular mechanisms of injury that could contribute to disease pathogenesis. The colored symbols represent various types of proteins, including charged, lipophilic, conformationally flexible proteins, catalytic enzymes with sensitive thiol groups, and proteins that crowd together or interact to form fibrils. Adapted with permission from Ref. [366]

3.8 Bio-nano Interface

To understand the dynamic interactions of the bio-nano interface, we have to consider the nanoparticle surface and its solid-liquid contact zone with the biological environment. Physicochemical characteristics of the nanoparticle like chemical composition, surface functionalization, angle of curvature, porosity and surface crystallinity are key properties to understand the bio-nano interaction. Other quantifiable properties, such as effective surface charge (zeta potential), particle aggregation, state of dispersion, stability/biodegradability and dissolution characteristics are determined by the characteristics of the suspending media [350, 365, 366, 378, 521, 615] (Fig. 3.7). To minimize possible side effects and to control the interaction of the nanosurface and the biological environment different strategies can be employed. One approach is to use polymer- based spacers to minimize interaction of the nanoparticle with the cell membrane. Another approach is to form an inorganic shell around the nanoparticle for anchoring different ligands and dye labels on the surface. The interaction of the nanoparticle with cells is a sum of different contributions, ranging from traditional forces for colloids (attractive VDW and repulsive electrostatic forces) to complex biomolecule interactions including receptor-ligand binding and membrane wrapping. These forces can be short-range (1–100 nm) for polymer bridging, steric and solvent interaction or long range for (up to 10^2–10^6 nm) for hydrodynamic interactions like Brownian diffusion and convective drag [365, 366]. Nanoparticles tend to aggregate when the energy for a single particle distribution in solution is smaller than the aggregation energy. PEG chains are able to sterically stabilize the nanoparticles in solution leaving a minimum interaction. Another approach to ensure colloidal stability is the electrokinetic stabilization of nanoparticles with a relative zeta potential of 20 mV. Surface concentration of the organic groups and steric stabilization also play an important role [34, 430]. The size of the carrier system is crucial for efficient cancer treatment (depth of penetration, aggregation) and needs to be adjusted according to the biological problems. Nanoparticles larger than 10 nm avoid renal clearance and do not accumulate in the kidney. Nanoparticles larger than 200 nm are filtered from the blood stream through sequestration by sinusoids in the spleen and fenestra of the liver (Fig. 3.8). Gold nanoparticles have shown a size and shape dependence for

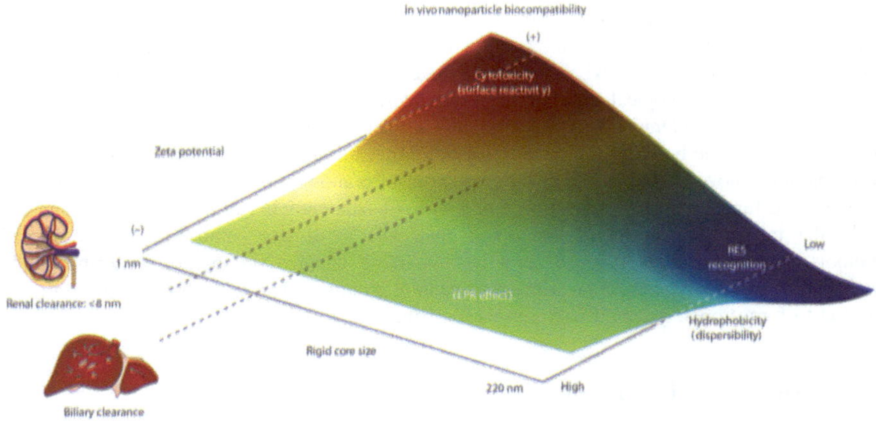

Fig. 3.8 Physical characteristics of nanoparticles determine in vivo biocompatibility. The three dimensional phase diagram displays the qualitative biocompatibility trends revealed after in vivo screening of around 130 nanoparticles intended for therapeutic use [340]. The main independent particle variables that determine the in vivo biocompatibility (color spectrum) are size, zeta potential (surface charge) and dispersibility (particularly the effect of hydrophobicity). Biocompatibility is reflected in the color spectrum, with red representing likely toxicity, blue likely safety and blue–green–yellow intermediate levels of safety (in the same order). Cationic particles or particles with high surface reactivity are more likely to be toxic (red hue) than the larger relatively hydrophobic or poorly dispersed particles, which are rapidly and safely (blue hue) removed by the RES. Particles that promote EPR effects – and are therefore optimal for chemotherapeutic drug delivery to cancers-generally have mid range sizes and relatively neutral surface charges. Reprinted with permission from Ref. [366]. Copyright Macmillan Publishers Ltd. 2009

uptake into mammalian cells [95]. Ultraviolet radiation of human skin revealed deep penetration of small quantum dots, but not of TiO_2 and polystyrene nanoparticles [357]. Sub-100 nm polymeric micelles showed a size-dependent accumulation in poorly permeable pancreatic tumors, which increases for larger micelles when transforming growth factor-β inhibitor-affected tumor permeability [55]. PEG-coated spherical and rod-shaped quantum dots deviate in their tumor penetration capabilities, suggesting the use of rod-shaped nanotherapeutics for efficient cancer therapy [76]. Additionally, microfluidic systems mimicking the tumor vasculature showed a shape-dependent enhancement of the endothelial targeting of antibody-coated polystyrene nanoparticles [258]. Real-time in vivo microscopic imaging was utilized to compare the extravasation of spherical quantum dots and rod-like single-walled carbon nanotubes (SWNTs) in three murine tumor models. It was concluded that despite similar surface coatings, area, and charge, the nanoparticle shape and physical parameters of the tumor endothelium are responsible for increased extravasation [467]. Multistage delivery systems are designed to utilize the TMI in order to overcome biological barriers by subsequent degradation. Larger particle constructs aim to prolong the blood circulation and exploit passive targeting via EPR. After TMI triggered fractioning, smaller particles and molecules are capable to penetrate deeper into the tumor tissue. MSN and polymeric clustered nanoparticles showed efficient delivery of chemotherapeutics and deep penetration into cancer cells in vivo. [290, 501, 568]

3.9 Electronic Properties of Nanomaterials

In recent years researchers have paid attention to explore the new electronic processes originating in artificial nanomaterial structures. Starting from luminescence in quantum dots (QDs) [30], complex interfacial electronic communication in heterodimers resulted in a variety of new applications. Owing to their fluorescent properties the heterodimers were utilized for biological detection and diagnostics [11, 69, 348]. The wide area of photocatalysis with nanomaterials mainly focuses on catalytic applications such as photovoltaics [224], photodegradation [531] and photogeneration [451]. Still, heterodimers containing noble metals and semiconductors are also of interest for many biochemical applications like photodynamic therapy or biomolecular detection/reaction [80] and diagnostics towards lab-on-a-particle architecture [529]. We will now take a closer look on the physical processes linked with QDs and their application in the area [545].

3.9.1 Luminescence of Quantum Dots

Semiconductor QDs absorb photons with energies larger than their band gap. This results in the generation of electron-hole pairs (excitons) by promotion of electrons from the valence band (VB) to the conduction band (CB). Light is emitted when electron-hole recombination takes place. The emitted photons belong to a narrow emission window due to the quantum confinement of the QDs. This results from the spatial confinement of the wave function of QDs which is localized in a quantum well. It can be observed for nanomaterials with a smaller radius than the Bohr exciton radius [69]. These materials exhibit quantized energy states because they are in an intermediate size between an atom/molecule with defined molecular orbitals and orbital energies and a bulk solid with continuous electron bands. They can be referred as artificial molecules and described by the "particle in a box" model. Resulting from this model, the energy of the exciton is inversely proportional to the size of the particle and an increase in particle size leads to a red shift in the absorption spectrum [545]. Therefore tuning the QD size and material allows a precise adjustment of the emission wavelength [8, 9]. The QDs show remarkable properties in biochemical imaging applications. They exhibit high brightness due to a high quantum yield in combination with large molecular extinction coefficients and a broad absorption spectrum [279]. Additionally they possess a high two-photon conversion cross section [275] for two-photon spectroscopy with NIR radiation. Photobleaching is usually negligible [9, 561]. In contrast, organic dye molecules for bioimaging usually have a narrow excitation, but a broad emission

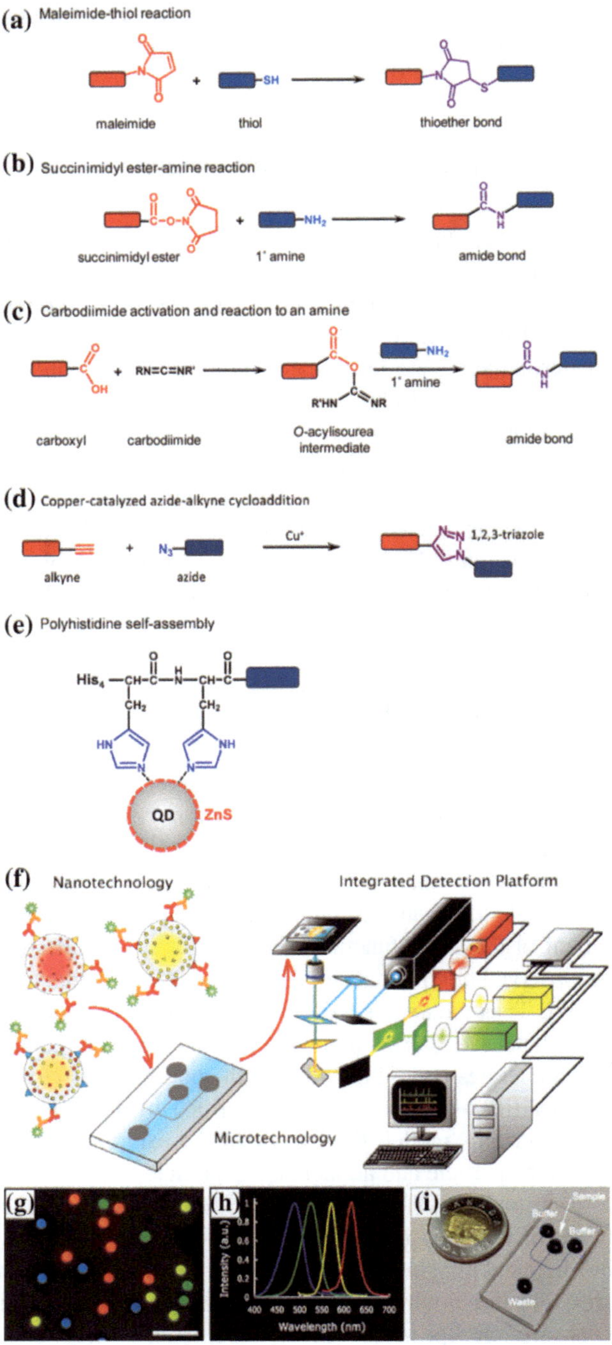
(a) Maleimide-thiol reaction
SH
maleimide
thiol
thioether bond
(b) Succinimidyl ester-amine reaction
NH2
succinimidyl ester
1° amine
amide bond
(c) Carbodiimide activation and reaction to an amine
OH
RN=C=NR'
NH2
1° amine
R'HN
NR
carboxyl
carbodiimide
O-acylisourea intermediate
amide bond
(d) Copper-catalyzed azide-alkyne cycloaddition
N3
Cu+
1,2,3-triazole
alkyne
azide
(e) Polyhistidine self-assembly
His4
CH2
HN
NH
QD
ZnS
(f) Nanotechnology
Integrated Detection Platform
Microtechnology
(g)
(h)
Intensity (a.u.)
Wavelength (nm)
(i)
Sample
Buffer
Waste

◂**Fig. 3.9** Bioconjugation reactions for QDs, including **a** maleimide-thiol, **b** succinimidyl ester-amine, **c** carbodiimide-mediated coupling between carboxyls and amines, **d** copper-catalyzed azide-alkyne cycloaddition and **e** self-assembly of polyhistidine to the inorganic ZnS surface of a QD *via* direct coordination of the imidazole moieties. Adapted with permission from Ref. [7]. **f** Diagram illustrating the integration of QDs, solution-based sandwich assay, microfluidics and fluorescence detection with custom software for high throughput, multiplexed blood-borne pathogen detection. **g** Fluorescence image of a collection of different color emitting, 5.0 μm diameter polystyrene QDs suitable for proteomic or genomic assays (scale bar 20 μm). **h** Normalized QD emission profiles corresponding to the QDs used for the barcodes in (**g**), all excited using 365 nm light. **i** Sample microfluidic chip, fabricated in polydimethylsiloxane with wells labeled. Channel dimensions are 100 μm wide by 15 μm high. Blue dye was used to visualize the channel intersection for electrokinetic focusing. Adapted with permission from Ref. [252]

range, and they are prone to photobleaching [38]. Compared to QDs they are less toxic and easy to conjugate to biomolecules without changing their biological function. In recent years surface functionalized QDs have been utilized for biomedical detection and labeling [11, 90].

Normally, the as-synthesized QDs are not water dispersable and need additional surface functionalization. The functionalization strategies are comparable to those mentioned above for metal and metal oxide nanoparticles. They are mainly based on chemisorption using the HSAB principle and tailored polymeric ligands. Because of the Pearson soft elements in QDs chemisorption typically relies on Pearson soft linkers like phosphines, phospholipids or thiols [596]. Dihydrolipoic acid with two sulfur anchor groups is a widely used bidentate linker which can be tailored subsequently with different PEG oligomers to achieve stability in aqueous dispersion independent of the pH or the ionic strength [489]. For the preparation of their nanobioconjugates, thiols, amines and carboxylate groups are used together with malimide, succinimidyl ester (NHS-ester) or carbodiimide activation for amide bonding (Fig. 3.9a–c) [7]. Click-chemistry like alkyne-azide cycloaddition to form 1,2,3-triazoles and self-assembly of polyhistidine ligands due to chemisorption of the imidazole side chains are also used on ZnS QDs (Fig. 3.9d, e) [7]. Biochemical applications are divided in many in vitro and some in vivo applications. Due to their excellent fluorescence properties QDs have been utilized for fast and highly sensitive diagnostic assays as biomarkers for infectious diseases which can be performed in microfluidic setups (lab-on-a-chip) [409]. The Chan group developed a QD based barcode assay for the detection of the hepatitis B, hepatitis C and HI viruses (Fig. 3.9f–i). The corresponding antigens are coupled to different fluorescent QDs and incubated with human serum containing the corresponding antibodies for viral antigens. A sandwich assay complex is formed by incubation with fluorophore-antibody conjugates. The QDs can now be traced with a microfluidic system on different emission wavelengths to target different viruses and antibody conjugations by the signal peak of the fluorophore. With this technique a 50 times higher sensitivity compared to FDA-approved methods was achieved for sample volumes less than 100 μL within a time span of less than one hour [172, 252]. In vitro drug targeting and bioimaging techniques have been developed, e.g. QD-aptamer-doxorubicin conjugates which simultaneously target prostate cancer and image the drug delivery process by activating the QD fluorescence when doxorubicin is released [24]. Further applications include the measurement of cell motility to evaluate the correlation between cell migration and their metastatic potential [395], detection and imaging of apoptotic cells upon binding of annexin V functionalized QDs on phosphatidylserine [163] and visualization of folic receptors overexpressed in many types of cancer cells by a fluorescence switch principle [609]. However, for in vivo applications the toxicity is mostly associated with the elements Cd and Se of the CdSe QDs, whereas coating, blood circulation, accumulation as well as degradation and clearance are equally important to estimate the potential for clinical use and the risks for the patients [42]. In a mouse model five different NIR-QDs were used for lymphangiography [256] and NIR-QDs were applied for multimodal NIR and positron emission tomography (PET) studies, where the QDs were functionalized with a ^{64}Cu-DOTA ligand for PET to enhance the quantitative tomographic image of PET by qualitative information of QD fluorescence [56]. Ding et al. utilized QDs for improving CT diagnostics by forming nanoemulsions of QDs and iodinated oil [124]. This dual contrast agent was subsequently used to visualize atherosclerosis in rabbits by macrophage uptake.

3.10 Nanoparticles as Enzyme Mimetics

Enzymes are known to catalyze different biochemical reactions with extraordinary high efficiency, velocity and specificity due to the induced-fit adaptability of the enzyme pocket with respect to the structure of the applied substrate [234, 547]. These catalytic properties have been applied e.g. for cellular imaging and H_2O_2 detection based on horse radish peroxidase- (HRP) functionalized TiO_2 nanorods [493]. However, when functionalized with a natural enzyme those nanomaterials still show the intrinsic disadvantages of enzymes like low stability and specific pH, temperature and substrate requirements [234]. Although alternative solutions for natural enzymes with higher stability were pursued by combining a metal atom as catalytic center with an artificial enzyme-like binding environment, it is still challenging to mimic enzymatic reactions while facing concomitant biochemical reactions in vitro [47, 547]. Important progress was made since 2007, when Gao et al. showed that Fe_3O_4 nanoparticles exhibit an intrinsic peroxidase-like activity [166]. Driven by the hypothesis that chemical redox processes occurring in natural peroxidases such as HRP [135], which contains a Fe^{2+} (haem co-factor) in the active center [120, 519], can be related to Fe^{2+}-Fe^{3+} oxidation state transitions in Fe_3O_4, other materials were tested and found to mimic natural enzymes. The following section discusses nanoparticle enzyme mimics and highlights possible applications ranging from biochemistry to materials science.

Since the detection of peroxidase-like activity of iron oxide nanoparticles many other nanomaterials with peroxidase-like activity such as MnO_2 [305], Ag/Pt [612] or V_2O_5 [17] have been reported. In general, peroxidases decompose H_2O_2 by reduction to H_2O while oxidizing at the same time another substrate [135]. Peroxidases are especially valuable in the respiratory chain of cells because incomplete reduction of O_2 in the mitochondria leads to the formation of reactive oxygen species (ROS) such as H_2O_2, hydroxide radicals ($OH^·$) and superoxides ($O_2^{-·}$) which are potentially harmful [312]. Most peroxidases contain Fe, Mn or V in their active sites [134]. One of the oldest known examples of peroxidase activity is the Fenton reaction [151]. The iron oxide nanoparticles have advantages in pH (0–12) and temperature stability (4–90 °C) compared to HRP [166]. They also show higher catalytic activity than HRP with 7.5–24.5 times higher catalytic turnover numbers (k_{cat}) for the TMB (3,3′,5,5′-tetramethylbenzidine) and H_2O_2 substrates, respectively [166]. Similarly important is the surface area of the nanoparticles as the smallest particles have the highest catalytic activity [166]. Also presented by the Yan group were applications for peroxidase mimicking nanoparticles by encapsulating iron oxide nanoparticles into a heavy-chain ferritin protein for targeting the transferrin receptor and visualizing the tumor tissue by the peroxidase-like activity of iron oxide [146] and the strip-based detection of the Ebola virus by conjugating anti-EBOV antibodies to iron oxide nanoparticles (Fig. 3.10a). These particles subsequently recognize the EBOV-glycoproteins bound on the antibody functionalized test-strip to form a sandwich complex. After oxidation of 3,3′-diaminobenzidine (DAB) through the enzyme mimicking activity of the iron oxide nanoparticles the naked-eye detection limit was increased by two order of magnitude compared to standard colloidal Au-strips [132].

Another application exploring nanoparticles as enzyme mimic is the glucose detection for food analysis or clinical purposes. Glucose can be oxidized to gluconic acid via glucose oxidase. O_2 is subsequently reduced to H_2O_2 (Fig. 3.10b). Thus, the glucose detection can be coupled to peroxidase mimicking nanoparticles for colorimetric detection [546, 547]. It was also demonstrated that both activities can be exploited by nanomaterials where Au nanoparticles are used as glucose oxidase and V_2O_5 nanowires as peroxidase mimic [412]. In addition to their peroxidase-like activity some nanoparticles also exhibit catalase-like activity. This was demonstrated by Chen et al. for magnetite (Fe_3O_4) and maghemite (γ-Fe_2O_3) nanoparticles (Fig. 3.10c–e) [84]. Catalase not only reduces H_2O_2 to H_2O but also oxidizes H_2O_2 to O_2 [362]. They reported a pH dependent enzyme mimicking function of the particles: (i) At a physiological pH of 7.4 where the particles are in cytosolic environment the decomposition of H_2O_2 produces $OH^·$ and $O_2^{-·}$, which quickly react with each other to form O_2 and H_2O as required for a catalase-like activity, and (ii) At lysosomal pH of 4.8 the formation of $HO_2^·$ as preliminary step for the production of superoxide is slower resulting in a more pronounced Fenton reaction and peroxidase-like activity. The generated $OH^·$ species have cellular toxicity which is more prominent for magnetite than for maghemite particles in harmony with their peroxidase-like activities [84].

Superoxide radicals as part of ROS are also responsible for oxidative stress under aerobic conditions. As a defense mechanism the enzyme superoxide dismutase (SOD) terminates $O_2^{-·}$ to H_2O_2 to O_2 [338]. Natural SOD contains Fe/Mn [14] or Cu/Zn [405] as co-factor and Mn^{2+} ions have been shown to exhibit

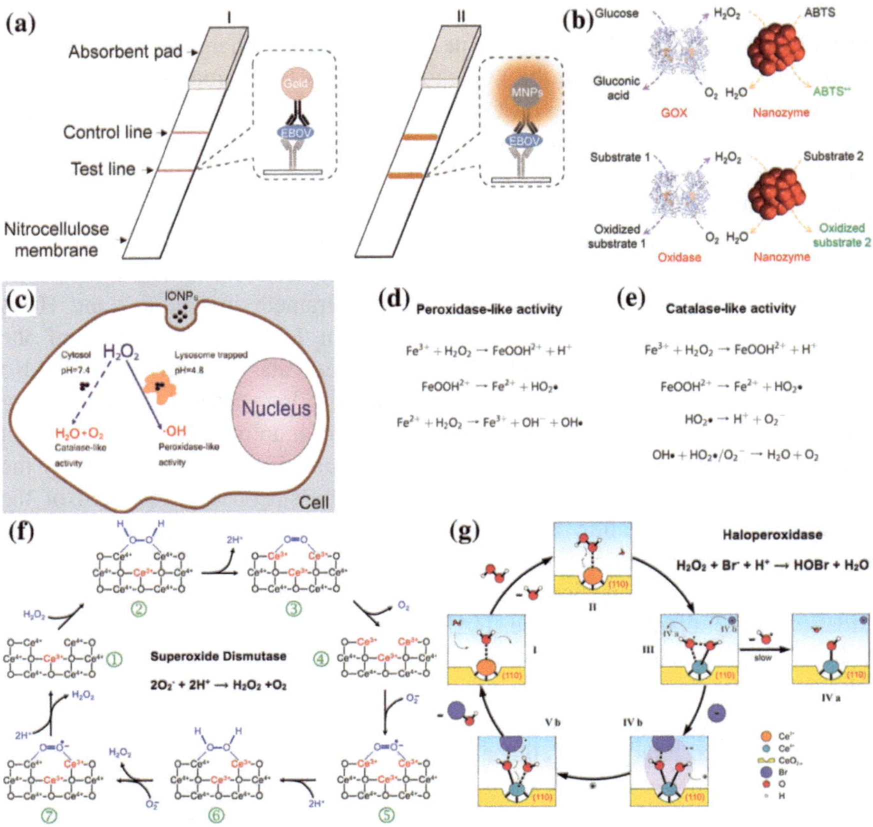

Fig. 3.10 **a** Strip based detection of Ebola virus with (I) standard colloidal gold strip and (II) nanozyme-strip employing iron oxide nanoparticles in place of colloidal gold. The probe with nanozyme activity generates a color reaction with substrates, which significantly enhances the signal so that it can be visualized by the naked-eye. Adapted with permission from Ref. [132]. **b** Nanozyme as peroxidase mimic for colorimetric sensing of H_2O_2 and glucose when combined with glucose oxidase. The sensing format could be extended to other targets (substrate 1) when combined with a appropriate oxidase. Adapted with permission from Ref. [547]. **c** Schematic illustration of peroxidase-like activity induced cytotoxicity by iron oxide nanoparticles (IONPs). IONPs are trapped in acidic lysosomes when internalized into cells, so they catalyze (**d**) H_2O_2 to produce hydroxyl radicals through peroxidase-like activity. However, in neutral cytosol, IONPs would decompose (**e**) H_2O_2 through catalase-like activity. Adapted with permission from Ref. [84]. **f** A model of the reaction mechanism for the oxidation of hydrogen peroxide by nanoceria and the regeneration via reduction by superoxide. Reduction of hydrogen peroxide to molecular oxygen takes places on initial Ce^{4+} sites, while superoxide reduction on Ce^{3+} sites restores initial Ce^{4+} sites. Adapted with permission from Ref. [67]. **g** Proposed catalytic bromination mechanism of the CeO_{2-x} nanorods. On a (110) model surface (I), the H_2O ligand can be exchanged against H_2O_2 (II). An oxidation of the Ce^{3+} site with OH^- anion and $OH^{\cdot}$ radical ligands (III) takes place. These two groups are bound to each other through a weak two-center three-electron bond. A release of an OH· radical from this species into solution may occur, but it represents a step "uphill" in Gibbs free energy and is therefore slow (IVa). A Br^- anion can add to one of the O atoms to form a species which is best described as an anionic surface site (with two hydroxide ligands) where one of the OH^- anions interacts with a $Br^{\cdot}$ radical (IVb). The other, noninteracting OH^- anion is protonated to restore a neutral surface site (Vb). Dissociation of the HOBr product finally regenerates the initial Ce^{3+} site (I). Adapted with permission from Ref. [197]

protection against oxygen radical induced damage [19] which is associated with the formation of a Mn^{4+} species (MnO^{2+}) which disproportionates in a follow-up reaction to Mn^{2+}, O_2 and H_2O_2 [28]. Therefore it is self-evident that materials that can reversibly switch oxidation states and display oxygen affinity are candidates for SOD mimetics. Nanoceria can switch easily between Ce^{3+} and Ce^{4}. Oxygen vacancies compensate the reduced positive charge of Ce^{3+} and therefore stabilize the trivalent state [237]. The Seal group reported that normal cells can be protected from radiation induced radical mediated damage by a Ce^{3+}/Ce^{4+} catalytic process [500]. Self and coworkers afterwards demonstrated SOD activity of vacancy engineered nanoceria [260]. The Ce^{3+}/Ce^{4+} ratio is important for the SOD activity. A higher Ce^{3+} content supports enhanced SOD activity [260], whereas a higher Ce^{4+} content supports the catalase reaction [406]. A mechanism was proposed that links H_2O_2 to two Ce^{4+} surface sites with subsequent reduction to Ce^{3+} and a release of O_2 (Fig. 3.10f). Now the Ce^{3+} sites can coordinate superoxide and reduce it to H_2O_2. After one repeat cycle both Ce^{3+} sites are reoxidized for the next catalytic cycle [67]. Hyeon and co-workers presented an in vivo application of ceria particles with diameters of 3 nm for ROS reduction after stroke. The antioxidant properties of the nanoceria were used to reduce the infarct volume and oxidative induced apoptosis as a neuroprotective effect [246].

In 2012 we presented the haloperoxidase-like activity of V_2O_5 nanowires which is suitable for marine paints with anti-biofouling properties [364]. Generally, haloperoxidases oxidize H_2O_2 in the presence of halides (Cl^-, Br^-, I^-) to form hypohalous acids (HOCl, HOBr, HOI) [417]. The hypohalous acid in turn acts as halogenation reagent for nucleophilic acceptor molecules involved in the intracellular communication of bacteria and therefore exhibit biocidal activity [40, 428]. We showed under laboratory and under seawater conditions (60 days in Atlantic Ocean as marine paint containing V_2O_5) that V_2O_5 nanowires follow a similar mechanism of forming HOBr and singlet oxygen as naturally occurring vanadium haloperoxidases. They are at the same time less toxic for marine biota than currently approved antifouling coatings [364]. Recently we showed that oxygen deficient ceria nanorods (dimensions 20–100 nm) also exhibited haloperoxidase-like activity suitable for antifouling applications (Fig. 3.10g) [197]. To evaluate the mechanism of HOBr formation, quantum chemical calculations were performed on the basis of X-ray photoemission spectroscopy to locate the Ce^{3+} surface sites. The results of the calculations suggest a mechanism of H_2O_2 adsorption on one Ce^{3+} (110) surface atom and a subsequent two-center three-electron bonding of OH^- and $OH^{\cdot}$ species at oxidized Ce^{4+} sites. Upon addition of Br^- a radical anionic surface site with two hydroxide ligands is formed which is degraded to water in the next step. HOBr, that is formed simultaneously, restores the Ce^{3+} surface site for the next catalytic cycle [197]. These examples highlight the importance of enzyme mimcs and nanomaterials for catalysis and material science because recombinant haloperoxidases have been proposed as additives for marine paints [191]. However, the lack of long-term stability and high production costs have prevented its use so far [469].

The hypothesis that nanomaterials containing the transition atoms in the same oxidation state as the active center of enzymes has served as a useful guide to identify other new enzymes mimics. Ragg et al. demonstrated that MoO_3 nanoparticles with diameters of approx. 2 nm mimics the reaction of sulfite oxidase and can be used to treat sulfite oxidase deficiency [415]. Patients suffering from this disease lack an adequate sulfite oxidase enzyme in the intermembrane space of liver and kidney cells which plays a major role in cellular detoxification and catabolism of sulfur containing amino acids containing [449]. Typically, sulfite oxidase catalyzes the oxidation of sulfite to sulfate which otherwise would cause severe neurological damage and early childhood death of patients [168, 226]. It has been shown that MoO_3 nanoparticles, functionalized with triphenylphosphonium units for mitochondria targeting [352], possess in vitro activity towards sulfite oxidase deficient cells. The functionalized MoO_3 nanoparticles were successfully internalized in the mitochondria and the sulfite oxidase activity could be fully recovered [415].

3.11 Protein Corona

After injecting DDSs into the bloodstream there are several obstacles and barriers to overcome for successful drug delivery. They must disseminate by tissue perfusion, penetrate across microvascular walls and then distribute throughout the tumor stroma [75]. The formation of a protein corona is a natural consequence of carrier-cell interactions in vitro and in vivo. They are caused by the adsorption of proteins on the high energy surface of the nanoparticle [350]. NP properties like, size, composition, shape, crystallinity and conformational changes of proteins upon binding affect the composition, thermodynamics and kinetics of the protein corona [326]. Solid and mesoporous silica nanoparticles with various sizes show a correlation of external surface area, surface curvature and porosity with the protein corona formation [101]. Silica nanoparticles show similar as gold nanoparticles an aggregation and size due to corona formation and phagocyte interaction, which is described as their "biological identity" [524]. Important physical properties that affect the composition of the protein corona are the surface charge of the DDS. Although surface charge is responsible for colloidal stabilization, interaction with counter charged solvent (medium) or ligands may lead to increased aggregation, e.g. via protein corona forming [524]. We can distinguish between a tightly bound monolayer of biomolecules ("hard" corona) and a more loosely associated and highly fluctuating layer ("soft" corona) [353, 419] (Fig. 3.11). In general, proteins like apolipoprotein-1, albumins, immunoglobulins and complement proteins constitute the majority of the corona [558]. Polystyrene NPs showed reduced clearance by the MPS in vivo due to a pre-coating of albumin on their surface. This lead to the conclusion that albumin coating prevents adsorption of opsonization-active serum proteins [379]. Mailänder and coworkers could verify apolipoprotein assisted

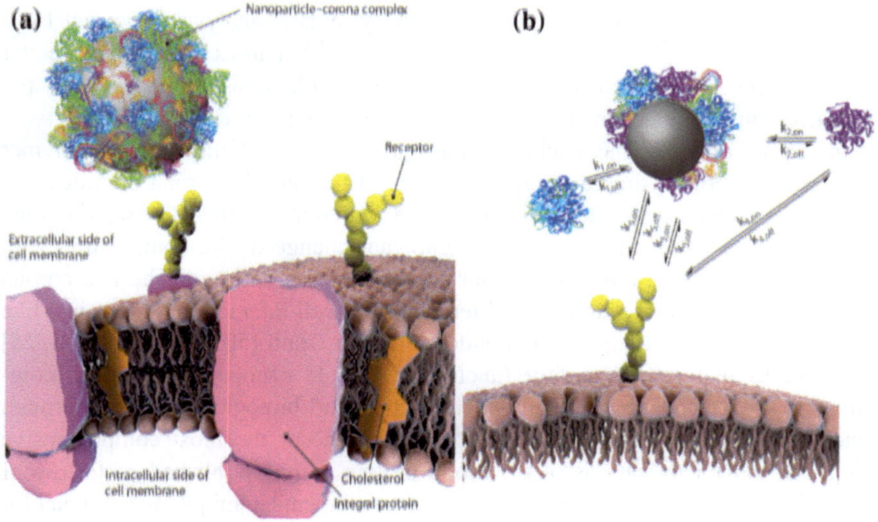

Fig. 3.11 The nanoparticle–corona complex in a biological environment. **a** It is the nanoparticle–corona complex, rather than the bare nanoparticle, that interacts with biological machinery, here with a cell membrane receptor. **b** Relevant processes (arrows), in both directions (on/off), for a nanoparticle interacting with a receptor. Biomolecules in the environment adsorb strongly to the bare nanoparticle surface (k_1), forming a tightly bound layer of biomolecules, the 'hard' corona, in immediate contact with the nanoparticle. Other biomolecules, the 'soft' corona, have a residual affinity to the nanoparticle–hard-corona complex (primarily to the hard corona itself), but this is much lower, so those molecules are in rapid exchange with the environment (k_2). If sufficiently long-lived in the corona, a biomolecule may lead to recognition of the nanoparticle–corona complex as a whole by a cell-membrane receptor (k_3). The same biomolecule alone can also be recognized by the receptor (k_4). If present, the bare surface of the nanoparticle may also interact with cell surface receptors (k_5) or other constituents of the cell membrane. Reprinted with permission from Ref. [353]. Copyright Macmillan Publishers Ltd. 2012

cellular uptake and identify ApoA4 or ApoC3 to be responsible for the decreased cellular uptake [424]. Polystyrene and silica nanoparticles incubated with disease-specific human plasma showed a "personalized protein corona", correlating medical conditions to plasma protein concentrations and structures [185]. Puntes et al. [61] described protein corona formation of metal (Au, Ag) and metal oxide (Fe_3O_4, CoO, CeO_2) nanoparticles that can reduce ROS formation. The hydrophobicity of zwitterionic nanoparticles could be tuned to avoid hard corona forming in serum [358], showing the interconnection between nanoparticle surface interaction (chemical motif), cellular uptake and hemolysis. Another problem of protein corona formation is that "active targeting ligands" on the particle surface can be hidden, thereby preventing receptor mediated endocytose. Solid fluorescent silica nanoparticles conjugated with human transferrin showed a loss of targeting capabilities when a protein corona is formed on their surface [440]. Despite many efforts to control nanoparticle blood circulation time by tuning corona formation, to enhance cellular uptake and to minimize off-target drug delivery, a crucial task is the quantification and characterization of the protein corona. A systematic analysis is needed for a range of carrier compositions, particle sizes and surface charges to allow for a better understanding of the protein corona in nanoparticle assisted cancer therapy. Isothermal titration calorimetry and surface plasmon resonance spectroscopy were employed to quantify the exchange rates and protein affinities on copolymer nanoparticles [66]. A time-dependent analysis of the protein corona on silica and polystyrene nanoparticles identified almost 300 different proteins. The rapid corona formation (<0.5 min) did not show a significant change of the composition with time, but an increased amount of adsorbed protein was detected. The fast corona formation could be correlated with different carrier-cell interaction mechanism like hemolysis, nanoparticle uptake and endothelial cell death [503]. Chan et al. [525] were able to analyze 105 surface functionalized gold nanoparticles with a bioinformatics approach to predict cellular interaction and biological response through protein corona fingerprints. This study includes one of the most comprehensive quantitative characterizations of the blood protein corona published so far. Mohr et al. [556] showed sample preparation and measurement technique to be crucial for corona formation, which indicates that protein corona identification and characterization must be handled with care. To compare protein corona formation in vitro and in vivo, superparamagnetic iron oxide NPs (SPIONs) were magnetically extracted after intravenous injection in rats. This ex situ analysis revealed pronounced differences in the hard corona formation and question the validity of previous in vitro experiments for screening toxicity or uptake mechanism of DDS [438].

3.12 Passive and Active Targeting

Drug delivery of chemotherapeutics to cancer cells can be achieved by passive and active targeting (Fig. 3.12). Passive targeting utilizes the abnormal vasculature formation of highly proliferated tissue. Fast tumor growth and chaotic vasculature lead to a preferred accumulation of polymers with high molecular weight and nanoparticles inside the tumor stroma (enhanced permeability and retention (EPR) effect) [334]. Neocarzinostatin (SMANCS) conjugated with maleic acid to polystyrene in a murine tumor model was used to show the preferential accumulation at locations of increased vascular permeability (solid hepatocellular carcinoma) [336]. Jain and coworkers [170] developed a microvascular permeability measurement technique to quantify the increased permeability and diffusivity of dextrane macromolecules in solid tumors. Passive targeting via EPR provides a higher accumulation of the transported drugs in the tumor tissue, thereby sparing healthy tissue and reducing off-target damage. There is a need to differentiate between the permeability of cancer blood vessels and the retention of nanoparticles or macromolecules. The biological characteristics of blood vessels consist of well-defined smooth muscle cells and pericytes, an organized vessel architecture, a homogeneously distributed junction of endothelial cells, a functional lymphatic drainage and an underlying basement membrane [268, 400, 401]. The permeability

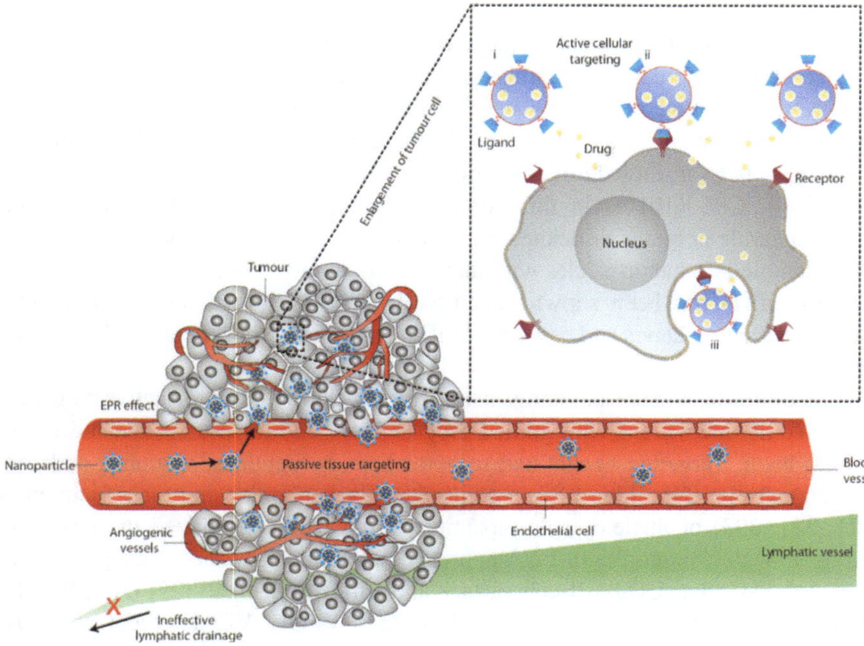

Fig. 3.12 Physical characteristics of nanoparticles determine in vivo biocompatibility. Schematic representation of different mechanisms by which nanocarriers can deliver drugs to tumors. Nanoparticles are shown as representative nanocarriers (circles). Passive tissue targeting is achieved by extravasation of nanoparticles through increased permeability of the tumor vasculature and ineffective lymphatic drainage (EPR effect). Active cellular targeting (inset) can be achieved by functionalizing the surface of nanoparticles with ligands that promote cell-specific recognition and binding. The nanoparticles can (i) release their contents in close proximity to the target cells; (ii) attach to the membrane of the cell and act as an extracellular sustained-release drug depot; or (iii) internalize into the cell. Reprinted with permission from Ref. [402]. Copyright Macmillan Publishers Ltd. 2007

of the tumor neovasculature develops from the necessity to maintain a sufficient supply of nutrients and oxygen (rapid tumor angiogenesis) [119]. This leads to aberrant vascular structures, adapted endothelial-cell-pericyte interactions, abnormal blood flow, increased permeability and delayed maturation, i.e. a complete change of the characteristics of normal blood vessels. An elevated microvessel density in the tumor is believed to advance the successful therapeutic outcome because of the EPR effect [158]. Depending on the tumor type, the pores of the tumor vasculature are typically in the 100–800 nm size range [484]. Matsumoto et al. showed that the permeability of the tumor blood vessels are based on a vigorous outward flow of fluid into the tumor interstitial space [335]. The retention of the nanoparticles or macromolecules can be attributed to defect or insufficient lymphatic drainage. Increased endothelial fenestration, architectural anarchy and dysfunctional or collapsed lymph vessels lead to decreased lymphatic drainage and, thus, to accumulation of nanoparticles inside the tumor stroma [119, 322, 366].

A major analysis of the EPR effect and the accumulation of nanomedicines in solid tumors was conducted in small animal models, while clinical data and experimental evidence for an increased accumulation via the EPR effect in humans are scarce. Tumor accumulation in humans was tracked with radiolabeled liposomes [190, 261]. Local colocalization of camptothecin and CRXL101 was demonstrated, however, only by a single human gastric tumor biopsy and for human breast cancer [140, 550]. CRLX101 is in this context a nanoparticle consisting of a cyclic oligosaccharide polymer associated with camptothecin. Approximately five strands self-assemble into nanoparticles with diameters of approx. 30 nm. The existence of the EPR effect has been shown conclusively only in animal models of human cancers. Recently, Davis and coworkers [100] showed it to be efficient in humans as well.

In addition to passive targeting, nanoparticles can be used for active targeting (cell specific interaction) with guided drug delivery and release on demand at the cellular level. Targeting agents like receptor ligands (peptides, vitamins, and carbohydrates) [48, 296, 328, 481], proteins (mainly antibodies and their fragments) [121, 321, 393] or nucleic acids (aptamers) [165, 614] are utilized in active targeting strategies. Examples of FDA approved antibodies for cancer therapy include rituximab [41] (anti-CD20, B-cell surface antigen), trastuzumab [29] (Anti-ERBB, ERBB2 receptor), and bevacizumab [153] (anti-VEGF, VEGF receptor) [119]. Active targeting is based on the ability of an antibody or protein to recognize cell specific receptors. The first example for the active targeting of nanoparticles is based on a fluorescent liposome antibody construct. The heterobifunctional cross-linker N-hydroxysuccinimidyl 3-(2-pyridyldithio)propionate (SDPD) was used to covalently couple mouse monoclonal antibody against human β2-microglublin to the liposomes [286]. Overexpression of small peptide receptors has been documented for many examples of cancer [422]. For active targeting, the receptor density should be in the range of 10^4–10^5 copies per cell. Lower densities in cell populations are less efficiently targeted [311]. Tumors in the millimeter size range (1–2 mm) appear to have no vasculature [157]. Without the proper pathophysiological characteristics (EPR effect) active targeting in small metastases (<100 mm^3) can be a suitable treatment strategy [92] by increasing the selectivity compared to passive targeting. The perquisites are a strong ligand–receptor interaction (affinity) or multiple ligand–receptor interactions (avidity) [12]. At the same time it is important to control the density and accessibility of the targeting ligands on the carrier. With transferrin-coated gold nanoparticles the tumor accumulation was enhanced by the nanoparticle avidity, the accumulation being 5 times faster and 2-times higher compared to passive targeting [490]. Dendrimer-bound folate molecules showed a 2500–170,000-fold increase of the dissociation constants (KD) compared to free folate [205]. Multivalent target binding is one approach to ensure antibody-antigen interaction, but increasing the amount of ligand on the nanoparticle surface does not necessarily increase the avidity due to steric considerations [450]. The mode of action for antibody therapy depends on the individual targeted tumor antigen and may range from a downregulation of the cell surface receptors to an activation of the antibody-dependent cell-mediated cytotoxicity (ADCC) or the complement-dependent cytotoxicity (CDC) [450]. For active targeting strategies we may distinguished between cancer cell and tumor endothelium targeting. These two strategies include receptor-mediated internalization of encapsulated chemotherapeutics for reduced off-target damage and direct cell killing strategies that exploit tumor angiogenesis by targeting VEGF receptors in the tumor-associated vasculature. Transferrin, folate, glycoproteins and epidermal growth factor (EGFR) are all suitable targeting receptors for active targeting of cancer cells. These receptors are highly overexpressed, they have an internalization-prone mechanism (endocytosis-prone surface receptors [251]) and their targeting ligands can easily be conjugated to the carrier [116].

Therapeutic targeting of the tumor microenvironment [229, 413] can be implemented in various ways, which makes the tumor stroma [359] a viable target for cancer therapy. Angiogenesis-associated targeting [221, 367] is effective in limiting the tumor growth. It is universal for all cancer types, it normalizes the tumor stroma, reduces the IFP and increases oxygen flow, blood and nutrient supply. This leads to a deep penetration of the nanocarriers that avoid the chaotic tumor vasculature and the associated physiological barriers [16, 75, 218]. IFP reduction by 65% through collagenase-1 induced type-1 collagen depletion led to a higher accumulation and gene expression for intravenous injected lipoplexes [239]. In addition, disrupting the growth of the blood supply cripples the metastatic capabilities of the primary tumor and reduces the secondary drug resistance due to smaller phenotypic variations in the neovascular endothelial cells [54]. Tumor vascular targeting through a neovascular proliferation biomarker ($\alpha_V\beta_3$-integrin) was utilized to deliver doxorubicin resulting in a reduced metastatic activity [361]. Another approach to normalize the tumor neovasculature is to block the VEGF. VEGF-specific antibodies like bevacizumab are able to normalize the tumor vessel and to reduce the interstitial fluid pressure in animal and human models [554]. Vascular normalization by the VEGF receptor-2 blocking antibody DC101 showed improved delivery of quantum dots in a size-dependent manner. It was suggested that 12 nm nanoparticles are optimum for deep penetration, owning to the DC101 induced reduction of the IFP [77]. Ribal et al. [118] demonstrated the capability of non-antibody targeting ligands with a human transferrin glycoprotein targeting ligand that induced an RNAi activity in humans via siRNA delivery. This study demonstrated for the first time a dose-dependent accumulation of targeted nanoparticles in human tumors. It showed the potential of active targeting to elucidate the EPR effect and the concept of passive targeting in humans.

3.13 Endocytose

The transport of chemotherapeutic drugs, proteins or fluorescent dyes into tumor cells occurs through endocytose (Fig. 3.13a). It is characterized by carrier-membrane interaction followed by an engulfment and endosomal

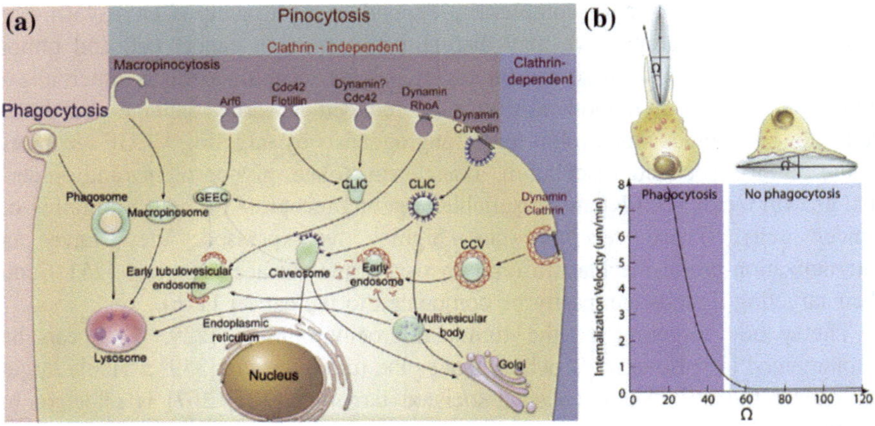

Fig. 3.13 **a** Multiple pathways for cellular entry of particles and solutes. In all cases the initial stage of endocytosis proceeds from the plasma membrane portals of cellular entry and involves engulfment of cargo into intracellular vesicles. The second stage involves sorting of the cargo through endosomes. It is followed by the final stage where the cargo is delivered to its final destination, recycled to the extracellular milieu or delivered across cells (not shown). Abbreviations are: CCV, clathrin coated vesicles, CLIC, clathrin-independent carriers; GEEC, GPI-anchored protein-enriched compartment; GPI, glycophosphatidylinositol, MVB, multivesicular body. **b** Effect of particle geometry on phagocytosis. The entry of nanoparticles inside macrophages depends on the angle between the membrane normal at the point of initial contact and the line defining the particle curvature at this point (Ω). The internalization velocity is positive at $\Omega \leq 45°$, which indicates that the particle undergoes internalization. As the angle exceeds critical value ($\approx 45°$) the internalization velocity is zero, the macrophages lose the ability to entrap particles and start spreading over the particle. Adapted with permission from Ref. [433]

distribution to different cell compartments. Phagocytose is primarily observed in cell populations of the MPS (macrophages, dendritic cell) while pinocytosis is classified by participating endocytosis proteins. One can differentiate between a clathrin-dependent and clathrin-independent mechanism [125, 433]. Nanoparticle uptake is based on factors like surface charge, particle size. Ligand density and composition are two key characteristics for carrier-cell interaction and determine the pathway for nanoparticle internalization [605]. Simulations of nanoparticle wrapping in cell membranes showed that next to particle size and geometry ligand density is an important factor for minimizing endocytic time [591] (Fig. 3.13b). Transferrin-coated gold NPs showed clathrin-mediated endocytosis [94], while QDs [502, 603], silica [465], iron oxide [3, 227] and silver NPs [207] showed variable endocytose mechanisms dependent on size, surface charge and functionalizing strategy.

3.14 Gatekeeper

Drug loading of nanoparticles is followed by encapsulation to prevent premature release and off-side host-drug interaction. This is particularly important for biologically sensitive agents which are degraded enzymatically or are inactivated in the blood stream. Additionally, highly toxic cargos need to be shielded to protect the body from lethal drug doses. One example is the highly-cytotoxic tumor necrosis factor-alpha [206, 468] (TNF-α, Beromun), whose short half-life and acute toxicity makes it applicable only for treating non-resectable tumors of the extremities via isolated limb perfusion (ILP) [200]. Important for drug delivery applications is the capability of zero premature release of the drug. Lin and coworkers [269] were the first to functionalize MCM-41 mesoporous silica particles with cadmium sulfide NPs to ensure zero premature release of vancomycin- and adenosine triphosphate (ATP) by physically blocking of the pore entrance. Gatekeeper systems have to release the cargo through tumor cell specific mechanisms and triggers by utilizing the TMI and intrinsic features of the tumor phenotype. These intrinsic properties of the TMI like pH increase, oxygen deficit, ROS level and intracellular glutathione concentration are utilized for an intracellularly triggered release of the drug. pH-triggered release systems depend on the low pH value within the lysosome and endosomal compartment inside tumor cells and the evaluated pH value outside the cell. A possible explanation for the increased external H^+ concentration is the high lactate production through increased glycolytic metabolism (aerobic glycolysis, Warburg effect [543] and carbon dioxide concentration (oxidative metabolism) [196, 476]. The hypothesis that mitochondrial defects cause insufficient oxidative phosphorylation and—as a result—cancer was modified some time ago when it was shown that the mitochondrial metabolism is reprogrammed by altered oncogenes and tumor suppressors [544]. Increased abnormal vasculature formation of high proliferation tissue, elevated interstitial pressure and decreased lymphatic drainage lead to the observed increased of the pH value. Endosomal fusing with lysosomal vesicles after endocytose exposed the delivery system to pH values up to 4.5 [380] to trigger pH induced drug release. This pH induced gate effect was first utilized by Latorre et al. [63] through polycondensation of amino silane onto mesoporous MCM-41 materials. pH-triggered release via biocompatible polycations like poly (2-vinylpyridine) (PVP) [195, 370] and poly(ethylene imine) (PEI) [123, 155, 188, 210, 216, 290, 293, 294, 320, 347, 371, 407, 429, 452, 482, 564, 597] is an efficient method to encapsulate and deliver various small molecules, small interfering RNA (siRNA) and chemotherapeutics in vitro and in vivo. The key was to control the toxicity of high charged polycations [85, 441, 523] onto cells and a successful release of the loaded drug into the cytoscl through lysosomal escape. The proton sponge effect of the polycation is a controversial issue. It is unclear, if the large buffer capacity of PEI and other polycations is the main factor responsible for a successful endosomal/lysosomal escape [32, 33, 592]. Inherent problems of siRNA therapeutics are intravascular nucleotide degradation, proper tissue penetrance, immune-mediated toxicities and complete release of the siRNA into the cytosol after endocytose [401]. The combination of modified PEI and MSN were used to co-deliver doxorubicin (inducing apoptosis) and siRNA (gene suppression of cellular anti-apoptotic defense and drug exporter) to overcome multidrug-resistant cancer cells [78, 345, 346]. Additionally, anti-angiogenic response in vitro and in vivo was possible through knockdown of the TWIST1 transcription factor [155] and suppression of neovascularization via VEGF [293]. Numerous applications of pH gated polymer systems [73, 81] have been used in drug delivery applications, transport bioimaging nanosystems [96] or anthracycline delivery chemotherapy agents [111]. Poly(N-succinimidyl acrylate) (PSA) combined with an acid-labile acetal linker was used to encapsulate the antineoplastic antibiotic doxorubicin, functionalized with PEG (in order to decrease non-cellular uptake) and decorated with folic acid for folate receptor targeting [576]. Redox-responsive release can be achieved by reducing unstable redox linkers (e.g. disulfide bridges) by the redox couple glutathione (GSH)/glutathione disulfide (GSSG) [436]. The intracellular glutathione concentration ranges from 0.5 to 10 mM, the corresponding extracellular values are one to three orders of magnitude lower [91, 325]. This anti-oxidative capacity of cells and the associated release potential can be utilized for surface coverage [390], drug encapsulation and transport [128, 542] or for an enhanced degradability of the carrier system [324]. Another approach to release the drug at the tumor site depends on external stimuli such as light, ultrasound and temperature. Temperature-dependent release of the loaded drug is possible through conformational change of the polymer chains by heating with light or utilizing metal-based nanocarriers for hypothermal treatment [181, 238]. Photothermal activation of the gate system via NIR [192, 376, 599, 608], UV [375], red light [68] or enzymatic degradation of a biodegradable gatekeeper are viable alternatives for drug release. Biodegradation is based on the nature of cellular proteases to degrade the gatekeeper inside the lysosome [35, 268, 572]. Fukumura and coworkers [560] were able to establish a multistage delivery system, enzymatically degradable by proteases that are highly expressed in the TMI to ensure long circulation half-life and deep tumor penetration. Tumor targeting and release triggered by the matrix metalloprotease 2 (MMP 2) is a potent concept for drug delivery [82, 624]. An alternative approach to encapsulate highly toxic chemotherapeutic agents in carrier systems is based on supramolecular interactions of proteins. The interaction between avidin or streptavidin and biotin [202] (strong non-covalent biological interaction ($K_d = 4 \times 10^{-14}$ M) [178] leads to a light-responsive protein shell, covering an MSN for successful doxorubicin delivery [333]. Aptamers [241] are nucleic acid ligands generated with SELEX (systematic evolution of ligands by exponential) [141]. Compared to siRNAs, where the therapeutic target is only intracellular, aptamers can be used for intracellular, extracellular [86] or cell-surface targets [150] and stimuli-responsive drug delivery [240, 298]. The protocell, a unique class of nanocarrier combining features of mesoporous silica particles and liposomes in a synergistic fashion was devised by the Brinker group several years ago. The combination of a high surface area and the rigid structure of the silica core combined with a lipid bilayer coverage for drug encapsulation and targeting moieties represented a 10^6 fold improvement in cancer cell killing over comparable liposomes [20]. Further developments made contributions to the treatment of neuromuscular disorders and leukemia via individual cell targeting [136, 175]. All strategies described above are individually capable of drug delivery and cell targeting. To ensure complete drug release, zero premature release co-delivery of different chemotherapeutics, site-specific drug release and possible exploiting of combined chemo –and photodynamic therapy [126], internal and external stimulus can be linked. pH/redox-responsive nanocomplexes [186, 244, 292, 540] or light- and pH-activated intracellular drug release [506, 580] are examples for dual-responsive systems.

3.15 Biorelated Degradation of the DDS

A biorelated degradation of nanomaterials is a key property for DDS to prevent severe and unpredictable side-effects caused by accumulating individual components during treatment. The International Union of Pure and Applied Chemistry (IUPAC) defined biodegradation as the "degradation caused by enzymatic processes resulting from the action of cells" [520]. In this context, most carrier systems should be addressed as "biorelated degradable" nanomaterials [108]. We must differentiate between complete degradation of the whole carrier system or fracturing into small pieces for on-site drug release. If the shell is composed of biodegradable polymers like poly(D,L-lactide-co-glycoside) (PLGA) or poly(ε-caprolactone) (PCL) [6], drug release can be achieved much faster. Typical examples for degradable inorganic nanoparticles are silicon [79, 189], iron oxide [276, 287], gold [420, 432] and QDs [97]. Silica, as food additive, has been "generally recognized as safe" by the FDA for over 50 years. Mesoporous silicon is biocompatible and has received FDA approval for pharmaceutical applications for radiotherapy and implant-associated drug delivery [259, 600]. Silicon-based nanomaterials are hydrolytically unstable and dissolve in a time-dependent manner into silicic acid ($Si(OH)_4$, pKa 9.6). The rate of dissolution of silicon NPs depends on the degradation medium, degree of saturation, temperature, pH, pore size, functionalization and degree and crystallinity [98, 300, 515]. In vivo degradation studies of mesoporous silica in different simulated body fluid helps to support the connection between dissolution kinetics and drug release on-demand in a biological context [46, 184, 193, 263].

3.16 Cancer Immunotherapy

The premise of an immune system-mediated intervention in cancer progression lies in the capacity of the immune system to distinguish between self and non-self. While highly equipped and effective in the eradication of pathogens, the ability of the immune system to effectively deal with transforming cancer cells is hampered by the fact that the cancer cell's origin is self [225]. Despite the fact that depletion of self-antigen reactive T lymphocytes during cancer progression combined with numerous genetic alterations and loss of cellular regulatory processes should lead to immune-recognition, cancer cells evade immunosurveillance through various immunosuppressive mechanism [463, 623]. Recognition of cancer cells is circumvented by down-regulating the expression of surface antigens, by recruiting immune suppressive cell types to the TMI and overall the cytokine-derived immunosuppressive environment [344, 448]. Cancer immunotherapy comprises several treatments in order to activate the patient′s own immune system to detect and eradicate cancer cells. Promising approaches in solid tumor treatment are cancer vaccines for active immunization [427], antibody-based cancer therapies [60] and immune-checkpoint blockade [396] for rewiring the immune system to fight the growing cancer. The activation of a tumor-associated immunosuppressive cell population (T- and dendritic cells [154, 437] and macrophage [318, 329]) is needed for recruiting an anti-tumor response and eradicating immunosuppression by the TMI. Nanoparticles can be utilized for the delivery of vaccines, adjuvants or immune-modulatory/stimulatory molecules, making use of advantages like passive accumulation, cell target delivery, stimuli-responsive drug release and protection of biodegradable cargos [439]. Chemotherapy itself can help to activate the immune system, but only if the associated cell death is immunogenic. Immunogenicity describes the ability of a substance to provoke an immune response in the body [512]. Immunogenic responses can be triggered by chemotherapy-induced cell death, release of pro-inflammatory factors and upregulation of antigen presentation [466]. Immunogenically rendered cells can effective stimulate the immune system and thus counteract the immune-suppressive TMI [18, 377]. Nevertheless, many cytotoxic anti-cancer agents can lead to an immunologically silent or tolerogenic immune response [62] due to apoptosis [270]. Apoptosis is not accompanied by the release of pro-inflammatory cytokines into the TMI that prevent an inflammatory response. In addition, apoptotic cells are quickly phagocytozed and do not release their cellular constituents into the tumor microenvironment [142]. As a result, this silent cell death does not trigger the immune system. For a combination of chemo- and immunotherapy one needs to tune the synergy of a chemotherapeutic agent and an immunotherapeutic stimulus to achieve an effective anti-tumor response. For example, gold nanoparticle transporting siRNA to a tumor-associated macrophage (TAM [329]) and lung cancer cells silencing the VEGF mRNA combine an immune modulation of the TMI with tumor suppressor effects in vivo [102]. Gold nanoparticles for cancer immunotherapy are applied in difference areas like cancer vaccines, gene therapy and adjuvant delivery [13]. Furthermore, hollow MSN could deliver cancer antigens and showed an intrinsic immune-modulatory activity, suggesting MSN to be an efficient adjuvant for cancer immunotherapy [539]. Another example for a combination therapy is the synergistic effect of cisplatin and synthetic peptides (SLP) of the human papillomavirus type 16 (HPV16) that lead to increased anti-cancer activity [466]. Still, examples of nanoparticles used for a combined chemo -and immunotherapy are scarce. Most of them focus on imaging and active targeting to a cell subpopulation [245]. MR imaging TAM with

FDA-approved iron oxide nanoparticles (ferumoxytol or Feraheme) showed possible applications as biomarker to develop new immune-targeted therapies [115]. Polymer-based nanoparticles were used to deliver [616] a fluorescent platinum(IV) pro-drug [349] and hydrazinocurcumin (HC) [607], a synthetic analogue of curcumin, to TAM. To circumvent immunosuppression, an immune-checkpoint blockade of cytotoxic T lymphocyte associated antigen 4 (CTLA4) and programmed death 1 (PD-1) receptors are explored in clinical trials [113, 557]. Blockade of these receptors leads to an enhanced antitumor immune response by diminishing suppressive activity of intratumoral regulatory T-cells (T_{reg}) [454]. This T_{reg} population is believed to represent a major immune resistance mechanism in the tumor to prevent an immune response of cancer patients. In addition, these two T-cell regulators are highly expressed on T_{regs} making them valuable targets for antibody-mediated cancer treatment via immune-checkpoint blockade [45, 396]. Examples for nanoparticle-induced elimination of T_{reg} and myeloid-derived suppressor cells (MDSCs [162]) are the transport of a broad spectrum of anti-inflammatory triterpenoids via lipid-coated calcium phosphate nanoparticles [604]. PD–L1 siRNA target delivery with PEI liposomes resulted in the formation of re-programmed tumor-associated dendritic cells [112].

3.17 Limitation of the DDS

There is a variety of difficulties for a successful delivery of chemotherapeutics to cancer cells ranging from restrictions of the carrier (blood circulation, aggregation, protein corona) to the journey in the body (influence TMI, MPS recognition) and cellular uptake. Physical barriers and the derived rules for drug delivery are complex because of the biological problem and the multicomponent system. The effects of size, shape and charge of the DDS on diffusion and Brownian motion, particle aggregation or in vivo flow have to be considered [156]. MPS recognition and reduced aggregation can be remedied by PEG stealth coatings. Drawbacks of surface passivation by PEGylation are the possibility of complement activation leading to hypersensitivity reactions, toxicity of side products, degradation under stress and biodegradability. Polymer costs, FDA approval, stealth behavior, prolonged blood circulation, diminished RES uptake and EPR effect outweigh all disadvantages [254]. Still, the immunological response and phenomena like accelerated blood clearance [1] must be considered. 30 years after the first carrier system for cancer treatment had been approved by the FDA Chan et al. analyzed 117 publications of the past decade and concluded that only 0.7% (median) of the administered nanoparticle dose was actually delivered to solid tumors [553]. They concluded that a successful transfer of nanomedicines into clinical trials is only possible after re-examining of all existing DDSs in order to establish a simple, effective and reproducible product. The underwhelming delivery efficiency of many DDSs can be linked to an incomplete understanding of passive and active delivery strategies (Fig. 3.14). Overestimation of the EPR effect combined with the

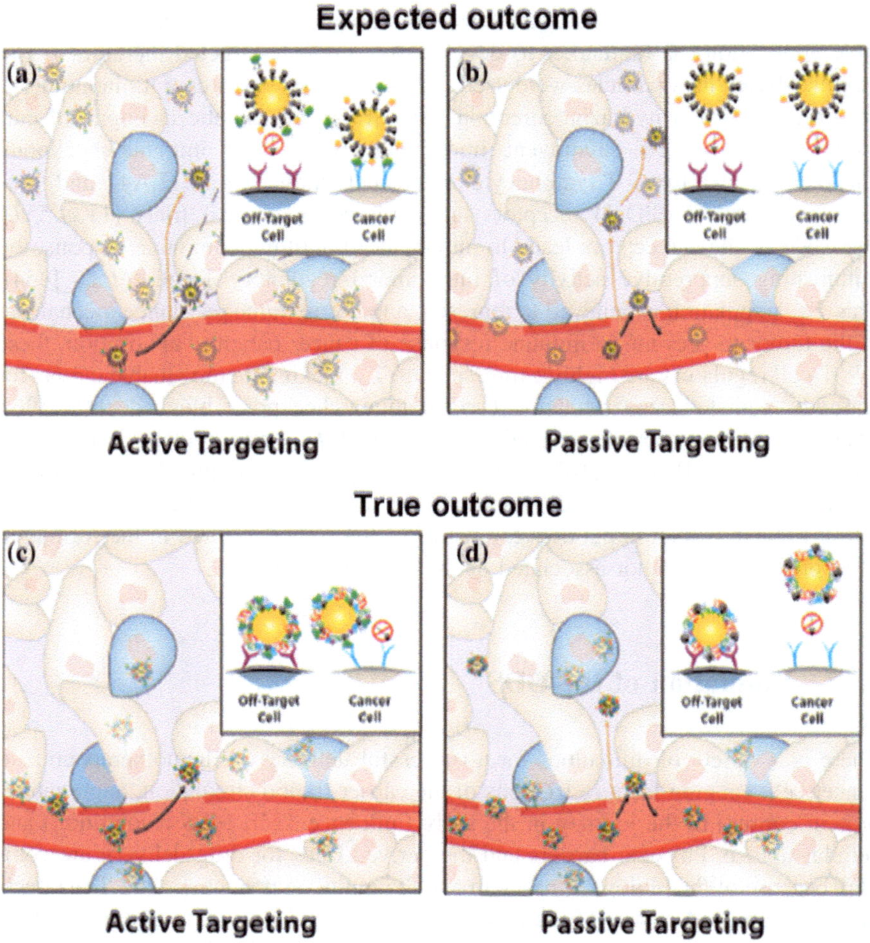

Fig. 3.14 Ideal versus actual nanoparticle-tumor targeting strategies. In tumors, the unique fenestrated vasculature facilitates nanoparticle extravasation from the blood and into the interstitium. The poor lymphatic drainage and high intratumoral pressure help retain the nanoparticles within the mass. In an ideal situation without nanoparticle–blood interactions: **a** nanoparticles are surface-functionalized with ligands, bind to cell surface receptors and undergo receptor mediated endocytosis for selective entry. **b** Nanoparticles are surface-passivated with PEG to increase half-life, and provide greater opportunity to extravasate through leaky tumor endothelium and penetrate into the tumor. The true nanoparticle–blood interactions in vivo are very different. Protein corona formation (**c**) sterically hinders and masks surface bound ligands, limits cancer cell receptor-mediated interactions and supports off-target cell binding (indicated by blue cells). **d** Nanoparticles increase in the size and limits their depth of penetration into the tumor (i.e., the particles are more likely to stay near the vessel). Of note, the green and orange structures on nanoparticle surface in (**a**) and (**b**) represents a targeting ligand and cancer agent. Adapted with permission from Ref. [278]

fundamental differences (size, metabolic and development rates, tumor biology) between animal tumors (specifically murine models) and human cancers makes progress in nanomedicine sparse [369]. Unlike artificially induced animal tumors, human tumors can take decades to develop, giving time for a more controlled angiogenic process [548]. Additionally, real human tumors are generally much smaller ranging from a few millimeters to a few centimeters at the time of diagnosis and treatment, in comparison to oversized tumors in a mouse model [59]. Active targeting strategies are hampered through the "binding-site barrier", resulting in decreased penetration of the carrier system due to high binding affinity [12]. Penetration depth is essential for efficient cancer treatment, and diffusion problems arise from the size difference between DDS and cells. Thus, large molecules like a monoclonal antibody (approximately 150,000 Da) need up to several months to reach a uniform concentration in a tumor (diameter 1 cm) by diffusion [217]. Nichols et al. [369] interpreted the diffusion process of a nanoparticle with a reduced effectivity due to physical barriers of the ECM and tightly packed cells. It is further decreased when chemical interactions between the particles and tumor components are considered. Despite these barriers and obstacles, multidrug resistance (MDR) is another important constraint in applied cancer therapy. MDR in cancer is defined as resistance of tumor cells to a single anticancer drug, which is accompanied by resistance to structurally and mechanistically completely unrelated drugs [613]. We can distinguish between pre-existing (de novo) and (acquired) resistance developed during treatment. De novo drug resistance like environment-mediated drug resistance (EMDR) protects cancer cells from apoptosis caused by receptor-mediated cell death, radiotherapy or chemotherapy. It is regulated by the TMI (soluble factors) and components of the ECM [341]. This intrinsic protection against cancer therapy leads to survival of a small subset of cancer cells (named minimal residual disease) and without eradication of the vast majority of cancer cells, a relapse in cancer patient is highly probable [404]. This survival can be explained by the highly molecular and genetic heterogeneity of cancer itself. This subset of tumor cells, originating from initially survived through de novo resistance, are now able to develop acquired resistance. This can be achieved through drug-mediated mutation and therapy-induced selection of a resistant subpopulation [203]. It leads to a highly resistant tumor, rendered nearly immune to additional cancer therapy. Cellular mechanisms of multidrug resistance are diverse, ranging from increased efflux of hydrophobic drugs, alterations in drug metabolism and cell cycle, increased repair of DNA damage, reduced apoptosis or adaption of the tumor through survival signaling pathways and inactivation of downstream death signaling pathways [176, 403, 491, 613]. The resistance mechanism of increased drug efflux (pump resistance) is caused by the adenosine triphosphate (ATP) binding cassette (ABC) transporters such as P-glycoprotein (Pgp) or the multidrug resistance associated-protein 1 (MRP1, also known as ABCC1) [171]. They are known transmembrane proteins regulating the flux of chemotherapeutic agents across the plasma membrane [137, 285], and valuable targets for nanoparticle-mediated delivery of ABC inhibitors to overcome pump resistance in cancer cells. Non-pump resistance is caused by activation of cellular anti-apoptotic defense (Bcl-2 protein) [583]. The simultaneous inhibition of both resistance pathways is key for successful cancer treatment [299, 388, 389] and was utilized for the co-delivery of doxorubicin and siRNA, targeting mRNA encoding Bcl-2 protein [78] and Pgb [345]. Dendrimeric nanoassemblies [295], MSNs [302] and gold nanoparticles [530] were functionalized to combat physiological barriers and regulate cellular factors of multidrug resistance. Additionally, size-dependent drug efflux resistance of gold nanoparticles led to increased cancer cell killing capabilities [533].

The final important area in nanotechnology development and safety is the toxicity of the nanoparticle-based formula. We have to distinguish between the different ways of exposure (ingestion, injection, transdermal delivery, and inhalation) [313] to accurately describe nanotoxicity. In the context of drug delivery, important factors are biorelated degradation byproducts of the base material, the colloidal stability of the formula and the clearance pathways from the body (MPS recognition, kidney). Additional factors to characterize nanotoxicity are the nature of the carrier system and nanoparticle properties like size, porosity, morphology, surface coating, thermal oxidation and surface functionalization [145]. These properties have an impact on cellular uptake, DNA synthesis and uptake, altered gene expression, immunogenicity, oxidative stress, and cell proliferation [195, 622]. Additional problems arise from interactions with red blood cells (RBCs) [610], cellular respiration [499] and membrane disruption [461]. To assess nanoparticle toxicity, QDs [26, 122], carbon nanotubes [308], gold [248] and silicon [397] nanoparticle have been analyzed in vivo. Wang et al. analyzed the toxicity mechanism (hemolysis mechanism at the molecular level) of difference QDs to RBCs [534, 537]. In vitro and in vivo studies of silica nanoparticles and their interactions with different parts of the body show ambiguous results. The results vary from high toxicity to general safety, probably due to differences in functionalization, size, morphology and crystallinity [194, 314, 423].

To quantify nanoparticle-induced toxicity, different analytical methods and viability assays were used to track metabolic activity, hemolysis, apoptosis and necrosis [267, 288, 331]. It is crucial to consider failure in the cell viability assessment of nanomaterials when using viability assays. Particular, high surface areas, crystallinity and dye-nanomaterial interactions of silicon microparticles, carbon nanomaterials and QDs led to wrong conclusions concerning cell viabilities [267, 354].

3.18 Abbreviations

AC	Alternating current
CB	Conduction band
CNT	Classical nucleation theory
CT	Computed tomography
CTAB	Cetyltrimethylammonium bromide

DDS	Drug delivery system
ECM	Extracellular matrix
EMDR	Environment-mediated drug resistance
EPR	Enhanced permeability and retention
FDA	Food and Drug Administration (USA)
FID	Free induction decay
HRP	Horse radish peroxidase
IFP	Interstitial fluid pressure
MCM	Mobile crystalline material
MDR	Multidrug resistance
MPS	Mononuclear phagocyte system
MRI	Magneto-resonance imaging
MSN	Mesoporous silica nanoparticle
NIR	Near infrared
NMR	Nuclear magnetic resonance
NP	Nanoparticle
OFDI	Optical frequency domain imaging
PEG	Polyethylene glycol
PET	Positron emission tomography
QD	Quantum dot
ROS	Reactive oxygen species
SAR	Specific absorption rate
SERS	Surface-enhanced Raman scattering
SLP	Specific loss power
SOD	Superoxide dismutase
SPR	Surface plasmon resonance
SWNT	Single-walled carbon nanotube
TAM	Tumor-associated macrophage
TMI	Tumor microenvironment
VB	Valence band
VDW	Van der Waals
VEGF	Vascular epidermal growth factor

References

1. Abu Lila AS, Kiwada H, Ishida T (2013) The accelerated blood clearance (ABC) phenomenon: clinical challenge and approaches to manage. J Control Release 172:38–47. https://doi.org/10.1016/j.jconrel.2013.07.026
2. Adiseshaiah PP, Hall JB, McNeil SE (2010) Nanomaterial standards for efficacy and toxicity assessment. WIREs Nanomed Nanobiotechnol 2:99–112. https://doi.org/10.1002/wnan.66
3. Ahrens ET, Feili-Hariri M, Xu H, Genove G, Morel PA (2003) Receptor-mediated endocytosis of iron-oxide particles provides efficient labeling of dendritic cells for in vivo MR imaging. Magn Reson Med 49:1006–1013. https://doi.org/10.1002/mrm.10465

4. Ai K, Liu Y, Liu J, Yuan Q, He Y, Lu L (2011) Large-scale synthesis of Bi2S3 nanodots as a contrast agent for in vivo X-ray computed tomography imaging. Adv Mater 23:4886–4891. https://doi.org/10.1002/adma.201103289
5. Aird WC (2007) Phenotypic heterogeneity of the endothelium: I. Structure, function, and mechanisms. Circ Res 100:158–173. https://doi.org/10.1161/01.res.0000255691.76142.4a
6. Alexis F, Pridgen E, Molnar LK, Farokhzad OC (2008) Factors affecting the clearance and biodistribution of polymeric nanoparticles. Mol Pharm 5:505–515. https://doi.org/10.1021/mp800051m
7. Algar WR, Prasuhn DE, Stewart MH, Jennings TL, Blanco-Canosa JB, Dawson PE, Medintz IL (2011) The controlled display of biomolecules on nanoparticles: a challenge suited to bioorthogonal chemistry. Bioconjug Chem 22:825–858. https://doi.org/10.1021/bc200065z
8. Algar WR, Susumu K, Delehanty JB, Medintz IL (2011) Semiconductor quantum dots in bioanalysis: crossing the valley of death. Anal Chem 83:8826–8837. https://doi.org/10.1021/ac201331r
9. Alivisatos AP (1996) Semiconductor clusters, nanocrystals, and quantum dots. Science 271:933–937
10. Alivisatos AP (2000) Naturally aligned nanocrystals. Science 289:736–737. https://doi.org/10.1126/science.289.5480.736
11. Alivisatos P (2004) The use of nanocrystals in biological detection. Nat Biotechnol 22:47–52. https://doi.org/10.1038/nbt927
12. Allen TM (2002) Ligand-targeted therapeutics in anticancer therapy. Nat Rev Cancer 2:750–763. https://doi.org/10.1038/nrc903
13. Almeida JPM, Figueroa ER, Drezek RA (2014) Gold nanoparticle mediated cancer immunotherapy. Nanomedicine 10:503–514. https://doi.org/10.1016/j.nano.2013.09.011
14. Amano A, Shizukuishi S, Tamagawa H, Iwakura K, Tsunasawa S, Tsunemitsu A (1990) Characterization of superoxide dismutases purified from either anaerobically maintained or aerated *Bacteroides gingivalis*. J Bacteriol 172:1457–1463
15. Amendola V, Scaramuzza S, Litti L, Meneghetti M, Zuccolotto G, Rosato A, Nicolato E, Marzola P, Fracasso G, Anselmi C, Pinto M, Colombatti M (2014) Magneto-plasmonic Au-Fe alloy nanoparticles designed for multimodal SERS-MRI-CT imaging. Small 10:2476–2486. https://doi.org/10.1002/smll.201303372
16. Anchordoquy TJ, Barenholz Y, Boraschi D, Chorny M, Decuzzi P, Dobrovolskaia MA, Farhangrazi ZS, Farrell D, Gabizon A, Ghandehari H, Godin B, La-Beck NM, Ljubimova J, Moghimi SM, Pagliaro L, Park J-H, Peer D, Ruoslahti E, Serkova NJ, Simberg D (2017) Mechanisms and barriers in cancer nanomedicine: addressing challenges, looking for solutions. ACS Nano 11:12–18. https://doi.org/10.1021/acsnano.6b08244
17. André R, Natálio F, Humanes M, Leppin J, Heinze K, Wever R, Schröder H-C, Müller WEG, Tremel W (2011) V_2O_5 nanowires with an intrinsic peroxidase-like activity. Adv Funct Mater 21:501–509. https://doi.org/10.1002/adfm.201001302
18. Apetoh L, Ghiringhelli F, Tesniere A, Obeid M, Ortiz C, Criollo A, Mignot G, Maiuri MC, Ullrich E, Saulnier P, Yang H, Amigorena S, Ryffel B, Barrat FJ, Saftig P, Levi F, Lidereau R, Nogues C, Mira J-P, Chompret A, Joulin V, Clavel-Chapelon F, Bourhis J, André F, Delaloge S, Tursz T, Kroemer G, Zitvogel L (2007) Toll-like receptor 4-dependent contribution of the immune system to anticancer chemotherapy and radiotherapy. Nat Med 13:1050–1059. https://doi.org/10.1038/nm1622
19. Archibald FS, Fridovich I (1981) Manganese, superoxide dismutase, and oxygen tolerance in some lactic acid bacteria. J Bacteriol 146:928–936
20. Ashley CE, Carnes EC, Phillips GK, Padilla D, Durfee PN, Brown PA, Hanna TN, Liu J, Phillips B, Carter MB, Carroll NJ, Jiang X, Dunphy DR, Willman CL, Petsev DN, Evans DG, Parikh AN, Chackerian B, Wharton W, Peabody DS, Brinker CJ (2011) The targeted delivery of multicomponent cargos to cancer cells by nanoporous particle-supported lipid bilayers. Nat Mater 10:389–397. https://doi.org/10.1038/nmat2992

21. Atabaev TS, Lee JH, Lee JJ, Han D-W, Hwang Y-H, Kim H-K, Hong NH (2013) Mesoporous silica with fibrous morphology: a multifunctional core–shell platform for biomedical applications. Nanotechnology 24:345603/1–345603/7. https://doi.org/10.1088/0957-4484/24/34/345603
22. Awwad HK, El Aggar M, Mocktar N, Barsoum M (1986) Intercapillary distance measurement as an indicator of hypoxia in carcinoma of the cervix uteri. Int J Radiat Oncol 12:1329–1333. https://doi.org/10.1016/0360-3016(86)90165-3
23. Baeza A, Colilla M, Vallet-Regí M (2015) Advances in mesoporous silica nanoparticles for targeted stimuli-responsive drug delivery. Expert Opin Drug Deliv 12:319–337. https://doi.org/10.1517/17425247.2014.953051
24. Bagalkot V, Zhang L, Levy-Nissenbaum E, Jon S, Kantoff PW, Langer R, Farokhzad OC (2007) Quantum dot—aptamer conjugates for synchronous cancer imaging, therapy, and sensing of drug delivery based on bi-fluorescence resonance energy transfer. Nano Lett 7:3065–3070. https://doi.org/10.1021/nl071546n
25. Ballauff M, Lu Y (2007) "Smart" nanoparticles: preparation, characterization and applications. Polymer 48:1815–1823. https://doi.org/10.1016/j.polymer.2007.02.004
26. Ballou B, Lagerholm BC, Ernst LA, Bruchez MP, Waggoner AS (2004) Noninvasive imaging of quantum dots in mice. Bioconjug Chem 15:79–86. https://doi.org/10.1021/bc034153y
27. Barenholz Y (2012) Doxil®–the first FDA-approved nano-drug: lessons learned. J Control Release 160:117–134. https://doi.org/10.1016/j.jconrel.2012.03.020
28. Barnese K, Gralla EB, Cabelli DE, Valentine JS (2008) Manganous phosphate acts as a superoxide dismutase. J Am Chem Soc 130:4604–4606. https://doi.org/10.1021/ja710162n
29. Baselga J (2001) Herceptin alone or in combination with chemotherapy in the treatment of HER2-positive metastatic breast cancer: pivotal trials. Oncology 61(Suppl 2):14–21. https://doi.org/10.1159/000055397
30. Bawendi MG, Steigerwald ML, Brus LE (1990) The quantum mechanics of larger semiconductor clusters ("quantum dots"). Annu Rev Phys Chem 41:477–496. https://doi.org/10.1146/annurev.pc.41.100190.002401
31. Beck JS, Vartuli JC, Roth WJ, Leonowicz ME, Kresge CT, Schmitt KD, Chu CTW, Olson DH, Sheppard EW (1992) A new family of mesoporous molecular sieves prepared with liquid crystal templates. J Am Chem Soc 114:10834–10843. https://doi.org/10.1021/ja00053a020
32. Behr J-P (1997) The Proton Sponge: a Trick to Enter Cells the Viruses Did Not Exploit. Chim Int J Chem 51:34–36
33. Benjaminsen RV, Mattebjerg MA, Henriksen JR, Moghimi SM, Andresen TL (2013) The possible "proton sponge " effect of polyethylenimine (PEI) does not include change in lysosomal pH. Mol Ther 21:149–157. https://doi.org/10.1038/mt.2012.185
34. Bergman L, Rosenholm J, Öst A-B, Duchanoy A, Kankaanpää P, Heino J, Lindén M (2008) On the complexity of electrostatic suspension stabilization of functionalized silica nanoparticles for biotargeting and imaging applications. J Nanomater 2008:1–9. https://doi.org/10.1155/2008/712514
35. Bernardos A, Mondragon L, Aznar E, Marcos MD, Martinez-Mañez R, Sancenon F, Soto J, Barat JM, Perez-Paya E, Guillem C, Amoros P (2010) Enzyme-responsive intracellular controlled release using nanometric silica mesoporous supports capped with "saccharides". ACS Nano 4:6353–6368. https://doi.org/10.1021/nn101499d
36. Berry CC, Wells S, Charles S, Curtis ASG (2003) Dextran and albumin derivatised iron oxide nanoparticles: Influence on fibroblasts in vitro. Biomaterials 24:4551–4557. https://doi.org/10.1016/s0142-9612(03)00237-0
37. Bertrand N, Leroux J-C (2012) The journey of a drug-carrier in the body: an anatomo-physiological perspective. J Control Release 161:152–163. https://doi.org/10.1016/j.jconrel.2011.09.098

38. Biju V, Itoh T, Ishikawa M (2010) Delivering quantum dots to cells: bioconjugated quantum dots for targeted and nonspecific extracellular and intracellular imaging. Chem Soc Rev 39:3031–3056. https://doi.org/10.1039/b926512k
39. Boal A, Ilhan F, DeRouchey J, Thurn-Albrecht T, Russell T, Rotello V (2000) Self-assembly of nanoparticles into structured spherical and network aggregates. Nature 404:746–748. https://doi.org/10.1038/35008037
40. Borchardt SA, Allain EJ, Michels JJ, Stearns GW, Kelly RF, McCoy WF (2001) Reaction of acylated homoserine lactone bacterial signaling molecules with oxidized halogen antimicrobials. Appl Environ Microbiol 67:3174–3179. https://doi.org/10.1128/aem.67.7.3174-3179.2001
41. Borisch B, Semac I, Soltermann A, Palomba C, Hoessli DC (2001) Anti-CD20 treatments and the lymphocyte membrane: pathology for therapy. Verh Dtsch Ges Pathol 85:161–166
42. Bottrill M, Green M (2011) Some aspects of quantum dot toxicity. Chem Commun 47:7039–7050. https://doi.org/10.1039/c1cc10692a
43. Boucher Y, Baxter LT, Jain RK (1990) Interstitial pressure gradients in tissue-isolated and subcutaneous tumors: implications for therapy. Cancer Res 50:4478–4484
44. Boucher Y, Kirkwood JM, Opacic D, Desantis M, Jain RK (1991) Interstitial hypertension in superficial metastatic melanomas in humans. Cancer Res 51:6691–6694
45. Brahmer JR, Tykodi SS, Chow LQM, Hwu W-J, Topalian SL, Hwu P, Drake CG, Camacho LH, Kauh J, Odunsi K, Pitot HC, Hamid O, Bhatia S, Martins R, Eaton K, Chen S, Salay TM, Alaparthy S, Grosso JF, Korman AJ, Parker SM, Agrawal S, Goldberg SM, Pardoll DM, Gupta A, Wigginton JM (2012) Safety and activity of anti-PD-L1 antibody in patients with advanced cancer. N Engl J Med 366:2455–2465. https://doi.org/10.1056/nejmoa1200694
46. Braun K, Pochert A, Beck M, Fiedler R, Gruber J, Lindén M (2016) Dissolution kinetics of mesoporous silica nanoparticles in different simulated body fluids. J Sol-Gel Sci Technol 79:319–327. https://doi.org/10.1007/s10971-016-4053-9
47. Breslow R, Overman LE (1970) An "artificial enzyme" combining a metal catalytic group and a hydrophobic binding cavity. J Am Chem Soc 92:1075–1077. https://doi.org/10.1021/ja00707a062
48. Brevet D, Gary-Bobo M, Raehm L, Richeter S, Hocine O, Amro K, Loock B, Couleaud P, Frochot C, Morère A, Maillard P, Garcia M, Durand J-O (2009) Mannose-targeted mesoporous silica nanoparticles for photodynamic therapy. Chem Commun 1475–1477. https://doi.org/10.1039/b900427k
49. Brinker C, Scherer W (1990) The physics and chemistry of sol-gel processing. Academic Press, San Diego
50. Bruchez M Jr, Moronne M, Gin P, Weiss S, Alivisatos AP (1998) Semiconductor nanocrystals as fluorescent biological labels. Science 281:2013–2016. https://doi.org/10.1126/science.281.5385.2013
51. Buck MR, Bondi JF, Schaak RE (2012) A total-synthesis framework for the construction of high-order colloidal hybrid nanoparticles. Nat Chem 4:37–44. https://doi.org/10.1038/nchem.1195
52. Burns A, Ow H, Wiesner U (2006) Fluorescent core-shell silica nanoparticles: towards "Lab on a Particle" architectures for nanobiotechnology. Chem Soc Rev 35:1028–1042. https://doi.org/10.1039/b600562b
53. Butcher NJ, Mortimer GM, Minchin RF (2016) Drug delivery: unravelling the stealth effect. Nat Nanotechnol 11:310–311. https://doi.org/10.1038/nnano.2016.6
54. Byrne JD, Betancourt T, Brannon-Peppas L (2008) Active targeting schemes for nanoparticle systems in cancer therapeutics. Adv Drug Deliv Rev 60:1615–1626. https://doi.org/10.1016/j.addr.2008.08.005
55. Cabral H, Matsumoto Y, Mizuno K, Chen Q, Murakami M, Kimura M, Terada Y, Kano MR, Miyazono K, Uesaka M, Nishiyama N, Kataoka K (2011) Accumulation of sub-100 nm polymeric micelles in poorly permeable tumours depends on size. Nat Nanotechnol 6:815–823. https://doi.org/10.1038/nnano.2011.166

56. Cai W, Chen K, Li Z-BB, Gambhir SS, Chen X (2007) Dual-function probe for PET and near-infrared fluorescence imaging of tumor vasculature. J Nucl Med 48:1862–1870. https://doi.org/10.2967/jnumed.107.043216
57. Caltagirone C, Bettoschi A, Garau A, Montis R (2014) Silica-based nanoparticles: a versatile tool for the development of efficient imaging agents. Chem Soc Rev 44:4645–4671. https://doi.org/10.1039/c4cs00270a
58. Carbone L, Cozzoli PD (2010) Colloidal heterostructured nanocrystals: synthesis and growth mechanisms. Nano Today 5:449–493. https://doi.org/10.1016/j.nantod.2010.08.006
59. Carter CL, Allen C, Henson DE (1989) Relation of tumor size, lymph node status, and survival in 24,740 breast cancer cases. Cancer 63:181–187. https://doi.org/10.1002/1097-0142(19890101)63:1<181:aid-cncr2820630129>3.0.co;2-h
60. Carter P (2001) Improving the efficacy of antibody-based cancer therapies. Nat Rev Cancer 1:118–129. https://doi.org/10.1038/35101072
61. Casals E, Pfaller T, Duschl A, Oostingh GJ, Puntes VF (2011) Hardening of the nanoparticle-protein corona in metal (Au, Ag) and oxide (Fe_3O_4, CoO, and CeO_2) nanoparticles. Small 7:3479–3486. https://doi.org/10.1002/smll.201101511
62. Casares N, Pequignot MO, Tesniere A, Ghiringhelli F, Roux S, Chaput N, Schmitt E, Hamai A, Hervas-Stubbs S, Obeid M, Coutant F, Métivier D, Pichard E, Aucouturier P, Pierron G, Garrido C, Zitvogel L, Kroemer G (2005) Caspase-dependent immunogenicity of doxorubicin-induced tumor cell death. J Exp Med 202:1691–1701. https://doi.org/10.1084/jem.20050915
63. Casasús R, Marcos MD, Martínez-Máñez R, Ros-Lis JV, Soto J, Villaescusa LA, Amorós P, Beltrán D, Guillem C, Latorre J (2004) Toward the development of ionically controlled nanoscopic molecular gates. J Am Chem Soc 126:8612–8613. https://doi.org/10.1021/ja048095i
64. Cauda V, Argyo C, Bein T (2010) Impact of different PEGylation patterns on the long-term bio-stability of colloidal mesoporous silica nanoparticles. J Mater Chem 20:8693–8699. https://doi.org/10.1039/c0jm01390k
65. Cauda V, Schlossbauer A, Bein T (2010) Bio-degradation study of colloidal mesoporous silica nanoparticles: Effect of surface functionalization with organo-silanes and poly (ethylene glycol). Microporous Mesoporous Mater 132:60–71. https://doi.org/10.1016/j.micromeso.2009.11.015
66. Cedervall T, Lynch I, Lindman S, Berggård T, Thulin E, Nilsson H, Dawson KA, Linse S (2007) Understanding the nanoparticle-protein corona using methods to quantify exchange rates and affinities of proteins for nanoparticles. Proc Natl Acad Sci 104:2050–2055. https://doi.org/10.1073/pnas.0608582104
67. Celardo I, Pedersen JZ, Traversa E, Ghibelli L (2011) Pharmacological potential of cerium oxide nanoparticles. Nanoscale 3:1411–1420. https://doi.org/10.1039/c0nr00875c
68. Chai S, Guo Y, Zhang Z, Chai Z, Ma Y, Qi L (2017) Cyclodextrin-gated mesoporous silica nanoparticles as drug carriers for red light-induced drug release. Nanotechnology 28:145101/1–145101/10. https://doi.org/10.1088/1361-6528/aa5e74
69. Chan WCW, Maxwell DJ, Gao X, Bailey RE, Han M, Nie S (2002) Luminescent quantum dots for multiplexed biological detection and imaging. Curr Opin Biotechnol 13:40–46. https://doi.org/10.1016/s0958-1669(02)00282-3
70. Chan WCW, Nie S (1998) Quantum dot bioconjugates for ultrasensitive nonisotopic detection. Science 281:2016–2018. https://doi.org/10.1126/science.281.5385.2016
71. Chang C-L, Fogler HS (1996) Kinetics of silica particle formation in nonionic W/O microemulsions from TEOS. AIChE J 42:3153–3163. https://doi.org/10.1002/aic.690421115
72. Chang C-L, Fogler HS (1997) Controlled formation of silica particles from tetraethyl orthosilicate in nonionic water-in-oil microemulsions. Langmuir 13:3295–3307. https://doi.org/10.1021/la961062z

73. Chang Y-J, Liu X-Z, Zhao Q, Yang X-H, Wang K-M, Wang Q, Lin M, Yang M (2015) P (VPBA-DMAEA) as a pH-sensitive nanovalve for mesoporous silica nanoparticles based controlled release. Chinese Chem Lett 26:1203–1208. https://doi.org/10.1016/j.cclet.2015.08.005
74. Chaudhuri RG, Paria S (2012) Core/shell nanoparticles: classes, properties, synthesis mechanisms, characterization, and applications. Chem Rev 112:2373–2433. https://doi.org/10.1021/cr100449n
75. Chauhan VP, Jain RK (2013) Strategies for advancing cancer nanomedicine. Nat Mater 12:958–962. https://doi.org/10.1038/nmat3792
76. Chauhan VP, Popović Z, Chen O, Cui J, Fukumura D, Bawendi MG, Jain RK (2011) Fluorescent nanorods and nanospheres for real-time in vivo probing of nanoparticle shape-dependent tumor penetration. Angew Chem Int Ed 50:11417–11420. https://doi.org/10.1002/anie.201104449
77. Chauhan VP, Stylianopoulos T, Martin JD, Popović Z, Chen O, Kamoun WS, Bawendi MG, Fukumura D, Jain RK (2012) Normalization of tumour blood vessels improves the delivery of nanomedicines in a size-dependent manner. Nat Nanotechnol 7:383–388. https://doi.org/10.1038/nnano.2012.45
78. Chen AM, Zhang M, Wei D, Stueber D, Taratula O, Minko T, He H (2009) Co-delivery of Doxorubicin and Bcl-2 siRNA by mesoporous silica nanoparticles enhances the efficacy of chemotherapy in multidrug-resistant cancer cells. Small 5:2673–2677. https://doi.org/10.1002/smll.200900621
79. Chen G, Teng Z, Su X, Liu Y, Lu G (2015) Unique biological degradation behavior of stöber mesoporous silica nanoparticles from their interiors to their exteriors. J Biomed Nanotechnol 11:722–729. https://doi.org/10.1166/jbn.2015.2072
80. Chen G, Wang J, Wu C, Li CZ, Jiang H, Wang X (2012) Photoelectrocatalytic oxidation of glutathione based on porous TiO_2-Pt nanowhiskers. Langmuir 28:12393–12399. https://doi.org/10.1021/la302355b
81. Chen T, Wu W, Xiao H, Chen Y, Chen M, Li J (2016) Intelligent drug delivery system based on mesoporous silica nanoparticles coated with an ultra-pH-sensitive gatekeeper and poly(ethylene glycol). ACS Macro Lett 5:55–58. https://doi.org/10.1021/acsmacrolett.5b00765
82. Chen W-H, Luo G-F, Lei Q, Jia H-Z, Hong S, Wang Q-R, Zhuo R-X, Zhang X-Z (2015) MMP-2 responsive polymeric micelles for cancer-targeted intracellular drug delivery. Chem Commun 51:465–468. https://doi.org/10.1039/c4cc07563c
83. Chen YS, Choi H, Kamat PV (2013) Metal-cluster-sensitized solar cells. A new class of thiolated gold sensitizers delivering efficiency greater than 2%. J Am Chem Soc 135:8822–8825. https://doi.org/10.1021/ja403807f
84. Chen Z, Yin JJ, Zhou YT, Zhang Y, Song L, Song M, Hu S, Gu N (2012) Dual enzyme-like activities of iron oxide nanoparticles and their implication for diminishing cytotoxicity. ACS Nano 6:4001–4012. https://doi.org/10.1021/nn300291r
85. Cheng H, Zhu J-L, Zeng X, Jing Y, Zhang X-Z, Zhuo R-X (2009) Targeted gene delivery mediated by folate-polyethylenimine-block-poly(ethylene glycol) with receptor selectivity. Bioconjug Chem 20:481–487. https://doi.org/10.1021/bc8004057
86. Cheng J, Teply BA, Sherifi I, Sung J, Luther G, Gu FX, Levy-Nissenbaum E, Radovic-Moreno AF, Langer R, Farokhzad OC (2007) Formulation of functionalized PLGA-PEG nanoparticles for in vivo targeted drug delivery. Biomaterials 28:869–876. https://doi.org/10.1016/j.biomaterials.2006.09.047
87. Cheng K, Frey NA, Sun S, Peng S, Cheng K, Sun S (2009) Magnetic nanoparticles: synthesis, functionalization, and applications in bioimaging and magnetic energy storage. Chem Soc Rev 38:2532–2542. https://doi.org/10.1039/b815548h
88. Cheng K, Yang M, Zhang R, Qin C, Su X, Cheng Z (2014) Hybrid nanotrimers for Dual T1 and T2-weighted magnetic resonance imaging. ACS Nano 8:9884–9896. https://doi.org/10.1021/nn500188y

89. Cheng L-C, Huang J-H, Chen HM, Lai T-C, Yang K-Y, Liu R-S, Hsiao M, Chen C-H, Her L-J, Tsai DP (2012) Seedless, silver-induced synthesis of star-shaped gold/silver bimetallic nanoparticles as high efficiency photothermal therapy reagent. J Mater Chem 22:2244–2253. https://doi.org/10.1039/c1jm13937a
90. Cheng MMC, Cuda G, Bunimovich YL, Gaspari M, Heath JR, Hill HD, Mirkin CA, Nijdam AJ, Terracciano R, Thundat T, Ferrari M (2006) Nanotechnologies for biomolecular detection and medical diagnostics. Curr Opin Chem Biol 10:11–19. https://doi.org/10.1016/j.cbpa.2006.01.006
91. Cheng R, Feng F, Meng F, Deng C, Feijen J, Zhong Z (2011) Glutathione-responsive nano-vehicles as a promising platform for targeted intracellular drug and gene delivery. J Control Release 152:2–12. https://doi.org/10.1016/j.jconrel.2011.01.030
92. Cheng Z, Al Zaki A, Hui JZ, Muzykantov VR, Tsourkas A (2012) Multifunctional nanoparticles: cost versus benefit of adding targeting and imaging capabilities. Science 338:903–910. https://doi.org/10.1126/science.1226338
93. Cheon J, Lee JH (2008) Synergistically integrated nanoparticles as multimodal probes for nanobiotechnology. Acc Chem Res 41:1630–1640. https://doi.org/10.1021/ar800045c
94. Chithrani BD, Chan WCW (2007) Elucidating the mechanism of cellular uptake and removal of protein-coated gold nanoparticles of different sizes and shapes. Nano Lett 7:1542–1550. https://doi.org/10.1021/nl070363y
95. Chithrani BD, Ghazani AA, Chan WCW (2006) Determining the size and shape dependence of gold nanoparticle uptake into mammalian cells. Nano Lett 6:662–668. https://doi.org/10.1021/nl052396o
96. Chiu H-Y, Deng W, Engelke H, Helma J, Leonhardt H, Bein T (2016) Intracellular chromobody delivery by mesoporous silica nanoparticles for antigen targeting and visualization in real time. Sci Rep 6:25019/1–25019/12. https://doi.org/10.1038/srep25019
97. Choi HS, Liu W, Misra P, Tanaka E, Zimmer JP, Itty Ipe B, Bawendi MG, Frangioni JV (2007) Renal clearance of quantum dots. Nat Biotechnol 25:1165–1170. https://doi.org/10.1038/nbt1340
98. Choi Y, Lee J-E, Lee JH, Jeong JH, Kim J (2015) A biodegradation study of SBA-15 microparticles in simulated body fluid and in vivo. Langmuir 31:6457–6462. https://doi.org/10.1021/acs.langmuir.5b01316
99. Chu YT, Chanda K, Lin PH, Huang MH (2012) Aqueous phase synthesis of palladium tripod nanostructures for sonogashira coupling reactions. Langmuir 28:11258–11264. https://doi.org/10.1021/la302284m
100. Clark AJ, Wiley DT, Zuckerman JE, Webster P, Chao J, Lin J, Yen Y, Davis ME (2016) CRLX101 nanoparticles localize in human tumors and not in adjacent, nonneoplastic tissue after intravenous dosing. Proc Natl Acad Sci 113:3850–3854. https://doi.org/10.1073/pnas.1603018113
101. Clemments AM, Botella P, Landry CC (2015) Protein adsorption from biofluids on silica nanoparticles: corona analysis as a function of particle diameter and porosity. ACS Appl Mater Interfaces 7:21682–21689. https://doi.org/10.1021/acsami.5b07631
102. Conde J, Bao C, Tan Y, Cui D, Edelman ER, Azevedo HS, Byrne HJ, Artzi N, Tian F (2015) Dual targeted immunotherapy via in vivo delivery of biohybrid RNAi-peptide nanoparticles to tumour-associated macrophages and cancer cells. Adv Funct Mater 25:4183–4194. https://doi.org/10.1002/adfm.201501283
103. Costi R, Saunders AE, Banin U (2010) Colloidal hybrid nanostructures: a new type of functional materials. Angew Chem Int Ed 49:4878–4897. https://doi.org/10.1002/anie.200906010
104. Courty A, Henry A-I, Goubet N, Pileni M-P (2007) Large triangular single crystals formed by mild annealing of self-organized silver nanocrystals. Nat Mater 6:900–907. https://doi.org/10.1038/nmat2004
105. Courty A, Mermet A, Albouy PA, Duval E, Pileni MP (2005) Vibrational coherence of self-organized silver nanocrystals in f.c.c. supra-crystals. Nat Mater 4:395–398. https://doi.org/10.1038/nmat1366

106. Cozzoli PD, Pellegrino T, Manna L (2006) Synthesis, properties and perspectives of hybrid nanocrystal structures. Chem Soc Rev 35:1195–1208. https://doi.org/10.1039/b517790c
107. Croissant J, Cattoën X, Man MWC, Gallud A, Raehm L, Trens P, Maynadier M, Durand J-O (2014) Biodegradable ethylene-bis(propyl)disulfide-based periodic mesoporous organosilica nanorods and nanospheres for efficient in-vitro drug delivery. Adv Mater 26:6174–6180. https://doi.org/10.1002/adma.201401931
108. Croissant JG, Fatieiev Y, Khashab NM (2017) Degradability and clearance of silicon, organosilica, silsesquioxane, silica mixed oxide, and mesoporous silica nanoparticles. Adv Mater 29:1604634/1–1604634/51. https://doi.org/10.1002/adma.201604634
109. Croissant JG, Fatieiev Y, Omar H, Anjum DH, Gurinov A, Lu J, Tamanoi F, Zink JI, Khashab NM (2016) Periodic mesoporous organosilica nanoparticles with controlled morphologies and high drug/dye loadings for multicargo delivery in cancer cells. Chemistry 22:9607–9615. https://doi.org/10.1002/chem.201600587
110. Croissant JG, Picard S, Aggad D, Klausen M, Mauriello Jimenez C, Maynadier M, Mongin O, Clermont G, Genin E, Cattoën X, Wong Chi Man M, Raehm L, Garcia M, Gary-Bobo M, Blanchard-Desce M, Durand J-O (2016) Fluorescent periodic mesoporous organosilica nanoparticles dual-functionalized via click chemistry for two-photon photodynamic therapy in cells. J Mater Chem B 4:5567–5574. https://doi.org/10.1039/c6tb00638h
111. Croissant JG, Zhang D, Alsaiari S, Lu J, Deng L, Tamanoi F, AlMalik AM, Zink JI, Khashab NM (2016) Protein-gold clusters-capped mesoporous silica nanoparticles for high drug loading, autonomous gemcitabine/doxorubicin co-delivery, and in-vivo tumor imaging. J Control Release 229:183–191. https://doi.org/10.1016/j.jconrel.2016.03.030
112. Cubillos-Ruiz JR, Engle X, Scarlett UK, Martinez D, Barber A, Elgueta R, Wang L, Nesbeth Y, Durant Y, Gewirtz AT, Sentman CL, Kedl R, Conejo-Garcia JR (2009) Polyethylenimine-based siRNA nanocomplexes reprogram tumor-associated dendritic cells via TLR5 to elicit therapeutic antitumor immunity. J Clin Invest 119:2231–2244. https://doi.org/10.1172/jci37716
113. Curran MA, Montalvo W, Yagita H, Allison JP (2010) PD-1 and CTLA-4 combination blockade expands infiltrating T cells and reduces regulatory T and myeloid cells within B16 melanoma tumors. Proc Natl Acad Sci 107:4275–4280. https://doi.org/10.1073/pnas.0915174107
114. Dai L, Zhang Q, Gu H, Cai K (2015) Facile synthesis of yolk–shell silica nanoparticles for targeted tumor therapy. J Mater Chem B 3:8303–8313. https://doi.org/10.1039/c5tb01620g
115. Daldrup-Link HE, Golovko D, Ruffell B, Denardo DG, Castaneda R, Ansari C, Rao J, Tikhomirov GA, Wendland MF, Corot C, Coussens LM (2011) MRI of tumor-associated macrophages with clinically applicable iron oxide nanoparticles. Clin Cancer Res 17:5695–5704. https://doi.org/10.1158/1078-0432.ccr-10-3420
116. Danhier F, Feron O, Préat V (2010) To exploit the tumor microenvironment: Passive and active tumor targeting of nanocarriers for anti-cancer drug delivery. J Control Release 148:135–146. https://doi.org/10.1016/j.jconrel.2010.08.027
117. Daniel MC, Astruc D (2004) Gold nanoparticles: assembly, supramolecular chemistry, quantum-size-related properties, and applications toward biology, catalysis, and nanotechnology. Chem Rev 104:293–346. https://doi.org/10.1021/cr030698+
118. Davis ME, Zuckerman JE, Choi CHJ, Seligson D, Tolcher A, Alabi CA, Yen Y, Heidel JD, Ribas A (2010) Evidence of RNAi in humans from systemically administered siRNA via targeted nanoparticles. Nature 464:1067–1070. https://doi.org/10.1038/nature08956
119. Dawidczyk CM, Kim C, Park JH, Russell LM, Lee KH, Pomper MG, Searson PC (2014) State-of-the-art in design rules for drug delivery platforms: lessons learned from FDA-approved nanomedicines. J Control Release 187:133–144. https://doi.org/10.1016/j.jconrel.2014.05.036
120. Dawson JH (1988) Probing structure-function relations in heme-containing oxygenases and peroxidases. Science 240:433–439. https://doi.org/10.1126/science.3358128

121. Deng Z, Zhen Z, Hu X, Wu S, Xu Z, Chu PK (2011) Hollow chitosan-silica nanospheres as pH-sensitive targeted delivery carriers in breast cancer therapy. Biomaterials 32:4976–4986. https://doi.org/10.1016/j.biomaterials.2011.03.050
122. Derfus AM, Chan WCW, Bhatia SN (2004) Probing the Cytotoxicity of Semiconductor Quantum Dots. Nano Lett 4:11–18. https://doi.org/10.1021/nl0347334
123. Desai D, Prabhakar N, Mamaeva V, Karaman DŞ, Lähdeniemi IAK, Sahlgren C, Rosenholm JM, Toivola DM (2016) Targeted modulation of cell differentiation in distinct regions of the gastrointestinal tract via oral administration of differently PEG-PEI functionalized mesoporous silica nanoparticles. Int J Nanomedicine 11:299–313. https://doi.org/10.2147/ijn.s94013
124. Ding J, Wang Y, Ma M, Zhang Y, Lu S, Jiang Y, Qi C, Luo S, Dong G, Wen S, An Y, Gu N (2013) CT/fluorescence dual-modal nanoemulsion platform for investigating atherosclerotic plaques. Biomaterials 34:209–216. https://doi.org/10.1016/j.biomaterials.2012.09.025
125. Doherty GJ, McMahon HT (2009) Mechanisms of endocytosis. Annu Rev Biochem 78:857–902. https://doi.org/10.1146/annurev.biochem.78.081307.110540
126. Dolmans DEJGJ, Fukumura D, Jain RK (2003) Photodynamic therapy for cancer. Nat Rev Cancer 3:380–387. https://doi.org/10.1038/nrc1071
127. Dong A, Jiao Y, Milliron DJ (2013) Electronically coupled nanocrystal superlattice films by in situ ligand exchange at the liquid-air interface. ACS Nano 7:10978–10984. https://doi.org/10.1021/nn404566b
128. Dong R, Su Y, Yu S, Zhou Y, Lu Y, Zhu X (2013) A redox-responsive cationic supramolecular polymer constructed from small molecules as a promising gene vector. Chem Commun 49:9845–9847. https://doi.org/10.1039/c3cc46123h
129. Dragovich T, Gordon M, Mendelson D, Wong L, Modiano M, Chow H-HS, Samulitis B, O'Day S, Grenier K, Hersh E, Dorr R (2007) Phase I trial of imexon in patients with advanced malignancy. J Clin Oncol 25:1779–1784. https://doi.org/10.1200/jco.2006.08.9672
130. Du X, He J (2010) Fine-tuning of silica nanosphere structure by simple regulation of the volume ratio of cosolvents. Langmuir 26:10057–10062. https://doi.org/10.1021/la100196j
131. Du X, Li X, Xiong L, Zhang X, Kleitz F, Qiao SZ (2016) Mesoporous silica nanoparticles with organo-bridged silsesquioxane framework as innovative platforms for bioimaging and therapeutic agent delivery. Biomaterials 91:90–127. https://doi.org/10.1016/j.biomaterials.2016.03.019
132. Duan D, Fan K, Zhang D, Tan S, Liang M, Liu Y, Zhang J, Zhang P, Liu W, Qiu X, Kobinger GP, Fu Gao G, Yan X (2015) Nanozyme-strip for rapid local diagnosis of Ebola. Biosens Bioelectron 74:134–141. https://doi.org/10.1016/j.bios.2015.05.025
133. Dubertret B, Skourides P, Norris DJ, Noireaux V, Brivanlou AH, Libchaber A (2002) In vivo imaging of quantum dots encapsulated in phospholipid micelles. Science 298:1759–1762. https://doi.org/10.1126/science.1077194
134. Dunford HB (2010) Peroxidases and catalases: biochemistry biophysics, biotechnology and physiology. Wiley, Hoboken
135. Dunford HB, Stillman JS (1976) On the function and mechanism of action of peroxidases. Coord Chem Rev 19:187–251. https://doi.org/10.1016/s0010-8545(00)80316-1
136. Durfee PN, Lin Y-S, Dunphy DR, Muñiz AJ, Butler KS, Humphrey KR, Lokke AJ, Agola JO, Chou SS, Chen I-M, Wharton W, Townson JL, Willman CL, Brinker CJ (2016) Mesoporous silica nanoparticle-supported lipid bilayers (protocells) for active targeting and delivery to individual leukemia cells. ACS Nano 10:8325–8345. https://doi.org/10.1021/acsnano.6b02819
137. Eckford PDW, Sharom FJ (2009) ABC efflux pump-based resistance to chemotherapy drugs. Chem Rev 109:2989–3011. https://doi.org/10.1021/cr9000226
138. Ehlerding EB, Chen F, Cai W (2016) Biodegradable and renal clearable inorganic nanoparticles. Adv Sci 3:1500223/1–1500223/8. https://doi.org/10.1002/advs.201500223

139. Elechiguerra JL, Reyes-Gasga J, Yacaman MJ (2006) The role of twinning in shape evolution of anisotropic noble metal nanostructures. J Mater Chem 16:3906–3919. https://doi.org/10.1039/b607128g
140. Eliasof S, Lazarus D, Peters CG, Case RI, Cole RO, Hwang J, Schluep T, Chao J, Lin J, Yen Y, Han H, Wiley DT, Zuckerman JE, Davis ME (2013) Correlating preclinical animal studies and human clinical trials of a multifunctional, polymeric nanoparticle. Proc Natl Acad Sci 110:15127–15132. https://doi.org/10.1073/pnas.1309566110
141. Ellington AD, Szostak JW (1990) In vitro selection of RNA molecules that bind specific ligands. Nature 346:818–822. https://doi.org/10.1038/346818a0
142. Elmore S (2007) Apoptosis: a review of programmed cell death. Toxicol Pathol 35:495–516. https://doi.org/10.1080/01926230701320337
143. Eskey CJ, Koretsky AP, Domach MM, Jain RK (1993) Role of oxygen vs. glucose in energy metabolism in a mammary carcinoma perfused ex vivo: direct measurement by 31P NMR. Proc Natl Acad Sci 90:2646–2650
144. Facciabene A, Peng X, Hagemann IS, Balint K, Barchetti A, Wang L-P, Gimotty PA, Gilks CB, Lal P, Zhang L, Coukos G (2011) Tumour hypoxia promotes tolerance and angiogenesis via CCL28 and T(reg) cells. Nature 475:226–230. https://doi.org/10.1038/nature10169
145. Fadeel B, Garcia-Bennett AE (2010) Better safe than sorry: Understanding the toxicological properties of inorganic nanoparticles manufactured for biomedical applications. Adv Drug Deliv Rev 62:362–374. https://doi.org/10.1016/j.addr.2009.11.008
146. Fan K, Cao C, Pan Y, Lu D, Yang D, Feng J, Song L, Liang M, Yan X (2012) Magnetoferritin nanoparticles for targeting and visualizing tumour tissues. Nat Nanotechnol 7:459–464. https://doi.org/10.1038/nnano.2012.90
147. Fang W, Tang S, Liu P, Fang X, Gong J, Zheng N (2012) Pd nanosheet-covered hollow mesoporous silica nanoparticles as a platform for the chemo-photothermal treatment of cancer cells. Small 8:3816–3822. https://doi.org/10.1002/smll.201200962
148. Faraday M (1857) The Bakerian lecture: experimental relations of gold (and other metals) to light. Philos Trans R Soc London 147:145–181. https://doi.org/10.1098/rstl.1857.0011
149. Farber S, Diamond LK (1948) Temporary remissions in acute leukemia in children produced by folic acid antagonist, 4-aminopteroyl-glutamic acid. N Engl J Med 238:787–793. https://doi.org/10.1056/nejm194806032382301
150. Farokhzad OC, Cheng J, Teply BA, Sherifi I, Jon S, Kantoff PW, Richie JP, Langer R (2006) Targeted nanoparticle-aptamer bioconjugates for cancer chemotherapy in vivo. Proc Natl Acad Sci 103:6315–6320. https://doi.org/10.1073/pnas.0601755103
151. Fenton H (1894) LXXIII.—Oxidation of tartaric acid in presence of iron. J Chem Soc Trans 65:899–910
152. Ferrando R, Jellinek J, Johnston RL (2008) Nanoalloys: from theory to applications of alloy clusters and nanoparticles. Chem Rev 108:845–910. https://doi.org/10.1021/cr040090g
153. Ferrara N, Hillan KJ, Novotny W (2005) Bevacizumab (Avastin), a humanized anti-VEGF monoclonal antibody for cancer therapy. Biochem Biophys Res Commun 333:328–335. https://doi.org/10.1016/j.bbrc.2005.05.132
154. Figdor CG, de Vries IJM, Lesterhuis WJ, Melief CJM (2004) Dendritic cell immunotherapy: mapping the way. Nat Med 10:475–480. https://doi.org/10.1038/nm1039
155. Finlay J, Roberts CM, Dong J, Zink JI, Tamanoi F, Glackin CA (2015) Mesoporous silica nanoparticle delivery of chemically modified siRNA against TWIST1 leads to reduced tumor burden. Nanomedicine 11:1657–1666. https://doi.org/10.1016/j.nano.2015.05.011
156. Florence AT (2012) "Targeting" nanoparticles: the constraints of physical laws and physical barriers. J Control Release 164:115–124. https://doi.org/10.1016/j.jconrel.2012.03.022
157. Folkman J (1971) Tumor angiogenesis: therapeutic implications. N Engl J Med 285:1182–1186. https://doi.org/10.1056/nejm197111182852108
158. Folkman J (1990) What is the evidence that tumors are angiogenesis dependent? J Natl Cancer Inst 82:4–7. https://doi.org/10.1093/jnci/82.1.4

159. Fortin JJ-P, Wilhelm C, Servais J, Menager C, Bacri J-CJ-C, Gazeau F, Ménager C, Bacri J-CJ-C, Gazeau Florence (2007) Size-sorted anionic iron oxide nanomagnets as colloidal mediators for magnetic hyperthermia. J Am Chem Soc 129:2628–2635. https://doi.org/10.1021/ja067457e
160. Fu PP, Xia Q, Hwang H-M, Ray PC, Yu H (2014) Mechanisms of nanotoxicity: generation of reactive oxygen species. Nanomater Toxicol Med Appl 22:64–75. https://doi.org/10.1016/j.jfda.2014.01.005
161. Fu WH, Guan Y, Wang YM, He M-Y (2016) A facile synthesis of monodispersed mesoporous silica nanospheres with Pm3n structure. Microporous Mesoporous Mater 220:168–174. https://doi.org/10.1016/j.micromeso.2015.09.004
162. Gabrilovich DI, Nagaraj S (2009) Myeloid-derived suppressor cells as regulators of the immune system. Nat Rev Immunol 9:162–174. https://doi.org/10.1038/nri2506
163. Gac LS, Vermes I, Van Den Berg A Den (2006) Quantum dots based probes conjugated to Annexin V for photostable apoptosis detection and imaging. Nano Lett 6:1863–1869. https://doi.org/10.1021/nl060694v
164. Gai S, Yang P, Ma P, Wang D, Li C, Li X, Niu N, Lin J (2011) Fibrous-structured magnetic and mesoporous Fe_3O_4/silica microspheres: synthesis and intracellular doxorubicin delivery. J Mater Chem 21:16420–16426. https://doi.org/10.1039/c1jm13357h
165. Gao L, Cui Y, He Q, Yang Y, Fei J, Li J (2011) Selective recognition of co-assembled thrombin aptamer and docetaxel on mesoporous silica nanoparticles against tumor cell proliferation. Chemistry 17:13170–13174. https://doi.org/10.1002/chem.201101658
166. Gao L, Zhuang J, Nie L, Zhang J, Zhang Y, Gu N, Wang T, Feng J, Yang D, Perrett S, Yan X (2007) Intrinsic peroxidase-like activity of ferromagnetic nanoparticles. Nat Nanotechnol 2:577–583. https://doi.org/10.1038/nnano.2007.260
167. Gao X, Cui Y, Levenson RM, Chung LWK, Nie S (2004) In vivo cancer targeting and imaging with semiconductor quantum dots. Nat Biotechnol 22:969–976. https://doi.org/10.1038/nbt994
168. Garrett RM, Johnson JL, Graf TN, Feigenbaum A, Rajagopalan KV (1998) Human sulfite oxidase R160Q: identification of the mutation in a sulfite oxidase-deficient patient and expression and characterization of the mutant enzyme. Proc Natl Acad Sci 95:6394–6398. https://doi.org/10.1073/pnas.95.11.6394
169. Gennes DP-G (1992) Soft matter (Nobel Lecture). Angew Chem Int Ed 31:842–845. https://doi.org/10.1002/anie.199208421
170. Gerlowski LE, Jain RK (1986) Microvascular permeability of normal and neoplastic tissues. Microvasc Res 31:288–305. https://doi.org/10.1016/0026-2862(86)90018-x
171. Gillet J-P, Gottesman MM (2010) Mechanisms of multidrug resistance in cancer. Methods Mol Biol 596:47–76. https://doi.org/10.1007/978-1-60761-416-6_4
172. Giri S, Sykes EA, Jennings TL, Chan WCW (2011) Rapid screening of genetic biomarkers of infectious agents using quantum dot barcodes. ACS Nano 5:1580–1587. https://doi.org/10.1021/nn102873w
173. Glaser N, Adams DJ, Bo A, Krausch G (2006) Janus particles at liquid–liquid interfaces. Langmuir 22:5227–5229. https://doi.org/10.1021/la060693i
174. Gomez-Grana S, Hubert F (2011) Surfactant (bi) layers on gold nanorods. Langmuir 1453–1459. https://doi.org/10.1021/la203451p
175. Gonzalez Porras MA, Durfee PN, Gregory AM, Sieck GC, Brinker CJ, Mantilla CB (2016) A novel approach for targeted delivery to motoneurons using cholera toxin-B modified protocells. J Neurosci Methods 273:160–174. https://doi.org/10.1016/j.jneumeth.2016.09.003
176. Gottesman MM, Fojo T, Bates SE (2002) Multidrug resistance in cancer: role of ATP-dependent transporters. Nat Rev Cancer 2:48–58. https://doi.org/10.1038/nrc706
177. Graf C, Vossen DLJ, Imhof A, van Blaaderen A (2003) A general method to coat colloidal particles with silica. Langmuir 19:6693–6700. https://doi.org/10.1021/la0347859
178. Green NM (1990) Avidin and streptavidin. Methods Enzymol 184:51–67

179. Gref R, Minamitake Y, Peracchia MT, Trubetskoy V, Torchilin V, Langer R (1994) Biodegradable long-circulating polymeric nanospheres. Science 263:1600–1603. https://doi.org/10.1126/science.8128245
180. Grzelczak M, Pérez-Juste J, Mulvaney P, Liz-Marzán LM (2008) Shape control in gold nanoparticle synthesis. Chem Soc Rev 37:1783–1791. https://doi.org/10.1039/b711490g
181. Guisasola E, Baeza A, Talelli M, Arcos D, Vallet-Regí M (2016) Design of thermoresponsive polymeric gates with opposite controlled release behaviors. RSC Adv 6:42510–42516. https://doi.org/10.1039/c6ra02260j
182. Guo L, Yan DD, Yang D, Li Y, Xiaodong W, Zalewski O, Yan B, Lu W (2015) Combinatorial photothermal and immuno cancer therapy using chitosan-coated hollow copper sulfide nanoparticles. ACS Nano 8:5670–5681. https://doi.org/10.1021/nn5002112
183. Haacke EM, Brown RW, Thomson MR, Venkatesan R (1999) Magnetic resonance imaging—physical principles and sequence design. Wiley, New York
184. Haartman E, Lindberg D, Prabhakar N, Rosenholm JM (2016) On the intracellular release mechanism of hydrophobic cargo and its relation to the biodegradation behavior of mesoporous silica nanocarriers. Eur J Pharm Sci 95:17–27. https://doi.org/10.1016/j.ejps.2016.06.001
185. Hajipour MJ, Laurent S, Aghaie A, Rezaee F, Mahmoudi M (2014) Personalized protein coronas: a "key" factor at the nanobiointerface. Biomater Sci 2:1210–1221. https://doi.org/10.1039/c4bm00131a
186. Han L, Tang C, Yin C (2015) Dual-targeting and pH/redox-responsive multi-layered nanocomplexes for smart co-delivery of doxorubicin and siRNA. Biomaterials 60:42–52. https://doi.org/10.1016/j.biomaterials.2015.05.001
187. Han M, Gao X, Su JZ, Nie S (2001) Quantum-dot-tagged microbeads for multiplexed optical coding of biomolecules. Nat Biotechnol 19:631–635. https://doi.org/10.1038/90228
188. Hanafi-Bojd MY, Jaafari MR, Ramezanian N, Xue M, Amin M, Shahtahmassebi N, Malaekeh-Nikouei B (2015) Surface functionalized mesoporous silica nanoparticles as an effective carrier for epirubicin delivery to cancer cells. Eur J Pharm Biopharm 89:248–258. https://doi.org/10.1016/j.ejpb.2014.12.009
189. Hao X, Hu X, Zhang C, Chen S, Li Z, Yang X, Liu H, Jia G, Liu D, Ge K, Liang X-J, Zhang J (2015) Hybrid mesoporous silica-based drug carrier nanostructures with improved degradability by hydroxyapatite. ACS Nano 9:9614–9625. https://doi.org/10.1021/nn507485j
190. Harrington KJ, Mohammadtaghi S, Uster PS, Glass D, Peters AM, Vile RG, Stewart JS (2001) Effective targeting of solid tumors in patients with locally advanced cancers by radiolabeled pegylated liposomes. Clin Cancer Res 7:243–254
191. Hasan Z, Renirie R, Kerkman R, Ruijssenaars HJ, Hartog AF, Wever R (2006) Laboratory-evolved vanadium chloroperoxidase exhibits 100-fold higher halogenating activity at alkaline pH: Catalytic effects from first and second coordination sphere mutations. J Biol Chem 281:9738–9744. https://doi.org/10.1074/jbc.m512166200
192. He Q, Kiesewetter DO, Qu Y, Fu X, Fan J, Huang P, Liu Y, Zhu G, Liu Y, Qian Z, Chen X (2015) NIR-responsive on-demand release of co from metal carbonyl-caged graphene oxide nanomedicine. Adv Mater 27:6741–6746. https://doi.org/10.1002/adma.201502762
193. He Q, Shi J, Zhu M, Chen Y, Chen F (2010) The three-stage in vitro degradation behavior of mesoporous silica in simulated body fluid. Microporous Mesoporous Mater 131:314–320. https://doi.org/10.1016/j.micromeso.2010.01.009
194. He Q, Zhang Z, Gao Y, Shi J, Li Y (2009) Intracellular localization and cytotoxicity of spherical mesoporous silica nano- and microparticles. Small 5:2722–2729. https://doi.org/10.1002/smll.200900923
195. Heidegger S, Gößl D, Schmidt A, Niedermayer S, Argyo C, Endres S, Bein T, Bourquin C (2015) Immune response to functionalized mesoporous silica nanoparticles for targeted drug delivery. Nanoscale 8:938–948. https://doi.org/10.1039/c5nr06122a

196. Helmlinger G, Sckell A, Dellian M, Forbes NS, Jain RK (2002) Acid production in glycolysis-impaired tumors provides new insights into tumor metabolism. Clin Cancer Res 8:1284–1291
197. Herget K, Hubach P, Pusch S, Deglmann P, Götz H, Gorelik TE, Gural'skiy IA, Pfitzner F, Link T, Schenk S, Panthöfer M, Ksenofontov V, Kolb U, Opatz T, André R, Tremel W (2016) Haloperoxidase mimicry by CeO_{2-x} nanorods combats biofouling. Adv Mater 29:1603823/1-1603823/8. https://doi.org/10.1002/adma.201603823
198. Herricks T, Chen J, Xia Y (2004) Polyol synthesis of platinum nanoparticles: control of morphology with sodium nitrate. Nano Lett 4:2367–2371. https://doi.org/10.1021/nl048570a
199. Hobson B, Denekamp J (1984) Endothelial proliferation in tumours and normal tissues: continuous labelling studies. Br J Cancer 49:405–413
200. Hoekstra HJ, van Ginkel RJ (2003) Hyperthermic isolated limb perfusion in the management of extremity sarcoma. Curr Opin Oncol 15:300–303
201. Hoffmann F, Cornelius M, Morell J, Fröba M (2006) Silica-based mesoporous organic-inorganic hybrid materials. Angew Chem Int Ed 45:3216–3251. https://doi.org/10.1002/anie.200503075
202. Holmberg A, Blomstergren A, Nord O, Lukacs M, Lundeberg J, Uhlén M (2005) The biotin-streptavidin interaction can be reversibly broken using water at elevated temperatures. Electrophoresis 26:501–510. https://doi.org/10.1002/elps.200410070
203. Holohan C, van Schaeybroeck S, Longley DB, Johnston PG (2013) Cancer drug resistance: an evolving paradigm. Nat Rev Cancer 13:714–726. https://doi.org/10.1038/nrc3599
204. Hong R, Fischer N (2005) Surface PEGylation and ligand exchange chemistry of FePt nanoparticles for biological applications. Chem Mater 4617–4621. https://doi.org/10.1021/cm0507819
205. Hong S, Leroueil PR, Majoros IJ, Orr BG, Baker JR, Banaszak Holl MM (2007) The binding avidity of a nanoparticle-based multivalent targeted drug delivery platform. Chem Biol 14:107–115. https://doi.org/10.1016/j.chembiol.2006.11.015
206. Horssen VR, Ten Hagen TLM, Eggermont AMM (2006) TNF-alpha in cancer treatment: molecular insights, antitumor effects, and clinical utility. Oncologist 11:397–408. https://doi.org/10.1634/theoncologist.11-4-397
207. Hsiao I-L, Hsieh Y-K, Chuang C-Y, Wang C-F, Huang Y-J (2017) Effects of silver nanoparticles on the interactions of neuron- and glia-like cells: toxicity, uptake mechanisms, and lysosomal tracking. Environ Toxicol. https://doi.org/10.1002/tox.22397
208. Hu C-MJ, Fang RH, Wang K-C, Luk BT, Thamphiwatana S, Dehaini D, Nguyen P, Angsantikul P, Wen CH, Kroll AV, Carpenter C, Ramesh M, Qu V, Patel SH, Zhu J, Shi W, Hofman FM, Chen TC, Gao W, Zhang K, Chien S, Zhang L (2015) Nanoparticle biointerfacing by platelet membrane cloaking. Nature 526:118–121. https://doi.org/10.1038/nature15373
209. Hu C-MJ, Zhang L, Aryal S, Cheung C, Fang RH, Zhang L (2011) Erythrocyte membrane-camouflaged polymeric nanoparticles as a biomimetic delivery platform. Proc Natl Acad Sci 108:10980–10985. https://doi.org/10.1073/pnas.1106634108
210. Hu H, You Y, He L, Chen T (2015) Rational design of NAMI-A-loaded mesoporous silica nanoparticles as antiangiogeneic nanosystem. J Mater Chem B 3:6346–6368. https://doi.org/10.1039/c5tb00612k
211. Hu JW, Li JF, Ren B, Wu DY, Sun SG, Tian ZQ (2007) Palladium-coated gold nanoparticles with a controlled shell thickness used as surface-enhanced raman scattering substrate. J Phys Chem C 111:1105–1112. https://doi.org/10.1021/jp0652906
212. Hu Y, Yang Y, Wang H, Du H (2015) Synergistic integration of layer-by-layer assembly of photosensitizer and gold nanorings for enhanced photodynamic therapy in the near infrared. ACS Nano 9:8744–8754. https://doi.org/10.1021/acsnano.5b03063

213. Huang J, Guo M, Ke H, Zong C, Ren B, Liu G, Shen H, Ma Y, Wang X, Zhang H, Deng Z, Chen H, Zhang Z (2015) Rational design and synthesis of γFe_2O_3@Au magnetic gold nanoflowers for efficient cancer theranostics. Adv Mater 27:5049–5056. https://doi.org/10.1002/adma.201501942
214. Huang X, Tang S, Mu X, Dai Y, Chen G, Zhou Z, Ruan F, Yang Z, Zheng N (2011) Freestanding palladium nanosheets with plasmonic and catalytic properties. Nat Nanotechnol 6:28–32. https://doi.org/10.1038/nnano.2010.235
215. Huynh WU, Dittmer JJ, Alivisatos AP (2002) Hybrid nanorod-polymer solar cells. Science 295:2425–2427. https://doi.org/10.1126/science.1069156
216. Hwang AA, Lee B-Y, Clemens DL, Dillon BJ, Zink JI, Horwitz MA (2015) pH-responsive isoniazid-loaded nanoparticles markedly improve tuberculosis treatment in mice. Small 11:5066–5078. https://doi.org/10.1002/smll.201500937
217. Jain RK (1996) 1995 Whitaker lecture: delivery of molecules, particles, and cells to solid tumors. Ann Biomed Eng 24:457–473. https://doi.org/10.1007/bf02648108
218. Jain RK (2005) Normalization of tumor vasculature: an emerging concept in antiangiogenic therapy. Science 307:58–62. https://doi.org/10.1126/science.1104819
219. Jain RK (2013) Normalizing tumor microenvironment to treat cancer: bench to bedside to biomarkers. J Clin Oncol 31:2205–2218. https://doi.org/10.1200/jco.2012.46.3653
220. Jain RK, Baxter LT (1988) Mechanisms of heterogeneous distribution of monoclonal antibodies and other macromolecules in tumors: significance of elevated interstitial pressure. Cancer Res 48:7022–7032
221. Jain RK, Stylianopoulos T (2010) Delivering nanomedicine to solid tumors. Nat Rev Clin Oncol 7:653–664. https://doi.org/10.1038/nrclinonc.2010.139
222. Jana N, Earhart C, Ying J (2007) Synthesis of water-soluble and functionalized nanoparticles by silica coating. Chem Mater 5:5074–5082. https://doi.org/10.1021/cm071368z
223. Jang J, Nah H, Lee J-H, Moon SH, Kim MG, Cheon J (2009) Critical enhancements of MRI contrast and hyperthermic effects by dopant-controlled magnetic nanoparticles. Angew Chem Int Ed 48:1234–1238. https://doi.org/10.1002/anie.200805149
224. Jang YH, Jang YJ, Kochuveedu ST, Byun M, Lin Z, Kim DH (2014) Plasmonic dye-sensitized solar cells incorporated with Au-TiO_2 nanostructures with tailored configurations. Nanoscale 6:1823–1832. https://doi.org/10.1039/c3nr05012b
225. Jiang H, Chess L (2009) How the immune system achieves self–nonself discrimination during adaptive immunity. In: Advances in immunology. Academic Press, New York, pp 95–133
226. Johnson JL, Rajagopalan KV, Wadman SK (1993) Human molybdenum cofactor deficiency. Adv Exp Med Biol 338:373–378
227. Jordan A, Scholz R, Wust P, Schirra H, Schiestel T, Schmidt H, Felix R (1999) Endocytosis of dextran and silan-coated magnetite nanoparticles and the effect of intracellular hyperthermia on human mammary carcinoma cells in vitro. J Magn Magn Mater 194:185–196. https://doi.org/10.1016/s0304-8853(98)00558-7
228. Josephson L, Tung C-H, Moore A, Weissleder R (1999) High-efficiency intracellular magnetic labeling with novel superparamagnetic-tat peptide conjugates. Bioconjug Chem 10:186–191. https://doi.org/10.1021/bc980125h
229. Joyce JA (2005) Therapeutic targeting of the tumor microenvironment. Cancer Cell 7:513–520. https://doi.org/10.1016/j.ccr.2005.05.024
230. Juarez JC, Manuia M, Burnett ME, Betancourt O, Boivin B, Shaw DE, Tonks NK, Mazar AP, Doñate F (2008) Superoxide dismutase 1 (SOD1) is essential for H_2O_2-mediated oxidation and inactivation of phosphatases in growth factor signaling. Proc Natl Acad Sci 105:7147–7152. https://doi.org/10.1073/pnas.0709451105
231. Jun Y-W, Huh Y-M, Choi J-S, Lee J-H, Song H-T, Kim S, Yoon S, Kim K-S, Shin J-S, Suh J-S, Cheon J (2005) Nanoscale size effect of magnetic nanocrystals and their utilization for cancer diagnosis via magnetic resonance imaging. J Am Chem Soc 127:5732–5733. https://doi.org/10.1021/ja0422155

232. Jun Y, Lee J-H, Cheon J (2008) Chemical design of nanoparticle probes for high-performance magnetic resonance imaging. Angew Chem Int Ed 47:5122–5135. https://doi.org/10.1002/anie.200701674
233. Jun YW, Choi JS, Cheon J (2006) Shape control of semiconductor and metal oxide nanocrystals through nonhydrolytic colloidal routes. Angew Chem Int Ed 45:3414–3439. https://doi.org/10.1002/anie.200503821
234. Kaim W, Schwederski B (2005) Bioanorganische Chemie. Springer, Wiesbaden
235. Kang Z, Yan X, Zhao L, Liao Q, Zhao K, Du H, Zhang X, Zhang X, Zhang Y (2015) Gold nanoparticle/ZnO nanorod hybrids for enhanced reactive oxygen species generation and photodynamic therapy. Nano Res 8:1–11. https://doi.org/10.1007/s12274-015-0712-3
236. Karakashev SV, Reginato MJ (2015) Progress toward overcoming hypoxia-induced resistance to solid tumor therapy. Cancer Manage Res 7:253–264. https://doi.org/10.2147/cmar.s58285
237. Karakoti A, Singh S, Dowding JM, Seal S, Self WT (2010) Redox-active radical scavenging nanomaterials. Chem Soc Rev 39:4422–4432. https://doi.org/10.1039/b919677n
238. Karesoja M, McKee J, Karjalainen E, Hietala S, Bergman L, Linden M, Tenhu H (2013) Mesoporous silica particles grafted with poly(ethyleneoxide- block—N-vinylcaprolactam). J Polym Sci A Polym Chem 51:5012–5020. https://doi.org/10.1002/pola.26928
239. Kato M, Hattori Y, Kubo M, Maitani Y (2012) Collagenase-1 injection improved tumor distribution and gene expression of cationic lipoplex. Int J Pharm 423:428–434. https://doi.org/10.1016/j.ijpharm.2011.12.015
240. Kavruk M, Celikbicak O, Ozalp VC, Borsa BA, Hernandez FJ, Bayramoglu G, Salih B, Arica MY (2015) Antibiotic loaded nanocapsules functionalized with aptamer gates for targeted destruction of pathogens. Chem Commun 51:8492–8495. https://doi.org/10.1039/c5cc01869b
241. Keefe AD, Pai S, Ellington A (2010) Aptamers as therapeutics. Nat Rev Drug Discov 9:537–550. https://doi.org/10.1038/nrd3141
242. Keith B, Johnson RS, Simon MC (2011) HIF1α and HIF2α: sibling rivalry in hypoxic tumour growth and progression. Nat Rev Cancer 12:9–22. https://doi.org/10.1038/nrc3183
243. Khanal S, Casillas G, Bhattarai N, Velázquez-Salazar JJ, Santiago U, Ponce A, Mejía-Rosales S, José-Yacamán M (2013) CuS(2)-passivated Au-core, Au(3)Cu-shell nanoparticles analyzed by atomistic-resolution C(s)-corrected STEM. Langmuir 29:9231–9239. https://doi.org/10.1021/la401598e
244. Khashab NM, Trabolsi A, Lau YA, Ambrogio MW, Friedman DC, Khatib HA, Zink JI, Stoddart JF (2009) Redox- and pH-controlled mechanized nanoparticles. Eur J Org Chem 2009:1669–1673. https://doi.org/10.1002/ejoc.200801300
245. Kievit FM, Zhang M (2011) Cancer nanotheranostics: improving imaging and therapy by targeted delivery across biological barriers. Adv Mater 23:H217–H247. https://doi.org/10.1002/adma.201102313
246. Kim CK, Kim T, Choi IY, Soh M, Kim D, Kim YJ, Jang H, Yang HS, Kim JY, Park HK, Park SP, Park S, Yu T, Yoon BW, Lee SH, Hyeon T (2012) Ceria nanoparticles that can protect against ischemic stroke. Angew Chem Int Ed 51:11039–11043. https://doi.org/10.1002/anie.201203780
247. Kim D, Park S, Lee JH, Jeong YY, Jon S (2007) Antibiofouling polymer-coated gold nanoparticles as a contrast agent for in vivo X-ray computed tomography imaging. J Am Chem Soc 129:7661–7665. https://doi.org/10.1021/ja071471p
248. Kim D, Park S, Lee JH, Jeong YY, Jon S (2007) Antibiofouling polymer-coated gold nanoparticles as a contrast agent for in vivo X-ray computed tomography imaging. J Am Chem Soc 129:7661–7665. https://doi.org/10.1021/ja071471p
249. Kim J, Kim HHS, Lee N, Kim T, Kim HHS, Yu T, Song IC, Moon WK, Hyeon T (2008) Multifunctional uniform nanoparticles composed of a magnetite nanocrystal core and a mesoporous silica shell for magnetic resonance and fluorescence imaging and for drug delivery. Angew Chem Int Ed 47:8438–8441. https://doi.org/10.1002/anie.200802469

250. Kim SW, Kim S, Tracy JB, Jasanoff A, Bawendi MG (2005) Phosphine oxide polymer for water-soluble nanoparticles. J Am Chem Soc 127:4556–4557. https://doi.org/10.1021/ja043577f
251. Kirpotin DB, Drummond DC, Shao Y, Shalaby MR, Hong K, Nielsen UB, Marks JD, Benz CC, Park JW (2006) Antibody targeting of long-circulating lipidic nanoparticles does not increase tumor localization but does increase internalization in animal models. Cancer Res 66:6732–6740. https://doi.org/10.1158/0008-5472.can-05-4199
252. Klostranec JM, Xiang Q, Farcas GA, Lee JA, Rhee A, Laffert EI, Perrault SD, Kain KC, Chan WCW (2007) Convergence of quantum dot barcodes with microfluidics and signal processing for multiplexed high-throughput infectious disease diagnostics. Nano Lett 7:2812–2818. https://doi.org/10.1021/nl071415m
253. Kluenker M, Tahir MN, Ragg R, Korschelt K, Simon P, Gorelik TE, Barton B, Shylin SI, Panthoefer M, Herzberger J, Frey H, Ksenofontov V, Möller A, Kolb U, Grin Y, Tremel W (2017) Pd@Fe_2O_3 superparticles with enhanced peroxidase activity by solution phase. Chem Mater 29:1134–1146. https://doi.org/10.1021/acs.chemmater.6b04283
254. Knop K, Hoogenboom R, Fischer D, Schubert US (2010) Poly(ethylene glycol) in drug delivery: pros and cons as well as potential alternatives. Angew Chem Int Ed 49:6288–6308. https://doi.org/10.1002/anie.200902672
255. Knopp D, Tang D, Niessner R (2009) Review: bioanalytical applications of biomolecule-functionalized nanometer-sized doped silica particles. Anal Chim Acta 647:14–30. https://doi.org/10.1016/j.aca.2009.05.037
256. Kobayashi H, Hama Y, Koyama Y, Barrett T, Regino CAS, Urano Y, Choyke PL (2007) Simultaneous multicolor imaging of five different lymphatic basins using quantum dots. Nano Lett 7:1711–1716. https://doi.org/10.1021/nl0707003
257. Kochuveedu ST, Jang YH, Kim DH (2013) A study on the mechanism for the interaction of light with noble metal-metal oxide semiconductor nanostructures for various photophysical applications. Chem Soc Rev 42:8467–8493. https://doi.org/10.1039/c3cs60043b
258. Kolhar P, Anselmo AC, Gupta V, Pant K, Prabhakarpandian B, Ruoslahti E, Mitragotri S (2013) Using shape effects to target antibody-coated nanoparticles to lung and brain endothelium. Proc Natl Acad Sci 110:10753–10758. https://doi.org/10.1073/pnas.1308345110
259. Korotchenkov GS (2015) Porous silicon: from formation to application. CRC Press, Boca Raton
260. Korsvik C, Patil S, Seal S, Self WT (2007) Superoxide dismutase mimetic properties exhibited by vacancy engineered ceria nanoparticles. Chem Commun 1056–1058. https://doi.org/10.1039/b615134e
261. Koukourakis MI, Koukouraki S, Giatromanolaki A, Archimandritis SC, Skarlatos J, Beroukas K, Bizakis JG, Retalis G, Karkavitsas N, Helidonis ES (1999) Liposomal doxorubicin and conventionally fractionated radiotherapy in the treatment of locally advanced non-small-cell lung cancer and head and neck cancer. J Clin Oncol 17:3512–3521. https://doi.org/10.1200/jco.1999.17.11.3512
262. Kresge CT, Leonowicz ME, Roth WJ, Vartuli JC, Beck JS (1992) Ordered mesoporous molecular sieves synthesized by a liquid-crystal template mechanism. Nature 359:710–712. https://doi.org/10.1038/359710a0
263. Kumar R, Roy I, Ohulchanskky TY, Vathy LA, Bergey EJ, Sajjad M, Prasad PN (2010) In vivo biodistribution and clearance studies using multimodal organically modified silica nanoparticles. ACS Nano 4:699–708. https://doi.org/10.1021/nn901146y
264. Kumar R, Roy I, Ohulchanskyy TY, Goswami LN, Bonoiu AC, Bergey EJ, Tramposch KM, Maitra A, Prasad PN (2008) Covalently dye-linked, surface-controlled, and bioconjugated organically modified silica nanoparticles as targeted probes for optical imaging. ACS Nano 2:449–456. https://doi.org/10.1021/nn700370b
265. Kuszyk BS, Corl FM, Franano FN, Bluemke DA, Hofmann LV, Fortman BJ, Fishman EK (2001) Tumor transport physiology: implications for imaging and imaging-guided therapy. Am J Roentgenol 177:747–753. https://doi.org/10.2214/ajr.177.4.1770747

266. Kwon SG, Hyeon T (2011) Formation mechanisms of uniform nanocrystals via hot-injection and heat-up methods. Small 7:2685–2702. https://doi.org/10.1002/smll.201002022
267. Laaksonen T, Santos H, Vihola H, Salonen J, Riikonen J, Heikkilä T, Peltonen L, Kumar N, Murzin DY, Lehto V-P, Hirvonen J (2007) Failure of MTT as a toxicity testing agent for mesoporous silicon microparticles. Chem Res Toxicol 20:1913–1918. https://doi.org/10.1021/tx700326b
268. Rica DLR, Aili D, Stevens MM (2012) Enzyme-responsive nanoparticles for drug release and diagnostics. Adv Drug Deliv Rev 64:967–978. https://doi.org/10.1016/j.addr.2012.01.002
269. Lai C-Y, Trewyn BG, Jeftinija DM, Jeftinija K, Xu S, Jeftinija S, Lin VS-Y (2003) A mesoporous silica nanosphere-based carrier system with chemically removable CdS nanoparticle caps for stimuli-responsive controlled release of neurotransmitters and drug molecules. J Am Chem Soc 125:4451–4459. https://doi.org/10.1021/ja028650l
270. Lake RA, van der Most RG (2006) A better way for a cancer cell to die. N Engl J Med 354:2503–2504. https://doi.org/10.1056/nejmcibr061443
271. LaMer VK (1952) Nucleation in phase transitions. Ind Eng Chem 44:1270–1277. https://doi.org/10.1021/ie50510a027
272. LaMer VK, Dinegar RRH (1950) Theory, production and mechanism of formation of monodispersed hydrosols. J Am Chem Soc 72:4847–4854. https://doi.org/10.1021/ja01167a001
273. Landgraf L, Christner C, Storck W, Schick I, Krumbein I, Dähring H, Haedicke K, Heinz-Herrmann K, Teichgräber U, Reichenbach JR, Tremel W, Tenzer S, Hilger I (2015) A plasma protein corona enhances the biocompatibility of Au@Fe_3O_4 Janus particles. Biomaterials 68:77–88. https://doi.org/10.1016/j.biomaterials.2015.07.049
274. Landgraf L, Ernst P, Schick I, Köhler O, Oehring H, Tremel W, Hilger I (2014) Anti-oxidative effects and harmlessness of asymmetric Au@Fe_3O_4 Janus particles on human blood cells. Biomaterials 35:6986–6997. https://doi.org/10.1016/j.biomaterials.2014.04.111
275. Larson DR, Zipfel WR, Williams RM, Clark SW, Bruchez MP, Wise FW, Webb WW (2003) Water-soluble quantum dots for multiphoton fluorescence imaging in vivo. Science 300:1434–1436. https://doi.org/10.1126/science.1083780
276. Lartigue L, Alloyeau D, Kolosnjaj-Tabi J, Javed Y, Guardia P, Riedinger A, Péchoux C, Pellegrino T, Wilhelm C, Gazeau F (2013) Biodegradation of iron oxide nanocubes: high-resolution in situ monitoring. ACS Nano 7:3939–3952. https://doi.org/10.1021/nn305719y
277. Lartigue L, Hugounenq P, Alloyeau D, Clarke SP, Lévy M, Bacri JC, Bazzi R, Brougham DF, Wilhelm C, Gazeau F (2012) Cooperative organization in iron oxide multi-core nanoparticles potentiates their efficiency as heating mediators and MRI contrast agents. ACS Nano 6:10935–10949. https://doi.org/10.1021/nn304477s
278. Lazarovits J, Chen YY, Sykes EA, Chan WCW (2015) Nanoparticle-blood interactions: the implications on solid tumour targeting. Chem Commun 51:2756–2767. https://doi.org/10.1039/c4cc07644c
279. Leatherdale CA, Woo W-K, Mikulec FV, Bawendi MG (2002) On the absorption cross section of CdSe nanocrystal quantum dots. J Phys Chem B 106:7619–7622. https://doi.org/10.1021/jp025698c
280. Lee BP, Dalsin JL, Messersmith PB (2002) Synthesis and gelation of DOPA-modified poly (ethylene glycol) hydrogels. Biomacromolecules 3:1038–1047. https://doi.org/10.1021/bm025546n
281. Lee J-H, Huh Y-M, Jun Y, Seo J, Jang J, Song H-T, Kim S, Cho E-J, Yoon H-G, Suh J-S, Cheon J (2007) Artificially engineered magnetic nanoparticles for ultra-sensitive molecular imaging. Nat Med 13:95–99. https://doi.org/10.1038/nm1467
282. Lee J-H, Jang J-T, Choi J-S, Moon SH, Noh S-H, Kim J-W, Kim J-G, Kim I-S, Park KI, Cheon J (2011) Exchange-coupled magnetic nanoparticles for efficient heat induction. Nat Nanotechnol 6:418–422. https://doi.org/10.1038/nnano.2011.95

283. Lee N, Kim H, Choi SH, Park M, Kim D, Kim H-C, Choi Y, Lin S, Kim BH, Jung HS, Kim H, Park KS, Moon WK, Hyeon T (2011) Magnetosome-like ferrimagnetic iron oxide nanocubes for highly sensitive MRI of single cells and transplanted pancreatic islets. Proc Natl Acad Sci 108:2662–2667. https://doi.org/10.1073/pnas.1016409108
284. Lee Y, Thompson DH (2017) Stimuli-responsive liposomes for drug delivery. WIREs Nanomed Nanobiotechnol e1450:1–40. https://doi.org/10.1002/wnan.1450
285. Leonessa F (2003) ATP binding cassette transporters and drug resistance in breast cancer. Endocr Relat Cancer 10:43–73. https://doi.org/10.1677/erc.0.0100043
286. Leserman LD, Barbet J, Kourilsky F, Weinstein JN (1980) Targeting to cells of fluorescent liposomes covalently coupled with monoclonal antibody or protein A. Nature 288:602–604. https://doi.org/10.1038/288602a0
287. Levy M, Luciani N, Alloyeau D, Elgrabli D, Deveaux V, Pechoux C, Chat S, Wang G, Vats N, Gendron F, Factor C, Lotersztajn S, Luciani A, Wilhelm C, Gazeau F (2011) Long term in vivo biotransformation of iron oxide nanoparticles. Biomaterials 32:3988–3999. https://doi.org/10.1016/j.biomaterials.2011.02.031
288. Lewinski N, Colvin V, Drezek R (2008) Cytotoxicity of nanoparticles. Small 4:26–49. https://doi.org/10.1002/smll.200700595
289. Li C, Sun L, Sun Y, Teranishi T (2013) One-pot controllable synthesis of Au@ Ag heterogeneous nanorods with highly tunable plasmonic absorption. Chem Mater 25:2580–2590. https://doi.org/10.1021/cm400392e
290. Li H-J, Du J-Z, Du X-J, Xu C-F, Sun C-Y, Wang H-X, Cao Z-T, Yang X-Z, Zhu Y-H, Nie S, Wang J (2016) Stimuli-responsive clustered nanoparticles for improved tumor penetration and therapeutic efficacy. Proc Natl Acad Sci 113:4164–4169. https://doi.org/10.1073/pnas.1522080113
291. Li P, Wei Z, Wu T, Peng Q, Li Y (2011) Au–ZnO hybrid nanopyramids and their photocatalytic properties. J Am Chem Soc 133:5660–5663. https://doi.org/10.1021/ja111102u
292. Li Q-L, Xu S-H, Zhou H, Wang X, Dong B, Gao H, Tang J, Yang Y-W (2015) pH and glutathione dual-responsive dynamic cross-linked supramolecular network on mesoporous silica nanoparticles for controlled anticancer drug release. ACS Appl Mater Interfaces 7:28656–28664. https://doi.org/10.1021/acsami.5b10534
293. Li X, Chen Y, Wang M, Ma Y, Xia W, Gu H (2013) A mesoporous silica nanoparticle–PEI–fusogenic peptide system for siRNA delivery in cancer therapy. Biomaterials 34:1391–1401. https://doi.org/10.1016/j.biomaterials.2012.10.072
294. Li X, Xie QR, Zhang J, Xia W, Gu H (2011) The packaging of siRNA within the mesoporous structure of silica nanoparticles. Biomaterials 32:9546–9556. https://doi.org/10.1016/j.biomaterials.2011.08.068
295. Li Y, Xu X, Zhang X, Li Y, Zhang Z, Gu Z (2017) Tumor-specific multiple stimuli-activated dendrimeric nanoassemblies with metabolic blockade surmount chemotherapy resistance. ACS Nano 11:416–429. https://doi.org/10.1021/acsnano.6b06161
296. Li Z-Y, Hu J-J, Xu Q, Chen S, Jia H-Z, Sun Y-X, Zhuo R-X, Zhang X-Z (2015) A redox-responsive drug delivery system based on RGD containing peptide-capped mesoporous silica nanoparticles. J Mater Chem B 3:39–44. https://doi.org/10.1039/c4tb01533a
297. Li Z, Li W, Camargo PHC, Xia Y (2008) Facile synthesis of branched Au nanostructures by templating against a self-destructive lattice of magnetic Fe nanoparticles. Angew Chem Int Ed 47:9653–9656. https://doi.org/10.1002/anie.200804634
298. Li Z, Liu Z, Yin M, Yang X, Yuan Q, Ren J, Qu X (2012) Aptamer-capped multifunctional mesoporous strontium hydroxyapatite nanovehicle for cancer-cell-responsive drug delivery and imaging. Biomacromolecules 13:4257–4263. https://doi.org/10.1021/bm301563q
299. Lima RT, Martins LM, Guimarães JE, Sambade C, Vasconcelos MH (2004) Specific downregulation of bcl-2 and xIAP by RNAi enhances the effects of chemotherapeutic agents in MCF-7 human breast cancer cells. Cancer Gene Ther 11:309–316. https://doi.org/10.1038/sj.cgt.7700706

300. Lin Y-S, Haynes CL (2010) Impacts of mesoporous silica nanoparticle size, pore ordering, and pore integrity on hemolytic activity. J Am Chem Soc 132:4834–4842. https://doi.org/10.1021/ja910846q
301. Liou G-Y, Storz P (2010) Reactive oxygen species in cancer. Free Radic Res 44:479–496. https://doi.org/10.3109/10715761003667554
302. Liu J, Li Q, Zhang J, Huang L, Qi C, Xu L, Liu X, Wang G, Wang L, Wang Z (2017) Safe and effective reversal of cancer multidrug resistance using sericin-coated mesoporous silica nanoparticles for lysosome-targeting delivery in mice. Small 139:1602567/1–1602567/14. https://doi.org/10.1002/smll.201602567
303. Liu J, Qiao SZ, Budi Hartono S, Lu GQM (2010) Monodisperse yolk-shell nanoparticles with a hierarchical porous structure for delivery vehicles and nanoreactors. Angew Chem Int Ed 49:4981–4985. https://doi.org/10.1002/anie.201001252
304. Liu J, Yu M, Zhou C, Zheng J (2013) Renal clearable inorganic nanoparticles: a new frontier of bionanotechnology. Mater Today 16:477–486. https://doi.org/10.1016/j.mattod.2013.11.003
305. Liu X, Wang Q, Zhao H, Zhang L, Su Y, Lv Y (2012) BSA-templated MnO_2 nanoparticles as both peroxidase and oxidase mimics. Analyst 137:4552–4558. https://doi.org/10.1039/c2an35700c
306. Liu Y, Ai K, Liu J, Yuan Q, He Y, Lu L (2012) A high-performance ytterbium-based nanoparticulate contrast agent for in vivo X-ray computed tomography imaging. Angew Chem Int Ed 51:1437–1442. https://doi.org/10.1002/anie.201106686
307. Liu Y, Ai K, Lu L (2012) Nanoparticulate X-ray computed tomography contrast agents: from design validation to in vivo applications. Acc Chem Res 45:1817–1827. https://doi.org/10.1021/ar300150c
308. Liu Z, Davis C, Cai W, He L, Chen X, Dai H (2008) Circulation and long-term fate of functionalized, biocompatible single-walled carbon nanotubes in mice probed by Raman spectroscopy. Proc Natl Acad Sci 105:1410–1415. https://doi.org/10.1073/pnas.0707654105
309. Lofgreen JE, Ozin GA (2014) Controlling morphology and porosity to improve performance of molecularly imprinted sol-gel silica. Chem Soc Rev 43:911–933. https://doi.org/10.1039/c3cs60276a
310. Loo C, Lin A, Hirsch L, Lee M-H, Barton J, Halas N, West J, Drezek R (2004) Nanoshell-enabled photonics-based imaging and therapy of cancer. Technol Cancer Res Treat 3:33–40. https://doi.org/10.1177/153303460400300104
311. Lopes de Menezes DE, Pilarski LM, Allen TM (1998) In vitro and in vivo targeting of immunoliposomal doxorubicin to human B-cell lymphoma. Cancer Res 58:3320–3330
312. López-Lázaro M (2007) Dual role of hydrogen peroxide in cancer: possible relevance to cancer chemoprevention and therapy. Cancer Lett 252:1–8. https://doi.org/10.1016/j.canlet.2006.10.029
313. Love SA, Maurer-Jones MA, Thompson JW, Lin Y-S, Haynes CL (2012) Assessing nanoparticle toxicity. Annu Rev Anal Chem 5:181–205. https://doi.org/10.1146/annurev-anchem-062011-143134
314. Lu J, Liong M, Li Z, Zink JI, Tamanoi F (2010) Biocompatibility, biodistribution, and drug-delivery efficiency of mesoporous silica nanoparticles for cancer therapy in animals. Small 6:1794–1805. https://doi.org/10.1002/smll.201000538
315. Lu N, Tian Y, Tian W, Huang P, Liu Y, Tang Y, Wang C, Wang S, Su Y, Zhang Y, Pan J, Teng Z, Lu G (2016) Smart cancer cell targeting imaging and drug delivery system by systematically engineering periodic mesoporous organosilica nanoparticles. ACS Appl Mater Interfaces 8:2985–2993. https://doi.org/10.1021/acsami.5b09585
316. Lu Y, Yin Y, Li Z, Xia Y (2002) Synthesis and self-assembly of Au@SiO_2 core–shell colloids. Nano Lett 2:785–788. https://doi.org/10.1021/nl025598i
317. Ludi B, Süess MJ, Werner IA, Niederberger M (2012) Mechanistic aspects of molecular formation and crystallization of zinc oxide nanoparticles in benzyl alcohol. Nanoscale 4:1982. https://doi.org/10.1039/c1nr11557j

318. Luo Y, Zhou H, Krueger J, Kaplan C, Lee S-H, Dolman C, Markowitz D, Wu W, Liu C, Reisfeld RA, Xiang R (2006) Targeting tumor-associated macrophages as a novel strategy against breast cancer. J Clin Invest 116:2132–2141. https://doi.org/10.1172/jci27648
319. Ma L, Wang C, Gong M, Liao L, Long R, Wang J, Wu D, Zhong W, Kim MJ, Chen Y, Xie Y, Xiong Y (2012) Control over the branched structures of platinum nanocrystals for electrocatalytic applications. ACS Nano 6:9797–9806. https://doi.org/10.1021/nn304237u
320. Ma X, Zhao Y, Ng KW, Zhao Y (2013) Integrated hollow mesoporous silica nanoparticles for target drug/siRNA co-delivery. Chemistry 19:15593–15603. https://doi.org/10.1002/chem.201302736
321. Mackowiak SA, Schmidt A, Weiss V, Argyo C, Schirnding C, Bein T, Bräuchle C (2013) Targeted drug delivery in cancer cells with red-light photoactivated mesoporous silica nanoparticles. Nano Lett 13:2576–2583. https://doi.org/10.1021/nl400681f
322. Maeda H, Tsukigawa K, Fang J (2016) A retrospective 30 years after discovery of the enhanced permeability and retention effect of solid tumors: next-generation chemotherapeutics and photodynamic therapy-problems, solutions, and prospects. Microcirculation 23:173–182. https://doi.org/10.1111/micc.12228
323. Magda D, Miller RA (2006) Motexafin gadolinium: a novel redox active drug for cancer therapy. Semin Cancer Biol 16:466–476. https://doi.org/10.1016/j.semcancer.2006.09.002
324. Maggini L, Cabrera I, Ruiz-Carretero A, Prasetyanto EA, Robinet E, Cola L (2016) Breakable mesoporous silica nanoparticles for targeted drug delivery. Nanoscale 8:7240–7247. https://doi.org/10.1039/c5nr09112h
325. Maher P (2005) The effects of stress and aging on glutathione metabolism. Ageing Res Rev 4:288–314. https://doi.org/10.1016/j.arr.2005.02.005
326. Mahmoudi M, Lynch I, Ejtehadi MR, Monopoli MP, Bombelli FB, Laurent S (2011) Protein-nanoparticle interactions: opportunities and challenges. Chem Rev 111:5610–5637. https://doi.org/10.1021/cr100440g
327. Malvi B, Sarkar BR, Pati D, Mathew R, Ajithkumar TG, Sen Gupta S (2009) "Clickable" SBA-15 mesoporous materials: synthesis, characterization and their reaction with alkynes. J Mater Chem 19:1409–1416. https://doi.org/10.1039/b815350g
328. Mamaeva V, Rosenholm JM, Bate-Eya LT, Bergman L, Peuhu E, Duchanoy A, Fortelius LE, Landor S, Toivola DM, Lindén M, Sahlgren C (2011) Mesoporous silica nanoparticles as drug delivery systems for targeted inhibition of notch signaling in cancer. Mol Ther 19:1538–1546. https://doi.org/10.1038/mt.2011.105
329. Mantovani A, Sica A (2010) Macrophages, innate immunity and cancer: balance, tolerance, and diversity. Curr Opin Immunol 22:231–237. https://doi.org/10.1016/j.coi.2010.01.009
330. Markov IV (2003) Crystal growth for beginners: fundamentals of nucleation, crystal growth and epitaxy. World Scientific, Singapore
331. Marquis BJ, Love SA, Braun KL, Haynes CL (2009) Analytical methods to assess nanoparticle toxicity. Analyst 134:425–439. https://doi.org/10.1039/b818082b
332. Martinez-Boubeta C, Simeonidis K, Makridis A, Angelakeris M, Iglesias O, Guardia P, Cabot A, Yedra L, Estradé S, Peiró F, Saghi Z, Midgley P, Conde-Leborán I, Serantes D, Baldomir D (2013) Learning from nature to improve the heat generation of iron-oxide nanoparticles for magnetic hyperthermia applications. Sci Rep 3:1652/1–1652/8. https://doi.org/10.1038/srep01652
333. Martinez-Carmona M, Baeza A, Rodriguez-Milla MA, García-Castro J, Vallet M (2015) Mesoporous silica nanoparticles grafted with light-responsive protein shell for highly cytotoxic antitumoral therapy. J Mater Chem B 3:5746–5752. https://doi.org/10.1039/c5tb00304k
334. Matsumoto K, Yamamoto T, Kamata R, Maeda H (1986) Enhancement of vascular permeability upon serratial infection: activation of Hageman factor–kallikrein–kinin cascade. Adv Exp Med Biol 198:71–78

335. Matsumoto Y, Nichols JW, Toh K, Nomoto T, Cabral H, Miura Y, Christie RJ, Yamada N, Ogura T, Kano MR, Matsumura Y, Nishiyama N, Yamasoba T, Bae YH, Kataoka K (2016) Vascular bursts enhance permeability of tumour blood vessels and improve nanoparticle delivery. Nat Nanotechnol 11:533–538. https://doi.org/10.1038/nnano.2015.342
336. Matsumura Y, Maeda H (1986) A new concept for macromolecular therapeutics in cancer chemotherapy: mechanism of tumoritropic accumulation of proteins and the antitumor agent smancs. Cancer Res 46:6387–6392
337. Maye MM, Luo J, Lim IIS, Han L, Kariuki NN, Rabinovich D, Liu T, Zhong CJ (2003) Size-controlled assembly of gold nanoparticles induced by a tridentate thioether ligand. J Am Chem Soc 125:9906–9907. https://doi.org/10.1021/ja0363866
338. McCord JM, Keele BB, Fridovich I (1971) An enzyme-based theory of obligate anaerobiosis: the physiological function of superoxide dismutase. Proc Natl Acad Sci 68:1024–1027. https://doi.org/10.1073/pnas.68.5.1024
339. McCurrie RA (1994) Ferromagnetic materials: structure and properties. Academic Press, San Diego
340. McNeil SE (2009) Nanoparticle therapeutics: a personal perspective. WIREs Nanomed Nanobiotechnol 1:264–271. https://doi.org/10.1002/wnan.6
341. Meads MB, Gatenby RA, Dalton WS (2009) Environment-mediated drug resistance: a major contributor to minimal residual disease. Nat Rev Cancer 9:665–674. https://doi.org/10.1038/nrc2714
342. Meka AK, Niu Y, Karmakar S, Hartono SB, Zhang J, Lin CXC, Zhang H, Whittaker A, Jack K, Yu M, Yu C (2016) Facile synthesis of large-pore bicontinuous cubic mesoporous silica nanoparticles for intracellular gene delivery. ChemNanoMat 2:220–225. https://doi.org/10.1002/cnma.201600021
343. Mekaru H, Lu J, Tamanoi F (2015) Development of mesoporous silica-based nanoparticles with controlled release capability for cancer therapy. Adv Drug Deliv Rev 95:40–49. https://doi.org/10.1016/j.addr.2015.09.009
344. Melief CJM (2008) Cancer immunotherapy by dendritic cells. Immunity 29:372–383. https://doi.org/10.1016/j.immuni.2008.08.004
345. Meng H, Liong M, Xia T, Li Z, Ji Z, Zink JI, Nel AE (2010) Engineered design of mesoporous silica nanoparticles to deliver doxorubicin and P-glycoprotein siRNA to overcome drug resistance in a cancer cell line. ACS Nano 4:4539–4550. https://doi.org/10.1021/nn100690m
346. Meng H, Mai WX, Zhang H, Xue M, Xia T, Lin S, Wang X, Zhao Y, Ji Z, Zink JI, Nel AE (2013) Codelivery of an optimal drug/siRNA combination using mesoporous silica nanoparticles to overcome drug resistance in breast cancer in vitro and in vivo. ACS Nano 7:994–1005. https://doi.org/10.1021/nn3044066
347. Meng H, Xue M, Xia T, Ji Z, Tarn DY, Zink JI, Nel AE (2011) Use of size and a copolymer design feature to improve the biodistribution and the enhanced permeability and retention effect of doxorubicin-loaded mesoporous silica nanoparticles in a murine xenograft tumor model. ACS Nano 5:4131–4144. https://doi.org/10.1021/nn200809t
348. Michalet X, Pinaud FF, Bentolila LA, Tsay JM, Doose S, Li JJ, Sundaresan G, Wu AM, Gambhir SS, Weiss S (2008) Quantum dots for live cells, in vivo imaging, and diagnostics. Science 538:538–544. https://doi.org/10.1126/science.1104274
349. Miller MA, Zheng Y-R, Gadde S, Pfirschke C, Zope H, Engblom C, Kohler RH, Iwamoto Y, Yang KS, Askevold B, Kolishetti N, Pittet M, Lippard SJ, Farokhzad OC, Weissleder R (2015) Tumour-associated macrophages act as a slow-release reservoir of nano-therapeutic Pt(IV) pro-drug. Nat Commun 6:8692. https://doi.org/10.1038/ncomms9692
350. Min Y, Akbulut M, Kristiansen K, Golan Y, Israelachvili J (2008) The role of interparticle and external forces in nanoparticle assembly. Nat Mater 7:527–538. https://doi.org/10.1038/nmat2206
351. Mitchell DG, Cohen MS (2004) MRI principles. Elsevier Saunders, Philadelphia

352. Modica-Napolitano JS, Aprille JR (2001) Delocalized lipophilic cations selectively target the mitochondria of carcinoma cells. Adv Drug Deliv Rev 49:63–70. https://doi.org/10.1016/s0169-409x(01)00125-9
353. Monopoli MP, Aberg C, Salvati A, Dawson KA (2012) Biomolecular coronas provide the biological identity of nanosized materials. Nat Nanotechnol 7:779–786. https://doi.org/10.1038/nnano.2012.207
354. Monteiro-Riviere NA, Inman AO, Zhang LW (2009) Limitations and relative utility of screening assays to assess engineered nanoparticle toxicity in a human cell line. Toxicol Appl Pharmacol 234:222–235. https://doi.org/10.1016/j.taap.2008.09.030
355. Morales MP, Verdangure SV-, Montero MI, Serna CJ, Roig A, Casas L, Martinez B, Sandiumenge F (1999) Surface and internal spin canting in γ-Fe_2O_3 nanoparticles. Chem Mater 11:3058–3064. https://doi.org/10.1021/cm991018f
356. Mornet S, Vasseur S, Grasset F, Duguet E (2004) Magnetic nanoparticle design for medical diagnosis and therapy. J Mater Chem 14:2161–2175. https://doi.org/10.1039/b402025a
357. Mortensen LJ, Oberdörster G, Pentland AP, Delouise LA (2008) In vivo skin penetration of quantum dot nanoparticles in the murine model: the effect of UVR. Nano Lett 8:2779–2787. https://doi.org/10.1021/nl801323y
358. Moyano DF, Saha K, Prakash G, Yan B, Kong H, Yazdani M, Rotello VM (2014) Fabrication of corona-free nanoparticles with tunable hydrophobicity. ACS Nano 8:6748–6755. https://doi.org/10.1021/nn5006478
359. Mueller MM, Fusenig NE (2004) Friends or foes—bipolar effects of the tumour stroma in cancer. Nat Rev Cancer 4:839–849. https://doi.org/10.1038/nrc1477
360. Mulvihill MJ, Ling XY, Henzie J, Yang P (2010) Anisotropic etching of silver nanoparticles for plasmonic structures capable of single-particle SERS. J Am Chem Soc 132:268–274. https://doi.org/10.1021/ja906954f
361. Murphy EA, Majeti BK, Barnes LA, Makale M, Weis SM, Lutu-Fuga K, Wrasidlo W, Cheresh DA (2008) Nanoparticle-mediated drug delivery to tumor vasculature suppresses metastasis. Proc Natl Acad Sci 105:9343–9348. https://doi.org/10.1073/pnas.0803728105
362. Murthy MRN, Reid TJ, Sicignano A, Tanaka N, Rossmann MG (1982) The structure of beef liver catalase. In: Dunford HB, Dolphin D, Raymond KN, Sieker L (eds) The biological chemistry of iron—a look at the metabolism of iron and its subsequent uses living organisms. Proceedings of NATO advanced study institute held at Edmonton, Alberta, Canada, 13 Aug 13–4 Sept 1981. Springer Netherlands, Dordrecht, pp 439–458
363. Myers CE, McGuire WP, Liss RH, Ifrim I, Grotzinger K, Young RC (1977) Adriamycin: the role of lipid peroxidation in cardiac toxicity and tumor response. Science 197:165–167. https://doi.org/10.1126/science.877547
364. Natalio F, André R, Hartog AF, Stoll B, Jochum KP, Wever R, Tremel W (2012) Vanadium pentoxide nanoparticles mimic vanadium haloperoxidases and thwart biofilm formation. Nat Nanotechnol 7:530–535. https://doi.org/10.1038/nnano.2012.91
365. Nel A, Xia T, Mädler L, Li N (2006) Toxic potential of materials at the nanolevel. Science 311:622–627. https://doi.org/10.1126/science.1114397
366. Nel AE, Mädler L, Velegol D, Xia T, Hoek EMV, Somasundaran P, Klaessig F, Castranova V, Thompson M (2009) Understanding biophysicochemical interactions at the nano-bio interface. Nat Mater 8:543–557. https://doi.org/10.1038/nmat2442
367. Neri D, Bicknell R (2005) Tumour vascular targeting. Nat Rev Cancer 5:436–446. https://doi.org/10.1038/nrc1627
368. Netti PA, Berk DA, Swartz MA, Grodzinsky AJ, Jain RK (2000) Role of extracellular matrix assembly in interstitial transport in solid tumors. Cancer Res 60:2497–2503
369. Nichols JW, Bae YH (2012) Odyssey of a cancer nanoparticle: from injection site to site of action. Nano Today 7:606–618. https://doi.org/10.1016/j.nantod.2012.10.010
370. Niedermayer S, Weiss V, Herrmann A, Schmidt A, Datz S, Müller K, Wagner E, Bein T, Bräuchle C (2015) Multifunctional polymer-capped mesoporous silica nanoparticles for pH-responsive targeted drug delivery. Nanoscale 7:7953–7964. https://doi.org/10.1039/c4nr07245f

371. Niemelä E, Desai D, Nkizinkiko Y, Eriksson JE, Rosenholm JM (2015) Sugar-decorated mesoporous silica nanoparticles as delivery vehicles for the poorly soluble drug celastrol enables targeted induction of apoptosis in cancer cells. Eur J Pharm Biopharm 96:11–21. https://doi.org/10.1016/j.ejpb.2015.07.009
372. Nikolic MS, Krack M, Aleksandrovic V, Kornowski A, Förster S, Weller H (2006) Maßgeschneiderte Liganden für biokompatible Nanopartikel. Angew Chem 118:6727–6731. https://doi.org/10.1002/ange.200602209
373. Niu W, Zhang W, Firdoz S, Lu X (2014) Controlled synthesis of palladium concave nanocubes with sub-10-nanometer edges and corners for tunable plasmonic property. Chem Mater 26:2180–2186. https://doi.org/10.1021/cm500210u
374. Noh SH, Na W, Jang JT, Lee JH, Lee EJ, Moon SH, Lim Y, Shin JS, Cheon J (2012) Nanoscale magnetism control via surface and exchange anisotropy for optimized ferrimagnetic hysteresis. Nano Lett 12:3716–3721. https://doi.org/10.1021/nl301499u
375. Noureddine A, Gary-Bobo M, Lichon L, Garcia M, Zink JI, Wong Chi Man M, Cattoën X (2016) Bis-clickable mesoporous silica nanoparticles: straightforward preparation of light-actuated nanomachines for controlled drug delivery with active targeting. Chem Eur J 22:9624–9630. https://doi.org/10.1002/chem.201600870
376. Noureddine A, Lichon L, Maynadier M, Garcia M, Gary-Bobo M, Zink JI, Wong Chi Man M, Cattoën X (2015) Controlled multiple functionalization of mesoporous silica nanoparticles: homogeneous implementation of pairs of functionalities communicating through energy or proton transfers. Nanoscale 7:11444–11452. https://doi.org/10.1039/c5nr02620b
377. Obeid M, Tesniere A, Ghiringhelli F, Fimia GM, Apetoh L, Perfettini J-L, Castedo M, Mignot G, Panaretakis T, Casares N, Métivier D, Larochette N, van Endert P, Ciccosanti F, Piacentini M, Zitvogel L, Kroemer G (2007) Calreticulin exposure dictates the immunogenicity of cancer cell death. Nat Med 13:54–61. https://doi.org/10.1038/nm1523
378. Oberdörster G, Maynard A, Donaldson K, Castranova V, Fitzpatrick J, Ausman K, Carter J, Karn B, Kreyling W, Lai D, Olin S, Monteiro-Riviere N, Warheit D, Yang H (2005) Principles for characterizing the potential human health effects from exposure to nanomaterials: elements of a screening strategy. Part Fibre Toxicol 2:8/1–8/35. https://doi.org/10.1186/1743-8977-2-8
379. Ogawara K, Furumoto K, Nagayama S, Minato K, Higaki K, Kai T, Kimura T (2004) Pre-coating with serum albumin reduces receptor-mediated hepatic disposition of polystyrene nanosphere: implications for rational design of nanoparticles. J Control Release 100:451–455. https://doi.org/10.1016/j.jconrel.2004.07.028
380. Ohkuma S, Poole B (1978) Fluorescence probe measurement of the intralysosomal pH in living cells and the perturbation of pH by various agents. Proc Natl Acad Sci 75:3327–3331
381. Olive PL, Vikse C, Trotter MJ (1992) Measurement of oxygen diffusion distance in tumor cubes using a fluorescent hypoxia probe. Int J Radiat Oncol 22:397–402. https://doi.org/10.1016/0360-3016(92)90840-e
382. Osaki F, Kanamori T, Sando S, Sera T, Aoyama Y (2004) A quantum dot conjugated sugar ball and its cellular uptake. On the size effects of endocytosis in the subviral region. J Am Chem Soc 126:6520–6521. https://doi.org/10.1021/ja048792a
383. Osseo-Asare K, Arriagada FJ (1990) Preparation of SiO_2 nanoparticles in a non-ionic reverse micellar system. Colloids Surf 50:321–339. https://doi.org/10.1016/0166-6622(90)80273-7
384. Ostwald W (1900) Über die vermeintliche Isomerie des roten und gelben Quecksilberoxyds und die Oberflächenspannung fester Körper. Zeitschrift für Phys Chemie 34:495–503
385. Owens DE, Peppas NA (2006) Opsonization, biodistribution, and pharmacokinetics of polymeric nanoparticles. Int J Pharm 307:93–102. https://doi.org/10.1016/j.ijpharm.2005.10.010
386. Ozbay E (2006) Plasmonics: merging photonics and electronics at nanoscale dimensions. Science 311:189–194. https://doi.org/10.1126/science.1114849
387. Padera TP, Stoll BR, Tooredman JB, Capen D, Di Tomaso E, Jain RK (2004) Pathology: cancer cells compress intratumour vessels. Nature 427:695. https://doi.org/10.1038/427695a

388. Pakunlu RI, Cook TJ, Minko T (2003) Simultaneous modulation of multidrug resistance and antiapoptotic cellular defense by MDR1 and BCL-2 targeted antisense oligonucleotides enhances the anticancer efficacy of doxorubicin. Pharm Res 20:351–359. https://doi.org/10.1023/a:1022687617318
389. Pakunlu RI, Wang Y, Tsao W, Pozharov V, Cook TJ, Minko T (2004) Enhancement of the efficacy of chemotherapy for lung cancer by simultaneous suppression of multidrug resistance and antiapoptotic cellular defense: novel multicomponent delivery system. Cancer Res 64:6214–6224. https://doi.org/10.1158/0008-5472.can-04-0001
390. Palanikumar L, Kim HY, Oh JY, Thomas AP, Choi ES, Jeena MT, Joo SH, Ryu J-H (2015) Noncovalent surface locking of mesoporous silica nanoparticles for exceptionally high hydrophobic drug loading and enhanced colloidal stability. Biomacromolecules 16:2701–2714. https://doi.org/10.1021/acs.biomac.5b00589
391. Palma DR, Peeters S, Van Bael MJ, Van den Rul H, Bonroy K, Laureyn W, Mullens J, Borghs G, Maes G (2007) Silane ligand exchange to make hydrophobic superparamagnetic nanoparticles water-dispersible. Chem Mater 19:1821–1831. https://doi.org/10.1021/cm0628000
392. Pan D, Roessl E, Schlomka JP, Caruthers SD, Senpan A, Scott MJ, Allen JS, Zhang H, Hu G, Gaffney PJ, Choi ET, Rasche V, Wickline SA, Proksa R, Lanza GM (2010) Computed tomography in color: Nanok-enhanced spectral CT molecular imaging. Angew Chem Int Ed 49:9635–9639. https://doi.org/10.1002/anie.201005657
393. Pan L, He Q, Liu J, Chen Y, Ma M, Zhang L, Shi J (2012) Nuclear-targeted drug delivery of TAT peptide-conjugated monodisperse mesoporous silica nanoparticles. J Am Chem Soc 134:5722–5725. https://doi.org/10.1021/ja211035w
394. Pang J, Zhou G, Liu R, Li T (2016) Esterification of oleic acid with methanol by immobilized lipase on wrinkled silica nanoparticles with highly ordered, radially oriented mesochannels. Mater Sci Eng C 59:35–42. https://doi.org/10.1016/j.msec.2015.09.088
395. Parak WJ, Boudreau R, Le Gros M, Gerion D, Zanchet D, Micheel CM, Williams SC, Alivisatos AP, Larabell C (2002) Cell motility and metastatic potential studies based on quantum dot imaging of phagokinetic tracks. Adv Mater 14:882–885. https://doi.org/10.1002/1521-4095(20020618)14:12<882:aid-adma882>3.0.co;2-y
396. Pardoll DM (2012) The blockade of immune checkpoints in cancer immunotherapy. Nat Rev Cancer 12:252–264. https://doi.org/10.1038/nrc3239
397. Park J-H, Gu L, Maltzahn G, Ruoslahti E, Bhatia SN, Sailor MJ (2009) Biodegradable luminescent porous silicon nanoparticles for in vivo applications. Nat Mater 8:331–336. https://doi.org/10.1038/nmat2398
398. Park J, Joo J, Soon GK, Jang Y, Hyeon T (2007) Synthesis of monodisperse spherical nanocrystals. Angew Chem Int Ed 46:4630–4660. https://doi.org/10.1002/anie.200603148
399. Parodi A, Quattrocchi N, van de Ven AL, Chiappini C, Evangelopoulos M, Martinez JO, Brown BS, Khaled SZ, Yazdi IK, Enzo MV, Isenhart L, Ferrari M, Tasciotti E (2013) Synthetic nanoparticles functionalized with biomimetic leukocyte membranes possess cell-like functions. Nat Nanotechnol 8:61–68. https://doi.org/10.1038/nnano.2012.212
400. Pearson RG (1963) Hard and soft acids and bases. J Am Chem Soc 85:3533–3539. https://doi.org/10.1021/ja00905a001
401. Pecot CV, Calin GA, Coleman RL, Lopez-Berestein G, Sood AK (2011) RNA interference in the clinic: challenges and future directions. Nat Rev Cancer 11:59–67. https://doi.org/10.1038/nrc2966
402. Peer D, Karp JM, Hong S, Farokhzad OC, Margalit R, Langer R (2007) Nanocarriers as an emerging platform for cancer therapy. Nat Nanotechnol 2:751–760. https://doi.org/10.1038/nnano.2007.387
403. Peer D, Margalit R (2006) Fluoxetine and reversal of multidrug resistance. Cancer Lett 237:180–187. https://doi.org/10.1016/j.canlet.2005.06.003
404. Persidis A (1999) Cancer multidrug resistance. Nat Biotechnol 17:94–95. https://doi.org/10.1038/5289

405. Piddington DL, Fang FC, Laessig T, Cooper M, Orme IM, Buchmeier N, Cooper AM (2001) Cu , zn superoxide dismutase of mycobacterium tuberculosis contributes to survival in activated macrophages that are generating an oxidative burst Cu, Zn superoxide dismutase of *Mycobacterium tuberculosis* contributes to survival in activated macrophages. Infect Immun 69:4980–4987. https://doi.org/10.1128/iai.69.8.4980
406. Pirmohamed T, Dowding JM, Singh S, Wasserman B, Heckert E, Karakoti AS, King JES, Seal S, Self WT (2010) Nanoceria exhibit redox state-dependent catalase mimetic activity. Chem Commun 46:2736–2738. https://doi.org/10.1039/b922024k
407. Prabhakar N, Zhang J, Desai D, Casals E, Gulin-Sarfraz T, Näreoja T, Westermarck J, Rosenholm JM (2016) Stimuli-responsive hybrid nanocarriers developed by controllable integration of hyperbranched PEI with mesoporous silica nanoparticles for sustained intracellular siRNA delivery. Int J Nanomedicine 11:6591–6608. https://doi.org/10.2147/ijn.s120611
408. Pradhan S, Ghosh D, Chen S (2009) Janus nanostructures based on Au-TiO_2 heterodimers and their photocatalytic activity in the oxidation of methanol. ACS Appl Mater Interfaces 1:2060–2065. https://doi.org/10.1021/am900425v
409. Pumera M (2011) Nanomaterials meet microfluidics. Chem Commun 47:5671–5680. https://doi.org/10.1039/c1cc11060h
410. Qin J, Jo YS, Muhammed M (2009) Coating nanocrystals with amphiphilic thermosensitive copolymers. Angew Chem Int Ed 48:7845–7849. https://doi.org/10.1002/anie.200900489
411. Qin J, Laurent S, Jo YS, Roch A, Mikhaylova M, Bhujwalla ZM, Müller RN, Muhammed M (2007) A high-performance magnetic resonance imaging T2 contrast agent. Adv Mater 19:1874–1878. https://doi.org/10.1002/adma.200602326
412. Qu K, Shi P, Ren J, Qu X (2014) Nanocomposite incorporating V_2O_5 nanowires and gold nanoparticles for mimicking an enzyme cascade reaction and its application in the detection of biomolecules. Chem Eur J 20:7501–7506. https://doi.org/10.1002/chem.201400309
413. Quail DF, Joyce JA (2013) Microenvironmental regulation of tumor progression and metastasis. Nat Med 19:1423–1437. https://doi.org/10.1038/nm.3394
414. Radu DR, Lai C, Jeftinija K, Rowe EW, Jeftinija S, Lin VS (2004) Communication A polyamidoamine dendrimer-capped mesoporous silica nanosphere-based gene transfection reagent a polyamidoamine dendrimer-capped mesoporous silica nanosphere-based. J Am Chem Soc 126:13216–13217. https://doi.org/10.1021/ja046275m
415. Ragg R, Natalio F, Tahir MN, Janssen H, Kashyap A, Strand D, Strand S, Tremel W (2014) Molybdenum trioxide nanoparticles with intrinsic sulfite oxidase activity. ACS Nano 8:5182–5189. https://doi.org/10.1021/nn501235j
416. Ragg R, Schilmann AM, Korschelt K, Wieseotte C, Kluenker M, Viel M, Völker L, Preiß S, Herzberger J, Frey H, Heinze K, Blümler P, Tahir MN, Natalio F, Tremel W (2016) Intrinsic superoxide dismutase activity of MnO nanoparticles enhances the magnetic resonance imaging contrast. J Mater Chem B 4:7423–7428. https://doi.org/10.1039/c6tb02078j
417. Ragg R, Tahir MN, Tremel W (2016) Solids Go Bio: inorganic nanoparticles as enzyme mimics. Eur J Inorg Chem 2016:1906–1915. https://doi.org/10.1002/ejic.201501237
418. Ramanujan S, Pluen A, McKee TD, Brown EB, Boucher Y, Jain RK (2002) Diffusion and convection in collagen gels: implications for transport in the tumor interstitium. Biophys J 83:1650–1660. https://doi.org/10.1016/s0006-3495(02)73933-7
419. Rana S, Yeh Y-C, Rotello VM (2010) Engineering the nanoparticle-protein interface: applications and possibilities. Curr Opin Chem Biol 14:828–834. https://doi.org/10.1016/j.cbpa.2010.10.001
420. Rengan AK, Bukhari AB, Pradhan A, Malhotra R, Banerjee R, Srivastava R, De A (2015) In vivo analysis of biodegradable liposome gold nanoparticles as efficient agents for photothermal therapy of cancer. Nano Lett 15:842–848. https://doi.org/10.1021/nl5045378
421. Renzo DF, Testa F, Chen J, Cambon H, Galarneau A, Plee D, Fajula F (1999) Textural control of micelle-templated mesoporous silicates: the effects of co-surfactants and alkalinity. Microporous Mesoporous Mater 28:437–446. https://doi.org/10.1016/s1387-1811(98)00315-1

422. Reubi JC (2003) Peptide receptors as molecular targets for cancer diagnosis and therapy. Endocr Rev 24:389–427. https://doi.org/10.1210/er.2002-0007
423. Rijt VS, Boeluekbas DA, Argyo C, Wipplinger K, Naureen M, Datz S, Eickelberg O, Meiners S, Bein T, Schmid O, Stoeger T (2016) Applicability of avidin protein coated mesoporous silica nanoparticles as drug carriers in the lung. Nanoscale 8:8058–8069. https://doi.org/10.1039/c5nr04119h
424. Ritz S, Schöttler S, Kotman N, Baier G, Musyanovych A, Kuharev J, Landfester K, Schild H, Jahn O, Tenzer S, Mailänder V (2015) Protein corona of nanoparticles: distinct proteins regulate the cellular uptake. Biomacromolecules 16:1311–1321. https://doi.org/10.1021/acs.biomac.5b00108
425. Rodriguez PL, Harada T, Christian DA, Pantano DA, Tsai RK, Discher DE (2013) Minimal "Self" peptides that inhibit phagocytic clearance and enhance delivery of nanoparticles. Science 339:971–975. https://doi.org/10.1126/science.1229568
426. Roduner E (2014) Understanding catalysis. Chem Soc Rev 2:8226–8239. https://doi.org/10.1039/c4cs00210e
427. Rosenberg SA, Yang JC, Restifo NP (2004) Cancer immunotherapy: moving beyond current vaccines. Nat Med 10:909–915. https://doi.org/10.1038/nm1100
428. Rosenhahn A, Schilp S, Kreuzer HJ, Grunze M (2010) The role of "'inert'" surface chemistry in marine biofouling prevention. Phys Chem Chem Phys 12:4275–4286. https://doi.org/10.1039/c004746p
429. Rosenholm JM, Meinander A, Peuhu E, Niemi R, Eriksson JE, Sahlgren C, Lindén M (2009) Targeting of porous hybrid silica nanoparticles to cancer cells. ACS Nano 3:197–206. https://doi.org/10.1021/nn800781r
430. Rosenholm JM, Sahlgren C, Lindén M (2010) Towards multifunctional, targeted drug delivery systems using mesoporous silica nanoparticles—opportunities & challenges. Nanoscale 2:1870–1883. https://doi.org/10.1039/c0nr00156b
431. Rosensweig RE (2002) Heating magnetic fluid with alternating magnetic field. J Magn Magn Mater 252:370–374. https://doi.org/10.1016/s0304-8853(02)00706-0
432. Sadauskas E, Danscher G, Stoltenberg M, Vogel U, Larsen A, Wallin H (2009) Protracted elimination of gold nanoparticles from mouse liver. Nanomedicine 5:162–169. https://doi.org/10.1016/j.nano.2008.11.002
433. Sahay G, Alakhova DY, Kabanov AV (2010) Endocytosis of nanomedicines. J Control Release 145:182–195. https://doi.org/10.1016/j.jconrel.2010.01.036
434. Sahoo JK, Tahir MN, Yella A, Schladt TD, Mugnaoli E, Kolb U, Tremel W (2010) Reversible self-assembly of metal chalcogenide/metal oxide nanostructures based on pearson hardness. Angew Chem Int Ed 49:7578–7582. https://doi.org/10.1002/anie.201000774
435. Sahoo JK, Tahir MN, Yella A, Schladt TD, Mugnaoli E, Kolb U, Tremel W (2010) Reversible Selbstorganisation von Metallchalkogenid-Metalloxid- Nanostrukturen basierend auf dem Pearson-Konzept. Angew Chem 122:7741–7745. https://doi.org/10.1002/ange.201000774
436. Saito G, Swanson JA, Lee K-D (2003) Drug delivery strategy utilizing conjugation via reversible disulfide linkages: role and site of cellular reducing activities. Adv Drug Deliv Rev 55:199–215. https://doi.org/10.1016/s0169-409x(02)00179-5
437. Sakaguchi S, Miyara M, Costantino CM, Hafler DA (2010) FOXP3+ regulatory T cells in the human immune system. Nat Rev Immunol 10:490–500. https://doi.org/10.1038/nri2785
438. Sakulkhu U, Maurizi L, Mahmoudi M, Motazacker M, Vries M, Gramoun A, Ollivier Beuzelin M-G, Vallée J-P, Rezaee F, Hofmann H (2014) Ex situ evaluation of the composition of protein corona of intravenously injected superparamagnetic nanoparticles in rats. Nanoscale 6:11439–11450. https://doi.org/10.1039/c4nr02793k
439. Saleh T, Shojaosadati SA (2016) Multifunctional nanoparticles for cancer immunotherapy. Hum Vaccin Immunother 12:1863–1875. https://doi.org/10.1080/21645515.2016.1147635

440. Salvati A, Pitek AS, Monopoli MP, Prapainop K, Bombelli FB, Hristov DR, Kelly PM, Åberg C, Mahon E, Dawson KA (2013) Transferrin-functionalized nanoparticles lose their targeting capabilities when a biomolecule corona adsorbs on the surface. Nat Nanotechnol 8:137–143. https://doi.org/10.1038/nnano.2012.237
441. Satchi-Fainaro R, Duncan R (2006) Advances in polymer science. Springer, Berlin
442. Schick I, Gehrig D, Montigny M, Balke B, Panthöfer M, Henkel A, Laquai F, Tremel W (2015) Effect of charge transfer in magnetic-plasmonic Au@MOx (M = Mn, Fe) heterodimers on the kinetics of nanocrystal formation. Chem Mater 27:4877–4884. https://doi.org/10.1021/acs.chemmater.5b01968
443. Schick I, Lorenz S, Gehrig D, Schilmann AM, Bauer H, Panthöfer M, Fischer K, Strand D, Laquai F, Tremel W (2014) Multifunctional two-photon active silica-coated Au@ MnO janus particles for selective dual functionalization and imaging. J Am Chem Soc 136:2473–2483. https://doi.org/10.1021/ja410787u
444. Schladt T, Graf T, Köhler O (2012) Synthesis and magnetic properties of FePt@ MnO nano-heteroparticles. Chem Mater 24:525–535. https://doi.org/10.1021/cm2030685
445. Schladt TD, Schneider K, Schild H, Tremel W (2011) Synthesis and bio-functionalization of magnetic nanoparticles for medical diagnosis and treatment. Dalt Trans 40:6315–6343. https://doi.org/10.1039/c0dt00689k
446. Schladt TD, Shukoor MI, Schneider K, Tahir MN, Natalio F, Ament I, Becker J, Jochum FD, Weber S, Köhler O, Theato P, Schreiber LM, Sönnichsen C, Schröder HC, Müller WEG, Tremel W (2010) Au@MnO-"Nanoblumen" - Hybrid-Nanokomposite zur selektiven dualen Funktionalisierung und Bildgebung. Angew Chem 122:4068–4072. https://doi.org/10.1002/ange.200906689
447. Schöttler S, Becker G, Winzen S, Steinbach T, Mohr K, Landfester K, Mailänder V, Wurm FR (2016) Protein adsorption is required for stealth effect of poly(ethylene glycol)- and poly(phosphoester)-coated nanocarriers. Nat Nanotechnol 11:372–377. https://doi.org/10.1038/nnano.2015.330
448. Schreiber RD, Old LJ, Smyth MJ (2011) Cancer immunoediting: integrating immunity's roles in cancer suppression and promotion. Science 331:1565–1570. https://doi.org/10.1126/science.1203486
449. Schwarz G, Mendel RR, Ribbe MW (2009) Molybdenum cofactors, enzymes and pathways. Nature 460:839–847. https://doi.org/10.1038/nature08302
450. Scott AM, Allison JP, Wolchok JD (2012) Monoclonal antibodies in cancer therapy. Cancer Immun 12:14/1–14/8.
451. Seh ZW, Liu S, Low M, Zhang SY, Liu Z, Mlayah A, Han MY (2012) Janus Au-TiO_2 photocatalysts with strong localization of plasmonic near-fields for efficient visible-light hydrogen generation. Adv Mater 24:2310–2314. https://doi.org/10.1002/adma.201104241
452. Şen Karaman D, Gulin-Sarfraz T, Hedström G, Duchanoy A, Eklund P, Rosenholm JM (2014) Rational evaluation of the utilization of PEG-PEI copolymers for the facilitation of silica nanoparticulate systems in biomedical applications. J Colloid Interface Sci 418:300–310. https://doi.org/10.1016/j.jcis.2013.11.080
453. Senyei A, Widder K, Czerlinski G (1978) Magnetic guidance of drug-carrying microspheres. J Appl Phys 49:3578–3583. https://doi.org/10.1063/1.325219
454. Sharma P, Allison JP (2015) The future of immune checkpoint therapy. Science 348:56–61. https://doi.org/10.1126/science.aaa8172
455. Sheldon MT, Trudeau PE, Mokari T, Wang LW, Paul Alivisatos A (2009) Enhanced semiconductor nanocrystal conductance via solution grown contacts. Nano Lett 9:3676–3682. https://doi.org/10.1021/nl902186v
456. Shen D, Chen L, Yang J, Zhang R, Wei Y, Li X, Li W, Sun Z, Zhu H, Abdullah AM, Al-Enizi A, Elzatahry AA, Zhang F, Zhao D (2015) Ultradispersed palladium nanoparticles in three-dimensional dendritic mesoporous silica nanospheres: toward active and stable heterogeneous catalysts. ACS Appl Mater Interfaces 7:17450–17459. https://doi.org/10.1021/acsami.5b04992

457. Shi H, Wang Z, Huang C, Gu X, Jia T, Zhang A, Wu Z, Zhu L, Luo X, Zhao X, Jia N, Miao F (2016) A functional CT contrast agent for in vivo imaging of tumor hypoxia. Small 12:3995–4006. https://doi.org/10.1002/smll.201601029
458. Shi S, Huang Y, Chen X, Weng J, Zheng N (2015) Optimization of surface coating on small Pd nanosheets for in vivo near-infrared photothermal therapy of tumor. ACS Appl Mater Interfaces 7:14369–14375. https://doi.org/10.1021/acsami.5b03106
459. Shi W, Sahoo Y, Zeng H, Ding Y, Swihart MT, Prasad PN (2006) Anisotropic growth of PbSe nanocrystals on Au-Fe_3O_4 hybrid nanoparticles. Adv Mater 18:1889–1894. https://doi.org/10.1002/adma.200600685
460. Shi W, Zeng H, Sahoo Y, Ohulchanskyy TY (2006) A general approach to binary and ternary hybrid nanocrystals. Nano Lett 6:875–881. https://doi.org/10.1021/nl0600833
461. Shinto H, Fukasawa T, Yoshisue K, Tezuka M, Orita M (2014) Cell membrane disruption induced by amorphous silica nanoparticles in erythrocytes, lymphocytes, malignant melanocytes, and macrophages. Adv Powder Technol 25:1872–1881. https://doi.org/10.1016/j.apt.2014.09.002
462. Shliomis MI, Pshenichnikov AF, Morozov KI, Shurubor IY (1990) Magnetic properties of ferrocolloids. J Magn Magn Mater 85:40–46. https://doi.org/10.1016/0304-8853(90)90013-g
463. Shukla S, Steinmetz NF (2016) Emerging nanotechnologies for cancer immunotherapy. Exp Biol Med 241:1116–1126. https://doi.org/10.1177/1535370216647123
464. Shultz MD, Ulises Reveles J, Khanna SN, Carpenter EE (2007) Reactive nature of dopamine as a surface functionalization agent in iron oxide nanoparticles. J Am Chem Soc 129:2482–2487. https://doi.org/10.1021/ja0651963
465. Slowing I, Trewyn BG, Lin VS-Y (2006) Effect of surface functionalization of MCM-41-type mesoporous silica nanoparticles on the endocytosis by human cancer cells. J Am Chem Soc 128:14792–14793. https://doi.org/10.1021/ja0645943
466. van der Sluis TC, van Duikeren S, Huppelschoten S, Jordanova ES, Beyranvand Nejad E, Sloots A, Boon L, Smit VTHBM, Welters MJP, Ossendorp F, van de Water B, Arens R, van der Burg SH, Melief CJM (2015) Vaccine-induced tumor necrosis factor-producing T cells synergize with cisplatin to promote tumor cell death. Clin Cancer Res 21:781–794. https://doi.org/10.1158/1078-0432.ccr-14-2142
467. Smith BR, Kempen P, Bouley D, Xu A, Liu Z, Melosh N, Dai H, Sinclair R, Gambhir SS (2012) Shape matters: intravital microscopy reveals surprising geometrical dependence for nanoparticles in tumor models of extravasation. Nano Lett 12:3369–3377. https://doi.org/10.1021/nl204175t
468. Smith CA, Farrah T, Goodwin RG (1994) The TNF receptor superfamily of cellular and viral proteins: activation, costimulation, and death. Cell 76:959–962. https://doi.org/10.1016/0092-8674(94)90372-7
469. Soedjak HS, Walker JV, Butler A (1995) Inhibition and inactivation of vanadium bromoperoxidase by the substrate hydrogen peroxide and further mechanistic studies. Biochemistry 34:12689–12696. https://doi.org/10.1021/bi00039a027
470. Soto-Cantu E, Cueto R, Koch J, Russo PS (2012) Synthesis and rapid characterization of amine-functionalized silica. Langmuir 28:5562–5569. https://doi.org/10.1021/la204981b
471. Spanhel L, Haase M, Weller H, Henglein A (1987) Photochemistry of colloidal semiconductors. 20. Surface modification and stability of strong luminescing CdS particles. J Am Chem Soc 109:5649–5655. https://doi.org/10.1021/ja00253a015
472. Steiner D, Mokari T, Banin U, Millo O (2005) Electronic structure of metal-semiconductor nanojunctions in gold CdSe nanodumbbells. Phys Rev Lett 95:056805/1–056805/4. https://doi.org/10.1103/physrevlett.95.056805
473. Stöber W, Fink A, Bohn E (1968) Controlled growth of monodisperse silica spheres in the micron size range. J Colloid Interface Sci 26:62–69. https://doi.org/10.1016/0021-9797(68)90272-5
474. Stohrer M, Boucher Y, Stangassinger M, Jain RK (2000) Oncotic pressure in solid tumors is elevated. Cancer Res 60:4251–4255
475. Storz P (2005) Reactive oxygen species in tumor progression. Front Biosci 10:1881–1896

476. Stubbs M, McSheehy PMJ, Griffiths JR, Bashford CL (2000) Causes and consequences of tumour acidity and implications for treatment. Mol Med Today 6:15–19. https://doi.org/10.1016/s1357-4310(99)01615-9
477. Stylianopoulos T, Poh M-Z, Insin N, Bawendi MG, Fukumura D, Munn LL, Jain RK (2010) Diffusion of particles in the extracellular matrix: the effect of repulsive electrostatic interactions. Biophys J 99:1342–1349. https://doi.org/10.1016/j.bpj.2010.06.016
478. Subramanian V, Wolf EE, Kamat PV (2004) Catalysis with TiO_2/gold nanocomposites. Effect of metal particle size on the fermi level equilibration. J Am Chem Soc 126:4943–4950. https://doi.org/10.1021/ja0315199
479. Sun C, Lee JSHJ, Zhang M (2008) Magnetic nanoparticles in MR imaging and drug delivery. Adv Drug Deliv Rev 60:1252–1265. https://doi.org/10.1016/j.addr.2008.03.018
480. Sun H, Yuan Q, Zhang B, Ai K, Zhang P, Lu L (2011) Gd(III) functionalized gold nanorods for multimodal imaging applications. Nanoscale 3:1990–1996. https://doi.org/10.1039/c0nr00929f
481. Sun J, Kim D-H, Guo Y, Teng Z, Li Y, Zheng L, Zhang Z, Larson AC, Lu G (2015) A c (RGDfE) conjugated multi-functional nanomedicine delivery system for targeted pancreatic cancer therapy. J Mater Chem B 3:1049–1058. https://doi.org/10.1039/c4tb01402b
482. Sun L, Liu Y-J, Yang Z-Z, Qi X-R (2015) Tumor specific delivery with redox-triggered mesoporous silica nanoparticles inducing neovascularization suppression and vascular normalization. RSC Adv 5:55566–55578. https://doi.org/10.1039/c5ra09633b
483. Sun S, Murray CB, Weller D, Folks L, Moser A (2000) Monodisperse FePt nanoparticles and ferromagnetic FePt nanocrystal superlattices. Science 287:1989–1992. https://doi.org/10.1126/science.287.5460.1989
484. Sun T, Zhang YS, Pang B, Hyun DC, Yang M, Xia Y (2014) Engineered nanoparticles for drug delivery in cancer therapy. Angew Chem Int Ed 53:12320–12364. https://doi.org/10.1002/anie.201403036
485. Sun X, Guo S, Chung C-S, Zhu W, Sun S (2013) A sensitive H_2O_2 assay based on dumbbell-like PtPd-Fe_3O_4 nanoparticles. Adv Mater 25:132–136. https://doi.org/10.1002/adma.201203218
486. Sun X, Klapper A, Su Y, Nemkovski K, Wildes A, Bauer H, Köhler O, Schilmann A, Tremel W, Petracic O, Brückel T (2017) Magnetism of monomer MnO and heterodimer FePt@MnO nanoparticles. Phys Rev B 95:134427/1–134427/8. https://doi.org/10.1103/physrevb.95.134427
487. Sun Y, Foley JJ, Peng S, Li Z, Gray SK (2013) Interfaced metal heterodimers in the quantum size regime. Nano Lett 13:3958–3964. https://doi.org/10.1021/nl402361b
488. Sun Y, Xia Y (2002) Shape-controlled synthesis of gold and silver nanoparticles. Science 298:2176–2180
489. Susumu K, Uyeda HT, Medintz IL, Pons T, Delehanty JB, Mattoussi H (2007) Enhancing the stability and biological functionalities of quantum dots via compact multifunctional ligands. J Am Chem Soc 129:13987–13996. https://doi.org/10.1021/ja0749744
490. Sykes EA, Chen J, Zheng G, Chan WCW (2014) Investigating the impact of nanoparticle size on active and passive tumor targeting efficiency. ACS Nano 8:5696–5706. https://doi.org/10.1021/nn500299p
491. Szakács G, Paterson JK, Ludwig JA, Booth-Genthe C, Gottesman MM (2006) Targeting multidrug resistance in cancer. Nat Rev Drug Discov 5:219–234. https://doi.org/10.1038/nrd1984
492. Szatrowski TP, Nathan CF (1991) Production of large amounts of hydrogen peroxide by human tumor cells. Cancer Res 51:794–798
493. Tahir MN, André R, Sahoo JK, Jochum FD, Theato P, Natalio F, Berger R, Branscheid R, Kolb U, Tremel W (2011) Hydrogen peroxide sensors for cellular imaging based on horse radish peroxidase reconstituted on polymer-functionalized TiO(2) nanorods. Nanoscale 3:3907–3914. https://doi.org/10.1039/c1nr10587f

494. Tahir MN, Herzberger J, Natalio F, Koehler O, Branscheid R, Mugnaioli E, Ksenofontov V, Panthoefer M, Kolb U, Frey H, Tremel W, Köhler O, Branscheid R, Mugnaioli E, Ksenofontov V, Panthöfer M, Kolb U, Frey H, Tremel W (2016) Hierachical Ni@Fe_2O_3 superparticles through epitaxial growth of γ-Fe_2O_3 nanorods on in situ formed Ni nanoplates. Nanoscale 8:9548–9555. https://doi.org/10.1039/b000000x
495. Tahir MN, Natalio F, Cambaz MA, Panthofer M, Branscheid R, Kolb U, Tremel W (2013) Controlled synthesis of linear and branched Au@ZnO hybrid nanocrystals and their photocatalytic properties. Nanoscale 5:9944–9949. https://doi.org/10.1039/c3nr02817h
496. Tang S, Chen M, Zheng N (2014) Sub-10-nm Pd nanosheets with renal clearance for efficient near-infrared photothermal cancer therapy. Small 10:3139–3144. https://doi.org/10.1002/smll.201303631
497. Tang S, Chen M, Zheng N (2014) Multifunctional ultrasmall Pd nanosheets for enhanced near-infrared photothermal therapy and chemotherapy of cancer. Nano Res 8:165–174. https://doi.org/10.1007/s12274-014-0605-x
498. Tao AR, Habas S, Yang P (2008) Shape control of colloidal metal nanocrystals. Small 4:310–325. https://doi.org/10.1002/smll.200701295
499. Tao Z, Morrow MP, Asefa T, Sharma KK, Duncan C, Anan A, Penefsky HS, Goodisman J, Souid A-K (2008) Mesoporous silica nanoparticles inhibit cellular respiration. Nano Lett 8:1517–1526. https://doi.org/10.1021/nl080250u
500. Tarnuzzer RW, Colon J, Patil S, Seal S (2005) Vacancy engineered ceria nanostructures for protection from radiation-induced cellular damage. Nano Lett 5:2573–2577. https://doi.org/10.1021/nl052024f
501. Tasciotti E, Liu X, Bhavane R, Plant K, Leonard AD, Price BK, Cheng MM-C, Decuzzi P, Tour JM, Robertson F, Ferrari M (2008) Mesoporous silicon particles as a multistage delivery system for imaging and therapeutic applications. Nat Nanotechnol 3:151–157. https://doi.org/10.1038/nnano.2008.34
502. Tekle C, van Deurs B, Sandvig K, Iversen T-G (2008) Cellular trafficking of quantum dot-ligand bioconjugates and their induction of changes in normal routing of unconjugated ligands. Nano Lett 8:1858–1865. https://doi.org/10.1021/nl0803848
503. Tenzer S, Docter D, Kuharev JJJJ, Musyanovych A, Fetz V, Hecht R, Schlenk F, Fischer D, Kiouptsi K, Reinhardt C, Landfester K, Schild HH, Maskos M, Knauer SK, Stauber RH (2013) Rapid formation of plasma protein corona critically affects nanoparticle pathophysiology. Nat Nano 8:772–781. https://doi.org/10.1038/nnano.2013.181
504. Thanh NTK, Maclean N, Mahiddine S (2014) Mechanisms of nucleation and growth of nanoparticles in solution. Chem Rev 3:7610–7630. https://doi.org/10.1021/cr400544s
505. Thomas A, Bauer H, Schilmann AM, Fischer K, Tremel W, Frey H (2014) The "needle in the haystack" Makes the difference: linear and hyperbranched polyglycerols with a single catechol moiety for metal oxide nanoparticle coating. Macromolecules 47:4557–4566. https://doi.org/10.1021/ma5003672
506. Tian Y, Kong Y, Li X, Wu J, Ko AC-T, Xing M (2015) Light- and pH- activated intracellular drug release from polymeric mesoporous silica nanoparticles. Colloids Surf B Biointerfaces 134:147–155. https://doi.org/10.1016/j.colsurfb.2015.04.069
507. Tong S, Hou S, Zheng Z, Zhou J, Bao G (2010) Coating optimization of superparamagnetic iron oxide nanoparticles for high T2 relaxivity. Nano Lett 10:4607–4613. https://doi.org/10.1021/nl102623x
508. Torchilin VP (2005) Recent advances with liposomes as pharmaceutical carriers. Nat Rev Drug Discov 4:145–160. https://doi.org/10.1038/nrd1632
509. Trachootham D, Alexandre J, Huang P (2009) Targeting cancer cells by ROS-mediated mechanisms: a radical therapeutic approach? Nat Rev Drug Discov 8:579–591. https://doi.org/10.1038/nrd2803
510. Traina CA, Schwartz J (2007) Surface Modification of Y_2O_3 Nanoparticles. Langmuir 23:9158–9161. https://doi.org/10.1021/la701653v
511. Tromsdorf U (2010) Magnetische Nanopartikel als Kontrastmittel für die MRT. Dissertation, University of Hamburg

512. Tsirikis P, Wilson K, Xiang S, Wei W, Ma G, Selomulya C, Plebanski M (2016) Immunogenicity and biodistribution of nanoparticles in vivo. J Immunol 196(75):28
513. Tsoi KM, MacParland SA, Ma X-Z, Spetzler VN, Echeverri J, Ouyang B, Fadel SM, Sykes EA, Goldaracena N, Kaths JM, Conneely JB, Alman BA, Selzner M, Ostrowski MA, Adeyi OA, Zilman A, McGilvray ID, Chan WCW (2016) Mechanism of hard-nanomaterial clearance by the liver. Nat Mater 15:1212–1221. https://doi.org/10.1038/nmat4718
514. Turkevich J, Stevenson PC, Hillier J (1951) A study of the nucleation and growth processes in the synthesis of colloidal gold. Discuss Faraday Soc 11:55–75. https://doi.org/10.1039/df9511100055
515. Tzur-Balter A, Shatsberg Z, Beckerman M, Segal E, Artzi N (2015) Mechanism of erosion of nanostructured porous silicon drug carriers in neoplastic tissues. Nat Commun 6:6208/1–6208/8. https://doi.org/10.1038/ncomms7208
516. Uyeda HT, Medintz IL, Jaiswal JK, Simon SM, Mattoussi H (2005) Synthesis of compact multidentate ligands to prepare stable hydrophilic quantum dot fluorophores. J Am Chem Soc 127:3870–3878. https://doi.org/10.1021/ja044031w
517. Vakoc BJ, Lanning RM, Tyrrell JA, Padera TP, Bartlett LA, Stylianopoulos T, Munn LL, Tearney GJ, Fukumura D, Jain RK, Bouma BE (2009) Three-dimensional microscopy of the tumor microenvironment in vivo using optical frequency domain imaging. Nat Med 15:1219–1223. https://doi.org/10.1038/nm.1971
518. Vallet-Regi M, Rámila A, del Real RP, Pérez-Pariente J (2001) A new property of MCM-41: drug delivery system. Chem Mater 13:308–311. https://doi.org/10.1021/cm0011559
519. Veitch NC (2004) Horseradish peroxidase: a modern view of a classic enzyme. Phytochemistry 65:249–259. https://doi.org/10.1016/j.phytochem.2003.10.022
520. Vert M, Doi Y, Hellwich K-H, Hess M, Hodge P, Kubisa P, Rinaudo M, Schué F (2012) Terminology for biorelated polymers and applications (IUPAC Recommendations 2012). Pure Appl Chem 84:377–410. https://doi.org/10.1351/pac-rec-10-12-04
521. Vertegel AA, Siegel RW, Dordick JS (2004) Silica nanoparticle size influences the structure and enzymatic activity of adsorbed lysozyme. Langmuir 20:6800–6807. https://doi.org/10.1021/la0497200
522. Vo DQ, Kim EJ, Kim S (2009) Surface modification of hydrophobic nanocrystals using short-chain carboxylic acids. J Colloid Interface Sci 337:75–80. https://doi.org/10.1016/j.jcis.2009.04.078
523. Wagner E, Kloeckner J (2006) Gene delivery using polymer therapeutics. Polymer Therapeutics I. https://doi.org/10.1007/12_023
524. Walkey CD, Olsen JB, Guo H, Emili A, Chan WCW (2012) Nanoparticle size and surface chemistry determine serum protein adsorption and macrophage uptake. J Am Chem Soc 134:2139–2147. https://doi.org/10.1021/ja2084338
525. Walkey CD, Olsen JB, Song F, Liu R, Guo H, Olsen DWH, Cohen Y, Emili A, Chan WCW (2014) Protein corona fingerprinting predicts the cellular interaction of gold and silver nanoparticles. ACS Nano 8:2439–2455. https://doi.org/10.1021/nn406018q
526. Walter A, Billotey C, Garofalo A, Ulhaq-Bouillet, Corinne Lefèvre C, Taleb J, Laurent S, Elst L Vander, Muller RN, Felder-Flesch D, Begin-Colin S (2014) Mastering the shape and composition of dendronized iron oxide nanoparticles to tailor magnetic resonance imaging and hyperthermia. Chem Mater 26:5252–5264. https://doi.org/10.1021/cm5019025
527. Wang C, Daimon H, Sun S (2009) Dumbbell-like Pt-Fe_3O_4 nanoparticles and their enhanced catalysis for oxygen reduction reaction. Nano Lett 9:1493–1496. https://doi.org/10.1021/nl8034724
528. Wang C, Xu C, Zeng H, Sun S (2009) Recent progress in syntheses and applications of dumbbell-like nanoparticles. Adv Mater 21:3045–3052. https://doi.org/10.1002/adma.200900320
529. Wang C, Xu Y, Deng C, Liu Z, Wang R, Zhao H (2016) Design and preparation of a recyclable microfluidic SERS chip with integrated Au@Ag/TiO_2 NTs. RSC Adv 6:113115–113122. https://doi.org/10.1039/c6ra14947b

530. Wang F, Wang Y-C, Dou S, Xiong M-H, Sun T-M, Wang J (2011) Doxorubicin-tethered responsive gold nanoparticles facilitate intracellular drug delivery for overcoming multidrug resistance in cancer cells. ACS Nano 5:3679–3692. https://doi.org/10.1021/nn200007z
531. Wang G, Wang X, Liu J, Sun X (2012) Mesoporous Au/TiO_2 nanocomposite microspheres for visible-light photocatalysis. Chem Eur J 18:5361–5366. https://doi.org/10.1002/chem.201101410
532. Wang J, Wang Y, Liu Q, Yang L, Zhu R, Yu C, Wang S (2016) Rational design of multifunctional dendritic mesoporous silica nanoparticles to load curcumin and enhance efficacy for breast cancer therapy. ACS Appl Mater Interfaces 8:26511–26523. https://doi.org/10.1021/acsami.6b08400
533. Wang R-H, Bai J, Deng J, Fang C-J, Chen X (2017) TAT-modified gold nanoparticle carrier with enhanced anticancer activity and size effect on overcoming multidrug resistance. ACS Appl Mater Interfaces 9:5828–5837. https://doi.org/10.1021/acsami.6b15200
534. Wang T, Jiang X (2015) Breaking of the phosphodiester bond: a key factor that induces hemolysis. ACS Appl Mater Interfaces 7:129–136. https://doi.org/10.1021/am503865g
535. Wang T, LaMontagne D, Lynch J, Zhuang J, Cao YC (2013) Colloidal superparticles from nanoparticle assembly. Chem Soc Rev 42:2804–2823. https://doi.org/10.1039/c2cs35318k
536. Wang T, Wang X, LaMontagne D, Wang Z, Wang Z, Cao YC (2012) Shape-controlled synthesis of colloidal superparticles from nanocubes. J Am Chem Soc 134:18225–18228. https://doi.org/10.1021/ja308962w
537. Wang T, Zhu S, Jiang X (2015) Toxicity mechanism of graphene oxide and nitrogen-doped graphene quantum dots in RBCs revealed by surface-enhanced infrared absorption spectroscopy. Toxicol Res 4:885–894. https://doi.org/10.1039/c4tx00138a
538. Wang T, Zhuang J, Lynch J, Chen O, Wang Z, Wang X, LaMontagne D, Wu H, Wang Z, Cao YC (2012) Self-assembled colloidal superparticles from nanorods. Science 338:358–363. https://doi.org/10.1126/science.1224221
539. Wang X, Li X, Yoshiyuki K, Watanabe Y, Sogo Y, Ohno T, Tsuji NM, Ito A (2016) Comprehensive mechanism analysis of mesoporous-silica-nanoparticle-induced cancer immunotherapy. Adv Healthc Mater 5:1169–1176. https://doi.org/10.1002/adhm.201501013
540. Wang Y, Cui Y, Huang J, Di D, Dong Y, Zhang X, Zhao Q, Han N, Gao Y, Jiang T, Wang S (2015) Redox and pH dual-responsive mesoporous silica nanoparticles for site-specific drug delivery. Appl Surf Sci 356:1282–1288. https://doi.org/10.1016/j.apsusc.2015.07.151
541. Wang Y, He J, Liu C, Chong WH, Chen H (2014) Thermodynamics versus kinetics in nanosynthesis. Angew Chem Int Ed 53:2–32. https://doi.org/10.1002/anie.201402986
542. Wang Y, Huang H-Y, Yang L, Zhang Z, Ji H (2016) Cetuximab-modified mesoporous silica nano-medicine specifically targets EGFR-mutant lung cancer and overcomes drug resistance. Sci Rep 6:25468. https://doi.org/10.1038/srep25468
543. Warburg O (1925) The Metabolism of Carcinoma Cells. J Cancer Res 9:148–163. https://doi.org/10.1158/jcr.1925.148
544. Ward PS, Thompson CB (2012) Metabolic reprogramming: a cancer hallmark even warburg did not anticipate. Cancer Cell 21:297–308. https://doi.org/10.1016/j.ccr.2012.02.014
545. Wegner KD, Hildebrandt N (2015) Quantum dots: bright and versatile in vitro and in vivo fluorescence imaging biosensors. Chem Soc Rev 44:4792–4834. https://doi.org/10.1039/c4cs00532e
546. Wei H, Wang E (2008) Fe_3O_4 magnetic nanoparticles as peroxidase mimetics and their applications in H_2O_2 and glucose detection hui. Anal Chem 80:2250–2254. https://doi.org/10.1016/j.ultsonch.2009.06.014
547. Wei H, Wang E (2013) Nanomaterials with enzyme-like characteristics (nanozymes): next-generation artificial enzymes. Chem Soc Rev 42:6060–6093. https://doi.org/10.1039/c3cs35486e
548. Weinberg RA (2007) The biology of cancer. Garland Science, New York

549. Weiner RG, Skrabalak SE (2015) Metal dendrimers: synthesis of hierarchically stellated nanocrystals by sequential seed—directed overgrowth. Angew Chem Int Ed 54:1181–1184. https://doi.org/10.1002/anie.201409966
550. Weiss GJ, Chao J, Neidhart JD, Ramanathan RK, Bassett D, Neidhart JA, Choi CHJ, Chow W, Chung V, Forman SJ, Garmey E, Hwang J, Kalinoski DL, Koczywas M, Longmate J, Melton RJ, Morgan R, Oliver J, Peterkin JJ, Ryan JL, Schluep T, Synold TW, Twardowski P, Davis ME, Yen Y (2013) First-in-human phase 1/2a trial of CRLX101, a cyclodextrin-containing polymer-camptothecin nanopharmaceutical in patients with advanced solid tumor malignancies. Invest New Drugs 31:986–1000. https://doi.org/10.1007/s10637-012-9921-8
551. Weissleder R, Stark DD, Engelstad BL, Bacon B, White DL, Jacobs P, Lewis J (1989) Superparamagnetic iron oxide: pharmacokinetics and toxicity. Am J Roentgenol 152:167–173
552. West CM, Cooper RA, Loncaster JA, Wilks DP, Bromley M (2001) Tumor vascularity: a histological measure of angiogenesis and hypoxia. Cancer Res 61:2907–2910
553. Wilhelm S, Tavares AJ, Dai Q, Ohta S, Audet J, Dvorak HF, Chan WCW (2016) Analysis of nanoparticle delivery to tumours. Nat Rev Mater 1:16014/1–16014/12. https://doi.org/10.1038/natrevmats.2016.14
554. Willett CG, Boucher Y, Di Tomaso E, Duda DG, Munn LL, Tong RT, Chung DC, Sahani DV, Kalva SP, Kozin SV, Mino M, Cohen KS, Scadden DT, Hartford AC, Fischman AJ, Clark JW, Ryan DP, Zhu AX, Blaszkowsky LS, Chen HX, Shellito PC, Lauwers GY, Jain RK (2004) Direct evidence that the VEGF-specific antibody bevacizumab has antivascular effects in human rectal cancer. Nat Med 10:145–147. https://doi.org/10.1038/nm988
555. Wilms VS, Bauer H, Tonhauser C, Schilmann AM, Müller MC, Tremel W, Frey H (2013) Catechol-initiated polyethers: multifunctional hydrophilic ligands for pegylation and functionalization of metal oxide nanoparticles. Biomacromolecules 14:193–199. https://doi.org/10.1021/bm3015889
556. Winzen S, Schoettler S, Baier G, Rosenauer C, Mailaender V, Landfester K, Mohr K (2015) Complementary analysis of the hard and soft protein corona: sample preparation critically effects corona composition. Nanoscale 7:2992–3001. https://doi.org/10.1039/c4nr05982d
557. Wolchok JD, Kluger H, Callahan MK, Postow MA, Rizvi NA, Lesokhin AM, Segal NH, Ariyan CE, Gordon R-A, Reed K, Burke MM, Caldwell A, Kronenberg SA, Agunwamba BU, Zhang X, Lowy I, Inzunza HD, Feely W, Horak CE, Hong Q, Korman AJ, Wigginton JM, Gupta A, Sznol M (2013) Nivolumab plus ipilimumab in advanced melanoma. N Engl J Med 369:122–133. https://doi.org/10.1056/nejmoa1302369
558. Wolfram J, Yang Y, Shen J, Moten A, Chen C, Shen H, Ferrari M, Zhao Y (2014) The nano-plasma interface: implications of the protein corona. Colloids Surf B Biointerfaces 124:17–24. https://doi.org/10.1016/j.colsurfb.2014.02.035
559. Wondrak GT (2007) NQO1-activated phenothiazinium redox cyclers for the targeted bioreductive induction of cancer cell apoptosis. Free Radic Biol Med 43:178–190. https://doi.org/10.1016/j.freeradbiomed.2007.03.035
560. Wong C, Stylianopoulos T, Cui J, Martin J, Chauhan VP, Jiang W, Popovic Z, Jain RK, Bawendi MG, Fukumura D (2011) Multistage nanoparticle delivery system for deep penetration into tumor tissue. Proc Natl Acad Sci 108:2426–2431. https://doi.org/10.1073/pnas.1018382108
561. Wu M, Lin Z, Wolfbeis OS (2003) Determination of the activity of catalase using a europium(III)-tetracycline-derived fluorescent substrate. Anal Biochem 320:129–135. https://doi.org/10.1016/s0003-2697(03)00356-7
562. Wu S-H, Mou C-Y, Lin H-P (2013) Synthesis of mesoporous silica nanoparticles. Chem Soc Rev 42:3862–3875. https://doi.org/10.1039/c3cs35405a
563. Wulff G (1901) Zur Frage der Geschwindigkeit des Wachsthums und der Auflösung der Kristallflächen. Zeitschrift für Krist 34:449

564. Xia T, Kovochich M, Liong M, Meng H, Kabehie S, George S, Zink JI, Nel AE (2009) Polyethyleneimine coating enhances the cellular uptake of mesoporous silica nanoparticles and allows safe delivery of siRNA and DNA constructs. ACS Nano 3:3273–3286. https://doi.org/10.1021/nn900918w
565. Xia Y, Xiong Y, Lim B, Skrabalak SE (2009) Shape-controlled synthesis of metal nanocrystals: simple chemistry meets complex physics? Angew Chem Int Ed 48:60–103. https://doi.org/10.1002/anie.200802248
566. Xie M, Shi H, Ma K, Shen H, Li B, Shen S, Wang X, Jin Y (2013) Hybrid nanoparticles for drug delivery and bioimaging: mesoporous silica nanoparticles functionalized with carboxyl groups and a near-infrared fluorescent dye. J Colloid Interface Sci 395:306–314. https://doi.org/10.1016/j.jcis.2013.01.001
567. Xu C, Yu M, Noonan O, Zhang J, Song H, Zhang H, Lei C, Niu Y, Huang X, Yang Y, Yu C (2015) Core-cone structured monodispersed mesoporous silica nanoparticles with ultra-large cavity for protein delivery. Small 25:5949–5955. https://doi.org/10.1002/smll.201501449
568. Xu R, Zhang G, Mai J, Deng X, Segura-Ibarra V, Wu S, Shen J, Liu H, Hu Z, Chen L, Huang Y, Koay E, Huang Y, Liu J, Ensor JE, Blanco E, Liu X, Ferrari M, Shen H (2016) An injectable nanoparticle generator enhances delivery of cancer therapeutics. Nat Biotechnol 34:414–418. https://doi.org/10.1038/nbt.3506
569. Yamada H, Ujiie H, Urata C, Yamamoto E, Yamauchi Y, Kuroda K (2015) Multifunctional role of trialkylbenzene for the preparation of aqueous colloidal mesostructured/mesoporous silica nanoparticles with controlled pore size, particle diameter, and morphology. Nanoscale 7:19557–19567. https://doi.org/10.1039/c5nr04465k
570. Yamauchi H, Ishikawa T, Kondo S (1989) Surface characterization of ultramicro spherical particles of silica prepared by w/o microemulsion method. Colloids Surf 37:71–80. https://doi.org/10.1016/0166-6622(89)80107-6
571. Yanagi M, Asano Y, Kandori K, Kon-on K (1986) 39th Symposium on Divison of Colloid and Interface Chemistry. Chemical Society pf Japan Abstracts, p 396.
572. Yang C, Guo W, An N, Cui L, Zhang T, Tong R, Chen Y, Lin H, Qu F (2015) Enzyme-sensitive magnetic core–shell nanocomposites for triggered drug release. RSC Adv 5:80728–80738. https://doi.org/10.1039/c5ra15026d
573. Yang H-J, He S-Y, Tuan H-Y (2014) Self-seeded growth of five-fold twinned copper nanowires: mechanistic study, characterization, and SERS applications. Langmuir 30:602–610. https://doi.org/10.1021/la4036198
574. Yang J, Chen W, Shen D, Wei Y, Ran X, Teng W, Fan J, Zhang W, Zhao D (2014) Controllable fabrication of dendritic mesoporous silica–carbon nanospheres for anthracene removal. J Mater Chem A 2:11045–11048. https://doi.org/10.1039/c4ta01516a
575. Yang J, Chng LL, Yang X, Chen X, Ying JY (2014) Multiply-twinned intermetallic AuCu pentagonal nanorods. Chem Commun 50:1141–1143. https://doi.org/10.1039/c3cc47254j
576. Yang K, Luo H, Zeng M, Jiang Y, Li J, Fu X (2015) Intracellular pH-triggered, targeted drug delivery to cancer cells by multifunctional envelope-type mesoporous silica nanocontainers. ACS Appl Mater Interfaces 7:17399–17407. https://doi.org/10.1021/acsami.5b04684
577. Yang M, Li J, Chen PR (2014) Transition metal-mediated bioorthogonal protein chemistry in living cells. Chem Soc Rev 43:6511–6526. https://doi.org/10.1039/c4cs00117f
578. Yang Y, Liu X, Lv Y, Herng TS, Xu X, Xia W, Zhang T, Fang J, Xiao W, Ding J (2015) Orientation mediated enhancement on magnetic hyperthermia of Fe_3O_4 nanodisc. Adv Funct Mater 25:812–820. https://doi.org/10.1002/adfm.201402764
579. Yang Y, Niu Y, Zhang J, Meka AK, Zhang H, Xu C, Lin CXC, Yu M, Yu C (2015) Biphasic synthesis of large-pore and well-dispersed benzene bridged mesoporous organosilica nanoparticles for intracellular protein delivery. Small 11:2743–2749. https://doi.org/10.1002/smll.201402779
580. Yao X, Chen X, He C, Chen L, Chen X (2015) Dual pH-responsive mesoporous silica nanoparticles for efficient combination of chemotherapy and photodynamic therapy. J Mater Chem B 3:4707–4714. https://doi.org/10.1039/c5tb00256g

581. Ye E, Regulacio MD, Zhang S-Y, Loh XJ, Han M-Y (2015) Anisotropically branched metal nanostructures. Chem Soc Rev 44:6001–6017. https://doi.org/10.1039/c5cs00213c
582. Yin Y, Alivisatos AP (2005) Colloidal nanocrystal synthesis and the organic-inorganic interface. Nature 437:664–670. https://doi.org/10.1038/nature04165
583. Yip KW, Reed JC (2008) Bcl-2 family proteins and cancer. Oncogene 27:6398–6406. https://doi.org/10.1038/onc.2008.307
584. Yoo D, Lee JH, Shin TH, Cheon J (2011) Theranostic magnetic nanoparticles. Acc Chem Res 44:863–874. https://doi.org/10.1021/ar200085c
585. Yoreo DJJ, Gilbert PUPA, Sommerdijk NAJM, Penn RL, Whitelam S, Joester D, Zhang H, Rimer JD, Navrotsky A, Banfield JF, Wallace AF, Michel FM, Meldrum FC, Colfen H, Dove PM (2015) Crystallization by particle attachment in synthetic, biogenic, and geologic environments. Science 349:6247/1–6247/9. https://doi.org/10.1126/science.aaa6760
586. Yoshii Y, Sugiyama K (1988) Intercapillary distance in the proliferating area of human glioma. Cancer Res 48:2938–2941
587. Yu H, Chen M, Rice PM, Wang SX, White RL, Sun S (2005) Dumbbell-like bifunctional Au–Fe_3O_4 nanoparticles. Nano Lett 5:379–382. https://doi.org/10.1021/nl035004r
588. Yu SB, Watson AD (1999) Metal-based X-ray contrast media. Chem Rev 99:2353–2377. https://doi.org/10.1021/cr980441p
589. Yu WW, Chang E, Falkner JC, Zhang J, Al-Somali a M, Sayes CM, Johns J, Drezek R, Colvin VL (2007) Forming biocompatible & nonaggregated nanocrystal FeNP in water using amphiphilic polymer. J Am Chem Soc 129:2871–2879. https://doi.org/10.1021/ja067184n
590. Yu Y, Sun K, Tian Y, Li X-Z, Kramer MJ, Sellmyer DJ, Shield JE, Sun S (2013) One-pot synthesis of urchin-like FePd-Fe_3O_4 and their conversion into exchange-coupled L1(0)-FePd-Fe nanocomposite magnets. Nano Lett 13:4975–4979. https://doi.org/10.1021/nl403043d
591. Yuan H, Zhang S (2010) Effects of particle size and ligand density on the kinetics of receptor-mediated endocytosis of nanoparticles. Appl Phys Lett 96:33704/1–33704/3. https://doi.org/10.1063/1.3293303
592. Yue Y, Jin F, Deng R, Cai J, Dai Z, Lin MCM, Kung H-F, Mattebjerg MA, Andresen TL, Wu C (2011) Revisit complexation between DNA and polyethylenimine–effect of length of free polycationic chains on gene transfection. J Control Release 152:143–151. https://doi.org/10.1016/j.jconrel.2011.03.020
593. Yun HJ, Lee H, Kim ND, Lee DM, Yu S, Yi J (2011) A combination of two visible-light responsive photocatalysts for achieving the Z-scheme in the solid state. ACS Nano 5:4084–4090. https://doi.org/10.1021/nn2006738
594. Zeng H, Li J, Liu JP, Wang ZL, Sun S (2002) Exchange-coupled nanocomposite magnets by nanoparticle self-assembly. Nature 420:395–398. https://doi.org/10.1038/nature01208
595. Zeng H, Sun S (2008) Syntheses, properties, and potential applications of multicomponent magnetic nanoparticles. Adv Funct Mater 18:391–400. https://doi.org/10.1002/adfm.200701211
596. Zhang F, Lees E, Amin F, Rivera-Gil P, Yang F, Mulvaney P, Parak WJ (2011) Polymer-coated nanoparticles: a universal tool for biolabelling experiments. Small 7:3113–3127. https://doi.org/10.1002/smll.201100608
597. Zhang J, Desai D, Rosenholm JM (2014) Tethered lipid bilayer gates: toward extended retention of hydrophilic cargo in porous nanocarriers. Adv Funct Mater 24:2352–2360. https://doi.org/10.1002/adfm.201302995
598. Zhang J, Langille MR, Personick ML, Zhang K, Li S, Mirkin CA (2010) Concave cubic gold nanocrystals with high-index facets. J Am Chem Soc 132:14012–14014
599. Zhang J, Rosenholm JM (2015) The viability of mesoporous silica nanoparticles for drug delivery. Ther Deliv 6:891–893. https://doi.org/10.4155/tde.15.46

600. Zhang K, Loong SLE, Connor S, Yu SWK, Tan S-Y, Ng RTH, Lee KM, Canham L, Chow PKH (2005) Complete tumor response following intratumoral 32P BioSilicon on human hepatocellular and pancreatic carcinoma xenografts in nude mice. Clin Cancer Res 11:7532–7537. https://doi.org/10.1158/1078-0432.ccr-05-0400
601. Zhang L, Chen Y, Li Z, Li L, Saint-Cricq P, Li C, Lin J, Wang C, Su Z, Zink JI (2016) Tailored synthesis of octopus-type janus nanoparticles for synergistic actively-targeted and chemo-photothermal therapy. Angew Chem Int Ed 55:2118–2121. https://doi.org/10.1002/anie.201510409
602. Zhang L, Dou Y-H, Gu H-C (2006) Synthesis of Ag-Fe_3O_4 heterodimeric nanoparticles. J Colloid Interface Sci 297:660–664. https://doi.org/10.1016/j.jcis.2005.11.009
603. Zhang LW, Monteiro-Riviere NA (2009) Mechanisms of quantum dot nanoparticle cellular uptake. Toxicol Sci 110:138–155. https://doi.org/10.1093/toxsci/kfp087
604. Zhang Q, Neoh KG, Xu L, Lu S, Kang ET, Mahendran R, Chiong E (2014) Functionalized mesoporous silica nanoparticles with mucoadhesive and sustained drug release properties for potential bladder cancer therapy. Langmuir 30:6151–6161. https://doi.org/10.1021/la500746e
605. Zhang S, Gao H, Bao G (2015) Physical principles of nanoparticle cellular endocytosis. ACS Nano 9:8655–8671. https://doi.org/10.1021/acsnano.5b03184
606. Zhang X, Chen YL, Liu R-S, Tsai DP (2013) Plasmonic photocatalysis. Reports Prog Phys 76:046401/1–046401/41. https://doi.org/10.1088/0034-4885/76/4/046401
607. Zhang X, Tian W, Cai X, Wang X, Dang W, Tang H, Cao H, Wang L, Chen T (2013) Hydrazinocurcumin Encapsuled nanoparticles "re-educate" tumor-associated macrophages and exhibit anti-tumor effects on breast cancer following STAT3 suppression. PLoS One 8: e65896/1–e65896/9. https://doi.org/10.1371/journal.pone.0065896
608. Zhang Y, Hou Z, Ge Y, Deng K, Liu B, Li X, Li Q, Cheng Z, Ma P, Li C, Lin J (2015) DNA-hybrid-gated photothermal mesoporous silica nanoparticles for NIR-responsive and aptamer-targeted drug delivery. ACS Appl Mater Interfaces 7:20696–20706. https://doi.org/10.1021/acsami.5b05522
609. Zhang Y, Liu JM, Yan XP (2013) Self-assembly of folate onto polyethyleneimine-coated CdS/ZnS quantum dots for targeted turn-on fluorescence imaging of folate receptor overexpressed cancer cells. Anal Chem 85:228–234. https://doi.org/10.1021/ac3025653
610. Zhao Y, Sun X, Zhang G, Trewyn BG, Slowing II, Lin VS-Y (2011) Interaction of mesoporous silica nanoparticles with human red blood cell membranes: size and surface effects. ACS Nano 5:1366–1375. https://doi.org/10.1021/nn103077k
611. Zhao Z, Zhou Z, Bao J, Wang Z, Hu J, Chi X, Ni K, Wang R, Chen X, Chen Z, Gao J (2013) Octapod iron oxide nanoparticles as high-performance T2 contrast agents for magnetic resonance imaging. Nat Commun 4:2266/1–2266/7. https://doi.org/10.1038/ncomms3266
612. Zheng C, Zheng A-X, Liu B, Zhang X-L, He Y, Li J, Yang H-H, Chen G (2014) One-pot synthesized DNA-templated Ag/Pt bimetallic nanoclusters as peroxidase mimics for colorimetric detection of thrombin. Chem Commun 50:13103–13106. https://doi.org/10.1039/c4cc05339g
613. Zhou J (2010) Multi-drug resistance in cancer. Humana Press, Totowa
614. Zhu C-L, Song X-Y, Zhou W-H, Yang H-H, Wen Y-H, Wang X-R (2009) An efficient cell-targeting and intracellular controlled-release drug delivery system based on MSN-PEM-aptamer conjugates. J Mater Chem 19:7765–7770. https://doi.org/10.1039/b907978e
615. Zhu M, Nie G, Meng H, Xia T, Nel A, Zhao Y (2013) Physicochemical properties determine nanomaterial cellular uptake, transport, and fate. Acc Chem Res 46:622–631. https://doi.org/10.1021/ar300031y
616. Zhu S, Niu M, O'Mary H, Cui Z (2013) Targeting of tumor-associated macrophages made possible by PEG-sheddable, mannose-modified nanoparticles. Mol Pharm 10:3525–3530. https://doi.org/10.1021/mp400216r

617. Zhuang J, Shaller AD, Lynch J, Wu H, Chen O, Li ADQ, Cao YC (2009) Cylindrical superparticles from semiconductor nanorods. J Am Chem Soc 131:6084–6085. https://doi.org/10.1021/ja9015183
618. Zhuang J, Wu H, Yang Y, Cao YC (2007) Supercrystalline colloidal particles from artificial atoms. J Am Chem Soc 129:14166–14167. https://doi.org/10.1021/ja076494i
619. Zhuang J, Wu H, Yang Y, Cao YC (2008) Controlling colloidal superparticle growth through solvophobic interactions. Angew Chem Int Ed 47:2208–2212. https://doi.org/10.1002/anie.200705049
620. Zimmer JP, Kim S, Ohnishi S, Tanaka E, Frangioni JV, Bawendi MG (2006) Size series of small indium arsenide–zinc selenide core-shell nanocrystals and their application to in vivo imaging. J Am Chem Soc 128:2526–2527. https://doi.org/10.1021/ja0579816
621. Zimmermann TS, Lee ACH, Akinc A, Bramlage B, Bumcrot D, Fedoruk MN, Harborth J, Heyes JA, Jeffs LB, John M, Judge AD, Lam K, McClintock K, Nechev LV, Palmer LR, Racie T, Röhl I, Seiffert S, Shanmugam S, Sood V, Soutschek J, Toudjarska I, Wheat AJ, Yaworski E, Zedalis W, Koteliansky V, Manoharan M, Vornlocher H-P, MacLachlan I (2006) RNAi-mediated gene silencing in non-human primates. Nature 441:111–114. https://doi.org/10.1038/nature04688
622. Zolnik BS, González-Fernández A, Sadrieh N, Dobrovolskaia MA (2010) Nanoparticles and the immune system. Endocrinology 151:458–465. https://doi.org/10.1210/en.2009-1082
623. Zou W (2006) Regulatory T cells, tumour immunity and immunotherapy. Nat Rev Immunol 6:295–307. https://doi.org/10.1038/nri1806
624. Zou Z, He X, He D, Wang K, Qing Z, Yang X, Wen L, Xiong J, Li L, Cai L (2015) Programmed packaging of mesoporous silica nanocarriers for matrix metalloprotease 2-triggered tumor targeting and release. Biomaterials 58:35–45. https://doi.org/10.1016/j.biomaterials.2015.04.034
625. Zürcher S, Wäckerlin D, Bethuel Y, Malisova B, Textor M, Tosatti S, Gademann K (2006) Biomimetic surface modifications based on the cyanobacterial iron chelator anachelin. J Am Chem Soc 128:1064–1065. https://doi.org/10.1021/ja056256s

Martin Kluenker studied chemistry at Johannes Gutenberg-University in Mainz, Germany and the University of Toledo in Toledo/Ohio, USA. He received his diploma in chemistry from Johannes Gutenberg-University Mainz in 2014 with Prof. W. Tremel. He finished his Ph.D. in 2018 under the supervision of Prof. W. Tremel at the Johannes Gutenberg-University Mainz in collaboration with the Max Planck Graduate Center Mainz. His research interests focus on the synthesis of inorganic–organic hybrid nanomaterials and their theranostic and catalytic applications.

Sven Kurch received his diploma degree in chemistry (2014) from Johannes Gutenberg University in Mainz, Germany. His diploma thesis focused on the synthesis of mesoporous silica nanoparticles for drug delivery. Currently he is a doctoral student in inorganic chemistry in the group of Prof. Tremel. In collaboration with the University Medical Center of Mainz, Mainz Research School of Translational Biomedicine (TransMed) and the CRC 1066 (Nanodimensional polymer therapeutics for tumor therapy) he completed his Ph.D. thesis at the interface between nanodimensional therapeutics and immunology in June 2018.

Muhammad Nawaz Tahir received his M.Phil. degree from Quaid-i-Azam University Islamabad Pakistan in 2001 and then joined Prof. Wolfgang Tremel's group at the Johannes Gutenberg University of Mainz, Germany. He was awarded a Ph.D. in 2006 from Johannes Gutenberg University of Mainz with distinction "*Summa Cum Laude*". Later he was part of research projects funded by DFG and the Max Planck Society. Since 2010 he works as senior scientist and group leader for nanomaterials synthesis at Johannes Gutenberg University of Mainz. His research involves the synthesis and characterization of innovative nanomaterials and their surface functionalization for biomedical and energy storage applications.

Wolfgang Tremel is an inorganic materials chemist. He earned a Ph.D. in chemistry in 1984 from the University of Münster with B. Krebs. Later he joined the group of R. Hoffman at Cornell University (1984–1986) and spent one term (1989) at Ames Laboratory with H. F. Franzen after having returned to Münster in 1987. In 1991 he was appointed associate professor and in 1996 full professor at the Johannes Gutenberg University in Mainz. His research focuses on the synthesis and structural and physical characterization as well as the design and potential applications of new materials, in particular the chemistry of organized matter.

Chapter 4
Moving Liposome Technology from the Bench to the Oncological Patient: Towards Performance-by-Design

Ana Filipa Cruz, Nuno A. Fonseca, Ana C. Gregório, Vera Moura, Sérgio Simões and João Nuno Moreira

Abstract A liposome is defined as a spherical vesicle having multiple or a single lipid bilayer surrounding an aqueous compartment enabling the entrapment of hydrophobic or water-soluble substances respectively, and thus suitable as a drug delivery system. Different methods for preparation of liposomes at small scale were developed throughout time, establishing the principles for the introduction of modern large scale industrial production processes. Associated with the adaptation and/or introduction of powerful characterization techniques, such as dynamic light scattering for particle size distribution analysis, novel therapeutic drug delivery systems were developed aiming at the oncology setting, many reaching the clinics, such as Doxil/Caelyx (liposomal PEGylated doxorubicin). Nevertheless, the approval of novel liposome-based medicines has not fulfilled the expectations, owing to the biological barriers they face, before reaching the cancer cell in the tumor microenvironment. Accordingly, the increased understanding on cancer biology, subsequently shaping the design of the delivery system, has become paramount for the successful development of novel generations of liposomes. Thus, herein we will address the technological accomplishments associated with the development of nanoliposomal formulations, focusing on their preparation and characterization as part of the design and quality control. We will further address the impact of liposome properties on their interaction with living organisms, upon systemic administration, and how this knowledge has benefited the oncological patient.

A. F. Cruz · N. A. Fonseca · A. C. Gregório · V. Moura · S. Simões · J. N. Moreira (✉)
CNC—Center for Neuroscience and Cell Biology, University of Coimbra, 3004-504 Coimbra, Portugal
e-mail: jmoreira@ff.uc.pt

A. F. Cruz · S. Simões · J. N. Moreira
FFUC—Faculty of Pharmacy, Pólo das Ciências da Saúde, University of Coimbra, 3000-548 Coimbra, Portugal

N. A. Fonseca · A. C. Gregório · V. Moura
TREAT U, SA—Parque Industrial de Taveiro, Lote 44, 3045-508 Coimbra, Portugal

H. G. Merkus et al. (eds.), *Particles and Nanoparticles in Pharmaceutical Products*, AAPS Advances in the Pharmaceutical Sciences Series 29,
https://doi.org/10.1007/978-3-319-94174-5_4

4.1 Introduction

A liposome is defined as a spherical vesicle having multiple or a single lipid bilayer surrounding an aqueous compartment, which enable the entrapment of hydrophobic or water-soluble substances, respectively. Thereby, liposomes are being developed as drug delivery systems since 1965 [9, 68–70, 72]. They may be designed to offer protection against enzyme degradation (for example siRNAs), and further change the pharmacokinetics and distribution profiles of encapsulated drugs. Furthermore, liposomes are highly flexible with respect to the nature of the payload [3]. Often building on years of development, several examples of liposomal formulations (mainly aiming at intravenous administration) have made their way into the clinics, each one bringing a different healthcare innovation breakthrough.

Accordingly, herein, we will address the technological accomplishments associated with the development of nanoliposomal formulations, approaching classical methodologies of preparation and their evolution into industrial up-scaling and manufacture. Furthermore, characterization techniques will be discussed focusing on their integration into liposome design and quality control. We will further address the impact of liposome properties on their interaction with living organisms, upon systemic administration, with emphasis on the oncology setting, and how this knowledge has been translated into the clinics.

4.2 Methods for Liposome Preparation

Several methods of liposome preparation have been developed and often adapted around the physicochemical features of a particular active agent to be encapsulated. Some methods are fit only for small scale preparation but have established the grounds from which modern industrial methods have been developed (Fig. 4.1) [209].

The developmental path of liposomes from the bench to the clinics is highly challenging. To achieve the intended target product profile, with the expected efficacy, particularly in the oncology setting, several hurdles either coming from the manufacture or from the biological point-of-view, should be overcome. Thus, successful translation of liposome technology to the clinical setting, is a dynamic process that bridges the knowledge acquired during the development with that of industrial up-scaling, with the critical assistance of characterization technologies.

4.2.1 Small-Scale Manufacture Methods

Properties of lipid formulations can vary depending on the composition. However, the same preparation method can be virtually used for all liposomal compositions, but may vary according to the nature of the active agent to be encapsulated [65, 85].

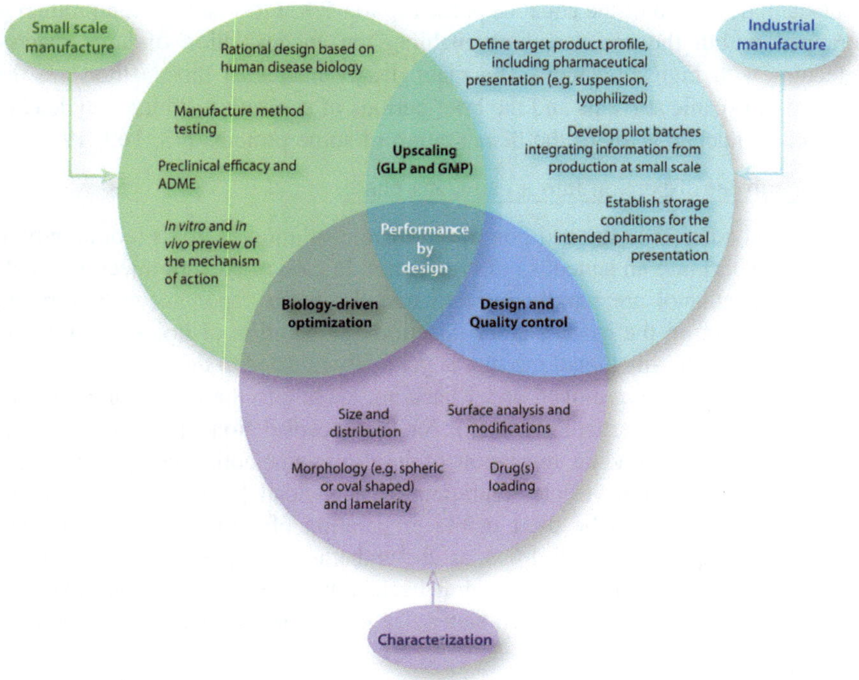

Fig. 4.1 The dynamic nature of performance-by-design associated with the development of liposomes

In general terms, the procedure involves preparation of the lipid vesicles using one of the methods described below, followed by homogenization of the dispersion.

Lipid-film hydration

Lipid-film hydration technique was one of the first ever developed [140]. In this process, the organic solvent (e.g. chloroform or ether), in which the lipids are dissolved, is evaporated until the formation of a thin dry film of lipids. The resulting film is then hydrated by adding an aqueous phase and dispersed by vigorous vortexing until completely suspended, spontaneously giving rise to multilamellar vesicles (MLVs) [9, 10, 26].

Reverse phase evaporation

Large unilamellar and oligolamellar vesicles are formed when an aqueous buffer is added to a solution of phospholipids in organic solvent, such as isopropyl ether, diethyl ether or a mixture of isopropyl ether and chloroform, forming a water-in-oil emulsion. The subsequent removal of the organic solvent by evaporation under reduced pressure, leads to the formation of a viscous gel. The liposomes are shaped when residual solvent is removed during continued rotary evaporation and the excess of phospholipids in the environment intercalate into the lipid bilayer.

A substantial fraction of the aqueous phase (up to 65% at low salt concentrations) is entrapped within the vesicles, thus enabling the encapsulation of large macromolecules with high efficiency. The main disadvantages of this method are the presence of organic solvents and the brief periods of sonication, which may lead to breakage of nucleic acids or the denaturation of some proteins [73, 169, 195].

Organic solvent dispersion into an aqueous phase

This method consists in the addition of an organic solution of lipids (commonly in ethanol or ether) into an aqueous buffer solution. In case of ethanol injection, lipids dissolved in ethanol are rapidly injected into the buffer, spontaneously forming MLVs [18]. Yet, in the case of ether injection, the solution of lipids dissolved in diethyl ether or ether-methanol mixture is gradually injected to the aqueous solution and the ether has to be removed under vacuum, leading to the formation of MLVs [37]. This method is particularly useful for the encapsulation of oligonucleotides since it enables the gradual formation of lipid-oligonucleotide complexes, often leading to multi-layered vesicles entrapping the payload [123]. Nonetheless, the current method can also be used in the preparation of liposomes encapsulating hydrophilic or lipophilic small drugs, as chemotherapeutics [58, 90]. The presence of the organic solvent represents a drawback, but it can be easily removed by dialysis or through buffer exchange using size exclusion chromatography [97].

Detergent dialysis

In this method, lipids are solubilized in ionic detergents, such as sodium cholate or cholic acid derivatives, or nonionic detergents (that may present different chemical structures), such as octyl glucoside, C12Es or Triton X-100. The latter are less harmful when in contact with proteins but often their use leads to larger vesicles. The subsequent removal of the detergent by dialysis from mixed micellar suspensions leads to the spontaneous formation of the vesicles, which are suspended in an aqueous buffer [5, 50, 133]. The detergent dialysis technique can be adjusted to encapsulate plasmids. In this case, the detergent solubilizes the preformed plasmid DNA-cationic lipid complex particles. The use of a PEG-modified lipid is important to prevent aggregation during dialysis and to stabilize the vesicles in aqueous suspension [54, 216].

4.2.2 Liposome Size Reduction and Homogenization Methods

After the formation of vesicles, it is extremely important to ensure that there is a homogeneous (near monodisperse) size distribution (with a small mean size, e.g. lower than 100 nm) and lamellarity of vesicles. Several methods have been used for this purpose, including French press, sonication and membrane extrusion. French press technique involves extrusion of vesicles at high pressures through a small orifice (French pressure cell) but the vesicles produced are larger than those obtained by other methods, and the technique has been disused [13, 75].

Sonication

Sonication is based on the exposure of the aqueous dispersions of vesicles to ultrasound waves rendering smaller unilamellar vesicles [82]. However, the vesicles formed are poorly reproducible and the lipid molecules can be damaged during the process. Furthermore, sonication results are heavily dependent on the careful control of conditions, like sonication time and/or power [176]. For example, it was demonstrated that short-time irradiation with strong ultrasound waves was more efficient than long-time irradiation with weak ultrasound, in terms of reduction of the average liposome diameter and the polydispersity index [165, 221]. Moreover, it was demonstrated that sonication depth (measured from the base of the vessel) also influences the rate and structure of the resulting vesicles [187].

Membrane extrusion

Alternatively, membrane extrusion is the most frequently used technique to reduce the mean size of liposomes, enabling the conversion of MLVs to small and large unilamellar vesicles. Vesicles are repetitively extruded through two stacked track-etched polycarbonate membrane filters (mounted in Avestin's LiposoFast extruder, for example), with defined pore sizes, resulting in a relatively homogeneous size distribution of vesicles [81, 114, 124]. The mean size of the extruded vesicles is a function of membrane pore size, lipid concentration, the number of extrusion cycles and pressure. Thus, larger membrane pores result in larger vesicles with higher trapping efficiencies. In turn, as the extrusion pressure is increased, the flow rate of vesicles through the pores increases and the mean size of the extruded vesicles decreases towards a given pore size [84, 163]. This relation is verified above a certain threshold, for extrusion pressure, that has been correlated with the lysis tension required to rupture the phospholipid membrane. Furthermore, the minimum pressure required for extrusion increases for higher lipid concentrations [113, 172].

4.2.3 Methods for Drug Loading

4.2.3.1 Nucleic Acids

Liposomes have been used to carry essentially two types of substances—small drugs and nucleic acids—and, depending on the loading and the purpose, liposomes may be prepared with different composition (cationic, anionic and neutral lipids) at different lipid ratios. Cationic lipids, such as N-[2,3-(dioleyloxy)propyl]-N,N, N-trimethylammonium chloride (DOTMA), are used in the preparation of cationic liposomes, adequate for the encapsulation of nucleic acids [136]. The preparation of these lipoplexes relies on an electrostatic interaction between the cationic lipid and the polyanionic nucleic acid, spontaneously forming the complexes [122]. Although the cationic lipid is a critical component for nucleic acid complexation and cellular internalization, it is often toxic to the target cells. Modulating charge, upon incorporation of polyethylene glycol (PEG) in the formulation, or reducing the cationic lipid content, may decrease the encapsulation and transfection efficiencies [183]. Therefore, ionizable cationic lipids, such as 1,2-diolyoxy-3-trimethylammonio propane (DOTAP), with pKa values below 7 have been developed [119]. The pKa of the ionizable cationic lipid determines the surface charge of the liposome under different pH conditions. Thus, at low pH values, the liposomes are positively charged, allowing the nucleic acids to be loaded into the vesicles but, at physiological pH values, the liposomes exhibit a low surface charge compatible with longer circulation times in the blood [16, 184]. Furthermore, encapsulation of nucleic acids is also dependent on the method of liposome preparation. To ensure that the nucleic acid is highly protected against degradation, the method of choice for production of nucleic acid-encapsulating liposomes is often based on ethanol injection [183, 211]. However, for small drugs different loading methods are available and may be adapted according to the physico-chemical properties of the drug.

4.2.3.2 Small Drugs

The loading method of small drugs depends largely on the physical characteristics of the drug. Hydrophobic drugs, such as paclitaxel, are often incorporated into the lipid bilayer during the formation of liposomes, but such entrapment method may lead to low encapsulation efficiencies or rapid leakage following systemic administration [15, 26]. On the other hand, water soluble drugs are mainly sequestered in aqueous core of the liposomes, through a method known as passive loading, characterized by low encapsulation efficiencies [125]. Between these two extremes are amphipathic drugs (weak bases such as doxorubicin). In this case, the drug is actively loaded into the liposome due to a transmembrane pH gradient, generated in response to internal acidic buffers or proton-generating dissociable salts, such as ammonium sulfate. Such method enables drug loading into preformed vesicles (remote loading) [24, 42, 126, 127]. This active loading strategy allows independent variation of any liposomal parameter and the achievement of much higher drug to lipid ratios than passive loading. Moreover, it is extremely versatile because it is not restricted to a specific drug class, as long as it has a primary, secondary or tertiary amine group [116]. Additionally, pH gradient-based loading techniques can also be used to solve the encapsulation and retention problems of hydrophobic (and uncharged) drugs such as docetaxel. In this case, the drug can be converted into a prodrug, a weak base derivate, which can be actively loaded into liposomes [229].

Overall, all the methods discussed above have enabled major breakthroughs in the liposome field and constituted the basis for many medicines currently used in the clinics or in clinical development. However, they were mainly engineered for laboratory scale, enabling only the production of small batches with the required features. Thus, more efficient methods enabling large batch manufacture were introduced, in response to market and regulatory demands.

4.2.4 Up-Scaling and Industrial Manufacture

When designing up-scaling of the liposomes manufacture, many considerations as discussed above should be taken. Despite useful at laboratory scale, many of the methods described above (e.g. lipid film hydration or detergent dialysis) are not suitable for industrial manufacture for several reasons.

Many of the organic solvents are unsuitable, even in residual amounts, as they may be hazardous for humans [83]. Often, these classical methods use gel resins for molecular exclusion chromatography for buffer exchange or purification, which are expensive and limit the preparation of large batches [83]. Furthermore, sterilization may be a challenge (as lipids are sensitive to temperature), and those methods are not readily adaptable to perform within a clean environment [83].

Addressing these issues, manufacturing processes involving new techniques were developed. For example, Peschka and colleagues developed the cross-flow (tangential) filtration detergent depletion method [167]. In this process, a lipid-detergent micellar solution is continuously filtered under pressure through a suitable membrane (with a defined exclusion limit), while detergent molecules are being removed, leading to the formation of unilamellar liposomes, virtually with unlimited up-scaling capacity [167].

Also based on tangential flow hollow fiber technology, and influenced by the ethanol injection technique, Charcosset and colleagues introduced the membrane contactor method [30]. This process enables the preparation of lipid nanoparticles by forcing an ethanolic lipid-mixture through a fixed-pore size membrane (using pressurized N_2 gas). The resulting small droplets then contact with an aqueous phase running within the hollow fiber module [30]. It can render liposome sizes up to 215 nm, according to process parameters (like temperature and flow rate), enabling a high encapsulation efficiency, at least for lipophilic drugs [30, 91, 101].

Another set of techniques based on dense gas processes to produce liposomes has been introduced as an attempt to solve some of the issues in classical methods, such as the use of organic solvents [130]. In this case, supercritical fluids are used as they present similar solubility to many other conventional solvents [130]. A good example is supercritical CO_2 which presents low critical temperature and pressure (31 °C and 74 bar) with solubility properties similar to organic solvents [83]. In this process, supercritical CO_2 is used to produce fine lipid particles within high pressure vessels, upon extraction of lipids from organic mixture and precipitation following to supersaturation of the supercritical phase. The liposomes are finally formed by adding the aqueous phase, presenting similar properties to those produced by other methods [105]. Despite the elevated pressures required and cost, it represents a single step process that may enable simultaneous encapsulation of lipophilic drugs and sterilization of the formulation (if the lipids used are stable to higher temperatures) [199].

Regarding encapsulation of hydrophilic or amphipathic drugs, more complex processes involving several steps (and employing some techniques above), may be necessary as they often involve buffer changes and active loading of the drug to encapsulate [11]. In these multi-step processes, industrial extrusion (for liposome size homogenization) and tangential flow filtration (for buffer exchange and non-encapsulated drug removal) are paramount to ensure the necessary scale of manufacture [167].

Independently of the method of preparation, the liposomal formulations (as medicines) are deemed to be sterile prior to human use. For example, filtration is commonly used owing to its low operating temperature, but requires aseptic conditions (increasing costs) and is limited to liposomes smaller than 200 nm. Other techniques, such as gamma irradiation or dry heat, despite highly effective in sterilization, often lead to degradation of liposomes [199]. As such, different sterilization processes should be used as a *per* case basis with final demonstration of sterility (with minimal degradation), as defined by regulatory agencies [199].

Overall, several processes exist to prepare liposomes that may be customized according to the needs and aims of a particular formulation. Still, as any other medicine, regulatory authorities (as FDA or EMA) will require reliability demonstration of all steps involved, including quality control characterization, as defined in guidelines issued in 2001 ("Liposome Drug Products: chemistry, manufacturing, and controls, human pharmacokinetics and bioavailability and labeling documentation") [209], and further updated in 2015.

4.3 Characterization of Liposomes: Towards Design and Quality Control

Every medicinal product requires a detailed characterization of its physico-chemical properties, including macroscopic (e.g. disintegration times of a compressed formulation in fluids) and microscopic (e.g. crystalline form of the active pharmaceutical ingredient). From a R&D perspective towards pharmaceutical development, this characterization provides relevant information that may be incorporated in the production process to optimize a given medicine formulation (Fig. 4.1). Such principle also applies to the development of innovative nanotechnology-based pharmaceutical formulations, including liposomes for advanced drug delivery.

The scale and complex nature of lipid-based nanoformulations represent an enormous challenge in terms of characterization of properties such morphology, particle mean size and polydispersity, associated electric charge, surface area and surface chemical modifications, as well as chemical composition, including drug loading [102]. Still, the acknowledgement that these properties affect many aspects of the interaction between the nanoformulation and the human organism, including its stability and therapeutic outcome, reflects the importance of their study [7, 164].

4.3.1 Assessing Liposomal Physical Properties: Morphology, Size, Homogeneity and Charge

Charge neutrality, mean size below 100 nm and surface modification (as PEG engraftment) are essential attributes that underlie the extended blood circulation features of recent liposome generations. Their precise characterization requires a vast number of principles and techniques that have been developed or translated from other scientific areas.

4.3.1.1 Microscopy-Based Techniques

Optical microscopy remains as a technique of choice to observe the "large" microscopic structures as mammalian cells. However, optical microscopes, which rely on 400–700 nm wavelengths (visible light spectrum), are not adequate to observe, with detail, objects at nanoscale (<100 nm), a limitation that arises from the low resolution (the distance between two distinguishable points) provided by the optical system, a parameter that is directly proportional to the wavelength of electromagnetic waves [102, 171]. Thus, to obtain detailed data on morphology and size of lipid-based nanoparticles, powerful tools providing higher resolutions are essential.

<u>Transmission electron microscopy</u>

In this respect, electron microscopy (EM) techniques, such as transmission electron microscopy (TEM), are particularly suitable for liposome characterization. For instance, TEM subnanometer resolution arises from the fact that high-energy electron beam (for example, 100 keV) present a theoretical wavelength of 0.0037 nm [43, 102]. Nevertheless, TEM requires a thin layered sample that enables non-interacting electrons to pass through [43]. Still, sample thickness may not be a constraint in the observation of nanoparticles, as they are often in liquid suspensions enabling the formation of a thin film. Furthermore, as analysis is performed in vacuum conditions, water cannot be present in the samples at its "natural" condition when loaded in the sample chamber. Accordingly, different methods of sample preparation were developed to overcome this problem [43]. Among others, cryogenic methods are common for liposomal EM imaging [102]. Cryo-TEM relies on imaging under freezing conditions in which frozen specimens are maintained in either liquid nitrogen or liquid helium [132]. This enables a vitreous state of water, preventing the formation of ice crystals that would disrupt the sample structure. As such, rapid freezing represents a good preparation method as compared to chemical treatment (as fixation and dehydration) and therefore sample alteration can be prevented [43, 132].

TEM has been paramount for the characterization of Doxil®, the first liposomal formulation containing doxorubicin (a chemotherapeutic anthracycline) to be approved by FDA for cancer treatment, presenting prolonged circulation times in the blood with significant drug retention [11, 155]. In this respect, cryo-TEM was used to characterize the ultrastructure of encapsulated doxorubicin. Crystallized $(Doxorubicin)_2SO_4$ rod-shaped fiber bundle structures in a gel-like phase were identified, only apparent for sulfate as a counter ion [11, 104]. These results were further confirmed by small X-ray scattering measurements [104].

TEM has also been playing an important role in the characterization of ligand-mediated targeted liposomes (as folate-targeted magnetoliposomes) as well as in the study of the underlying mechanism of action at the level of the target (tumor) cell. In this respect, TEM supported the accumulation of liposomes targeted by Sialyl X Lewis X (SLX) carbohydrate, a ligand of E-selectin, and loaded with colloidal nanogold (8 nm), within and in the abluminal side of the tumor endothelial cells. This was in contrast with the observation for the non-targeted counterpart [135].

Visualization of a protein corona deposited around liposomes, following interaction with plasmatic proteins [14, 180], has been also characterized by TEM, upon confirmation of superior electron density on liposome/PC biological entity [35]. The authors further confirmed the top 20 proteins in the protein corona, which included albumin and apolipoproteins [35]. Despite a phenomenon that may be attenuated using PEGylated liposomes, the formation of a protein corona should be considered when designing liposomes as systemic delivery systems.

<u>Surface analysis: scanning electron microscopy and atomic force microscopy</u>

Besides the ultrastructure of the liposomes, visualization of the surface also provides useful information about their physical properties, even when immobilized in 3D systems of polymeric nature. Scanning electron microscopy (SEM) and atomic force microscopy (AFM) are important methodologies to probe the liposome surface [102]. With SEM, a detailed image of the liposome's surface is obtained by scanning the sample with low-energy electron beam (up to 30 keV) and detecting the scattered electrons reflected from the surface [43]. For example, SEM was instrumental to define the structure of Exparel® (Pacira Pharmaceuticals), a novel bupivacaine extended-release liposomal formulation for local pain management [175]. Using freeze-fracture SEM, it was possible to visualize the honeycomb-like structure of each liposome particle (10–30 μm) with several chambers, with an inner aqueous core, encapsulating bupivacaine [175]. In another context, SEM was used to characterize and define the immobilization of dexamethasone-loaded liposomes on the surface of polycaprolactone electrospun nanomeshes for advanced tissue engineering [137, 139]. In this respect, the slow release of dexamethasone from liposomes, over 21 days, was reported. It enabled the osteogenic differentiation of human bone-marrow derived mesenchymal stem cells (hBMSC) as efficiently as cells cultured in osteogenic medium [137]. Aiming at the same purpose, a similar system was described for gene delivery. The system consisted on the immobilization of RUNX2 plasmid-loaded liposomes (with approx. 120 nm) into polycaprolactone nanofiber meshes, as confirmed by SEM [138]. It enabled prolonged gene expression in hBMSCs further promoting osteogenic differentiation [138].

Alternatively, AFM uses an extremely sharp probe tip (cantilever) to scan the surface structure of the sample at its natural environment with atomic resolution [60]. AFM was used to demonstrate the importance of maleimide groups attached at the terminal end of PEG chains for the specific coupling of transferrin, aiming at designing targeted pH-sensitive liposomes [56]. Targetability was confirmed upon correlating these results with cellular association experiments [56]. Furthermore, AFM can be useful to measure stiffness. In this respect, the technique demonstrated that the protein corona around the liposomes increased their stiffness (measured by Young's module) as compared to plain liposomes [35].

Overall, the microscopy techniques described above are paramount for the characterization of lipid-based nanoparticles providing valuable details about their ultrastructure and surface. Nevertheless, they present some disadvantages. TEM and SEM are time-demanding and require a high level of specialization, especially in terms of sample preparation (including staining, that may introduce artifacts) [178]. Furthermore, despite enabling the probing of non-conductive samples, AFM is limited by geometry and size of the cantilever, which may lead to overestimated lateral dimensions [60, 102]. Moreover, it is not easily used on nanoformulations with highly fluid membranes but still relevant to characterize elasticity and flexibility of liposomes by force-mapping [178]. Thus, complementary alternatives are often used to characterize liposomes including spectroscopy-based techniques.

4.3.1.2 Spectroscopy-Based and Separation Techniques

Another set of techniques, complementary to the above, rely on the interaction of light with the nanoparticles, as liposomes, or on their physical properties (e.g. surface charge and size) for further characterization [102].

Dynamic light scattering

Dynamic light scattering (DLS) explores the interaction of different light sources (laser, commonly used, or X-rays) with liposomes to establish their mean size (in fact the hydrodynamic radius, R_H) and infer on their homogeneity [22]. Determination of R_H (theoretical radius of a hard sphere) through DLS is based on the scattering of light intensity over time, owed to constructive and destructive interference triggered by the (Brownian) motion of the liposomes [22]. Accordingly, DLS data depend on several factors including viscosity and refractive index of the solvents, the concentration of the liposomes, as well as temperature [22]. One important limitation is the low resolution, particularly, in samples with a high (>0.3) polydispersity index (PDI), a measure of the width of the size distribution [22]. Thus, despite enabling the processing of large numbers of particles (with robust data), it may be helpful to use, concomitantly, another technique as TEM or SEM [22, 102].

Nanoparticle tracking analysis

Alternatively, a nanoparticle tracking analysis (NTA or PTA) system, as the Malvern's NanoSight, combines the laser light scattering with microscopy (with CCD camera detection) to track the Brownian motion of nanoparticles in suspension and define the size distribution [55]. Each particle is individually analyzed (acting as a point light scatter rather being visualized) by direct observation and measurement of diffusion events, which depend on their size (larger particles move slowly than smaller ones). This provides higher resolution data as compared to the bulk analysis of a DLS system, which turns it adequate for analysis of polydisperse samples [55]. However, data interpretation should be made carefully, as the movement of a very large number of particles must be measured to obtain a good size distribution and it was shown that the sample flow speed (specially associated with sample loading pumps) may influence the apparent size and concentration of particles measured by NTA [201].

Nevertheless, besides the structure, characterization of other physical properties is essential to estimate their impact in the performance of the liposomes, thus ensuring an adequate design.

Zeta-potential and capillary electrophoresis

The charge properties of nanoparticles, as liposomes, are important as they reflect the composition and critical attributes for production design. Zeta-potential (ZP) measurements and capillary electrophoresis are important methods to assess those properties.

ZP is the electric potential at the shear plane (interface particle/fluid) of a particle moving under an electric field [93]. It is deduced after analyzing the frequency's Doppler shift of scattered light (relative to the frequency of the original laser light) under electrophoretic movement [22]. As it is based in the principle of light scattering, ZP measurement may be used in conjunction with DLS in equipment such as the Malvern Zetasizer. ZP measurements are affected by conditions as pH, ionic strength of measuring media and concentration of nanoparticles [22]. As it only deals with the surface potential rather than charge density, one should regard the magnitude of ZP and cautiously consider the positive/negative finding (given by the identification of the electrode to which the particles are moving to), as it only provides indicative evidence of the nature of surface charge owing to the influence of the mentioned conditions at the shear plane [22].

Capillary electrophoresis (CE) also relies on charge and mass of the nanoparticle [215]. Separation occurs as the result of differences in the mass of charged particles moving in an electric field. However, owing to the large surface-to-volume ratio within the capillary, a phenomenon known as electro-osmosis creates a constant flow that may drag the charged particles in a different direction other than the one predicted solely through electrophoresis [215]. Thus, the adequate control of electro-osmosis combined with the improved control of temperature (in larger tubes, temperature leads to band-mixing), renders CE as a high resolution separation technique [202, 215]. Thus, CE enables accurate characterization of nanoparticles in terms of electrophoretic mobility, size and charge distribution, ZP, surface functionality or colloidal stability [202]. For example, Li and colleagues demonstrated, through CE, the coupling of nucleolin-binding AS1411 aptamer to the surface of PEGylated liposomes for the delivery of anti-BRAF siRNA [108]. Recently, Furst and colleagues demonstrated a CE method to determine the efficiency of siRNA complexation with cationic liposomes [61]. In contrast with other methods (e.g. RiboGreen assay), the CE method was optimized for separation and quantitation under automated conditions in the absence of carcinogenic molecules [61].

4.3.2 High Performance Chromatography: Analyzing Chemical Composition and Encapsulation Stability of Lipid-Based Nanoparticles

Besides the physical properties, the study of the chemical composition and the extent of drug leakage are critical parameters towards the successful development of lipid-based nanoparticles. In this respect, analytical methods, as those based on high performance liquid chromatography (HPLC), are essential tools to accurately determine the qualitative and quantitative composition of a given liposomal formulation [102, 186]. These methods can be further extended to study the stability of liposomes either in vitro or in vivo, owing to their high resolution, selectivity and adaptability to the sample matrix under analysis [186]. Furthermore, the possibility of combining those with different types of detectors, as UV, fluorescence or mass spectrometry, render HPLC methods highly flexible in terms of the target analytes, time and cost [186].

Among others, HPLC methods employing evaporative light scattering detection (ELSD) are becoming common, especially as an inexpensive alternative to UV detectors in the analysis of molecules without chromophores [4, 41]. In ELSD, the mobile phase (pumped from the HPLC) is vaporized in a nebulizing chamber (using N_2) and removed before the analytes cross an optical cell where they collide with a beam of light, scattering the light that is detected and processed as a signal [177]. Accordingly, a HPLC-ELSD method was developed for rapid and simultaneous quantification of four lipids in liposomal systems with high sensitivity and linearity [177]. Similarly, a method for the analysis of targeted liposomes composition was developed, aiming at simultaneous quantification of lipids including DSPE-PEG_{2000} and the maleimide derivative (DSPE-PEG_{2000}-maleimide), used for bioconjugation [159].

In a different perspective, analyzing the stability of drug encapsulation is paramount in terms of drug delivery, either in the quality control setting or in liposome design. Once again, HPLC methods are invaluable tools in those studies. For example, an in vitro HPLC assay to assess doxorubicin leakage, enabling the measurement of encapsulated and free drug, was established using a strategy of column switching [157]. It was demonstrated that the method did not have a rate-limiting step in the release of doxorubicin from the liposomes, contrary to conventional dialysis-based assay. In the latter, drug release profile is the result of both liposome membrane and dialysis membrane permeation kinetics [157]. In another example, Cipolla et al., developed an in vitro release assay for liposomal ciprofloxacin in bovine serum at pH 7.4, and centrifugal filtration to separate free and encapsulated ciprofloxacin, quantified by HPLC [33].

4.3.3 Liposomal Physicochemical Properties and Their Impact on Efficacy upon Intravenous Administration

In the previous sections, several examples of techniques aiming at characterizing liposomal formulations have been discussed. This characterization applied to liposomes, or every nanoparticle for the matter, is crucial from a quality control perspective. More importantly, it strongly contributes to the desired performance-by-design at the developmental stage, especially in terms of therapeutic activity [57]. In an oncology setting, and aiming at therapeutic intervention in solid tumors, tailoring the physicochemical properties of lipid nanoparticles, as a function of the biological challenges faced from the site of administration until the efficient delivery into the target tumor cell, is paramount from a pharmacodynamic standpoint [7, 57].

A first challenge is to ensure that liposomes present high circulation times in the blood to promote tumor accumulation. It was demonstrated that lower bilayer stiffness and surface charge (e.g. cationic lipids), both dependent on the lipid composition, lead to rapid blood elimination [79, 203]. Classical liposome formulations faced high extent of clearance by the mononuclear phagocyte system (MPS), as the result of adsorption of opsonins, a process highly dependent on the liposomal formulation [85, 212]. Polyethyleneglycol (PEG)-modified lipids were introduced in the lipid bilayer to improve circulation times (used in Doxil/Caelyx as discussed above). It was reasoned that the improved pharmacokinetics profile of PEGylated liposomes was owed to the hydrophilic cloud enabled by PEG, that minimized opsonization and further clearance by MPS [103, 161, 181].

Once reaching the tumor vasculature, characterized by an irregular network (with fenestrae up to 600 nm, depending on the tumor) of newly formed vessels through angiogenesis [76, 117, 164], liposomes have to exit the vascular compartment to the tumor parenchyma. Despite prolonged circulation times, enabled by PEGylation, improved tumor accumulation [62, 161] might not always be achieved [203]. In fact, mean size (hydrodynamic diameter) above 200 nm is a critical parameter inversely influencing the liposomal pharmacokinetics and tumor accumulation (in line with the fenestrae cutoff) [145, 203]. Furthermore, surface charge (defined by composition) also impacts vascular extravasation, as demonstrated by the inability of cationic liposomes to cross the vascular wall as compared, namely, with neutral liposomes in a melanoma mouse model [96].

Finally, in the parenchyma, liposomes may diffuse through the interstitial space reaching, for example, cancer cells. However, as in the brain [115], a size above 200 nm, a cationic surface or absence of PEG (both dependent on lipid composition) may hinder the diffusion of liposomes within the tumor. Further hindrance may arise from poor lymphatic drainage, supporting the enhanced permeability and retention (EPR) effect, which might limit liposomes diffusion owing to the increased tumor interstitial pressure resulting from the same EPR effect [117].

Even though, successful overcoming of those hurdles does not guarantee that the drug payload would become bioavailable to exert its action. This is the case of a highly stable formulation that reaches the tumor but releases its payload inefficiently at tumor site or within cells [7, 182]. To improve bioavailability, formulations enabling triggered drug release at the tumor site, as a response to pH or temperature change, have been developed [87, 94, 107]. Furthermore, it has been recognized that receptor-mediated targeting enabling intracellular delivery of drug payloads, by functionalizing the surface of liposomes with internalizing targeting ligands, may have a significant impact in terms of therapeutic outcome [57]. Different examples have emerged aiming at targeting cancer cells, exploring the overexpression of given receptors in cancer cells, such as HER2, using antibodies, aptamers or peptides specifically recognizing those receptors. Alternatively, targeting tumor vasculature (different in many aspects from mature blood vessels), has gained significant interest owing to its accessibility (relative to cancer cells). This enables one to avoid major barriers on drug delivery while hindering an essential framework for tumor maintenance and metastasis [57, 76].

Overall, the integration between development of nanoliposomes and tumor biology has enabled the conception of innovative, rationally designed, medicines that are under clinical development or already available for the oncological patients.

4.4 Current Application of Liposomes in the Oncological Setting

Nanomedicines have been investigated as therapeutic products for the treatment of a range of diseases, including cancer, infectious diseases, or inflammatory/immune disorders, among others [51]. Despite the number of approved nanomedicines being similar across the different categories, cancer treatment has received about two-thirds of the research attention [51]. In this section, the application of liposomes in oncology therapy will be addressed.

The unique ability of liposomes to encapsulate hydrophilic molecules within their aqueous environment, or to incorporate hydrophobic molecules in their phospholipid bilayer membrane, has been extensively explored for the development of drug delivery platforms. Liposomal formulations enabled the stabilization of the therapeutic compounds, improvement of the biodistribution of compounds, as well as tumor uptake, including at the cellular level.

4.4.1 Liposomes in Clinical Use

The concept of making "Old drugs in new clothing", as described by Gregoriadis in the 1980s, has opened a window of opportunity to add new therapeutic value to drugs already approved [71]. One of the best examples is the encapsulation of the cytotoxic anticancer drug doxorubicin into PEGylated liposomes (Doxil® in the United States or Caelyx® in Europe) [12]. In oncology, PEGylated liposomal doxorubicin was the first liposome-based nanomedicine to obtain FDA approval, in 1995, for the treatment of AIDS-associated Kaposi's sarcoma [12, 44]. Subsequent clinical studies supported approval in ovarian cancer refractory to paclitaxel and platinum-based chemotherapy [66, 67, 141, 205]. Later, Doxil® was also approved for the treatment of metastatic breast cancer (2004) and, in combination with bortezomib, for the treatment multiple myeloma (2007) [23, 44, 155, 158, 194]. Notwithstanding its superior therapeutic efficacy against both Kaposi´s sarcoma and ovarian cancer, the PEGylated formulation did not demonstrate superior efficacy but rather a better safety profile relative to the corresponding standard of care, in the setting of metastatic breast cancer and multiple myeloma [155, 158].

Nevertheless, treatment with Doxil® led to a shift of doxorubicin's safety profile. The extended circulation times in the blood of PEGylated liposomes resulted in palmar-plantar erythrodysesthesia (PPE), but significantly reduced the dose-limiting cardiotoxicity associated with free doxorubicin [155]. Therefore, the main therapeutic advantage of Doxil® results primarily from its enhanced safety, as compared to the standard therapies.

Following the approval of Doxil®, other liposome-based nanomedicines were also granted clinical application for cancer treatment (Table 4.1). Among them are Myocet® and DaunoXome®, both non-PEGylated liposome formulations encapsulating the anthracyclines doxorubicin and daunorubicin, respectively. Furthermore, and following the need to address a Doxil® shortage in the USA, LipoDox produced by Sun Pharma Global FZE was approved by the FDA in 2013 as the first generic of Doxil® [204].

In addition to anthracyclines, other anticancer drugs have also been delivered by liposomal formulations and approved for clinical use (Table 4.1). The most recent include vincristine (Marqibo®) and irinotecan (Onivyde®).

Table 4.1 Liposome nanomedicines on the market or in clinical trial for cancer therapy

Name	Drug	Drug formulation	Indication	Status
Doxil®/ Caelyx®	Doxorubicin	Doxorubicin hydrochloride encapsulated in long-circulating PEGylayed liposomes (STEALTH® liposomes) [44]	Kaposi's sarcoma, advanced ovarian cancer, metastatic breast cancer, multiple myeloma	Approved (1995)
Myocet®	Doxorubicin	Doxorubicin-citrate encapsulated in non-PEGylated liposomes [45]	Metastatic breast cancer	Approved (2000)
LipoDox	Doxorubicin	Doxorubicin hydrochloride encapsulated in PEGylated liposomes	Kaposi's sarcoma, breast cancer, ovarian cancer	Approved (2013)
Thermodox®	Doxorubicin	Doxorubicin encapsulated in lysolipid thermally sensitive liposomes [147, 148]	Primary hepatocellular carcinoma, recurrent chest wall breast cancer	**Phase I–III** NCT00441376[c] NCT02112656[b] NCT02181075[b] NCT00617981[c] NCT00826085[c] NCT02850419[a]
2B3-101	Doxorubicin	Doxorubicin encapsulated in glutathione-targeted PEGylated liposomes [63]	Breast cancer, meningeal carcinomatosis, solid tumors with brain metastases or recurrent malignant glioma	**Phase I/II** NCT01386580[c] NCT01818713[d]
DaunoXome®	Daunorubicin	Daunorubicin citrate encapsulated in non-PEGylated liposomes [59]	Kaposi's sarcoma	Approved (1996)
Marqibo® (formerly referred as Onco-TCS)	Vincristine	Vincristine sulfate encapsulated in sphingomyelin/cholesterol liposomes (OPTISOME®) [156, 214]	Acute lymphoblastic leucemia	Approved (2012)
Onivyde® (MM-398)	Irinotecan	Irinotecan sucroseoctasulfate salt encapsulated in pegylated liposomes [48]	Metastatic adenocarcinoma of the pancreas	Approved (2015)

(continued)

Table 4.1 (continued)

Name	Drug	Drug formulation	Indication	Status
NL CPT-11	Irinotecan	Liposomal irinotecan (or CPT-11) [42]	Recurrent high-grade gliomas	**Phase I** NCT00734682[c]
IHL-305	Irinotecan	PEGylated liposomal irinotecan [86, 218]	Advanced solid tumors	**Phase I** NCT00364143[d] NCT02631733[e]
CPX-1	Irinotecan:floxuridine (1:1 ratio)	Liposomal formulation of irinotecan and floxuridine encapsulated at a fixed 1:1 molar ratio [128]	Advanced colorectal cancer	**Phase II** NCT00361842[c]
Lipusu®	Paclitaxel	Liposomal paclitaxel	Solid tumors	Approved in China (2006)
PICN®	Paclitaxel	Polyoxyethylated castor oil- and albumin-free self-assembly nanoparticle formulation of paclitaxel stabilized with a polymer (polyvinylpyrrolidone) and lipid (cholesteryl sulfate and caprylic acid) using Nanotecton® technology [92]	Breast cancer [Solid tumors]	Approved in India [Phase I, NCT01304303[c]]
EndoTAG-1	Paclitaxel	Paclitaxel encapsulated in cationic liposomes [8, 52, 111]	Breast cancer, pancreatic cancer, liver cancer	**Phase II–III** NCT01537536[c] NCT00448305[c] NCT03002103[b] NCT00377936[c] NCT00542048[c]
LEP-ETU	Paclitaxel	Paclitaxel-containing liposomes [228]	Breast cancer	**Phase I/II** NCT00100139[c] NCT00080418[c] NCT01190982[c]

(continued)

Table 4.1 (continued)

Name	Drug	Drug formulation	Indication	Status
LE-DT	Docetaxel	Non-PEGylated liposomes containing docetaxel [38]	Pancreatic cancer, [prostate cancer]	**Phase I/II** NCT01151384[c] NCT01186731[c] [NCT01188408][e]
ATI-1123	Docetaxel	Docetaxel formulated in protein-stabilized liposomes (PSL™ technology) [118]	Solid tumors	**Phase I** NCT01041235[c]
Lipoplatin	Cisplatin	Cisplatin encapsulated in liposomes composed of dipalmitoyl phosphatidyl glycerol, soy phosphatidyl choline (SPC-3), cholesterol and methoxypolyethylene glycol-distearoyl phosphatidylethanolamine (mPEG2000-DSPE) [25]	Pleural malignancies, non-small cell lung cancer [143, 189, 190]	**Phase I–III** NCT02702700[b]
SPI-077	Cisplatin	Long-circulating pegylated (Stealth®) liposome-encapsulated cisplatin [152]	Recurrent ovarian cancer	**Phase II** NCT00004083[c]
LiPlaCis	Cisplatin	Cisplatin encapsulated in PEGylated liposomes (LiPlasomes technology) [110]	Advanced or refractory solid tumors	**Phase I/II** NCT01861496[b]
Aroplatin (also L-NDDP)	Cisplatin	Liposome-entrapped cis-bis-neodecanoate-trans-R, R-1,2-diaminocyclohexane platinum(II) [166]	Malignant pleural mesothelioma, advanced solid malignancies or B-cell lymphoma	**Phase I–II** NCT00004033[c] NCT00316511[c]
MBP-426	Oxaliplatin	Transferrin (Tf)-targeted N-glutaryl phosphatidylethanolamine liposomal formulation of oxaliplatin [192]	Advanced/metastatic solid tumors, metastatic gastric, gastro-esophageal junction or esophageal adenocarcinoma	**Phase I–I/II** NCT00355888[c] NCT00964080[d]

(continued)

Table 4.1 (continued)

Name	Drug	Drug formulation	Indication	Status
OSI-211	Lutortecan	Lutortecan encapsulated in unilamellar liposomes composed of 2:1 ratio of fully hydrogenated soy phosphatidylcholine:cholesterol [49]	Epithelial ovarian cancer, Advanced/metastatic solid tumors, Head and neck, Recurrent small cell lung cancer	**Phase I–II** NCT00046800[c] NCT00010179[c] NCT00003891[c] NCT00006036[c] NCT00022594[c] NCT00046787[c]
Brakiva™ (also topotecan liposomal injection, TLI)	Topotecan	Topotecan hydrochloride encapsulated in sphingomyelin/cholesterol liposomes (OPTISOME®) [225]	Small cell lung cancer, ovarian cancer and other advanced solid tumors	**Phase I** NCT00765973[a]
Alocrest™	Vinorelbine	Vinorelbine tartrate encapsulated in sphingomyelin/cholesterol liposomes (OPTISOME®) [39]	Advanced solid tumors, non-Hodgkin's lymphoma, or Hodgkin's disease	**Phase I** NCT00364676[c]
DepoCyt®	Cytabirine	Slow-release formulation of cytarabine encapsulated in spherical multivesicular particles (DepoFoam® technology) [46, 168]	Lymphomatous meningitis	Approved (1999/2007)
VYXEOS® (a.k.a. CPX-351)	Cytarabine:Daunorubicin (5:1 ratio)	Liposomal formulation of cytarabine and daunorubicin encapsulated at a fixed 5:1 molar ratio [128]	Acute myeloid leukemia	**Phase I–IV** NCT02019069[a] NCT01804101[a] NCT01696084[a] NCT02286726[b] NCT01943682[b] NCT02269579[b] NCT02642965[b] NCT02533115[b] NCT02533115 (EAP)

(continued)

Table 4.1 (continued)

Name	Drug	Drug formulation	Indication	Status
Mepact® (also Mifamurtide)	Liposomal muramyl tripeptide phosphatidylethanolamine	Muramyl tripeptide phosphatidylethanolamine (MTP-PE) intercalated in the lipid bilayer of multilamellar vesicles [47, 146]	High-grade resectable non-metastatic osteosarcoma	Approved (2009)
SGT-53	Plasmid encoding human *TP53*	Liposomes targeted by an anti-transferrin single-chain antibody fragment (TfRscFv) delivering human *TP53 gene* [219, 220]	Various solid tumors	**Phase I/II** NCT00470613[a] NCT02354547[b] NCT02340117[b] NCT02340156[b]
L9-NC	9-nitrocamptothecin	Dilauroylphosphatidylcholine (DLPC) liposome aerosols containing 9-nitrocamptothecin (9-NC) [95, 207]	Ewing's sarcoma and other solid tumors with lung involvement, non-small-cell lung cancer, metastatic or recurrent cancer of the endometrium	**Phase I/II–II** NCT00492141[c] NCT00250068[c] NCT00250016[c] NCT00249990[c] NCT00277082[c]
TKM-080301	Anti-*PLK1* siRNA	Anti-*PLK1* small interference RNA (siRNA) encapsulated in a liposomal formulation incorporating the ionizable cationic lipid dilinoleoyl-DMA (DLinDMA) [154]	Advanced hepatocellular carcinoma, neuroendocrine tumors and adrenocortical carcinoma, primary or secondary liver cancer	**Phase I–I/II** NCT02191878[c] NCT01262235[c] NCT01437007[c]
Atu027	Anti-*PKN3* siRNA	Anti-*PKN3* siRNA encapsulated in a formulation based on cationic lipids and incorporating neutral fusogenic and PEG-modified lipid components [2]	Advanced or metastatic pancreatic cancer	**Phase I/II** NCT00938574[c] NCT01808638[c]
MTL-CEBPA (former MTL-501)	*CEBPA* saRNA	Short activating RNA (saRNA) targeting the CCAAT enhancer binding protein-alpha (CEBPA; C/EBP-a) gene, encapsulated in a SMARTICLES® delivery platform [134]	Advanced liver cancer	**Phase I** NCT02716012[b]

(continued)

Table 4.1 (continued)

Name	Drug	Drug formulation	Indication	Status
LE-SN38	SN-38	7-ethyl-10-hydroxycamptothecin (SN-38) encapsulated in a liposome formulation composed of DOPC, cholesterol and cardiolipin [160]	Metastatic colorectal cancer	**Phase II** NCT00311610[c]
Allovectin-7®	Plasmid DNA encoding the genes *HLA-B7* and *β2-microglobulin*	Plasmid DNA encoding the genes *HLA-B7* and *β2-microglobulin* complexed with the cationic lipid N-(1-(2,3-dimyristyloxypropyl)-N,N-dimethyl-(2-hydroxyethyl) ammonium bromide/ dioleoyl phosphatidylethanolamine (DMRIE/DOPE) [34, 191]	Melanoma [Head and neck cancer]	**Phase II–III** NCT00395070[c] NCT00003647[c] NCT00003646[c] NCT00044356[c] [NCT00050388[c]]
Atragen®	All-*trans*-retinoic acid	All-*trans*-retinoic acid encapsulated in liposomes composed of diphosphatidyl palmitoylcholine (DPPC) and stearylamine (SA; 9:1, w/w) [162, 210, 213]	Metastatic kidney cancer	**Phase II** NCT00003656[c]
DCR-MYC	*MYC* siRNA	Anti-MYC DsiRNA formulated in EnCore lipid nanoparticles [1, 6, 200]	Refractory locally advanced or metastatic solid tumor malignancies, multiple myeloma, or lymphoma; hepatocellular carcinoma	**Phase I–I/II** NCT02110563[a] NCT02314052[a]
DOTAP: Chol-fus1	*FUS1* plasmid DNA	FUS1 plasmid DNA complexed to DOTAP: Chol liposome [88, 89]	Non-small cell lung cancer	**Phase I–I/II** NCT00059605[c] NCT01455389[b]

(continued)

Table 4.1 (continued)

Name	Drug	Drug formulation	Indication	Status
OSI-7904L	(*S*)-2-[5-[1,2-dihydro-3-methyl-1-oxobenzo[*f*]-quinazolin-9-yl)methyl] amino-1-oxo-2-isoindolynl]-glutaric acid (thymidylate synthase inhibitor)	OSI-7904 encapsulated in 4:1 fully hydrogenated soy phosphatidylcholine (HSPC)/cholesterol liposomes [21, 40]	Gastric or gastroesophageal cancer; advanced solid tumors; biliary tract cancer; head and neck cancer; colorectal cancer	**Phase I–II** NCT00073502[c] NCT00116896[c] NCT00088270[c] NCT00116909[c] NCT00081237[c]
S-CKD 602	(20S)-7-(2-isopropylamino)-ethylcamptothecin hydrochloride (topoisomerase inhibitor)	S-CKD 602 encapsulated in STEALTH® liposomes composed of MPEG-DSPE and 1,2-distearoyl-sn-glycero-3-phosphocholine (DSPC) in a 5:95 molar ratio [224, 226, 227]	Advanced malignancies	**Phase I** NCT00177281[c]
MM-302	Doxorubicin	HER2-targeted antibody–liposomal doxorubicin [131]	Advanced/metastatic breast cancer, HER2+ solid cancers with brain metastases	**Phase I–III** NCT02213744[e] NCT01304797[a] NCT02735798[a]

[a]Active but not recruiting; [b]Recruiting; [c]Completed; [d]Unknown status; [e]Suspended; *EAP* Expanded Access Protocol

4.4.2 Liposome-Based Nanoparticles in Clinical Trial

4.4.2.1 Multidrug-Loaded Liposomes

The combination of anticancer drugs is often the standard of care for the treatment of neoplastic diseases. Tumor heterogeneity leads to a variety of potential targets amenable to chemotherapy. Multidrug regimens address this issue upon encompassing cytotoxic agents with different mechanisms of action and non-overlapping toxicities. The underlying rationale of such regimens is to increase the therapeutic response and tolerability, while decreasing the chance of a patient developing drug resistance [74]. In this scenario, liposomes emerged as a promising strategy for the co-encapsulation of anticancer drugs.

As carriers of drug combinations, liposomes enable synchronization of their pharmacokinetics, selective localization and controlled release at tumor sites at the desired synergistic drug ratio [170]. Examples of such strategy are the CPX-1 and VYXEOS® (a.k.a. CPX-351) (Table 4.1) liposomal formulations, encapsulating the synergistic drug ratios of irinotecan:floxuridine (1:1) or cytarabine:daunorubicin (5:1), respectively, with proven therapeutic efficacy in preclinical tumor models [77, 128, 197, 198]. It was further demonstrated that the molar ratios of irinotecan to floxuridine (1:1) and cytarabine to daunorubicin (5:1) were also maintained in the plasma of patients for all doses tested, and for 8–12 h or 0.75–48 h, respectively [17, 53]. VYXEOS® is the most studied formulation as reflected by the extensive number of clinical trials, either completed or ongoing, performed in the setting of acute myeloid leukemia (AML) (Table 4.1). Results from a randomized, controlled phase II trial evaluating the efficacy of VYXEOS® as first line treatment for elderly patients with untreated AML, suggested a clinical benefit of the liposomal formulation over the non-encapsulated cytarabine plus daunorubicin drug combination [99]. The activity of VYXEOS® was particularly promising in a subset of patients with secondary AML, providing the rationale for a phase III study to confirm these results, with the primary endpoint being overall survival (NCT01696084). Additionally, the FDA has also approved a phase IV expanded access protocol (EAP) of VYXEOS®, in patients with secondary AML who are suitable for treatment with intensive chemotherapy (NCT02533115).

4.4.2.2 Stimulus-Responsive Liposomes

The development of stimuli-responsive nanoparticles has attained considerable advancements, particularly in the last decade. The exploitation of stimuli-responsive materials has led researchers to design delivery systems equipped with a dynamic sensitivity to the local microenvironment [142]. The concept of stimulus-responsive liposomes has been first introduced in 1978 by Yatvin and colleagues, in an experiment where the triggered release of thermosensitive liposome content was achieved through the application of mild hyperthermia [222]. The lipidic composition of thermosensitive liposomes often includes phospholipids with a melting phase transition temperature around 40–42 °C. At these temperatures, the lipid bilayer suffers conformational changes that render the liposomal membrane more permeable, facilitating the release of the encapsulated drug, while at the normal body temperature (37 °C) the drug load is retained [196]. This concept has been explored in the oncological setting, being ThermoDox® (Celsion Corporation) the most prominent example.

ThermoDox® is a lysolipid thermosensitive liposomal formulation of doxorubicin that releases the entrapped drug upon tumor application of mild hyperthermia [100]. The efficacy and safety of this liposomal formulation has been recently assessed in a double-blinded, placebo-controlled phase III study, in combination with radiofrequency ablation (RFA) compared to RFA-alone, for the treatment of non-resectable hepatocellular carcinoma (NCT00617981). ThermoDox® was well tolerated, although the primary outcome of the study (33% improvement in progression-free survival) was not attained [27]. In this study, the total length of RFA procedure was set to be proportional to the tumor size (patients with 3–7 cm single lesions), and estimated to range between 12 and 60 min for each lesion. Interestingly, further data analysis revealed a median survival benefit of 25.4 months in a large subgroup of patients treated with ThermoDox® combined with standardized RFA (sRFA > 45 min) [28]. The median of 80 months in overall survival observed in this subgroup of patients was considered a curative treatment for hepatocellular carcinoma, and prompted an additional evaluation trial. Thus, ThermoDox® is presently being evaluated in a global phase III study (OPTIMA) for the treatment of non-resectable hepatocellular carcinoma in combination with radiofrequency ablation standardized to a minimum of 45 min (NCT02112656). Ongoing is also a phase II trial to assess the efficacy and tolerability of ThermoDox® in combination with hyperthermia and radiotherapy for the treatment of loco-regional relapse in breast cancer patients (NCT02850419). In addition, the maximum tolerated dose, and recommended phase II dose, of ThermoDox® administered in combination with magnetic resonance high intensity focused ultrasound (MR-HIFU) is being evaluated in a phase I study in the setting of pediatric refractory solid tumors (NCT02536183). Lastly, a proof of concept study is also addressing the enhanced intra-tumoral delivery of doxorubicin by ThermoDox® upon application of focused ultrasound in liver tumors (NCT02181075).

The shaping of drug release profiles in a spatial and temporal manner could be achieved through the application of other exogenous stimuli, such as the magnetic field, ultrasound, light, or weak electric fields. Moreover, liposomes can also be tailored to respond to endogenous stimuli, such as pH variations, the altered expression profile of certain enzymes, or changes in the microenvironment redox status [142].

4.4.2.3 Ligand-Mediated Targeted Liposomes

Notwithstanding the large number of publications in this area, only a few ligand-mediated targeted nanoparticles have reached clinical evaluation [19]. This small list of targeted systems includes both polymeric- (BIND-014 and CALAA-01, both withdrawn from trials) and lipid-based (SGT-53 and MM-302) formulations (Table 4.1).

SGT-53 (SynerGene Therapeutics, Inc) consists of a liposomal formulation targeted with the anti-transferrin single-chain antibody fragment (scFv). It was designed to target and specifically deliver a plasmid encoding wild-type human *TP53* to cancer cells overexpressing transferrin, upon for systemic administration [219, 220]. The loss of p53 tumor suppressor function, via mutation or the partial inactivation of its pathway, is a feature reported in the majority of human cancers [31, 120, 206]. Moreover, the restoration of p53 function in human cancers has been supported by the results of several clinical trials assessing the impact of adenoviral-based delivery of wild-type *TP53* to cancer cells [149–151, 193]. SGT-53 thus emerged as a way to intravenously deliver p53, bypassing the previous need of intra-tumoral administration [185]. Presently, SGT-53 is being evaluated in four clinical studies (Table 4.1). So far, SGT-53 infusion in patients with advanced solid tumors (NCT00470613) was well tolerated and exhibited anticancer activity. Importantly, an accumulation of the transgene in metastatic sites, but not in normal skin tissue, was observed, suggesting that SGT-53 targets metastatic lesions as well as the primary site [185].

MM-302 (Merrimack Pharmaceuticals) is liposomal doxorubicin functionalized with an anti-HER2 monoclonal antibody. Data from the phase I study evaluating MM-302, alone or in combination with trastuzumab, in patients with advanced HER2-positive breast cancer, encouraged further evaluation in a randomized phase II trial (HERMIONE) [112]. In this study, stratification relied on HER2-positive breast cancer patients previously treated with trastuzumab, pertuzumab and T-DM1 (NCT02213744). However, this phase II study was terminated due to lack of benefit over the comparator, as confirmed through a futility analysis [129].

4.4.2.4 Liposomes Encapsulating Nucleic Acids

The modulation of gene expression in pathological conditions has grown into a new paradigm in medicine with tremendous therapeutic potential, as reflected by the number of ongoing or approved clinical studies [64]. RNA-based molecules emerged as a new class of drugs with therapeutic potential. Particularly, small interfering RNAs are being used to specifically silence genes with relevant pathological roles, such as oncogenes in cancer [29, 179, 217]. However, the vulnerability of these molecules to degradation by serum endonucleases, and their specific delivery to target tissues, still remain as major challenges for RNA drug development [223]. Hence, the encapsulation of RNA molecules in delivery systems became an important strategy to overcome these barriers. Delivery systems based on cationic lipids are the most commonly used nonviral vectors for RNA delivery [80, 106, 188], with several liposomal formulations currently under clinical investigation (Table 4.1).

Allovectin-7® is a HLA-B7/β2-microglobulin DNA-cationic liposome complex. In an initial first in human study, the intra-tumor administration of Allovectin-7® promoted the successful transfection of HLA-B7 DNA into tumor cells from cutaneous melanoma lesions, and elicited a melanoma-associated antigen-specific immune response in patients [144]. The encouraging clinical trial results that followed led to the evaluation of its therapeutic efficacy in a randomized phase III trial, comparing the combination Allovectin-7/dacarbazine against dacarbazine alone in chemotherapy-naive patients with advanced melanoma (NCT00003647). However, the former treatment did not demonstrate a significant clinical benefit: a median survival time of 10.75 months relative to 9.24 months, and a median time to disease progression of 1.9 months compared to 1.6 months for the combination therapy or dacarbazine monotherapy, respectively [20]. In a phase III study with patients with recurrent stage 3 or stage 4 melanoma (NCT00395070), Allovectin-7® performed similarly to Dacarbazine or Temozolomide, for both the primary outcome measure of objective response rate at ≥ 24 weeks after randomization, or the secondary endpoint of overall survival [208].

4.5 Concluding Remarks

In face of the small number of nanomedicines reaching clinical trials, when compared to the high number of publications in the field, it is clear that the pre-clinical thrive has largely failed to translate into the clinic. Over the recent years, overestimation of the EPR effect has been increasingly recognized as one of the major pitfalls in the nanomedicine field [36, 78]. For a long time, the preferential accumulation of nanomedicines in neoplastic tissues has been ascribed to the presence of fenestrations in the tumor blood vessels and the poor lymphatic drainage at these sites [121, 173]. Nevertheless, characteristics such as vascular density, localization of blood vessels within the tumor, or stromal morphology and content, determine the accessibility of nanoparticles to cancer cells and their capacity to diffuse and penetrate the tumor mass [32, 36, 78, 98, 109]. Intra- and inter-tumor variability with respect to these features also dictate the heterogeneity of the EPR phenomenon observed in patients. However, the recognition of the EPR effect as a highly heterogeneous phenomenon in human neoplastic tissues has been largely neglected, as well as failure to recognize differences between human tumors and those recapitulated in animal models. In rodent animal models, the EPR effect is well documented and is far superior than in human tumors [153]. This can lead to an overestimation of the therapeutic potential of nanoparticle formulations relying mainly on this phenomenon.

Another bottleneck in the translation of new nanopharmaceuticals relates with the scale-up of manufacturing under Good Manufacturing Practices (GMP) [174].

Translation from a lab to an industrial scale usually requires subtle adjustments in the manufacturing procedure that might result in changes of the nanoparticle's characteristics [174].

Altogether, these challenges often delay the progression of nanomedicines into the clinical setting. A better understanding of the (patho)physiological variability amongst individuals, and a better insight into the nanoparticles/biological system interaction, will assist in the design of nanomedicines with improved therapeutic efficacy. Moreover, adopting an industry-like rationale since early stages of development might reduce the risk and improve successful clinical translation.

References

1. Abrams M, Ganesh S, Ying B, Chopda G, Saxena U, Shah A, Koser M, Arvan R, Chen D, Shui S (2015) Abstract B20: EnCore-LNP mediated tumor delivery of MYC and CTNNB1 Dicer Substrate RNAs (DsiRNAs). AACR
2. Aleku M, Schulz P, Keil O, Santel A, Schaeper U, Dieckhoff B, Janke O, Endruschat J, Durieux B, Röder N (2008) Atu027, a liposomal small interfering RNA formulation targeting protein kinase N3, inhibits cancer progression. Can Res 68(23):9788–9798
3. Allen TM, Cullis PR (2004) Drug delivery systems: entering the mainstream. Science 303 (5665):1818–1822
4. Almeling S, Holzgrabe U (2010) Use of evaporative light scattering detection for the quality control of drug substances: influence of different liquid chromatographic and evaporative light scattering detector parameters on the appearance of spike peaks. J Chromatogr A 1217 (14):2163–2170
5. Alpes H et al (1986) Formation of large unilamellar vesicles using alkyl maltoside detergents. Biochim Biophys Acta 862:294–302
6. Amarzguioui M, Rossi JJ (2008) Principles of Dicer substrate (D-siRNA) design and function. In: RNAi: design and application. Humana Press, USA, pp 3–10
7. Anchordoquy TJ, Barenholz Y, Boraschi D, Chorny M, Decuzzi P, Dobrovolskaia MA, Farhangrazi ZS, Farrell D, Gabizon A, Ghandehari H, Godin B, La-Beck NM, Ljubimova J, Moghimi SM, Pagliaro L, Park JH, Peer D, Ruoslahti E, Serkova NJ, Simberg D (2017) Mechanisms and barriers in cancer nanomedicine: addressing challenges, looking for solutions. ACS Nano 11(1):12–18
8. Awada A, Bondarenko IN, Bonneterre J, Nowara E, Ferrero JM, Bakshi AV, Wilke C, Piccart M, Group CS (2014) A randomized controlled phase II trial of a novel composition of paclitaxel embedded into neutral and cationic lipids targeting tumor endothelial cells in advanced triple-negative breast cancer (TNBC). Ann Oncol 25(4):824–831
9. Bangham AD, Standish MM, Watkins JC (1965) Diffusion of univalent ions across the lamellae of swollen phospholipids. J Mol Biol 13(1):238–252
10. Bangham AD, Hill MW, Miller NGA (1974) Preparation and use of liposomes as models of biological membranes. Springer, Boston, pp 1–68
11. Barenholz Y (2012) Doxil(R)–the first FDA-approved nano-drug: lessons learned. J Control Release 160(2):117–134
12. Barenholz YC (2012) Doxil®—the first FDA-approved nano-drug: lessons learned. J Controlled Release 160(2):117–134
13. Barenholzt Y, Amselem S, Lichtenberg D (1979) A new method for preparation of phospholipid vesicles (liposomes)—French press. FEBS Lett 99(1):210–214

14. Barran-Berdon AL, Pozzi D, Caracciolo G, Capriotti AL, Caruso G, Cavaliere C, Riccioli A, Palchetti S, Lagana A (2013) Time evolution of nanoparticle-protein corona in human plasma: relevance for targeted drug delivery. Langmuir 29(21):6485–6494
15. Bartoli MH, Boitard M, Fessi H, Beriel H, Devissaguet JP, Picot F, Puisieux F (1990) In vitro and in vivo antitumoral activity of free, and encapsulated taxol. J Microencapsul 7 (2):191–197
16. Basha G, Novobrantseva TI, Rosin N, Tam YY, Hafez IM, Wong MK, Sugo T, Ruda VM, Qin J, Klebanov B, Ciufolini M, Akinc A, Tam YK, Hope MJ, Cullis PR (2011) Influence of cationic lipid composition on gene silencing properties of lipid nanoparticle formulations of siRNA in antigen-presenting cells. Mol Ther 19(12):2186–2200
17. Batist G, Gelmon KA, Chi KN, Miller WH, Chia SK, Mayer LD, Swenson CE, Janoff AS, Louie AC (2009) Safety, pharmacokinetics, and efficacy of CPX-1 liposome injection in patients with advanced solid tumors. Clin Cancer Res 15(2):692–700
18. Batzri S, Korn ED (1973) Single bilayer liposomes prepared without sonication. Biochim Biophys Acta 298(4):1015–1019
19. Bazak R, Houri M, El Achy S, Kamel S, Refaat T (2015) Cancer active targeting by nanoparticles: a comprehensive review of literature. J Cancer Res Clin Oncol 141(5):769–784
20. Bedikian AY, Del Vecchio M (2008) Allovectin-7 therapy in metastatic melanoma. Expert Opin Biol Ther 8(6):839–844
21. Beutel G, Glen H, Schöffski P, Chick J, Gill S, Cassidy J, Twelves C (2005) Phase I study of OSI-7904L, a novel liposomal thymidylate synthase inhibitor in patients with refractory solid tumors. Clin Cancer Res 11(15):5487
22. Bhattacharjee S (2016) DLS and zeta potential—what they are and what they are not? J Control Release 235:337–351
23. Bladé J, Sonneveld P, San Miguel JF, Sutherland HJ, Hajek R, Nagler A, Spencer A, Robak T, Lantz KC, Zhuang SH (2011) Efficacy and safety of pegylated liposomal doxorubicin in combination with bortezomib for multiple myeloma: effects of adverse prognostic factors on outcome. Clin Lymphoma Myeloma Leuk 11(1):44–49
24. Bolotin EM, Cohen R, Bar LK, Emanuel N, Ninio S, Lasic DD, Barenholz Y (1994) Ammonium sulfate gradients for efficient and stable remote loading of amphipathic weak bases into liposomes and ligandoliposomes. J Liposome Res 4(1):455–479
25. Boulikas T (2009) Clinical overview on lipoplatin: a successful liposomal formulation of cisplatin. Expert Opin Investig Drugs 18(8):1197–1218
26. Cabanes A, Briggs KE, Gokhale PC, Treat JA, Rahman A (1998) Comparative in vivo studies with paclitaxel and liposome-encapsulated paclitaxel. Int J Oncol 12(5):1035–1040
27. Celcion-Corporation (2013) Celsion announces results of Phase III HEAT study of ThermoDox® in primary liver cancer. 30 Mar 2017. Available from: http://investor.celsion.com/releasedetail.cfm?releaseid=737033
28. Celsion-Corporation (2016) Celsion announces presentation highlighting Phase III OPTIMA study at the Asia-Pacific primary liver cancer expert meeting. 30 Mar 2017. Available from: http://investor.celsion.com/releasedetail.cfm?releaseid=978853
29. Chakraborty C, Wen Z-H, Agoramoorthy G, Lin C-S (2016) Therapeutic microRNA delivery strategies with special emphasis on cancer therapy and tumorigenesis: current trends and future challenges. Curr Drug Metab 17(5):469–477
30. Charcosset C, El-Harati A, Fessi H (2005) Preparation of solid lipid nanoparticles using a membrane contactor. J Control Release 108(1):112–120
31. Cheok CF, Verma CS, Baselga J, Lane DP (2011) Translating p53 into the clinic. Nat Rev Clin Oncol 8(1):25–37
32. Choi J, Credit K, Henderson K, Deverkadra R, He Z, Wiig H, Vanpelt H, Flessner MF (2006) Intraperitoneal immunotherapy for metastatic ovarian carcinoma: resistance of intratumoral collagen to antibody penetration. Clin Cancer Res 12(6):1906
33. Cipolla D, Wu H, Eastman S, Redelmeier T, Gonda I, Chan HK (2014) Development and characterization of an in vitro release assay for liposomal ciprofloxacin for inhalation. J Pharm Sci 103(1):314–327

34. Clark PR, Stopeck AT, Ferrari M, Parker SE, Hersh EM (2000) Studies of direct intratumoral gene transfer using cationic lipid-complexed plasmid DNA. Cancer Gene Ther 7(6):853
35. Corbo C, Molinaro R, Taraballi F, Toledano Furman NE, Sherman MB, Parodi A, Salvatore F, Tasciotti E (2016) Effects of the protein corona on liposome-liposome and liposome-cell interactions. Int J Nanomed 11:3049–3063
36. Danhier F (2016) To exploit the tumor microenvironment: since the EPR effect fails in the clinic, what is the future of nanomedicine? J Control Release 244(Pt A):108–121
37. Deamer D, Bangham AD (1976) Large volume liposomes by an ether vaporization method. Biochim Biophys Acta 443(3):629–634
38. Deeken JF, Slack R, Weiss GJ, Ramanathan RK, Pishvaian MJ, Hwang J, Lewandowski K, Subramaniam D, He AR, Cotarla I, Rahman A, Marshall JL (2013) A phase I study of liposomal-encapsulated docetaxel (LE-DT) in patients with advanced solid tumor malignancies. Cancer Chemother Pharmacol 71(3):627–633
39. Deitcher SR, Cullis P, Wong M, Choy GS (2007) Vinorelbine liposomes injection results in greater tumor drug exposure compared to conventional vinorelbine in tumor-bearing nude mice. Liver 1668(1016):1–6
40. Desjardins J, Emerson DL, Colagiovanni DB, Abbott E, Brown EN, Drolet DW (2004) Pharmacokinetics, safety, and efficacy of a liposome encapsulated thymidylate synthase inhibitor, OSI-7904L [(S)-2-[5-[(1, 2-dihydro-3-methyl-1-oxobenzo [f] quinazolin-9-yl) methyl] amino-1-oxo-2-isoindolynl]-glutaric acid] in mice. J Pharmacol Exp Ther 309 (3):894
41. Douville V, Lodi A, Miller J, Nicolas A, Clarot I, Prilleux B, Megoulas N, Koupparis M (2006) Evaporative light scattering detection (ELSD): a tool for improved quality control of drug substances. Pharmeur Sci Notes 2006(1):9–15
42. Drummond DC, Noble CO, Guo Z, Hong K, Park JW, Kirpotin DB (2006) Development of a highly active nanoliposomal irinotecan using a novel intraliposomal stabilization strategy. Cancer Res 66(6):3271–3277
43. Dudkiewicz A, Tiede K, Loeschner K, Jensen LHS, Jensen E, Wierzbicki R, Boxall ABA, Molhave K (2011) Characterization of nanomaterials in food by electron microscopy. TrAC Trends Anal Chem 30(1):28–43
44. EMA (2009) Summary of product characteristics: Caelyx (doxorubicin hydrochloride in a pegylated liposomal formulation) 2 mg/mL concentrate for infusion. http://www.ema.europa.eu/ema/index.jsp?curl=pages/medicines/human/medicines/000089/human_med_000683.jsp&mid=WC0b01ac058001d124
45. EMA (2009) Summary of product characteristics: Myocet (Liposome-encapsulated doxorubicin-citrate complex) 50 mg powder, dispersion and solvent for concentrate for dispersion for infusion. http://www.ema.europa.eu/ema/index.jsp?curl=pages/medicines/human/medicines/000297/human_med_000916.jsp&mid=WC0b01ac058001d124
46. EMA (2009) Summary of product characteristics: DepoCyt (cytarabine suspension) 50 mg suspension for injection. http://www.ema.europa.eu/ema/index.jsp?curl=pages/medicines/human/medicines/000317/human_med_000740.jsp&mid=WC0b01ac058001d124
47. EMA (2009) Summary of product characteristics: Mepact (Mifamurtide) 4 mg powder for concentrate for dispersion for infusion. http://www.ema.europa.eu/ema/index.jsp?curl=pages/medicines/human/medicines/000802/human_med_000899.jsp&mid=WC0b01ac058001d124
48. EMA (2016) Summary of product characteristics: Onivyde (irinotecan hydrochloride trihydrate [as irinotecan sucrosofate salt in a pegylated liposomal formulation]) 5 mg/ml concentrate for solution for infusion. http://www.ema.europa.eu/ema/index.jsp?curl=pages/medicines/human/medicines/004125/human_med_002022.jsp&mid=WC0b01ac058001d124
49. Emerson DL, Bendele R, Brown E, Chiang S, Desjardins JP, Dihel LC, Gill SC, Hamilton M, LeRay JD, Moon-McDermott L (2000) Antitumor efficacy, pharmacokinetics,

and biodistribution of NX 211: a low-clearance liposomal formulation of lurtotecan. Clin Cancer Res 6(7):2903–2912
50. Enoch HG, Strittmatter P (1979) Formation and properties of 1000-A-diameter, single-bilayer phospholipid vesicles. Proc Natl Acad Sci U S A 76(1):145–149
51. Etheridge ML, Campbell SA, Erdman AG, Haynes CL, Wolf SM, McCullough J (2013) The big picture on small medicine: the state of nanomedicine products approved for use or in clinical trials. Nanomed Nanotechnol Biol Med 9(1):1
52. Fasol U, Frost A, Büchert M, Arends J, Fiedler U, Scharr D, Scheuenpflug J, Mross K (2011) Vascular and pharmacokinetic effects of EndoTAG-1 in patients with advanced cancer and liver metastasis. Ann Oncol 23(4):1030–1036
53. Feldman E, Kolitz J, Trang J, Liboiron B, Swenson C, Chiarella M, Mayer L, Louie A, Lancet J (2012) Pharmacokinetics of CPX-351; a nano-scale liposomal fixed molar ratio formulation of cytarabine: daunorubicin, in patients with advanced leukemia. Leuk Res 36 (10):1283–1289
54. Fenske DB, MacLachlan I, Cullis PR (2002) Stabilized plasmid-lipid particles: a systemic gene therapy vector. Methods Enzymol 346:36–71
55. Filipe V, Hawe A, Jiskoot W (2010) Critical evaluation of Nanoparticle Tracking Analysis (NTA) by NanoSight for the measurement of nanoparticles and protein aggregates. Pharm Res 27(5):796–810
56. Fonseca C, Moreira JN, Ciudad CJ, Pedroso de Lima MC, Simoes S (2005) Targeting of sterically stabilised pH-sensitive liposomes to human T-leukaemia cells. Eur J Pharm Biopharm 59(2):359–366
57. Fonseca NA, Gregorio AC, Valerio-Fernandes A, Simoes S, Moreira JN (2014) Bridging cancer biology and the patients' needs with nanotechnology-based approaches. Cancer Treat Rev 40(5):626–635
58. Fonseca NA, Rodrigues AS, Rodrigues-Santos P, Alves V, Gregorio AC, Valerio-Fernandes A, Gomes-da-Silva LC, Rosa MS, Moura V, Ramalho-Santos J, Simoes S, Moreira JN (2015) Nucleolin overexpression in breast cancer cell sub-populations with different stem-like phenotype enables targeted intracellular delivery of synergistic drug combination. Biomaterials 69:76–88
59. Forssen EA, Ross ME (1994) Daunoxome® treatment of solid tumors: preclinical and clinical investigations. J Liposome Res 4(1):481–512
60. Fu W, Zhang W (2017) Hybrid AFM for nanoscale physicochemical characterization: recent development and emerging applications. Small 13(11)
61. Furst T, Bettonville V, Farcas E, Frere A, Lechanteur A, Evrard B, Fillet M, Piel G, Servais AC (2016) Capillary electrophoresis method to determine siRNA complexation with cationic liposomes. Electrophoresis 37(20):2685–2691
62. Gabizon A, Catane R, Uziely B, Kaufman B, Safra T, Cohen R, Martin F, Huang A, Barenholz Y (1994) Prolonged circulation time and enhanced accumulation in malignant exudates of doxorubicin encapsulated in polyethylene-glycol coated liposomes. Cancer Res 54(4):987–992
63. Gaillard PJ, Appeldoorn CCM, Dorland R, van Kregten J, Manca F, Vugts DJ, Windhorst B, van Dongen GAMS, de Vries HE, Maussang D (2014) Pharmacokinetics, brain delivery, and efficacy in brain tumor-bearing mice of glutathione pegylated liposomal doxorubicin (2B3-101). PLoS ONE 9(1):e82331
64. Ginn SL, Alexander IE, Edelstein ML, Abedi MR, Wixon J (2013) Gene therapy clinical trials worldwide to 2012—an update. J Gene Med 15(2):65–77
65. Gomes-da-Silva LC, Santos AO, Bimbo LM, Moura V, Ramalho JS, Lima MC, Simoes S, Moreira JN (2012) Towards a siRNA-containing nanoparticle targeted to breast cancer cells and the tumor microenvironment. Int J Pharm 434(1–2):9–19
66. Gordon AN, Granai CO, Rose PG, Hainsworth J, Lopez A, Weissman C, Rosales R, Sharpington T (2000) Phase II study of liposomal doxorubicin in platinum-and paclitaxel-refractory epithelial ovarian cancer. J Clin Oncol 18(17):3093–3100

67. Gordon AN, Fleagle JT, Guthrie D, Parkin DE, Gore ME, Lacave AJ (2001) Recurrent epithelial ovarian carcinoma: a randomized phase III study of pegylated liposomal doxorubicin versus topotecan. J Clin Oncol 19(14):3312–3322
68. Gregoriadis G (1973) Drug entrapment in liposomes. FEBS Lett 36(3):292–296
69. Gregoriadis G (1976) The carrier potential of liposomes in biology and medicine (first of two parts). N Engl J Med 295(13):704–710
70. Gregoriadis G (1976) The carrier potential of liposomes in biology and medicine (second of two parts). N Engl J Med 295(14):765–770
71. Gregoriadis G (1984) Drug delivery: old drugs in new clothing. Nature 310:186–187
72. Gregoriadis G, Ryman BE (1971) Liposomes as carriers of enzymes or drugs: a new approach to the treatment of storage diseases. Biochem J 124(5):58P
73. Gruner SM, Lenk RP, Janoff AS, Ostro MJ (1985) Novel multilayered lipid vesicles: comparison of physical characteristics of multilamellar liposomes and stable plurilamellar vesicles. Biochemistry 24(12):2833–2842
74. Hait WN, Eder JP (2016) Principles of dose, schedule, and combination therapy, in Holland-Frei cancer medicine. Wiley, USA
75. Hamilton RL Jr, Goerke J, Guo LS, Williams MC, Havel RJ (1980) Unilamellar liposomes made with the French pressure cell: a simple preparative and semiquantitative technique. J Lipid Res 21(8):981–992
76. Hanahan D, Weinberg RA (2011) Hallmarks of cancer: the next generation. Cell 144 (5):646–674
77. Harasym TO, Tardi PG, Harasym NL, Harvie P, Johnstone SA, Mayer LD (2006) Increased preclinical efficacy of irinotecan and floxuridine coencapsulated inside liposomes is associated with tumor delivery of synergistic drug ratios. Oncol Res Featuring Preclinical Clin Cancer Ther 16(8):361–374
78. Hare JI, Lammers T, Ashford MB, Puri S, Storm G, Barry ST (2017) Challenges and strategies in anti-cancer nanomedicine development: an industry perspective. Adv Drug Deliv Rev 108:25–38
79. Ho EA, Ramsay E, Ginj M, Anantha M, Bregman I, Sy J, Woo J, Osooly-Talesh M, Yapp DT, Bally MB (2010) Characterization of cationic liposome formulations designed to exhibit extended plasma residence times and tumor vasculature targeting properties. J Pharm Sci 99(6):2839–2853
80. Hope MJ (2014) Enhancing siRNA delivery by employing lipid nanoparticles. Ther Deliv 5 (6):663–673
81. Hope MJ, Bally MB, Webb G, Cullis PR (1985) Production of large unilamellar vesicles by a rapid extrusion procedure: characterization of size distribution, trapped volume and ability to maintain a membrane potential. Biochim Biophys Acta 812(1):55–65
82. Huang C (1969) Studies on phosphatidylcholine vesicles. Formation and physical characteristics. Biochemistry 8(1):344–352
83. Huang Z, Li X, Zhang T, Song Y, She Z, Li J, Deng Y (2014) Progress involving new techniques for liposome preparation. Asian J Pharm Sci 9(4):176–182
84. Hunter DG, Frisken BJ (1998) Effect of extrusion pressure and lipid properties on the size and polydispersity of lipid vesicles. Biophys J 74(6):2996–3002
85. Immordino ML, Dosio F, Cattel L (2006) Stealth liposomes: review of the basic science, rationale, and clinical applications, existing and potential. Int J Nanomedicine 1(3):297–315
86. Infante JR, Keedy VL, Jones SF, Zamboni WC, Chan E, Bendell JC, Lee W, Wu H, Ikeda S, Kodaira H (2012) Phase I and pharmacokinetic study of IHL-305 (PEGylated liposomal irinotecan) in patients with advanced solid tumors. Cancer Chemother Pharmacol 70(5):699–705
87. Ishida T, Okada Y, Kobayashi T, Kiwada H (2006) Development of pH-sensitive liposomes that efficiently retain encapsulated doxorubicin (DXR) in blood. Int J Pharm 309(1–2):94–100
88. Ito I, Began G, Mohiuddin I, Saeki T, Saito Y, Branch CD, Vaporciyan A, Stephens LC, Yen N, Roth JA (2003) Increased uptake of liposomal–DNA complexes by lung metastases following intravenous administration. Mol Ther 7(3):409–418

89. Ito I, Ji L, Tanaka F, Saito Y, Gopalan B, Branch CD, Xu K, Atkinson EN, Bekele BN, Stephens LC (2004) Liposomal vector mediated delivery of the 3p FUS1 gene demonstrates potent antitumor activity against human lung cancer in vivo. Cancer Gene Ther 11(11):733–739
90. Jaafar-Maalej C, Diab R, Andrieu V, Elaissari A, Fessi H (2010) Ethanol injection method for hydrophilic and lipophilic drug-loaded liposome preparation. J Liposome Res 20(3):228–243
91. Jaafar-Maalej C, Charcosset C, Fessi H (2011) A new method for liposome preparation using a membrane contactor. J Liposome Res 21(3):213–220
92. Jain MM, Gupte SU, Patil SG, Pathak AB, Deshmukh CD, Bhatt N, Haritha C, Babu KG, Bondarde SA, Digumarti R (2016) Paclitaxel injection concentrate for nanodispersion versus nab-paclitaxel in women with metastatic breast cancer: a multicenter, randomized, comparative phase II/III study. Breast Cancer Res Treat 156(1):125–134
93. Kaszuba M, Corbett J, Watson FM, Jones A (1927) High-concentration zeta potential measurements using light-scattering techniques. Philos Trans A Math Phys Eng Sci 2010 (368):4439–4451
94. Koning GA, Eggermont AM, Lindner LH, ten Hagen TL (2010) Hyperthermia and thermosensitive liposomes for improved delivery of chemotherapeutic drugs to solid tumors. Pharm Res 27(8):1750–1754
95. Koshkina NV, Gilbert BE, Waldrep JC, Seryshev A, Knight V (1999) Distribution of camptothecin after delivery as a liposome aerosol or following intramuscular injection in mice. Cancer Chemother Pharmacol 44(3):187–192
96. Krasnici S, Werner A, Eichhorn ME, Schmitt-Sody M, Pahernik SA, Sauer B, Schulze B, Teifel M, Michaelis U, Naujoks K, Dellian M (2003) Effect of the surface charge of liposomes on their uptake by angiogenic tumor vessels. Int J Cancer 105(4):561–567
97. Kremer JM, Esker MW, Pathmamanoharan C, Wiersema PH (1977) Vesicles of variable diameter prepared by a modified injection method. Biochemistry 16(17):3932–3935
98. Krol A, Maresca J, Dewhirst MW, Yuan F (1999) Available volume fraction of macromolecules in the extravascular space of a fibrosarcoma: implications for drug delivery. Can Res 59(16):4136
99. Lancet JE, Cortes JE, Hogge DE, Tallman MS, Kovacsovics TJ, Damon LE, Komrokji R, Solomon SR, Kolitz JE, Cooper M, Yeager AM, Louie AC, Feldman EJ (2014) Phase 2 trial of CPX-351, a fixed 5:1 molar ratio of cytarabine/daunorubicin, vs cytarabine/daunorubicin in older adults with untreated AML. Blood 123(21):3239–3246
100. Landon CD, Park J-Y, Needham D, Dewhirst MW (2011) Nanoscale drug delivery and hyperthermia: the materials design and preclinical and clinical testing of low temperature-sensitive liposomes used in combination with mild hyperthermia in the treatment of local cancer. Open Nanomed J 3:38–64
101. Laouini A, Jaafar-Maalej C, Sfar S, Charcosset C, Fessi H (2011) Liposome preparation using a hollow fiber membrane contactor–application to spironolactone encapsulation. Int J Pharm 415(1–2):53–61
102. Lapresta-Fernández A, Salinas-Castillo A, Anderson de la Llana S, Costa-Fernández JM, Domínguez-Meister S, Cecchini R, Capitán-Vallvey LF, Moreno-Bondi MC, Marco MP, Sánchez-López JC, Anderson IS (2014) A general perspective of the characterization and quantification of nanoparticles: imaging, spectroscopic, and separation techniques. Crit Rev Solid State Mater Sci 39(6):423–458
103. Lasic DD, Martin FJ, Gabizon A, Huang SK, Papahadjopoulos D (1991) Sterically stabilized liposomes: a hypothesis on the molecular origin of the extended circulation times. Biochim Biophys Acta 1070(1):187–192
104. Lasic DD, Frederik PM, Stuart MCA, Barenholz Y, McIntosh TJ (1992) Gelation of liposome interior A novel method for drug encapsulation. FEBS Lett 312(2–3):255–258
105. Lesoin L, Boutin O, Crampon C, Badens E (2011) CO_2/water/surfactant ternary systems and liposome formation using supercritical CO_2: a review. Colloids Surf A 377(1–3):1–14
106. Leung A, Tam Y, Cullis PR (2014) Lipid nanoparticles for short interfering RNA delivery. Adv Genet 88:71–110

107. Li L, Ten Hagen TL, Hossann M, Suss R, van Rhoon GC, Eggermont AM, Haemmerich D, Koning GA (2013) Mild hyperthermia triggered doxorubicin release from optimized stealth thermosensitive liposomes improves intratumoral drug delivery and efficacy. J Control Release 168(2):142–150
108. Li L, Hou J, Liu X, Guo Y, Wu Y, Zhang L, Yang Z (2014) Nucleolin-targeting liposomes guided by aptamer AS1411 for the delivery of siRNA for the treatment of malignant melanomas. Biomaterials 35(12):3840–3850
109. Lieleg O, Baumgärtel RM, Bausch AR (2009) Selective filtering of particles by the extracellular matrix: an electrostatic bandpass. Biophys J 97(6):1569–1577
110. LiPlasome Pharma ApS (2017) LiPlaCis. Cited 2017 26 Apr 2017. Available from: http://www.liplasome.com/
111. Löhr JM, Haas SL, Bechstein W-O, Bodoky G, Cwiertka K, Fischbach W, Fölsch UR, Jäger D, Osinsky D, Prausova J (2012) Cationic liposomal paclitaxel plus gemcitabine or gemcitabine alone in patients with advanced pancreatic cancer: a randomized controlled phase II trial. Ann Oncol 23(5):1214–1222
112. LoRusso P, Krop I, Miller K, Ma C, Siegel BA, Shields AF, Molnar I, Wickham T, Reynolds J, Campbell K, Hendriks B, McClure T, Moyo V, Munster P (2015) Abstract CT234: A phase I study of MM-302, a HER2-targeted PEGylated liposomal doxorubicin, in patients with HER2+ metastatic breast cancer. Cancer Res 75(15 Supplement):CT234
113. Lu WL, Qi XR, Zhang Q, Li RY, Wang GL, Zhang RJ, Wei SL (2004) A pegylated liposomal platform: pharmacokinetics, pharmacodynamics, and toxicity in mice using doxorubicin as a model drug. J Pharmacol Sci 95(3):381–389
114. MacDonald RC, MacDonald RI, Menco BP, Takeshita K, Subbarao NK, Hu LR (1991) Small-volume extrusion apparatus for preparation of large, unilamellar vesicles. Biochim Biophys Acta 1061(2):297–303
115. MacKay JA, Deen DF, Szoka FC Jr (2005) Distribution in brain of liposomes after convection enhanced delivery; modulation by particle charge, particle diameter, and presence of steric coating. Brain Res 1035(2):139–153
116. Madden TD, Harrigan PR, Tai LC, Bally MB, Mayer LD, Redelmeier TE, Loughrey HC, Tilcock CP, Reinish LW, Cullis PR (1990) The accumulation of drugs within large unilamellar vesicles exhibiting a proton gradient: a survey. Chem Phys Lipids 53(1):37–46
117. Maeda H (2010) Tumor-selective delivery of macromolecular drugs via the EPR effect: background and future prospects. Bioconjug Chem 21(5):797–802
118. Mahalingam D, Nemunaitis JJ, Malik L, Sarantopoulos J, Weitman S, Sankhala K, Hart J, Kousba A, Gallegos NS, Anderson G, Charles J, Rogers JM, Senzer NN, Mita AC (2014) Phase I study of intravenously administered ATI-1123, a liposomal docetaxel formulation in patients with advanced solid tumors. Cancer Chemother Pharmacol 74(6):1241–1250
119. Martin B, Sainlos M, Aissaoui A, Oudrhiri N, Hauchecorne M, Vigneron JP, Lehn JM, Lehn P (2005) The design of cationic lipids for gene delivery. Curr Pharm Des 11(3):375–394
120. Martins CP, Brown-Swigart L, Evan GI (2006) Modeling the therapeutic efficacy of p53 restoration in tumors. Cell 127(7):1323–1334
121. Matsumura Y, Maeda H (1986) A new concept for macromolecular therapeutics in cancer chemotherapy: mechanism of tumoritropic accumulation of proteins and the antitumor agent smancs. Cancer Res 46(12 Part 1):6387–6392
122. Maurer N, Mori A, Palmer L, Monck MA, Mok KW, Mui B, Akhong QF, Cullis PR (1999) Lipid-based systems for the intracellular delivery of genetic drugs. Mol Membr Biol 16 (1):129–140
123. Maurer N, Wong KF, Stark H, Louie L, McIntosh D, Wong T, Scherrer P, Semple SC, Cullis PR (2001) Spontaneous entrapment of polynucleotides upon electrostatic interaction with ethanol-destabilized cationic liposomes. Biophys J 80(5):2310–2326
124. Mayer LD, Hope MJ, Cullis PR (1986) Vesicles of variable sizes produced by a rapid extrusion procedure. Biochim Biophys Acta 858(1):161–168
125. Mayer LD, Bally MB, Hope MJ, Cullis PR (1986) Techniques for encapsulating bioactive agents into liposomes. Chem Phys Lipids 40(2–4):333–345

126. Mayer LD, Bally MB, Cullis PR (1986) Uptake of adriamycin into large unilamellar vesicles in response to a pH gradient. Biochim Biophys Acta 857(1):123–126
127. Mayer LD, Tai LC, Bally MB, Mitilenes GN, Ginsberg RS, Cullis PR (1990) Characterization of liposomal systems containing doxorubicin entrapped in response to pH gradients. Biochim Biophys Acta 1025(2):143–151
128. Mayer LD, Harasym TO, Tardi PG, Harasym NL, Shew CR, Johnstone SA, Ramsay EC, Bally MB, Janoff AS (2006) Ratiometric dosing of anticancer drug combinations: controlling drug ratios after systemic administration regulates therapeutic activity in tumor-bearing mice. Mol Cancer Ther 5(7):1854–1863
129. Merrimack Pharmaceuticals (2016) I. Merrimack stops the Phase 2 HERMIONE trial of MM-302 in HER2-positive metastatic breast cancer patients. 21 Dec 2016. Cited 26 Apr 2017. Available from: http://investors.merrimackpharma.com/releasedetail.cfm?ReleaseID=1005127
130. Meure LA, Foster NR, Dehghani F (2008) Conventional and dense gas techniques for the production of liposomes: a review. AAPS PharmSciTech 9(3):798–809
131. Miller K, Cortes J, Hurvitz SA, Krop IE, Tripathy D, Verma S, Riahi K, Reynolds JG, Wickham TJ, Molnar I (2016) HERMIONE: a randomized Phase 2 trial of MM-302 plus trastuzumab versus chemotherapy of physician's choice plus trastuzumab in patients with previously treated, anthracycline-naïve, HER2-positive, locally advanced/metastatic breast cancer. BMC Cancer 16(1):352
132. Milne JL, Borgnia MJ, Bartesaghi A, Tran EE, Earl LA, Schauder DM, Lengyel J, Pierson J, Patwardhan A, Subramaniam S (2013) Cryo-electron microscopy—a primer for the non-microscopist. FEBS J 280(1):28–45
133. Milsmann MH, Schwendener RA, Weder HG (1978) The preparation of large single bilayer liposomes by a fast and controlled dialysis. Biochim Biophys Acta 512(1):147–155
134. MiNA Therapeutics (2017) OUTREACH—the first-in-human clinical study of a saRNA therapeutic. Cited 2017 26 Apr 2017. Available from: http://minatx.com/clinical_development/
135. Minematsu H, Otani T, Oohashi T, Hirai M, Oie K, Igarashi K, Ohtsuka A (2011) Development of an active targeting liposome encapsulated with high-density colloidal gold for transmission electron microscopy. J Electron Microsc (Tokyo) 60(1):95–99
136. Mok KW, Cullis PR (1997) Structural and fusogenic properties of cationic liposomes in the presence of plasmid DNA. Biophys J 73(5):2534–2545
137. Monteiro N, Martins A, Pires R, Faria S, Fonseca NA, Moreira JN, Reis RL, Neves NM (2014) Immobilization of bioactive factor-loaded liposomes on the surface of electrospun nanofibers targeting tissue engineering. Biomater Sci 2(9):1195
138. Monteiro N, Ribeiro D, Martins A, Faria S, Fonseca NA, Moreira JN, Reis RL, Neves NM (2014) Instructive nanofibrous scaffold comprising runt-related transcription factor 2 gene delivery for bone tissue engineering. ACS Nano 8(8):8082–8094
139. Monteiro N, Martins A, Ribeiro D, Faria S, Fonseca NA, Moreira JN, Reis RL, Neves NM (2015) On the use of dexamethasone-loaded liposomes to induce the osteogenic differentiation of human mesenchymal stem cells. J Tissue Eng Regen Med 9(9):1056–1066
140. Mozafari MR (2005) Liposomes: an overview of manufacturing techniques. Cell Mol Biol Lett 10(4):711–719
141. Muggia FM, Hainsworth JD, Jeffers S, Miller P, Groshen S, Tan M, Roman L, Uziely B, Muderspach L, Garcia A (1997) Phase II study of liposomal doxorubicin in refractory ovarian cancer: antitumor activity and toxicity modification by liposomal encapsulation. J Clin Oncol 15(3):987–993
142. Mura S, Nicolas J, Couvreur P (2013) Stimuli-responsive nanocarriers for drug delivery. Nat Mater 12(11):991–1003
143. Mylonakis N, Athanasiou A, Ziras N, Angel J, Rapti A, Lampaki S, Politis N, Karanikas C, Kosmas C (2010) Phase II study of liposomal cisplatin (Lipoplatin™) plus gemcitabine versus cisplatin plus gemcitabine as first line treatment in inoperable (stage IIIB/IV) non-small cell lung cancer. Lung Cancer 68(2):240–247

144. Nabel GJ, Nabel EG, Yang ZY, Fox BA, Plautz GE, Gao X, Huang L, Shu S, Gordon D, Chang AE (1993) Direct gene transfer with DNA-liposome complexes in melanoma: expression, biologic activity, and lack of toxicity in humans. Proc Natl Acad Sci 90 (23):11307–11311
145. Nagayasu A, Uchiyama K, Kiwada H (1999) The size of liposomes: a factor which affects their targeting efficiency to tumors and therapeutic activity of liposomal antitumor drugs. Adv Drug Deliv Rev 40(1–2):75–87
146. Nardin A, Lefebvre ML, Labroquere K, Faure O, Abastado JP (2006) Liposomal muramyl tripeptide phosphatidylethanolamine: targeting and activating macrophages for adjuvant treatment of osteosarcoma. Curr Cancer Drug Targets 6(2):123–133
147. Needham D, Dewhirst MW (2001) The development and testing of a new temperature-sensitive drug delivery system for the treatment of solid tumors. Adv Drug Deliv Rev 53(3):285–305
148. Needham D, Anyarambhatla G, Kong G, Dewhirst MW (2000) A new temperature-sensitive liposome for use with mild hyperthermia: characterization and testing in a human tumor xenograft model. Cancer Res 60(5):1197–1201
149. Nemunaitis J, Nemunaitis J (2011) Head and neck cancer: response to p53-based therapeutics. Head Neck 33(1):131–134
150. Nemunaitis J, Swisher S, Timmons T, Connors D, Mack M, Doerksen L, Weill D, Wait J, Lawrence D, Kemp B (2000) Adenovirus-mediated p53 gene transfer in sequence with cisplatin to tumors of patients with non–small-cell lung cancer. J Clin Oncol 18(3):609
151. Nemunaitis J, Clayman G, Agarwala SS, Hrushesky W, Wells JR, Moore C, Hamm J, Yoo G, Baselga J, Murphy BA (2009) Biomarkers predict p53 gene therapy efficacy in recurrent squamous cell carcinoma of the head and neck. Clin Cancer Res 15(24):7719–7725
152. Newman MS, Colbern GT, Working PK, Engbers C, Amantea MA (1999) Comparative pharmacokinetics, tissue distribution, and therapeutic effectiveness of cisplatin encapsulated in long-circulating, pegylated liposomes (SPI-077) in tumor-bearing mice. Cancer Chemother Pharmacol 43(1):1–7
153. Nichols JW, Bae YH (2014) EPR: evidence and fallacy. J Control Release 190:451–464
154. Northfelt DW, Hamburg SI, Borad MJ, Seetharam M, Curtis KK, Lee P, Crowell B, Vocila L, Fredlund P, Gilbert MJ, Mast CP, Semple SC, Judge AD, MacLachlan I, Ramanathan RK (2013) A phase I dose-escalation study of TKM-080301, a RNAi therapeutic directed against polo-like kinase 1 (PLK1), in patients with advanced solid tumors: expansion cohort evaluation of biopsy samples for evidence of pharmacodynamic effects of PLK1 inhibition. J Clin Oncol 31(15)
155. O'Brien MER, Wigler N, Inbar M, Rosso R, Grischke E, Santoro A, Catane R, Kieback DG, Tomczak P, Ackland SP, Orlandi F, Mellars L, Alland L, Tendler C (2004) Reduced cardiotoxicity and comparable efficacy in a phase III trial of pegylated liposomal doxorubicin HCl (CAELYX/Doxil) versus conventional doxorubicin for first-line treatment of metastatic breast cancer. Ann Oncol 15(3):440–449
156. O'Brien S, Schiller G, Lister J, Damon L, Goldberg S, Aulitzky W, Ben-Yehuda D, Stock W, Coutre S, Douer D (2012) High-dose vincristine sulfate liposome injection for advanced, relapsed, and refractory adult Philadelphia chromosome–negative acute lymphoblastic leukemia. J Clin Oncol 31(6):676–683
157. Ohnishi N, Tomida H, Ito Y, Tahara K, Takeuchi H (2014) Characterization of a doxorubicin liposome formulation by a novel in vitro release test methodology using column-switching high-performance liquid chromatography. Chem Pharm Bull 62(6):538–544
158. Orlowski RZ, Nagler A, Sonneveld P, Bladé J, Hajek R, Spencer A, San Miguel J, Robak T, Dmoszynska A, Horvath N (2007) Randomized phase III study of pegylated liposomal doxorubicin plus bortezomib compared with bortezomib alone in relapsed or refractory multiple myeloma: combination therapy improves time to progression. J Clin Oncol 25 (25):3892–3901

159. Oswald M, Platscher M, Geissler S, Goepferich A (2016) HPLC analysis as a tool for assessing targeted liposome composition. Int J Pharm 497(1–2):293–300
160. Pal A, Khan S, Wang Y-F, Kamath N, Sarkar AK, Ahmad A, Sheikh S, Ali S, Carbonaro D, Zhang A (2005) Preclinical safety, pharmacokinetics and antitumor efficacy profile of liposome-entrapped SN-38 formulation. Anticancer Res 25(1A):331–341
161. Papahadjopoulos D, Allen TM, Gabizon A, Mayhew E, Matthay K, Huang SK, Lee KD, Woodle MC, Lasic DD, Redemann C et al (1991) Sterically stabilized liposomes: improvements in pharmacokinetics and antitumor therapeutic efficacy. Proc Natl Acad Sci U S A 88(24):11460–11464
162. Parthasarathy R, Sacks PG, Harris D, Brock H, Mehta K (1994) Interaction of liposome-associated all-trans-retinoic acid with squamous carcinoma cells. Cancer Chemother Pharmacol 34(6):527–534
163. Patty PJ, Frisken BJ (2003) The pressure-dependence of the size of extruded vesicles. Biophys J 85(2):996–1004
164. Peer D, Karp JM, Hong S, Farokhzad OC, Margalit R, Langer R (2007) Nanocarriers as an emerging platform for cancer therapy. Nat Nanotechnol 2(12):751–760
165. Pereira-Lachataignerais J, Pons R, Panizza P, Courbin L, Rouch J, Lopez O (2006) Study and formation of vesicle systems with low polydispersity index by ultrasound method. Chem Phys Lipids 140(1–2):88–97
166. Perez-Soler R, Lautersztain J, Stephens L, Wright K, Khokhar A (1989) Preclinical toxicity and pharmacology of liposome-entrapped cis-bis-neodecanoato-trans-R, R-1, 2-diaminocyclohexane platinum (II). Cancer Chemother Pharmacol 24(1):1–8
167. Peschka R, Purmann T, Schubert R (1998) Cross-flow filtration—an improved detergent removal technique for the preparation of liposomes. Int J Pharm 162(1–2):177–183
168. Phuphanich S, Maria B, Braeckman R, Chamberlain M (2007) A pharmacokinetic study of intra-CSF administered encapsulated cytarabine (DepoCyt®) for the treatment of neoplastic meningitis in patients with leukemia, lymphoma, or solid tumors as part of a phase III study. J Neurooncol 81(2):201–208
169. Pidgeon C, McNeely S, Schmidt T, Johnson JE (1987) Multilayered vesicles prepared by reverse-phase evaporation: liposome structure and optimum solute entrapment. Biochemistry 26(1):17–29
170. Pinto AC, Moreira JN, Simoes S (2011) Liposomal imatinib-mitoxantrone combination: formulation development and therapeutic evaluation in an animal model of prostate cancer. Prostate 71(1):81–90
171. Placzek M, Kosela M (2016) Microscopic methods in analysis of submicron phospholipid dispersions. Acta Pharm 66(1):1–22
172. Popa R, Vranceanu M, Nikolaus S, Nirschl H, Leneweit G (2008) Entrance effects at nanopores of nanocapsules functionalized with poly(ethylene glycol) and their flow through nanochannels. Langmuir 24(22):13030–13036
173. Prabhakar U, Maeda H, Jain RK, Sevick-Muraca EM, Zamboni W, Farokhzad OC, Barry ST, Gabizon A, Grodzinski P, Blakey DC (2013) Challenges and key considerations of the enhanced permeability and retention effect for nanomedicine drug delivery in oncology. Cancer Res 73(8):2412–2417
174. Ragelle H, Danhier F, Préat V, Langer R, Anderson DG (2016) Nanoparticle-based drug delivery systems: a commercial and regulatory outlook as the field matures. Expert Opin Drug Delivery 1–14
175. Richard BM, Ott LR, Haan D, Brubaker AN, Cole PI, Nelson KG, Ross PE, Rebelatto MC, Newton PE (2011) The safety and tolerability evaluation of DepoFoam bupivacaine (bupivacaine extended-release liposome injection) administered by incision wound infiltration in rabbits and dogs. Expert Opin Investig Drugs 20(10):1327–1341
176. Richardson ES, Pitt WG, Woodbury DJ (2007) The role of cavitation in liposome formation. Biophys J 93(12):4100–4107

177. Roces CB, Kastner E, Stone P, Lowry D, Perrie Y (2016) Rapid quantification and validation of lipid concentrations within liposomes. Pharmaceutics 8(3):29
178. Ruozi B, Tosi G, Leo E, Vandelli MA (2007) Application of atomic force microscopy to characterize liposomes as drug and gene carriers. Talanta 73(1):12–22
179. Sahin U, Karikó K, Türeci Ö (2014) mRNA-based therapeutics—developing a new class of drugs. Nat Rev Drug Discov 13(10):759–780
180. Salvati A, Pitek AS, Monopoli MP, Prapainop K, Bombelli FB, Hristov DR, Kelly PM, Aberg C, Mahon E, Dawson KA (2013) Transferrin-functionalized nanoparticles lose their targeting capabilities when a biomolecule corona adsorbs on the surface. Nat Nanotechnol 8 (2):137–143
181. Sapra P, Allen TM (2003) Ligand-targeted liposomal anticancer drugs. Prog Lipid Res 42 (5):439–462
182. Seetharamu N, Kim E, Hochster H, Martin F, Muggia F (2010) Phase II study of liposomal cisplatin (SPI-77) in platinum-sensitive recurrences of ovarian cancer. Anticancer Res 30 (2):541–545
183. Semple SC, Klimuk SK, Harasym TO, Dos Santos N, Ansell SM, Wong KF, Maurer N, Stark H, Cullis PR, Hope MJ, Scherrer P (2001) Efficient encapsulation of antisense oligonucleotides in lipid vesicles using ionizable aminolipids: formation of novel small multilamellar vesicle structures. Biochim Biophys Acta 1510(1–2):152–166
184. Semple SC, Akinc A, Chen J, Sandhu AP, Mui BL, Cho CK, Sah DW, Stebbing D, Crosley EJ, Yaworski E, Hafez IM, Dorkin JR, Qin J, Lam K, Rajeev KG, Wong KF, Jeffs LB, Nechev L, Eisenhardt ML, Jayaraman M, Kazem M, Maier MA, Srinivasulu M, Weinstein MJ, Chen Q, Alvarez R, Barros SA, De S, Klimuk SK, Borland T, Kosovrasti V, Cantley WL, Tam YK, Manoharan M, Ciufolini MA, Tracy MA, de Fougerolles A, MacLachlan I, Cullis PR, Madden TD, Hope MJ (2010) Rational design of cationic lipids for siRNA delivery. Nat Biotechnol 28(2):172–176
185. Senzer N, Nemunaitis J, Nemunaitis D, Bedell C, Edelman G, Barve M, Nunan R, Pirollo KF, Rait A, Chang EH (2013) Phase I study of a systemically delivered p53 nanoparticle in advanced solid tumors. Mol Ther 21(5):1096–1103
186. Siddiqui MR, AlOthman ZA, Rahman N (2017) Analytical techniques in pharmaceutical analysis: a review. Arab J Chem 10:S1409–S1421
187. Silva R, Ferreira H, Little C, Cavaco-Paulo A (2010) Effect of ultrasound parameters for unilamellar liposome preparation. Ultrason Sonochem 17(3):628–632
188. Singh Y, Tomar S, Khan S, Meher JG, Pawar VK, Raval K, Sharma K, Singh PK, Chaurasia M, Reddy BS (2015) Bridging small interfering RNA with giant therapeutic outcomes using nanometric liposomes. J Controlled Release 220:368–387
189. Stathopoulos GP, Antoniou D, Dimitroulis J, Michalopoulou P, Bastas A, Marosis K, Stathopoulos J, Provata A, Yiamboudakis P, Veldekis D, Lolis N, Georgatou N, Toubis M, Pappas C, Tsoukalas G (2010) Liposomal cisplatin combined with paclitaxel versus cisplatin and paclitaxel in non-small-cell lung cancer: a randomized phase III multicenter trial. Ann Oncol 21(11):2227–2232
190. Stathopoulos G, Antoniou D, Dimitroulis J, Stathopoulos J, Marosis K, Michalopoulou P (2011) Comparison of liposomal cisplatin versus cisplatin in non-squamous cell non-small-cell lung cancer. Cancer Chemother Pharmacol 68(4):945–950
191. Stopeck AT, Jones A, Hersh EM, Thompson JA, Finucane DM, Gutheil JC, Gonzalez R (2001) Phase ii study of direct intralesional gene transfer of allovectin-7, an HLA-B7/β2-microglobulin DNA-liposome complex, in patients with metastatic melanoma. Clin Cancer Res 7(8):2285
192. Suzuki R, Takizawa T, Kuwata Y, Mutoh M, Ishiguro N, Utoguchi N, Shinohara A, Eriguchi M, Yanagie H, Maruyama K (2008) Effective anti-tumor activity of oxaliplatin encapsulated in transferrin-PEG-liposome. Int J Pharm 346(1–2):143–150

193. Swisher SG, Roth JA, Komaki R, Gu J, Lee JJ, Hicks M, Ro JY, Hong WK, Merritt JA, Ahrar K (2003) Induction of p53-regulated genes and tumor regression in lung cancer patients after intratumoral delivery of adenoviral p53 (INGN 201) and radiation therapy. Clin Cancer Res 9(1):93–101
194. Symon Z, Peyser A, Tzemach D, Lyass O, Sucher E, Shezen E, Gabizon A (1999) Selective delivery of doxorubicin to patients with breast carcinoma metastases by stealth liposomes. Cancer 86(1):72–78
195. Szoka F Jr, Papahadjopoulos D (1978) Procedure for preparation of liposomes with large internal aqueous space and high capture by reverse-phase evaporation. Proc Natl Acad Sci U S A 75(9):4194–4198
196. Ta T, Porter TM (2013) Thermosensitive liposomes for localized delivery and triggered release of chemotherapy. J Control Release 169(1–2):112–125
197. Tardi PG, Gallagher RC, Johnstone S, Harasym N, Webb M, Bally MB, Mayer LD (1768) Coencapsulation of irinotecan and floxuridine into low cholesterol-containing liposomes that coordinate drug release in vivo. Biochim Biophys Acta 1768(3):678–687
198. Tardi P, Johnstone S, Harasym N, Xie S, Harasym T, Zisman N, Harvie P, Bermudes D, Mayer L (2009) In vivo maintenance of synergistic cytarabine: daunorubicin ratios greatly enhances therapeutic efficacy. Leuk Res 33(1):129–139
199. Toh M-R, Chiu GNC (2013) Liposomes as sterile preparations and limitations of sterilisation techniques in liposomal manufacturing. Asian J Pharm Sci 8(2):88–95
200. Tolcher AW, Papadopoulos KP, Patnaik A, Rasco DW, Martinez D, Wood DL, Fielman B, Sharma M, Janisch LA, Brown BD (2015) Safety and activity of DCR-MYC, a first-in-class Dicer-substrate small interfering RNA (DsiRNA) targeting MYC, in a phase I study in patients with advanced solid tumors. American Society of Clinical Oncology
201. Tong M, Brown OS, Stone PR, Cree LM, Chamley LW (2016) Flow speed alters the apparent size and concentration of particles measured using NanoSight nanoparticle tracking analysis. Placenta 38:29–32
202. Trapiella-Alfonso L, d'Orlye F, Varenne A (2016) Recent advances in the development of capillary electrophoresis methodologies for optimizing, controlling, and characterizing the synthesis, functionalization, and physicochemical, properties of nanoparticles. Anal Bioanal Chem 408(11):2669–2675
203. Uchiyama K, Nagayasu A, Yamagiwa Y, Nishida T, Harashima H, Kiwada H (1995) Effects of the size and fluidity of liposomes on their accumulation in tumors: a presumption of their interaction with tumors. Int J Pharm 121(2):195–203
204. US Food Drug Administration (2017) FDA approval of generic version of cancer drug Doxil is expected to help resolve shortage. Accessed 1 Feb 2017. Available from: http://www.fda.gov/newsevents/newsroom/pressannouncements/ucm337872.htm
205. Uziely B, Jeffers S, Isacson R, Kutsch K, Wei-Tsao D, Yehoshua Z, Libson E, Muggia FM, Gabizon A (1995) Liposomal doxorubicin: antitumor activity and unique toxicities during two complementary phase I studies. J Clin Oncol 13(7):1777–1785
206. Ventura A, Kirsch DG, McLaughlin ME, Tuveson DA, Grimm J, Lintault L, Newman J, Reczek EE, Weissleder R, Jacks T (2007) Restoration of p53 function leads to tumour regression in vivo. Nature 445(7128):661–665
207. Verschraegen C, Gilbert B, Huaringa A, Newman R, Harris N, Leyva F, Keus L, Campbell K, Nelson-Taylor T, Knight V (2000) Feasibility, Phase I, and pharmacological study of aerosolized liposomal 9-nitro-20 (S)-camptothecin in patients with advanced malignancies in the lungs. Ann N Y Acad Sci 922(1):352–354
208. Vical Incorporated (2013) Vical Phase 3 trial of Allovectin® fails to meet efficacy endpoints
209. Wagner A, Vorauer-Uhl K (2011) Liposome technology for industrial purposes. J Drug Deliv 2011:591325
210. Wallace TL, Larson JL, Bazemore SA, Wilson CW, Cossum PA (2000) The nonclinical safety evaluation of the anticancer drug Atragen® (liposomal all-trans-retinoic acid). Int J Toxicol 19(1):33–42

211. Wan C, Allen TM, Cullis PR (2014) Lipid nanoparticle delivery systems for siRNA-based therapeutics. Drug Deliv Transl Res 4(1):74–83
212. Wang AZ, Langer R, Farokhzad OC (2012) Nanoparticle delivery of cancer drugs. Annu Rev Med 63:185–198
213. Wasan KM, Ramaswamy M, Ng SP, Wong W, Parrott SC, Ojwang JO, Wallace T, Cossum PA (1998) Differences in the lipoprotein distribution of free and liposome-associated all-trans-retinoic acid in human, dog, and rat plasma are due to variations in lipoprotein lipid and protein content. Antimicrob Agents Chemother 42 (7):1646–1653
214. Webb MS, Harasym TO, Masin D, Bally MB, Mayer LD (1995) Sphingomyelin-cholesterol liposomes significantly enhance the pharmacokinetic and therapeutic properties of vincristine in murine and human tumour models. Br J Cancer 72(4):896–904
215. Whatley H (2001) Basic principles and modes of capillary electrophoresis, in clinical and forensic applications of capillary electrophoresis. Humana Press, Totowa, pp 21–58
216. Wheeler JJ, Palmer L, Ossanlou M, MacLachlan I, Graham RW, Zhang YP, Hope MJ, Scherrer P, Cullis PR (1999) Stabilized plasmid-lipid particles: construction and characterization. Gene Ther 6(2):271–281
217. Wittrup A, Lieberman J (2015) Knocking down disease: a progress report on siRNA therapeutics. Nat Rev Genet 16(9):543–552
218. Wu H, Infante JR, Keedy VL, Jones SF, Chan E, Bendell JC, Lee W, Kirschbrown WP, Zamboni BA, Ikeda S (2015) Factors affecting the pharmacokinetics and pharmacodynamics of PEGylated liposomal irinotecan (IHL-305) in patients with advanced solid tumors. Int J Nanomed 10:1201
219. Xu L, Tang WH, Huang CC, Alexander W, Xiang LM, Pirollo KF, Rait A, Chang EH (2001) Systemic p53 gene therapy of cancer with immunolipoplexes targeted by anti-transferrin receptor scFv. Mol Med 7(10):723–734
220. Xu L, Huang C-C, Huang W, Tang W-H, Rait A, Yin YZ, Cruz I, Xiang L-M, Pirollo KF, Chang EH (2002) Systemic tumor-targeted gene delivery by anti-transferrin receptor scFv-immunoliposomes. Mol Cancer Ther 1(5):337–346
221. Yamaguchi T, Nomura M, Matsuoka T, Koda S (2009) Effects of frequency and power of ultrasound on the size reduction of liposome. Chem Phys Lipids 160(1):58–62
222. Yatvin MB, Weinstein JN, Dennis WH, Blumenthal R (1978) Design of liposomes for enhanced local release of drugs by hyperthermia. Science 202(4374):1290–1293
223. Yin H, Kanasty RL, Eltoukhy AA, Vegas AJ, Dorkin JR, Anderson DG (2014) Non-viral vectors for gene-based therapy. Nat Rev Genet 15(8):541–555
224. Yu NY, Conway C, Pena RL, Chen JY (2007) STEALTH® liposomal CKD-602, a topoisomerase I inhibitor, improves the therapeutic index in human tumor xenograft models. Anticancer Res 27(4B):2541–2545
225. Zamboni WC, Strychor S, Sidone BJ, Santory M, Berryman D, Abutin RM, Deitcher SR (2007) A pharmacokinetic study of a novel sphingomyelin/cholesterol liposomal topotecan and non-liposomal topotecan in rats. In: Molecular cancer therapeutics. American Association for Cancer Research, Philadelphia
226. Zamboni WC, Strychor S, Joseph E, Walsh DR, Zamboni BA, Parise RA, Tonda ME, Ning YY, Engbers C, Eiseman JL (2007) Plasma, tumor, and tissue disposition of STEALTH liposomal CKD-602 (S-CKD602) and nonliposomal CKD-602 in mice bearing A375 human melanoma xenografts. Clin Cancer Res 13(23):7217–7223
227. Zamboni WC, Strychor S, Maruca L, Ramalingam S, Zamboni B, Wu H, Friedland D, Edwards R, Stoller R, Belani C (2009) Pharmacokinetic study of pegylated liposomal CKD-602 (S-CKD602) in patients with advanced malignancies. Clin Pharmacol Ther 86 (5):519–526
228. Zhang JA, Anyarambhatla G, Ma L, Ugwu S, Xuan T, Sardone T, Ahmad I (2005) Development and characterization of a novel Cremophor EL free liposome-based paclitaxel (LEP-ETU) formulation. Eur J Pharm Biopharm 59(1):177–187

229. Zhigaltsev IV, Winters G, Srinivasulu M, Crawford J, Wong M, Amankwa L, Waterhouse D, Masin D, Webb M, Harasym N, Heller L, Bally MB, Ciufolini MA, Cullis PR, Maurer N (2010) Development of a weak-base docetaxel derivative that can be loaded into lipid nanoparticles. J Control Release 144(3):332–340

Ana Filipa Cruz has a master degree in Pharmaceutical Sciences from the Faculty of Pharmacy, University of Coimbra. Currently, she is a graduate student at CNC focused on the study of the molecular mechanisms underlying the activity of targeted delivery nanosystems encapsulating synergistic drug combinations.

Nuno A. Fonseca has a degree in Pharmaceutical Sciences and Ph.D. in Pharmaceutical Technology by the Faculty of Pharmacy of the University of Coimbra. His research focuses on bridging the fundamentals of cancer biology and the development of lipid-based nanoparticles for targeted delivery of synergistic drug combinations. He is author/co-author of several manuscripts in peer-reviewed journals and co-inventor of a patent application.

Ana C. Gregório has degree in Biochemistry by the University of Évora and a Ph.D. in Oncobiology from the University of Coimbra. Her research interests spanned different including immunology, neurosciences and oncobiology. Currently, she is a researcher in TREAT U, a spin-off of the University of Coimbra dedicated to the development of targeted nanotechnology-based platforms for the specific delivery of drugs. She is author/co-author of several manuscripts in peer-reviewed journals.

Vera Moura has a degree in Pharmaceutical Sciences and a Ph.D. in Pharmaceutical Technology by the Faculty of Pharmacy of the University of Coimbra. Her activity focuses on the development of lipid-based nanosystems for targeted drug delivery towards the tumor microenvironment. She is co-author of three patent applications granted in the USA and co-founder of TREAT U, a spin-off from the University of Coimbra.

Sérgio Simões is Tenured Associate Professor with Habilitation at the University of Coimbra and a contributing lecturer for the MIT-Portugal Program. His research activities include the participation in several national/international projects and private companies.

He is the author/co-author of over 11 book chapters, 110 scientific articles and co-inventor of 8 international patents. Moreover, his activities span to the entrepreneurial sector where he is member of the board and Vice-President of Bluepharma and the co-founder and shareholder of other pharmaceutical technology-based companies operating in the field of research, development and production of medicines, such as Pharmagen, Luzitin and TREAT U.

João Nuno Moreira is Tenured Assistant Professor with Habilitation at the University of Coimbra. Research activity is focused on the design of lipid-based nanoparticles for (small) drugs and nucleic acid targeted delivery, addressing the impact on the tumor microenvironment, both at the cellular and molecular level. He has been coordinator of several funded research projects as principal investigator, both at the national and European level, and has lectured over one hundred oral communications as invited lecturer. He has co-authored several publications in peer-reviewed journals, book chapters and patents (three of which granted in the USA). He is Co-founder of TREAT U, SA, a nanotechnology-based spin-off from the UC/CNC.

Chapter 5
Fundamentals of Dry Powder Inhaler Technology

Anthony J. Hickey

Abstract Dry powders and their delivery devices are an alternative to pressurized metered dose inhalers for the administration of aerosols to the lungs. Use of powders would reduce the necessity of employing global warming propellants; however further developments are needed in the technology required to disperse dry powders as aerosols if there is to be an alternative to pMDI as the primary means of aerosol delivery. For progress to occur the following factors influencing dry powder dispensing and dispersion must be considered. (1) The way in which particles are prepared may influence their subsequent behavioral characteristics. These preparative techniques fall in the general categories of constructive and destructive methods. (2) Specific morphological characteristics of particles such as size, shape, and rugosity may influence their interaction. In addition, particles may interact with each other or with surrounding surfaces. Among the underlying forces giving rise to these interactions are electrostatic, van der Waals, capillary and mechanical interlocking forces. (3) Powder may be dispersed by laboratory equipment, dry dust and fluidized bed generators, or small hand held devices designed for delivery of drugs to the lung. A development program requires selection of optimal conditions for each of these steps in the delivery of drugs to the lungs. The following description of dry-powder production, dispersion in air and particulate interactions discusses some of the difficulties that must be overcome for systems to advance as alternative to MDIs.

5.1 Introduction

Dry powder inhalers (DPIs) have been a component of therapy for asthma for over fifty years and for chronic obstructive lung disease for over a decade [20]. The effectiveness of the pharmaceutical aerosol powder delivered from these

A. J. Hickey (✉)
RTI International, Research Triangle Park, NC, USA
e-mail: ahickey@rti.org

H. G. Merkus et al. (eds.), *Particles and Nanoparticles in Pharmaceutical Products*,
AAPS Advances in the Pharmaceutical Sciences Series 29,
https://doi.org/10.1007/978-3-319-94174-5_5

devices in alleviating symptoms or treating the underlying cause of disease depends on the dose delivered and the aerodynamic particle size distribution. It is generally accepted that particles in the 1–5 μm size range are capable of entering the lungs whereas those 10 μm and greater have a low probability of entering the lungs and are predominantly captured in the oropharynx [16]. Bolus delivery of drug to achieve this objective has been effectively achieved by both pressurized metered dose inhalers (pMDIs) and DPIs. However, since the turn of the millennium there has been an increase in the use of DPIs for this purpose.

Until the development of Advair Diskus, the enormously successful dry powder inhaler (DPI) for the delivery of salmeterol and fluticasone in the treatment of asthma, pressurized metered dose inhalers (pMDI) were by far the dominant pharmaceutical inhalation aerosol product. Since the transition from chlorofluorocarbon propellants in pMDI due to their role in ozone depletion to the alternative hydrofluoroalkane (HFA) propellants the market for these systems has begun to increase with the development of new drugs and combinations delivered from them [17, 30]. However, the recent decision to include HFAs in the Montreal Protocol due to their global warming effects raises the specter again of a need for alternatives the most prominent of which is the dry powder inhaler [26].

Key product development activities for dry powder inhalers are identified in Fig. 5.1a. In order to address these activities certain processes and properties must be addressed as shown in Fig. 5.1b. These processes and properties relate to the components and system variables, shown in Fig. 5.1c, that must be understood and controlled for the product to exhibit robust, efficient and reproducible performance. The majority of commercially available DPIs consist of micronized drug powder mixed with an inert carrier-usually lactose. Hard capsules, formerly of gelatin but more frequently of hydroxypropyl methyl cellulose in current products, and multi-unit dose blister disks or strips are prominent examples of metering systems. The device is manipulated to pierce or rupture the capsule or blister and the patient's inspiratory airflow causes deaggregation and aerosolization of the powder. For these DPIs powder deagglomeration depends solely on the energy input provided by the patient's inspiratory flow. The extent of the dose delivered to the lung can change because of variability in inhalation characteristics between patients. Aerosol generators that rely on inspired airflow for dry powder dispersion are also subject to variability because of differences in the physical and chemical properties of the powder and the operating principles of the device. Poor dose reproducibility, and proportion of the dose in a size range capable of depositing in the lungs, may be improved by devices that employ a means of dispersion independent of the patients' inspiratory flow [7, 9, 15].

Considering the connection between the development activities, processes, properties, components and system variables outlined in Fig. 5.1 it is evident that global understanding and capacity to monitor and control variables that contribute to product performance is consistent with quality by design principles encouraged by regulators and adopted by the industry [19, 42].

A micronized powder requires an input of energy to overcome the interparticulate cohesive and particle-surface adhesive forces to produce airborne particles.

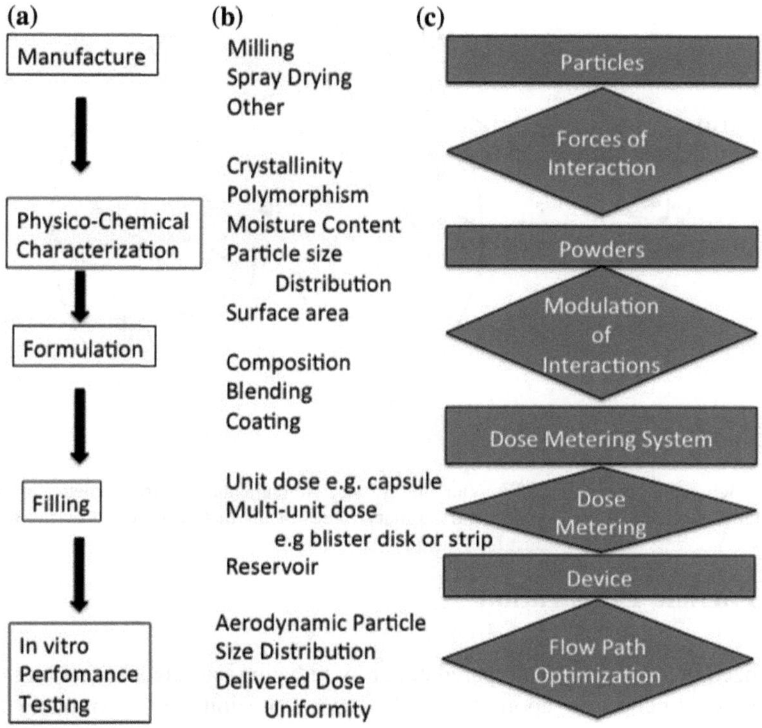

Fig. 5.1 Dry powder inhaler product: **a** development activities; **b** processes and properties; **c** components (filled rectangle) and system variables (filled diamond)

These interactions are primarily dictated by van der Waals, shown in Fig. 5.2a, electrostatic and capillary forces [25, 39], shown in Fig. 5.2b, and are affected by the size, shape, rugosity, hardness and chemical composition of the particle. Adhesive interactions between particles and surrounding surfaces are governed by the roughness, composition and contamination of the contacting surface, the duration of the contact, the initial contact velocity, and the environmental humidity [7]. Particles that deviate from sphericity experience additional interparticle interactions due largely to increased surface contact and mechanical interlocking, as illustrated in Fig. 5.2b. Consequently, more energy is required for dispersion and aerosolization. These additional interactions may be caused by either microscopic of macroscopic properties [22, 23]. Particle size, shape, crystalline state, and solid-solid interactions contribute to particle aggregation and are important microscopic properties. Macroscopic powder properties include packing density and equilibrium moisture content [22, 23]. An increase in the equilibrium moisture content and packing density generally reduces the ease of powder dispersibility.

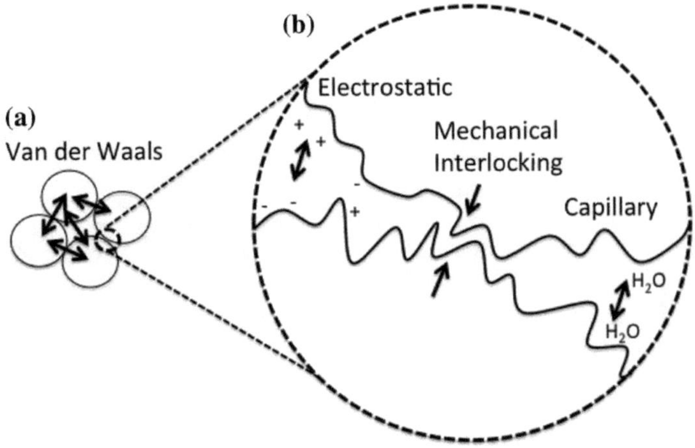

Fig. 5.2 Schematic depiction of **a** Van der Waals, fundamental electromagnetic; **b** surface electrostatic, mechanical interlocking and capillary forces of interaction between particles

5.2 Particle Preparation

Micronized powders can be prepared either by reducing particle size (milling) or by particle construction (evaporation/condensation, precipitation). Arguably spray drying is both a destructive (spray formation) and constructive (molecular association during drying) method while supercritical fluid manufacture is a constructive method of producing particles by a method that is a well-controlled mimic of precipitation.

5.2.1 Size Reduction

Attrition mills are used to produce particles in a size range suitable for aerosol delivery to the lungs (1–5 μm) [18]. Fluid energy mills and opposing jet mills are examples of attrition mills. Opposing jets of compressed gas (commonly air or nitrogen) at difference pressures force large particles to impact against one another and the walls of the mill. This violent impact results in reduction of particle size as energy is expended in cleaving particle usually along planes of the crystal system. Cyclone separation is used to collect the micronized powder. Particles generated by this method are charged. The property of greatest concern with respect to charge effects due to its impact on product performance is cohesion due to opposite, or complementary, charging. During the micronization process, the drug is exposed to minimum heating, and efficient size reduction and aerodynamic classification removes large particles. The reduction in particle size by jet milling brings about a commensurate increase in specific surface area. The increase in electrostatic charge on the particle surface may adversely influence dispersion.

5.2.2 Construction: Condensation

Condensation is preceded by evaporation to create a vapor. This process is driven by temperature elevation or pressure reduction according to principles of thermodynamics. The condensation process can proceed as a homogeneous phenomenon, where the vapor is condensed onto spontaneously forming nuclei or heterogeneously, where nucleating particles are introduced into the vapor. The process of condensation usually requires a supersaturated vapor and in the case of heterogeneous nucleation is initiated by the presence of small nucleating particles that serve as sites for particle growth [24]. The concentration of the vapor and nuclei are carefully controlled, and condensation occurs under slow and controlled conditions. The desired vapor concentration can be achieved by passing air through or over a heated reservoir of liquid or by nebulizing the liquid and heating the generated aerosol. A heat exchange process causes condensation to take place. Under controlled conditions, condensation and growth are the same for each nucleate particle. Differences in the size of the nuclei have minimum effect on the final droplet size, because nuclei are very much smaller than the particles formed after condensation. The process of condensation can produce particles in the nanometer size range and larger. Nucleated condensation can occur at supersaturations of only a few percent and is initiated by nuclei that may be soluble or insoluble in the condensed vapor.

Homogeneous or self-nucleation is a process of particle formation and growth from a supersaturated vapor without the aid of condensation nuclei. Homogeneous nucleation rarely occurs naturally and usually requires saturation ratios, of actual vapor pressure with respect to saturation vapor pressure, in the range of 2–10 [24].

The expression for the saturation ratio (p/p_s) required for equilibrium and the droplet size for pure liquids is given by the Kelvin-Gibbs equation [21]:

$$p/p_s = \exp[(4\gamma M)/(\rho RTd^*)] \tag{5.1}$$

where p, p_s, γ, M, ρ, and d^* are the partial and saturation vapor pressure, surface tension, molecular weight, density of the liquid, and the Kelvin diameter, respectively. At a particular saturation ratio, only particles having a diameter d^* are stable and will neither grow nor evaporate. Those smaller than d^* will evaporate and larger ones will grow.

Equation 5.1 suggests that for a given saturation ratio, a particle must first reach a diameter d^*. Only then can it grow to become a stable droplet. Initially, molecular clusters are formed because of intermolecular attractions. In a supersaturated vapor the number of these clusters increases and they collide with each other, forming stable agglomerates having diameters exceeding d^*, which grow by condensation to form large particles. The supersaturation supporting this phenomenon occurs at a fixed point called the 'critical saturation ratio'.

The principles of evaporation/condensation have recently been adopted to facilitate the production of dry particles for delivery through a nasal cannula for ventilated patients [27].

5.2.3 Construction: Evaporation

The process of evaporation is involved in spray drying, where, as the name suggests, spraying of droplets is followed by evaporative drying. The feed for spray drying is a liquid, and drying is accomplished by atomizing the feed into a hot drying medium. This is a one-step, continuous process, and the feed can be a solution, suspension, emulsion or paste [29]. The characteristics of the powder obtained by a spray-drying process depend on several factors, some of which include the initial solute concentration, the size distribution of droplets, and the rate of solvent removal [43].

The weight (W) of a given solid particle is determined by the volume (V) of the droplet from which it was derived and by the solute fraction (X, w/v). The solid particle diameter (d') and density (ρ) can also be calculated from the initial droplet diameter (d). The V of an initial droplet having a diameter d is given by

$$V = (4\pi/3)(d/2)^3 \tag{5.2}$$

The W of the dry particle is

$$W = X/V \tag{5.3}$$

And d' is

$$d'/2 = [(3XV)/(4\rho\pi)]^{1/3} \tag{5.4}$$

Combining (5.2) and (5.4)

$$d'/2 = \left[(X/\rho)(d/2)^3\right] \tag{5.5}$$

Spherical solid particles of uniform size are produced by the spray drying process. Particle size and other powder characteristics can be controlled, and either fine or coarse powder, agglomerates, or granules can be prepared by changing the physico-chemical properties of the feed and the design and operational conditions of the dryer. The above expression assumes that the particle is a non-porous solid. For consideration of low density, porous particles other considerations are required [43]. The use of low density particles in which the aerodynamic size is much smaller than the geometric size according to Stokes' law, for spherical particles greater than 1 μm in size as follows [2],

$$dae = d' \cdot \sqrt{\rho_1} \tag{5.6}$$

where *dae* is aerodynamic diameter and ρ_1 is density of the particle which for very low densities approaches the tapped or bulk density.

5.2.4 Construction: Precipitation and Crystallization

Precipitation and crystallization separate solid substances from solution by physical or chemical changes. A compound may precipitate or crystallize from solution when its solubility in the solvent is exceeded, i.e. the solution becomes supersaturated. Supersaturation can be achieved in several ways, including evaporation of the solvent, cooling of the solution, or production of additional solute as a result of a chemical reaction or a change in the solvent system (addition of a poor solvent for the precipitating material).

The rate of precipitation and the particle size and distribution depend on the composition of the nuclei and their growth rate. The particle size of the crystals also depends to a great extent on the concentration of the precipitating substance. The rate of addition of a poor solvent for the precipitating material, used to induce supersaturation in a cosolvent system, will influence the characteristics of the product. Very low and high supersaturations produce colloidal particle, whereas intermediate supersaturations tend to produce coarse crystals.

5.3 Characteristics Influencing Dispersion

The dispersion of dry powders is greatly influenced by particle size. Fine particles exhibit interparticulate forces that are difficult to overcome because the particle mass varies to the third power of particle size. As the particle size decreases, dispersion becomes increasingly difficult because of mechanical, electrostatic, capillary and van der Waals forces that lead to increased energies of association. The overall particle-size distribution also affects powder dispersion.

Particle morphology, such as crystal habit, surface texture, and porosity, also influences particle interactions. For irregularly shaped particles, surface roughness will cause an increase in the separation distance between neighboring particles. This may decrease the van der Waals attraction to some extent, which does not occur for surfaces that allow close contact. The force of adhesion is enhanced with plastic deformations. This may not be true for elastic deformations that retain their original shape on separation.

Particle shape and geometry are also of great importance. Flat and elongated crystals are less readily dispersed than spherical particles because of the greater area of contact between particles, which contributes to strong interparticulate forces. Mechanical forces die to the surface contact and interlocking play a role in particle-particle, cohesive and particle-surface, adhesive interactions. These forces are not easily delineated and will not be discussed in detail here.

In addition to particle size and shape, the chemical composition of particles also influences the flow behavior of the powder and its ease of dispersion. The electrical properties of a substance—i.e. whether it is metallic, a semiconductor, or an insulator—determine the charge on its particles. Factors such as elasticity and hygroscopicity are also important. If powders are too dry the presence of high triboelectric charges affects their dispersibility.

The interparticulate forces governing the dispersion of dry particles are mostly attractive and cause cohesion of particles and adhesion to the surfaces they encounter. For fine particles, such as those required for aerosol drug delivery to the lung, these forces are more pronounced and extremely difficult to overcome. This leads to increased aggregation, which affects powder flow and dispersion properties.

The most important mechanisms of cohesion between particles or adhesion of particles and a contacting surface are van der Waals, electrostatic and capillary forces [10, 12, 25, 31, 35, 39].

5.3.1 Van der Waals Forces

These forces are the major cause of particle-particle cohesion. They are a manifestation of the attraction between component molecules within each particle. Although an atom or molecule is neutral the random movement of electrons causes the formation of momentary dipoles, which in turn are attracted to complementary dipoles induced in nearby materials.

The energy of interaction between two molecules (V_{ss}) can be derived from London's theory to give the following expression [25, 32]

$$V_{ss} = -(3/4)\left[\left(h\nu_0\alpha^2\right)/\left(d^6\right)\right] \quad (5.7)$$

where α is the polarizability, h the Planck's constant, and ν_0 the characteristic frequency. Because ν_0 is found in the ultraviolet region of the absorption spectra and plays a key role in optical dispersion, the intermolecular London-van der Waals forces are also called dispersion forces.

These forces operate only at very short ranges but can attain values many times the weight of the particles when integrated over all the molecules in the two contacting particles. The interactions between dipoles and the induced dipoles in neighboring atoms, which is expressed as the Hamaker constant (A). For two perfectly spherical particles with diameters d_1 and d_2,

$$U_m = (A/12z)(d_1 d_2/(d_1 + d_2)) \tag{5.8}$$

where, U_m is the total molecular potential of the system, and z is the shortest interparticle distance. The Hamaker constant (A) is given by

$$A = \pi^2 \rho_1 \rho_2 C \tag{5.9}$$

where, ρ_1 and ρ_2 are the densities of interacting particles and C is the coefficient in the particle-particle interaction.

For two dissimilar materials, with Hamaker constants, A_{11} and A_{22}, the relationship is given by

$$A_{12} = \sqrt{(A_{11} \cdot A_{22})} \tag{5.10}$$

Equation 5.8 holds good where $z << D = d_1 d_2/(d_1 + d_2)$, which is not a practical limitation. A practical equation can be obtained by differentiating Eq. 5.8 with respect to z

$$F = \delta(U_m)/\delta z = (AD)/12z^2 \tag{5.11}$$

The attractive force (F) is dependent on the Hamaker constant (A) and the shortest distance between the particles, z. To bring about a decrease in F one must decrease A and increase z. Referring to Eq. 5.9 we can theoretically decrease the Hamaker constant (A) by decreasing the densities of the two interacting particles. Because the separation distance between particles is the single most important factor governing the van der Waals attraction (Eq. 5.11), any means to increase this distance will reduce the attractive force and increase the ease of dispersion. Surface roughening and the use of spacers can increase interparticulate separation.

The interactive energy for the van der Waals attraction is of the order of 0.1 eV and decreases with the sixth powder of the distance between the molecules. This provides an attractive force that decreases with the seventh power of the distance. The van der Waals forces have a smaller effect as the particle size increases.

In addition to the van der Waals force of attraction, electrostatic and capillary forces also operate between particles. For conducting particles these forces are smaller than the van der Waals force; however, most pharmaceutical powders are poor conductors, hence electrostatic charge plays an important role in particle interactions.

5.3.2 *Electrostatic Forces*

When two solid surfaces come in contact as described above, a contact potential is generated, which gives rise to electrostatic attractive forces that increase powder aggregation. Charges that arise because of frequent particle collisions with one another or against surfaces–called contact charging—also cause additional attractive forces. Triboelectric charging, a potential difference that arises between two interacting particles possessing different work functions, gives rise to electrostatic forces. This difference causes electrons to migrate from the body that has a smaller work function to one that has a larger one until an equilibrium charge is reached. A reduction in the charge does not always occur when a charged particle encounters another particle of surface, be it charged or uncharged. If the charge persists and remains on the particle surface, it will induce an equal and opposite charge on the neighboring particle or surface. The induced electrical force on a particle is given by [34]

$$F_1 = \varepsilon q^2/h^2 \tag{5.12}$$

where q is the charge on the particle and h is the separation distance between the adhering particles in a dielectric medium, ε.

In the case of triboelectric charging, the interaction force (F_c) between a spherical particle and an adjacent uncharged particle is given by the Coulomb equation

$$F_c = Q^2\left[1 - \left(\left[H/\left(R^2+H^2\right)^{1/2}\right]/16\pi\varepsilon_0 H^2\right)\right] \tag{5.13}$$

where R is the particle radius, Q the charge, H the separation distance, and ε_0 the permittivity of vacuum. In a humid environment, this Coulomb attraction reduces to zero because of a discharging of the system but is complicated by capillary forces.

When a potential difference ΔU arises particles of different work functions are brought in contact, the force of attraction F_w is given by

$$F_w = \left(\pi\varepsilon_0 R(\Delta U)^2\right)/H \tag{5.14}$$

The magnitude of electrostatic forces on particles on particles in quite small, but they play a large role in interparticulate cohesion.

Electrostatic charge effects are significant for poor conductors, which include most pharmaceutical powders. Methods that could be used for determining the electrostatic charge on dry powders have been reported [34, 40, 41].

5.3.3 Capillary Forces (Liquid Bridges)

Interparticulate cohesion usually increases as the relative humidity of the air increases. At humidities >65% fluid condenses in the space between particles that are close together. This can lead to liquid bridges between neighboring particles, and the effect of surface tension gives rise to attractive forces. These attractive forces can be quite strong, sometimes having a component several times the mass of the individual particles when the vapor pressure of the surrounding air approaches the saturation pressure.

For two smooth, spherical particles the force experienced is

$$F_H = 2\pi\gamma R \tag{5.15}$$

where R is the particle radius and γ the surface tension of the fluid completely wetting the surface of the particles.

If liquid is present between the surfaces of two solid bodies the surface tension of the liquid gives rise to boundary forces. A pressure difference called the capillary pressure (Pc) arises, which can be calculated from the Laplace equation

$$P_c = \gamma([1/R_1] + [1/R_2]) \tag{5.16}$$

where γ is the surface tension at the liquid-gas interface, R_1 and R_2 are the principal radii of curvature of the interface.

5.4 Surface Forces (Adhesion)

5.4.1 Forces of Adhesion

The importance mechanisms of particle-particle interactions (cohesion), that is, van der Waals, electrostatic and capillary forces, also apply to adhesion between particles and surfaces [18]. In addition to these forces, mechanical interlocking, double-layer repulsion, and hydrogen and metallic bonds also contribute to adhesion.

The van der Waals adhesive forces (F_v) between a spherical particle and a flat surface are given by

$$F_v \propto d/x^2 \tag{5.17}$$

where d is the particle diameter and x is the distance to the surface.

Most particles >0.1 μm carry electrostatic charge, q. The presence of this charge exerts an electrostatic force (F_E) in the presence of an opposite charge

$$F_E \propto q^2/x^2 \tag{5.18}$$

The van der Waals and electrostatic forces cause particle-surface adhesion and hence increase the area of contact: this force of attraction balances the forces resisting deformation. The hardness of the materials determines the final contact area and hence the strength of the adhesive force. The effect of hardness on the adhesive force has been noted [32]. The effect of particle geometry on adhesion is also an important consideration [37].

In addition to van der Waals and electrostatic forces, liquid bridges (capillary forces) also contribute to particle-surface adhesion. If a film of liquid is present between a sphere and a contacting plane a force of attraction is experience as a result of the surface tension γ of the liquid. This adhesive capillary force is given by

$$F = 4\pi\gamma r \tag{5.19}$$

where r is the radius of the particle.

5.4.2 Measurement of Adhesion

Experimental measurements of adhesive forces determine the force required to separate a particle from a surface. The theoretical forces of adhesion described above are for smooth surfaces and do not take into account the effect of surface roughness. The forces required to dislodge particles of different sizes will vary. Even for a number of practically identical particles on a surface, the force needed to dislodge each particle differ, due to variable inherent in the physical adhesion and the measurement error. The ratio of the number of particles retained on a surface after applying a force to the number that was originally present is defined as the 'adhesion number'. Estimates of the force needed to remove half the number of particles present are more practical for measuring adhesive force [37].

For particles >1.0 μm, several methods have been developed to estimate the force of adhesion of single particles on surfaces [32]. These methods include the variation of the slope of a surface, pendulum, microbalance, centrifuge, aerodynamic or hydrodynamic, and vibration methods. Only three methods, centrifugal, aerodynamic or hydrodynamic and vibrational are applicable to 0.1–50 μm diameter particles. The force necessary to detach a particle from a surface can be measured by subjecting particles to a centrifugal force, perpendicular to the surface, and measuring the speed of rotation required for detachment. In the aerodynamic method the velocity of the air required to remove a proportion of monodisperse particles from a surface is determined.

Adhesive forces are usually proportional to d, but forces needed for particle removal are proportional to d^3 for centrifugal and vibrational forces and d^2 for the aerodynamic method. Therefore, as the size of the particles decrease, it becomes increasingly difficult to remove them from surfaces.

Examples of adhesion measurement methods may be cited including centrifugal and vibrational [6, 8, 36, 37]. Two general principle of impact force and centrifugal force detachment are illustrated in Fig. 5.3. A new approach described as a drop test has also recently been reported [44].

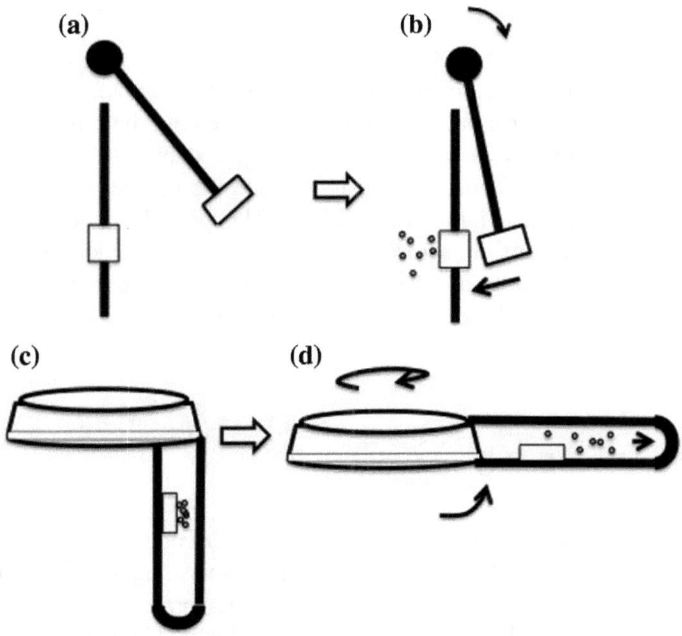

Fig. 5.3 Schematic depiction of **a** an impact force detachment apparatus and its use; **b** by detaching particles following impact on one side of a hard surface transferred to the opposite side (calibrated by a linear accelerometer) and; **c** a centrifugal apparatus and its use; **d** when under rotation a force of detachment of particles is achieved. Both methods are used for evaluation adhesion forces of particles

5.5 Powder Dispersion Devices

Disintegration and dispersion of bulk solids is an important mechanistic sequence for aerosol production. The ease of dispersibility of a powder depends on the composition, particle size and distribution, shape and moisture content. Hydrophobic materials are frequently more easily dispersed than hydrophilic materials because the smaller hydrophobic interactions dominate. Dry powders are usually easier to disperse than moist ones, although extremely dry powders are difficult to aerosolize because of the presence of high electrostatic surface charges that lead to increased aggregation. To completely disperse a powder sufficient energy should be supplied to overcome the interparticulate attractive forces. The composition and physicochemical properties of dry powders can be manipulated to aid in dispersion [33].

5.5.1 Dust Generators

Dry dispersion dust generators require some means of continuously feeding a powder into the generator at a constant rate and dispersing the powder to form an aerosol. Various methods are available to develop a stable dust [1, 33, 38]. The simplest dust metering systems use gravity to feed the powder into an airstream, which in many cases uses vibrators to aid flow. This provides an uneven delivery, which leads to variations in aerosol concentration. A cylinder of compressed powder that is scraped away at a fixed rate gives rise to a stable metering method and is exemplified in the Wright dust feed. Shear forces created by a high velocity airstream disperse the powder.

5.5.2 Fluidized Bed Generators

An alternative approach to powder dispersion is the introduction of powder into a fluidized bed containing 100–200 μm beads that disperse aggregates and suspend the powder particles in an airstream. Large agglomerates are prevented from escaping the fluidized bed through controlled air velocity. Commercial systems are available for this purpose [38].

Charging of particles during dispersion has a significant influence on the efficiency of dry dispersion aerosol generators because of contact charging as the particles impact and bounce off surfaces in the generator. These effects are of greatest magnitude for powders that have low moisture content. Extremely dry air with a relative humidity below 5% can also cause strong electrostatic forces between particles, thus reducing dispersibility.

5.5.3 Dry Powder Inhalers (DPIs)

Powder inhalers consist of three components: the drug formulation, the reservoir or metering system, and the dispersion mechanism to generate the aerosol. The airborne product generated by a powder inhaler should contain a significant proportion of particles <5 μm [14]. However, the cohesive nature of fine powders hinders motion within the device, leading to poor dispersibility and poor reproducibility of the dose delivered. The fraction of the dose delivered as respirable particles depends on the patient, the device, and the powder formulation. The design of inhalers varies greatly. These devices should optimally be capable of generating a high turbulence at a relatively low pressure drop. This can be brought about by manipulation of the geometry of the device and use of baffles, constrictions and grids [12, 39]. The adoption of computational fluid dynamic methods to predict the influence of these

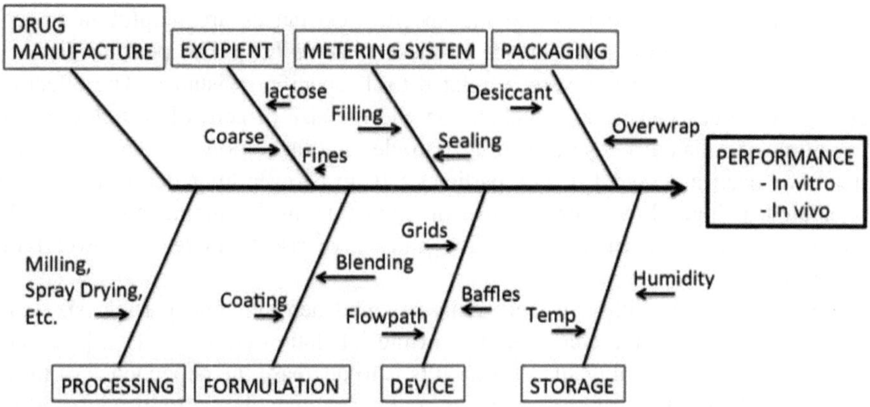

Fig. 5.4 Dry powder inhaler critical components, processes and conditions contributing to product performance

features in powder dispersion has been a valuable addition to the tools at the disposal of device designers [16].

There are three major types of DPIs: unit dose (capsules or blisters), multi-unit dose (disks or strips) and reservoirs. Each of these rely on the patient's inspired airflow to aerosolize the drug particles. DPIs often require inspiratory flow rates of ≥60 L/min to effectively deaggregate powder. It is now thought that the airflow rate requirements are modulated by the ability to induce shear and turbulence by incorporating a desirable pressure drop into the device [5]. Lactose is often included in the formulation as a large particle size carrier acting as both a diluent and dispersing agent. The carrier's purpose is to convey surface-associated drug particles into the airstream as 'controlled aggregates', where they are stripped off as individual respirable particles.

Figure 5.4 illustrates the sequence and relationship between components, processes and environment of a DPI that ultimately dictate the response to airflow and the efficiency and reproducibility of delivery. As described earlier, the drug manufacture, its processing into particles with desired properties, formulation with excipients (if necessary), filling into a dose metering system and combination with a device create the product that with adequate packaging and environmental controls will perform efficiently and reproducibly in the service of disease management. Therefore, control of these factors dictates the quality, performance and efficacy of drug delivered from the dosage form and conforms to quality by design principles [19, 42].

Powder deaggregation and aerosolization in all currently available commercial DPIs depend entirely on the energy input from the patient's inspiratory flow. A high flow rate increases the efficiency of the dose delivered. There are wide variations in the resistance to airflow occurring in DPIs [28].

The fraction of drug deposited in the lung depends on the airflow rate with which a patient inhales through the DPI. Reports relating the design of a DPI, the pressure drops and flow rates attained and the specific resistances are helpful in understanding the anticipated performance of these systems [28]. The peak inspiratory flow rate though a DPI is inversely related to its specific resistance. The effect of varying flow rates generated by inhalation and the size of aerosol particles on the deposition of inhaled aerosolized drug particles within regions of the respiratory tract have been assessed [4]. For particles >10 μm lower lung deposition is not anticipated at high flow rates because of impaction in the mouth and throat. In general, for smaller particles lower flow rates give rise to increased lower lung deposition.

Consequently, the mass of drug delivered and aerodynamic particle size can change depending on the characteristics of the inhalation effort [3, 4, 28]. Hence, one approach to developing successful DPIs is to decrease the dependence of these devices solely on the patient's inhalation. Energy may be introduced by electromechanical, mechanical, or compressed air assist methods to deaggregate the powder and disperse drug particles into the airstream. These systems may provide accurate dose delivery and an energy input that is independent of the patient's inhalation [8, 11, 13].

5.6 Conclusions

Hydrofluoroalkane propellants are under increasing scrutiny as global warming agents and steps are being taken to add these to the list of banned chemicals under the Montreal Protocol. This will inevitably lead to another period of innovation similar to that following the ban of chlorofluorocarbon propellants due to their involvement in atmospheric ozone depletion.

There will be increasing urgency to provide novel methods of powder production that scrutinize the forces of interaction and dispersion of particulates to allow for rapid advancement of innovative dry powder inhalers to meet the future needs for pharmaceutical aerosol delivery to the lungs. The effective treatment of pulmonary and systemic diseases depends upon rapid advances in powder technology and device design.

5.7 Definitions, Abbreviations and Symbols

Aerodynamic diameter	Diameter of a sphere with density 1000 kg/m^3 having the same aerodynamic property as the particle
Envelop density	Density of a particle including present pores
Particle size	Diameter of some defined equivalent sphere
True particle density	Density of a particle without pores
COPD	Chronic obstructive pulmonary disease
DPI	Dry powder inhalers

HFA	Hydrofluoroalkane
MDI	Metered dose inhalers
pMDI	Pressurized metered dose inhalers
A	Hamaker constant
A_{11}, A_{22}	Hamaker constants of two dissimilar materials
C	The coefficient in the particle-particle interaction
D	$d_1 d_2/(d_1 + d_2)$
d	Initial droplet diameter
d^*	Kelvin diameter of a particle
d'	Solid particle diameter
d_1, d_2	Diameters of two interacting spherical particles
dae	Aerodynamic diameter
dV_1, dV_2	Volume elements of two interacting particles
F	Attractive force
F	Adhesive capillary force
F_c	Interaction force between a spherical particle and an adjacent uncharged particle
F_E	Electrostatic force in the presence of an opposite charge
F_v	Van der Waals adhesive forces between a spherical particle and a flat surface
F_w	Force of attraction between particles of different work functions
H	Separation distance
h	Planck's constant
h	Separation distance between adhering particles
M	Molecular weight
ρ_1, ρ_2	Molecule densities of two interacting particles
p	Partial vapor pressure
p_s	Saturation vapor pressure
Q	Particle charge
q	Charge on a particle
R	Particle radius
R_1, R_2	Principal radii of curvature of the interface
r	Particle radius
U_m	Total molecular potential of the system
ΔU	Potential difference between particles of different work functions
X	Solute concentration (w/v)
V	Volume of a droplet
V_{ss}	Energy of interaction between two molecules
W	Weight of a given solid particle
Z	Shortest interparticle distance
α	Polarizability
γ	Surface tension of a liquid

ε	Dielectric constant
ε_0	Permittivity of vacuum
ν_0	Characteristic frequency
ρ	True density of a particle
ρ_e	Envelop density of a porous particle

Acknowledgements The author is grateful to Drs. Neville Concessio, Michiel Van Oort and Robert Platz who contributed to a paper published in Pharmaceutical Technology in 1994 from which the structure of this chapter was developed.

References

1. Carpenter RL, Yerkes K (1980) Relationship between fluid bed aerosol generator operation and the aerosol produced. Am Ind Hyg Assoc J 41:888–894
2. Cipolla D, Shekunov B, Blanchard J, Hickey A (2014) Lipid-based carriers for pulmonary products: preclinical development and case studies in humans. Adv Drug Deliv Rev 75:53–80
3. Clark AR, Egan M (1994) Modelling the deposition of inhaled powder drug aerosols. J Aerosol Sci 25:175–186
4. Clark AR, Hollingworth AM (1993) The relationship between powder inhaler resistance and peak inspiratory conditions in healthy volunteers—implications for in vitro testing. J Aerosol Med 6:99–110
5. Coates MS, Fletcher DF, Chan HK, Roper JA (2004) Effect of design on the performance of a dry powder inhaler using computational fluid dynamics. Part 1: Grid structure and mouthpiece length. J Pharm Sci 93:2863–2876
6. Concessio NM, van Oort MM, Hickey AJ (1996) Detachment of excipient particles from planar surfaces by applied impact force. Pharm Res 13:S158
7. Cordts E, Steckel H (2014) Formulation considerations for dry powder inhalers. Ther Del 5:675–689
8. Crowder T, Hickey A (2006) Powder specific active dispersion for generation of pharmaceutical aerosols. Int J Pharm 327:65–72
9. Crowder TM, Donovan MJ (2011) Science and technology of dry powder inhalers. In: Smyth HDC, Hickey AJ (eds) Controlled pulmonary drug delivery. Springer, New York, NY, pp 203–222
10. Crowder TM, Hickey AJ, Louey M, Orr N (2003) Behavior of particles in a guide to pharmaceutical particulate science. Interpharm/CRC, Boca Raton, FL, pp 77–121
11. Donovan MJ, Gibbons A, Herpin MJ, Marek S, McGill SL, Smyth HDC (2011) Novel dry powder inhaler particle-dispersion systems. Ther Deliv 2:1295–1311
12. Dunbar C, Hickey AJ, Holzner P (1998) Dispersion and characterization of pharmaceutical dry powder aerosols. KONA, Powder and Particle 167:7–44
13. Dunbar CA, Morgan B, Van Oort M, Hickey AJ (2000) A comparison of dry powder inhaler dose delivery characteristics using a power criterion. PDA J Pharm Sci Technol 54:478–484
14. Durham PG, Young EF, Braunstein MS, Welch JT, Hickey AJ (2016) A dry powder combination of pyrazinoic acid and its n-propyl ester for aerosol administration to animals. Int J Pharm 14:384–391
15. Frijlink HW, De Boer AH (2004) Dry powder inhalers for pulmonary drug delivery. Expert oppin Drug Deliv 1:67–86
16. Hickey A (2004) Summary of common approaches to pharmaceutical aerosol administration. In: Hickey AJ (ed) Pharmaceutical inhalation aerosol technology. Marcel Dekker Inc, New York, NY, pp 385–421

17. Hickey AJ (2013) Back to the future: inhaled drug products. J Pharm Sci 102(4):1165–1172
18. Hickey AJ (1993) Lung deposition and clearance: what can be learned from inhalation toxicology and industrial hygiene? Aerosol Sci Technol 18:290–304
19. Hickey AJ, Ganderton D (2011) Quality by design in pharmaceutical process engineering, 2nd edn. Informa Healthcare, New York, pp 193–196
20. Hickey AJ, Xu Z (2014) Dry powder inhalers. In: Merkus HG, Meesters GMH (eds) Particulate products—tailoring properties for optimal performance. Springer, NY, pp 295–332
21. Hickey AJ, Gonda I, Irwin WJ, Fildes FJ (1990) Effect of hydrophobic coating on the behavior of a hygroscopic aerosol powder in an environment of controlled temperature and relative humidity. J Pharm Sci 79:1009–1014
22. Hickey AJ, Mansour HM, Telko MJ, Xu Z, Smyth HD, Mulder T, McLean R, Langridge J, Papadopoulos D (2007) Physical characterization of component particles included in dry powder inhalers. I. Strategy review and static characteristics. J Pharm Sci 96:1282–1301
23. Hickey AJ, Mansour HM, Telko MJ, Xu Z, Smyth HD, Mulder T, McLean R, Langridge J, Papadopoulos D (2007) Physical characterization of component particles included in dry powder inhalers. II. Dynamic characteristics. J Pharm Sci 96:1302–1319
24. Hinds WC (1999) Aerosol technology properties, behavior, and measurement of airborne particles, 2nd edn. John Wiley and Sons, New York, p 1998
25. Israelachvili J (1991) Intermolecular and surface forces. Academic Press, London
26. Kigali Climate Change Agreement (2015) Nations fighting powerful refrigerant that warms planet reach landmark deal. New York Times, October 2015
27. Longest PW, Walenga RL, Son J-Y, Hindle M (2013) High efficiency generation and delivery of aerosols through nasal cannula during non-invasive ventilation. JAMPDD 26:266–279
28. Louey MD, Van Oort M, Hickey AJ (2006) Standardized entrainment tubes for the evaluation of pharmaceutical dry powder dispersion. J Aerosol Sci 37:1520–1531
29. Masters K (1991) Spray drying handbook. Longman Scientific and Technical, Harlow, England, pp 1–35
30. Molina MJ, Rowland FS (1974) Stratospheric sink for chorofluoromethanes: chlorine atom-catalysed destruction of ozone. Nature 249:810–812
31. Podczeck F (1998) Particle-particle adhesion in pharmaceutical powder handling. Imperial College Press, London, UK
32. Ranade MB (1987) Adhesion and removal of fine particles on surfaces. Aerosol Sci Tech 7:161–176
33. Rasmeijer F, Lexmond AJ, Maarten V, Hagedoorn P, Hickey AJ, Frijlink HW, de Boer AH (2014) New mechanisms to explain the effects of added lactose fines on the dispersion performance of adhesive mixtures for inhalation. PLoS ONE 9:1
34. Rietema K (1991) Theoretical derivation of interparticle forces. In: The dynamics of fine powders. Elsevier Applied Science, New York, pp 65–91
35. Rumpf H (1990) Particle Technology. Chapman and Hall, New York
36. Selvam P, Marek S, Truman CR, McNair D, Smyth HDC (2011) Micronized drug adhesion and detachment from surfaces: effect of loading conditions. Aerosol Sci Tech 45:81–87
37. Soltani M, Ahmadi G, Bayer RG, Gaynes MA (1995) Particle detachment mechanisms from rough surfaces under substrate acceleration. J Adhesion Sci Technol 9:453–473
38. Teague SV, Veranth JM, Aust AE, Pinkerton KE (2007) Dust generator for inhalation studies with limited amounts of archived materials. Aerosol Sci Technol 39:85–91
39. Telko M, Hickey AJ (2005) Dry powder inhaler formulation. Respir Care 50:1209–1227
40. Telko MJ, Hickey AJ (2014) Aerodynamic and electrostatic properties of model dry powder aerosols: a comprehensive study of formulation factors. AAPS PharmSciTech 15(6):1378–1397
41. Telko MJ, Kujanpaa J, Hickey AJ (2007) Investigation of triboelectric charging in dry powder inhalers using electrical low pressure impactor (ELPI). Int J Pharm 336:352–360
42. US DHHS, FDA (2009) Guidance for industry, Q8(R2) Pharmaceutical development

43. Vehring R (2007) Pharmaceutical particle engineering via spray drying. Pharm Res 25:1000–1022
44. Zafar U, Hare C, Hassanpou A, Ghadiri M (2014) Drop test: a new method to measure the particle adhesion force. Powder Technol 264:236–241

Anthony J. Hickey is Distinguished RTI Fellow, at the Research Triangle Institute, Emeritus Professor of Molecular Pharmaceutics of the Eshelman School of Pharmacy (2010–present, Professor 1993–2010), and Adjunct Professor Biomedical Engineering in the School of Medicine, at the University of North Carolina at Chapel Hill. He obtained Ph.D. (1984) and D.Sc. (2003) degrees in pharmaceutical sciences from Aston University, Birmingham, UK. Following postdoctoral positions, at the University of Kentucky (1984–1988) Dr. Hickey joined the faculty at the University of Illinois at Chicago (1988–1993). In 1990 he received the AAPS Young Investigator Award in Pharmaceutics and Pharmaceutical Technology. He is a Fellow of the Royal Society of Biology (2000), the American Association of Pharmaceutical Scientists (2003) and the American Association for the Advancement of Science (2005). He received the Research Achievement Award of the Particulate Presentations and Design Division of the Powder Technology Society of Japan (2012), the Distinguished Scientist Award of the American Association of Indian Pharmaceutical Scientists (2013) and the David W Grant Award in Physical Pharmacy of the American Association of Pharmaceutical Scientists (2015). He has published numerous papers and chapters in the pharmaceutical and biomedical literature, one of which received the AAPS Meritorious Manuscript Award in 2001. He has edited five texts on pharmaceutical inhalation aerosols and co-authored three others on 'Pharmaceutical Process Engineering', pharmaceutical particulate science and 'Pharmaco-complexity'. He is founder (1997, and formerly President and CEO, 1997–2013) of Cirrus Pharmaceuticals, Inc., which was acquired by Kemwell Pharma in 2013; founder (2001, and formerly CSO, 2002–2007) of Oriel Therapeutics, Inc, which was acquired by Sandoz in 2010 and founder and CEO of Astartein, Inc. (2013–present); member of the Pharmaceutical Dosage Forms Expert Committee of the United States Pharmacopeia (USP, 2010–2015, Chair of the sub-committee on Aerosols) and formerly Chair of the Aerosols Expert Committee of the USP (2005–2010). Dr. Hickey conducts a multidisciplinary research program in the field of pulmonary drug and vaccine delivery for treatment and prevention of a variety of diseases.

Chapter 6
Blending and Characterization of Pharmaceutical Powders

Carl A. Anderson and Natasha L. Velez

Abstract Blending is a critical part of pharmaceutical manufacturing. It consists of two parts, the actual mixing of powdered ingredients to a desired degree of homogeneity and the analytical testing to confirm the same. Blending equipment and formulation considerations are discussed, along with potential mechanisms for a loss of quality of blends. A general description of the role of blending and its impact on the quality of finished dosage forms is presented. Traditional and on-line methods for assessing the quality of blends are described. Though widely accepted, traditional blend uniformity testing (thief sampling) has a number of potential weaknesses. On-line techniques such as near-infrared spectroscopy, Raman spectroscopy and light-induced fluorescence are considered as alternatives to traditional testing. The characteristics of the technology and potential challenges of analysis and implementation are discussed for these on-line methods.

6.1 Introduction

Blending of powders is a critical unit operation in pharmaceutical manufacturing and is used to produce large quantities of materials. While the basic principles of blending powders are applicable to a wide range of industries, the highly regulated nature of pharmaceutical manufacturing creates specific needs for this unit operation. Direct compression tablet manufacturing is an example of the use of blending, where the characteristics of the blend have an important and direct influence on the quality of the product (tablets). A direct compression pharmaceutical manufacturing process will be used as an example throughout this chapter. Without a successful blending unit operation, it is highly unlikely that a drug product of adequate quality can be manufactured. Poor blending is a frequently cited root cause of content uniformity problems in finished dosage forms [49]. Due to statistical fluctuations and disturbances in the production process and material transport, the mixtures do

C. A. Anderson (✉) · N. L. Velez
Center for Pharmaceutical Technology, Duquesne University, Pittsburgh, USA
e-mail: andersonca@duq.edu

H. G. Merkus et al. (eds.), *Particles and Nanoparticles in Pharmaceutical Products*,
AAPS Advances in the Pharmaceutical Sciences Series 29,
https://doi.org/10.1007/978-3-319-94174-5_6

not have a constant composition and size distribution. Their quality has the potential to vary with time during the manufacturing process. Pharmaceutical products are usually composed of an active material and a combination of excipients. All materials considered here will be dry particles. The ultimate goal of the blending unit operation is to reach a homogeneous composition. However, reaching homogeneity poses a number of challenges for particulate mixtures.

Sampling of a product from a larger batch is executed to measure or control its quality. For pharmaceutical products, producing tablets is, essentially, a sampling exercise. Product quality may refer to a multitude of aspects, such as particle size distribution, particle shape, powder flowability, dispersion viscosity, chemical composition and degree and type of crystallinity. Most often, the collection of the test sample requires several steps, viz. a primary sampling step from the process lines or the stored product and, since this primary sample usually is too large for testing, one or more secondary sampling steps to reach the required test sample amount. Note that the limited size of the volume of interrogation of an instrument may introduce a third sampling step. The type of analysis and particle size determine the amount of test sample needed: this may range from milligrams to more than one hundred grams. Typical measurements in pharmaceutical applications range from tens of milligrams to grams. Test samples must be representative within known limits for the larger product bulk to allow an adequate quality statement for that product. The heterogeneous nature of particulate products poses significant challenges to the process of representative sampling. As a result, sampling is often a major source of error in the final measurement of the quality of a blend. On-line measurements have emerged over the past decade and are being increasingly used for control of the blending process. These measurements are based on a variety of rapid spectroscopic techniques such as near-infrared, Raman or light-induced fluorescence. This chapter discusses the challenges of the blending process and their potential solutions.

6.2 Particulate Mixtures

Powders processed to make pharmaceutical tablets (active ingredients and excipients) are typically composed of particles in the size range of 25–1000 μm. Typical pharmaceutical particles exhibit a dispersity in their size distribution, have a substantial range of different shapes (e.g., from spheres to high aspect ratio needles), and vary in their density. Because of these factors, the potential for segregation exists during all phases of processing, especially when the above parameters are mismatched and the blend is free flowing.

As a consequence, most often, particulates are present in **segregated mixtures**. Here, particles of one type have a greater or smaller probability of being found in a given part of the mixture than corresponding to their mean concentration. Segregation, i.e. separation of particles in a mixture (e.g. in a powder) due to differences in particle size, shape and/or density, or fluctuating production conditions and loading sequences may cause this type of mixtures (see Sect. 6.4).

Free-flowing powders commonly exhibit segregation, which may be severe after extensive exposure to vibration, transport and storage. The presence of segregation in both particulate components and multicomponent products asks for adequate blending before further processing.

Liquids represent an **ideal homogeneous mixture** of molecules, where the composition at any location is the same at the normal scale of scrutiny of the measurement. This contrast to powder mixtures, where such mixtures do not exist when we look at the scale of individual particles and the scale at which a sample requires homogeneity, the dosage form. [39] On the other hand, solutions, colloidal dispersions and stabilized gels may show such ideal homogeneity.

There are multiple descriptions of an optimum blend of powders. The two most prominent are the ordered mixture and the completely random mixture (CRM) [61]. A description of an **ideally ordered mixture** for a two component system is that each particle is next to a particle of a different type. However, blends of this type are not formed using commercial blenders in the pharmaceutical industry [62]. They can only exist if all particles have identical characteristics. An analogous system is **interactive ordered mixtures**, where particles of one type are coated with particles of a different type (often having a smaller size) through specific chemical and physical interactions. Such systems may e.g. be used to prevent particle agglomeration and to decrease powder cohesivity, for example for application in inhalation powders. As in completely random mixtures, the number of small particles per large particle usually varies according to statistics.

The optimum mixture for particulates in powders is a **completely random mixture** (CRM). The CRM is defined by a uniform probability of finding an element of a constituent throughout the mixture [36], or the probability of finding a particle of any type is the same at all locations and equal to the proportion of that particle type in the bulk. However, the exact proportion varies according to statistical laws. These mixtures may be found in well-mixed cohesive powders. Achieving such a result from blending typically requires similar particle shape, size and density for all constituents.

Materials used in formulating pharmaceutical products often do not meet the standards required to achieve a CRM. Fortunately, pharmaceutical blends used to make drug products do not require a CRM to be of acceptable quality [20, 48]. A better description of most pharmaceutical blends is a **partially randomized mixture**. This state of blending is described as having a standard deviation of the content of samples (s) greater than that of a CRM (s_R), where the standard deviation of the mean content in subsequent samples approaches zero. Again, a quality pharmaceutical product does not require a CRM; rather, it requires that the ingredients are distributed on a scale that is appropriate for further processing (e.g., encapsulation or tableting).

The size of sample, or scale of scrutiny, is a critical parameter in characterizing a blend. If we consider that the measure of homogeneity of a blend is the standard deviation of the contents of multiple samples from that blend, then the size of the sample becomes critical. Smaller samples require homogeneity on a smaller scale; while a larger sample requires less blending to achieve a similar standard deviation across samples. Consider that at the scale of a blender (the whole blender) the contents could be considered homogenous at the time the blender is charged. However, at the scale that is relevant to a patient (i.e., a patient will not ingest an entire blend) this blend would be highly heterogeneous. The smaller the scale of scrutiny, the more blended the constituents must be. If we consider only the active pharmaceutical ingredient (API) of a tablet, the API must be evenly distributed across the blend at a scale similar to, or less than, the mass of a tablet. This could ultimately be characterized by measuring the amount of API in tablets (the content uniformity of the tablets). However, contemporary practice in pharmacy now allows (if not encourages) the splitting of tablets. This places a greater burden on the blending process, as the mixture now needs to be homogenous on a ½ tablet scale.

The preceding example is limited to the distribution of the API and does not consider other formulation considerations such as disintegrants, compression aids, binders, boundary lubricants and other important elements of a formulation. These ingredients may have an even smaller scale of scrutiny requirement. For example, consider sodium croscarmellose; a super-disintegrant. Its role in a formulation is to wick water into the tablet and expand causing the tablet to mechanically fail in solution. If this ingredient is aggregated in a small region of a single tablet, then only that small region of the tablet will disintegrate. This leaves the bulk of the tablet still intact, slowing the ultimate dissolution of the API. However, if this disintegrant were properly distributed, it would effectively cause the tablet to disintegrate into small particles that would more easily dissolve, releasing the active ingredient. Thus, each constituent in a pharmaceutical formulation has a unique scale at which homogeneity is achieved.

A successful pharmaceutical blending operation requires two phases. The first is the manufacturing unit operation wherein powders such as the API and excipients are mixed together. The second is the confirmation that this process has achieved the required level of homogeneity. Both phases must be successful for the blending process to be considered complete. Consideration of the actual state of blending of the material vs. the measured state of the material can be a serious potential issue. Mechanisms by which the measurement of the state of the blend interfere with the blend itself will be discussed later in this chapter. This, when convoluted with analytical uncertainty, leads to one of the many complicating factors for deciding that a blend is finished, or not.

6.3 Blending as a Unit Operation

The objective of any mixing process is to produce a homogeneous blend. The basis of powder blending is a pharmaceutical unit operation for mixing a number of powdered ingredients. The goal is to distribute each of the ingredients throughout the blend such that samples of the blend will contain similar relative quantities of each ingredient. The quality of blending can be expressed as variance or standard deviation coming from a series of measurements of composition [60].

For some pharmaceutical processes, the unit operation that typically follows blending is tableting. The homogeneity of the ingredients in the tablets is the ultimate test of the success and stability of the blend. The process of tableting can be considered an elaborate sampling of the entire blend; wherein each tablet is a sample representative of a fraction of the blend. Comparisons of the blend homogeneity and the content uniformity of the blend have been conducted [49]. In general, mixing of particulates poses severe challenges, which relate to particle size distribution, density differences and particle shape [60]. Reasons are:

- The particles in free-flowing materials—equant particles having a size greater than about 30 μm and an aspect ratio close to 1—can move relatively easily under the influence of shaking or stirring. This means that not only mixing but also segregation proceeds fairly fast under influence of gravity and vibration.
- Elongated or flaky particles in dry powders mix and segregate much slower that their equant counterparts. The great differences between the three dimensions cause major mechanical hindrance for particle movement. Thus, both mixing and segregation proceed very slowly.
- In cohesive powders—where the attractive forces between particles are on the order of their mass, cause the primary particles not to move by themselves, but to form indistinct larger lumps that must be disrupted by external forces. Mixing is slower than in case of free-flowing materials, and stronger external forces are required to reach good mixing of the primary particles. On the other hand, the tendency for segregation after blending is reduced for cohesive powders.
- Environmental humidity often increases the cohesivity of particles, leading to issues associated with cohesive mixtures.
- Sub-micrometer particles may form agglomerates, which can be very stable and which also may form larger lumps in dry powders. These agglomerates require substantial external forces to dissociate, if it is possible to do so.

Several mechanisms play a role in mixing of dry powders [60]:

- *Convection or macro-mixing*: continuous movement of large groups of particles relative to each other. It is caused by rotation of the powder in a bin or by stirring it.
- *Diffusion or micro-mixing*: random changing of place by individual particles. General movement of the powder and vibration promote dispersion.
- *Shearing* by paddles or choppers may break agglomerates and, thus, improve dispersion.
- *Particle agglomeration* can make dispersion of primary particles impossible if the bonding is strong.
- *Segregation* leads to a lack of homogeneity (see Sect. 6.4 on segregation).

In general, it is advantageous to use as little mechanical force as possible to achieve the desired state of mixing. Application of excessive force to a blend can result in attrition of particles that is typically undesirable. Diffusion is a random, localized redistribution of blend constituents. It only occurs in a powder bed when

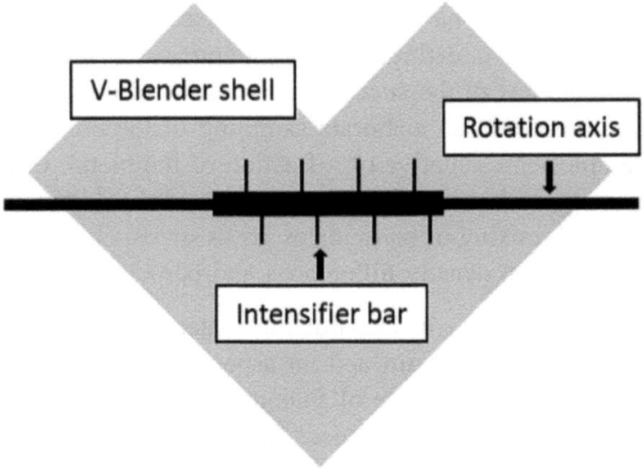

Fig. 6.1 Schematic of a V-blender with an intensifier bar

powders are allowed to move freely; this occurs when a bed is fluidized or during cascades in a tumble blender (diffusion would typically be parallel to the axis of rotation). Diffusion is typically most useful for causing blending on a small scale. Convection causes movement of particles on a larger scale. It is a mechanism that moves groups of particles from one location to another via mechanical force. An example of convection is the cascading of a powder bed in a tumble blender (convection occurs perpendicular to the axis of rotation) or the blade of a ribbon blender. When shear is applied to a powder bed, the ingredients are relocated via the formation of slip planes through the blend. The application of shear is often useful to break up agglomerates of material within the blender.

It is the design of the blender that determines the relative contribution of each mechanism described above. The success of the blending operation is a combination of the equipment used to blend the powders and the physical properties of the powders. The blender controls the forces the powder will experience and thus the dominant mechanism employed in blending. For example, tumble blenders mix by the mechanisms of diffusion and convection; while a planetary blender causes blending via convection and shear. Typical parameters for a bin-type blender are the speed of rotation and the time of blending. Additional elements may be added to the blender to enhance the degree of shear applied to the blend through the use of intensifier bars (Fig. 6.1). In general, the blender and its parameters are established and the blending process is controlled by time.

Various types of blenders exist for powders from a general design point of view, although for each type different specific designs are present [18, 60]:

- Tumbling blenders, where the whole vessel, or shell, containing a batch of powder rotates (e.g. drum mixers and V-blenders) [50]. The mixing is caused here by the fact that the powder falls at small time intervals due to gravity during rotation of the vessel, which promotes convection. They are only applicable for free-flowing powders, since particles in these blends disperse easily. Their energy consumption is typically low, viz. less than 1 kW/m^3. A further advantage is that they are typically easily cleaned. A drawback is that mixing is often incomplete for overly cohesive or free-flowing powders.

- Batch mixers with internal baffles or stirrers (e.g. ribbon blenders and paddle mixers). Here, either the vessel or the internals rotate while the other part is stagnant. Typically convection is increased in comparison with the tumbling blenders; also, dispersion is promoted, while also some shear is exerted to the powder to break up lumps. This type of equipment can be applied for both free-flowing and cohesive powders. Their energy consumption is at maximum about 5 kW/m^3.
- Bunker or silo mixers are comparable to batch mixers with internals. Only, their size is usually much greater in relation to the size of the vessel. As their name suggests, they are applied for mixing the content of e.g. a silo. The internals often consist of a screw. Also, air jets or fluid beds are used.
- Continuous mixers are similar to batch mixers in that they usually contain internal elements for powder blending. The major difference is their continuous operation. Material is continually fed into and leaves the mixer. Segregation in a continuous mixer is less common in the equipment itself.

The exact choice of equipment depends very strongly on the particulate properties of the materials involved, their (sensitivity to) humidity, the degree of agglomeration of the particles, the amounts of ingredients to be mixed, the mode of operation (batch or continuous) and the quality requirements set to the final product. Since good rules for scaling-up of equipment are a matter of considerable debate, extensive empirical investigation is required to find the optimum equipment and conditions. Such research should include investigations to best control of the mixing process and the final product. Typically, continuous blending processes give fewer chances to segregate than batch processes since their product can directly be applied in further processing.

6.4 Segregation

Typically, individual particles take different positions in dry powders due to different properties upon vibration (transport) and deposition (storage). Moreover, process fluctuations and loading sequences can contribute to segregation. Relevant properties of particles for segregation are size, shape, density and tendency to agglomerate. Mechanisms for segregation include:

- Percolation of small particles through the voids in between larger particles induced by gravity and vibration.
- Floating, where large particles move upwards during vibration and small particles fill the space underneath the large particles during their movement.
- Trajectory segregation during transport in gas or air, where particles follow different trajectories in relation to their impulse and mass. This holds for their

behavior at the end of the conveyor belt during conveyance into hoppers, in pipeline bends, during fluidization and in cyclones.

- Elutriation of small particles, where small/light particles are entrained in an upward gas flow and large/heavy particles are not.
- Different tendencies of ingredients to form agglomerates may also lead to segregation during vibration, when agglomerates are formed that stay together and move more slowly than the primary particles (often smaller and more mobile).

Several of these segregation processes occur during product transport. Thus, transport between the mixing process and subsequent manufacturing steps, such as tableting, should be as short as possible. Intermediate storage of particulate products also promotes segregation and, thus, should be avoided.

Although the range of selection of excipients for blending may be limited by larger formulation concerns, considerations of the particle size, shape, density and interaction potential are important for blending. These properties have a direct impact on the time and energy required for blending and on the potential for a blend to segregate immediately after blending and in subsequent processing steps. The size of particles plays a significant role in the potential for mixtures to segregate following the blending process. Specifically, small particles have the ability to percolate down through larger particles causing an increase in the concentration of the small particles near the bottom of the container after blending. This can be particularly detrimental to the quality of a drug product if it is the API or a key excipient for drug release that has segregated. The density of particles has the potential to elongate blend times and to be a source of segregation. However, these effects typically require density differences on the order of $4\times$ or greater before significant segregation is observed [50]. The shape of particles plays a significant role in blending largely through the effect of flow. Spherical or other similar equant particles tend to flow well, enhancing the rate at which blending occurs. However, the advantageous flow properties increase the potential for size and density based segregation. Irregular and large aspect ratio particles such as plates and needles decrease flow, increase blending times but decrease the potential for downstream segregation.

An additional factor that is often important is the cohesivity of the blend, as discussed above. This can be discussed as a function of the adhesive forces of the particles relative to their mass. If these adhesive forces are significantly greater than the mass of the particles then the cohesivity of the blend may be regarded as significant. The adhesive interactions include mechanical interlocking, electrostatic, Van der Waals and surface tension. The cohesivity of a blend has a substantial impact on the characteristic flow of that system. Highly cohesive blends have poor flow and may be difficult to mix and/or transfer between containers. Blends with these characteristics typically require significant shear to finish. However, they can be blended to a very high degree of homogeneity (blended to a very small scale of scrutiny) and are less susceptible to segregation in downstream processing of the mixture. However, non-cohesive blends can also be challenging to blend because of the free-flowing nature of these blends; this leads to a propensity to segregate when a size or density disparity of the particles exists. The combination of the flow properties of the blend and the matching of size, density and shape determine the requirements for blending and the propensity for the mixture to segregate in subsequent processing.

The relationship between blend uniformity and content uniformity (the uniformity of the content of API between tablets) has been studied in a limited number of accounts [6]. The ultimate objective of blending powders in the pharmaceutical industry is to produce tablets of consistent quality. Specifically, the content uniformity defines this attribute by the relative standard deviation of the API contents of a sample of tablets. While the relationship between blend uniformity and content uniformity has not been well defined, it is apparent that achieving appropriate content uniformity requires a certain degree of blend uniformity.

6.5 Blending Control and Sampling Methods

Confirmation that a blend has reached an adequate state of homogeneity is a subject of substantial debate in the pharmaceutical industry. There are two primary approaches to determining the state of a blend. These approaches have in common the goal of determining the distribution of the API throughout the volume of the blend. Another commonality is that both are quite sensitive to the scale of scrutiny (sample size) of the measurement of homogeneity. Both approaches must consider the analytical error associated with a given measurement. The traditional method for evaluating a blend is to stop it at various time points and analyze the contents of samples removed from specific locations within the volume. Each sample from the blend is then subject to wet chemical analysis to determine the concentration of the constituent of interest (typically the API). The blend uniformity (BU) is then defined as the relative standard deviation of those samples. A second approach to determining the progression of the blending process is to make measurements of the powder throughout the blending process. These measurements are typically restricted to a single point on a surface of the blend. This approach is commonly referred to, in the pharmaceutical industry, as process analytical technology, or PAT. Both of these methods seek to determine the state of the blend at a given point in time.

The preceding methods for characterizing the state of a pharmaceutical blend are used for both blend validation and for blend monitoring. In the case of blend validation, characterization techniques are used determine an appropriate blend time for the given formulation and blending parameters. This validation is completed as part of the development process for a drug product. In instances where the blend is considered low-risk, the blend validation will stand as the evidence needed to demonstrate that the product has been adequately blended (for future batches). That is, the materials will be blended in the prescribed manner and time; when the time has elapsed, the blending will stop and the mixture will be considered adequately homogenous. Examples of low-risk blends include high-drug dose blends (>35% w/w API) and boundary lubricant blending (i.e., magnesium stearate). Process analytical technology (on-line measurements) are not typically employed in this low-risk scenario. Routine blend monitoring may be required for products not considered to be low-risk. In this situation, each blend must be confirmed as properly blended by one of the two approaches to blend confirmation. When the traditional approach is used, samples are removed at the end of the blending time, analyzed in the laboratory, and processing of the material progresses only after a passing result is obtained from the testing. If PAT is applied, the blend will be stopped when a pre-determined end-point criterion (or criteria) is reached. The blend is considered to be appropriate when passing either of these tests.

In relation to tight specifications for pharmaceutical products, adequate control of both the blending process and its product is necessary. The progress of the blending process can be measured by new spectroscopic techniques, using near-infrared, Raman or light-induced fluorescence and chemometric methods. Note that the measurement depth is limited to several millimeters, restricting their application to relatively thin layers. On the other hand, frequent measurements are possible so that the progress of the blend can be continuously monitored and process control is possible.

Conventionally, both the blending progress and the final product quality are assessed by stopping the process and collecting physical samples for measurement.

The basic rule for correct sampling of a mixture for representative, quantitative analysis is that **all constituent elements have equal probability of being taken into the sample, given their proportions**. It is clear that this is not feasible if the product is segregated. Thus, chances of major errors increase with decreasing blend quality. These risks must be mitigated through application of adequate sampling procedures and equipment. Below, the conventional way of collecting samples will be discussed, although similar principles are applied for inspection by spectroscopic methods.

Most often, several sampling steps are required to reach a test sample for a product from a large batch. The primary sampling step is executed from a large amount of product, typically directly from the blend container, although sampling from a process stream is preferred. Secondary sampling is often required to reduce the large primary sample to a relatively small test sample.

6.5.1 Primary Sampling from a Process Line or a Stored Batch

The advantage of sampling from a process stream is that the particulate product is in movement, limiting the possibilities for segregation. Still, segregation due to differences in flow rate in a pipeline or to percolation in conveyor belts is possible. This means that the full trajectory of the pipeline or the conveyor belt has to be sampled in best practice. Examples of such samplers are the flexible hose sampler and the Vezin type sampler for pipelines and the cross-cutter sampler for conveyor belts [39].

Sampling from a stored batch (processing container or dedicated storage container) is less attractive since stored batches are often subject to severe segregation and, thus, sampling may result in large errors. Typically, different samples are collected at different times or locations and combined in a composite primary sample to take regard of segregation. The amounts of these sample increments should be about the same for statistical reasons (see Sect. 6.8.4).

Particulate products can be sampled in three different manners [57].

6.5.2 Random Sampling

Random sampling from a moving product stream or product boxes or tablets means that the time of sampling is determined by chance. Sampling times are typically based on random number generators or tables. For batch sampling, it means that the locations of sampling are determined by chance. This also means that random sampling of moving product streams, which are fairly homogeneously distributed over time, can be expected to yield better results than random sampling of segregated heaps. As can be expected, differences between samples may increase when the interval of time or location for sampling is increased. As a result, measurement errors may also increase. Thus, small intervals are recommended in the general case that segregation is to be expected.

6.5.3 Systematic Sampling

Sampling is named systematic if fixed intervals of time or distance are used for sampling. It is easier from an organizational point of view. However, it requires that the frequencies or distances are not correlated with process fluctuations or loading sequences.

6.5.4 Random Stratified Sampling

In this type of sampling, the sample is taken within fixed intervals but at a time or location determined by chance. Its advantage is that correlation with process fluctuations is avoided.

6.6 Sampling Protocol and General Rules

In order to ensure good quality of measurement results it is required that a detailed *sampling protocol* and/or *sampling plan* be written for each situation and material. Before writing, all details of the procedures should be investigated and demonstrated to have acceptable precision at a given confidence interval for all relevant product quality aspects of both excipients and API. The sampling protocol should regard the potential for product segregation and describe all details of both equipment and procedures. It should address all aspects that are relevant to the characterization of the particulate materials and the validation of the procedures:

- Product identification: typical composition; type (e.g., powders, tablets, suspensions, emulsions or aerosols) and properties (size distribution, shape, composition, cohesiveness, expected degree of segregation, hygroscopicity, suspension stability, toxicity, etc.)
- All results of preliminary investigations and reasoning for their consequences in relation to blending and its control, sampling and measurement of all relevant material and product properties
- Identification of specific sampling locations at conveyor belts, pipelines, silos, heaps, bags, boxes, etc. in relation to expected or measured segregation of materials
- Description of the purpose for sampling and characterization as well as of analysis goals: parameter(s) and required precision
- Requirements for personal protection and safety in view of dust, sound, toxicity, flammability, explosion risk, radiation, etc.
- Amount of test sample needed in relation to the method(s) to be used for characterization and the characteristics to be measured
- Number and size of primary sample increments to be taken
- Specification of frequency/time(s) or place(s) for primary sampling
- Sampling device(s) to be used
- Means for storage of samples
- Sample labeling
- Procedure(s) for sample treatment to arrive at test sample, e.g. combination of sample increments to a composite sample, splitting of composite sample, drying, etc.
- Procedures to be followed for calculation of sampling error, e.g. spreadsheet program
- Criteria for acceptance/rejection of a lot
- Statistical reasoning for sampling design
- Names of persons qualified for execution of sampling protocol
- Reporting of sampling and analysis
- Disposal of waste.

The sampling protocol should minimize hazards during sampling as well as contains the results of quantitative evaluation of (errors in) the total sampling procedure.

The *sampling plan* usually has somewhat limited contents, viz.:

- Product identification
- The number of units of product from each lot to be inspected (sample size and number of samples/increments)
- The procedure(s) followed to arrive from the primary sample to the test sample
- The criteria for acceptance/rejection of a lot
- The requirements for safety and personal protection.

Many of the parameters associated with confirmation of a blend are focused on appropriate sampling. The size of the samples (scale of scrutiny), number of samples, location from which samples are removed, and manner in which they are removed are all of substantial importance. The general principle for the size of samples is that they must be representative of the finished unit dose. While this is not directly reflected in the sparse regulatory guidance on this topic, it is self-evident. The number of samples, location and sampling technique should, ideally, be as close to probabilistic as possible. That is each volume within the full container should have an equal probability of being sampled [27, 28].

General 'golden' rules for both primary and secondary sampling of dry powders are:

- Sample where the material is well-mixed
- Sample when material is in motion
- Sample the whole cross section of particulate flow; do not stop the flow
- Design the sampling container large enough and without constraints
- Do not overfill the sampling container
- Prepare representative samples
- Use a rotary sample splitter in the laboratory
- Investigate and quantify the precision of measurement of material quality attributes through measurement of different samples
- Combine sufficient increments during primary sampling and split them later to form a test sample that matches the required precision for the quality attributes
- Use validated procedures for all sampling steps
- Make for each product a specific and well-documented sampling protocol.

Following these principles is complicated by traditional batch-wise blending in a closed container. The preferred *design of equipment* for primary sampling should take following aspects into account:

- The preferred design should collect the sample over the whole particulate stream for part of the time and not some fixed part of the stream continuously. Otherwise, segregation may lead to samples that give poor representation of product quality.

- Sampling times and locations should have no correlation to fluctuations in the production, transport or storage process. Thus, for example, the frequency of sampling should not coincide with the frequency of eventual process fluctuations or loading sequences.
- The sampler and sample container should have no constraints, so that even the largest particles are easily accepted. The opening diameter of sampling tubes and the distance between cutter edges should be at least 3 times the size of the largest particle.
- The sample container should have sufficient volume to easily receive the full sample. In case of nearly complete filling, usually the large particles are more easily rejected than the small particles.
- Sampler tubes and sample containers should have thin walls or sharp edges, to promote good control of acceptance of particles based on their center of gravity.
- Local sampling tubes in pipelines should be isoaxial to the pipe, so that the direction of the flow in pipeline and sampler remains unchanged.
- Local sampling in pipelines should be done isokinetically, while keeping the linear flow rate about the same. The sampler walls should not disturb the flow in the pipeline significantly.

Altogether, considering segregation of powders, maintaining optimum procedures to reach a representative test sample for a lot of material, assessing the ultimate error of sampling and blending and their confidence interval can lead to a situation where good decisions can be made on product quality and acceptance or rejection of produced product. Moreover, it is important that sampling is executed near to the final point of application of the product mixture (e.g. tableting), so that between the sampling point and the point of application no significant segregation occurs.

An example of a minimum level of homogeneity of a powder blend to be used to manufacture tablets is a relative standard deviation of less than 5% in the quantity of active ingredient present in a sample of powder equivalent to the mass of a unit dose (tablet). This is a typical means of describing the homogeneity of a blend. It incorporates the idea of the variance observed and (equally important) the mass of the sample. A minimum variation in the quantity of API in the tablet will represent a significant quality parameter to the patient. This is the ultimate objective of the blending process, to mix the powders to a degree that the API will be evenly distributed across the unit doses. Although it is not typically considered a critical quality attribute, the appropriate homogeneity of excipients critical to the formation of tablets, or of excipients that control the release of API can be important. However, the homogeneity (or lack thereof) is typically not measured directly during, or after blending. Effects of this type of heterogeneity are typically only observed indirectly. For example, heterogeneity of this type might be detected while testing the whole tablet (e.g., dissolution testing or tablet crushing strength).

The major factors that affect the blending process come from the formulation and the blending parameters. From the formulation, the most important elements are the size, shape and density of the particles. The blending parameters were discussed above and determine the relative importance of the three blending mechanisms, the total energy imparted upon the system and the total time over which that energy is imparted. Additional factors affecting blends are cohesivity, environmental conditions and triboelectric effects.

6.7 Formulation Considerations

In general, it can be said that the use of ingredients of uniform size and the application of slightly cohesive powders are advantageous for reaching good blends. On the other hand, blending of larger batches of cohesive products requires more time. In these cases, strong shear forces may be necessary to break agglomerates.

Choices made regarding the formulation are sometimes influenced by the sensitivities of the blending unit operation. Formulation of a product is typically not focused on blending. However, selection of materials can have a substantial influence on its success rate. The primary focus of formulating blends is to avoid segregation. The propensity for segregation is typically based on the size, shape, density, and interactions of the particles (as previously discussed). Other considerations, such as the quantity of API or the specific physical properties of the API are typically beyond the purview of the formulator. However, the formulation must accommodate these characteristics.

Formulations that require low percentage API place an additional burden on dry-powder blending. Strategies to overcome this burden include geometric dilution; a strategy where the API is combined with an equal mass of excipients to make a 50% blend. This blend is then combined with, again, an equal mass of excipients to make a 25% blend. The process is repeated until the desired API concentration has been reached. A slight modification to the geometric dilution strategy is to create a pre-blend that contains API at a level that is blended, with excipients conducive to this process. For example, a pre-blend of 10% API in microcrystalline cellulose could be prepared. This blend could then be added to the remaining excipients to make a low-concentration API blend in fewer steps than geometric dilution. In cases where a specific attractive interaction between the API and an excipient can be engineered, an interactive mixture can be prepared. This type of mixture, typically employed as part of a pre-blend strategy, has an advantage of minimizing segregation due to the interaction of the API and excipient. However, it requires careful particle characteristics control over both the API and excipient. The disadvantage of all of these processes is that they typically require multiple steps and containers to complete a single blend.

Other strategies for low percentage API blends require the use of solvents, or specific particle interactions, to distribute the API and prevent segregation. High-shear wet granulation can be used to combine low concentrations of API with excipients to form granules that have a very low potential to segregate after drying. Another strategy is to dissolve the API in a suitable solvent and spray it onto

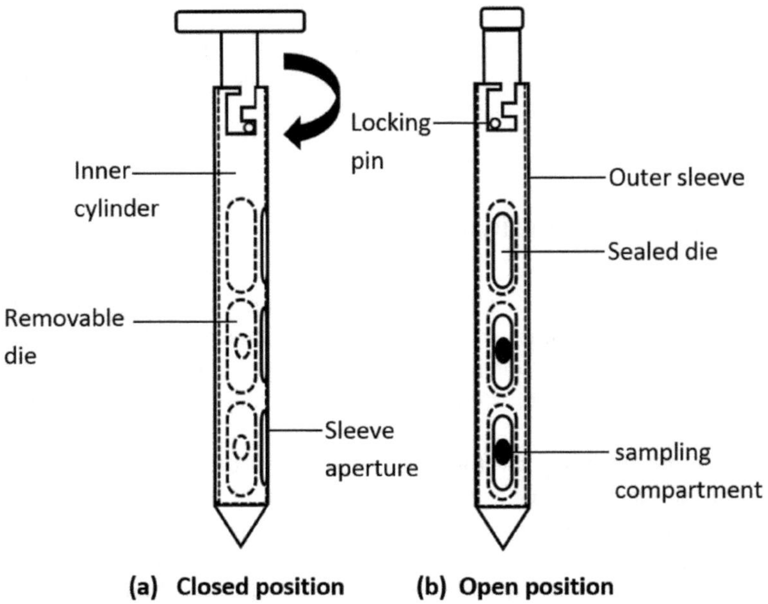

Fig. 6.2 Schematic side-sampling thief probe in **a** closed insertion/extraction position and **b** open sample retrieving position

excipients or particles. This process of drug loading immobilizes the API on a major excipient in the blend and prevents segregation of the API in the absence of attrition of the drug loaded particles. These strategies are suitable only when the API does not react with the solvent chemically, or physically and is adequately robust to withstand subsequent drying steps.

6.8 Traditional Thief Sampling and Blend Uniformity Assessment

Sampling from pharmaceutical blends is an essential step in the confirmation of blending quality. The conventional sampling method for the assessment of blend homogeneity involves the use of thief probes (Fig. 6.2). Thief sampling consists in inserting a cylindrically shaped probe into a stationary powder bed to extract samples from one or multiple locations. While several sampling tools exist, side-sampling and end-sampling thieves are predominantly used in industry for the evaluation of pharmaceutical powders.

Side-sampling thieves are probes with one or multiple sampling compartments built along the inner body of the device. The compartments are enclosed by an outer sleeve which prevents material flowing into the chambers prior to reaching the desired sampling location. The end of the probe has a conical shape to facilitate the insertion of the device into the powder bed. Once inserted, the thief is turned to the open position, allowing sampling compartments to coincide with apertures in the sleeve. Samples (ideally, similar in mass to a unit dose) are collected as the powder flows and fills the open compartments. The inner tube is then rotated to the closed position and the probe is pulled out from the powder bed. The sampling performance of this thief is significantly sensitive to material flow properties. Potential segregation can arise during sample collection as components with different properties flow selectively into the cavity [6, 7, 24, 44, 45]. Moreover, the poor flowability of cohesive blends typically results in unfilled or partially filled sampling compartments. As a result, side-sampling probes are limited to free-flowing blend formulations.

End-sampling probes are designed to facilitate the collection of samples from cohesive powders. These thieves are built with a single aperture, commonly located at the end or bottom of the device. A widely used end-sampling probe is the plug thief, which consists of a plunger inside an outer rod with a bottom orifice. This device is inserted in closed position into the powder bed to the desired depth and sampling location. The inner rod that plugs the tip opening is pulled back to a predefined length to create an open volume space. The thief is then inserted deeper into the powder bed to collect a stationary slug of powder inside the cavity, which is ejected by returning the device to its closed position. This probe does not have a feature to seal the open cavity and prevent material loss prior to sample ejection. Therefore, the performance of a plug thief mainly relies on material mechanical properties to form a plug of enough cohesive strength to withstand sequential handling. Attempts to provide external closing to the end cavity were reported, but these resulted in impractical alterations to the original shape of the probe and significantly increased perturbations to the blend during probe insertion [45].

The selection of the type of thief is of critical importance as sampling results can vary depending on the type of probe used. For example, the plug thief is designed to collect samples from undisturbed blend zones given the strategic location of the sampling cavity. While this represents an advantage over side-sampling thieves, it has been shown that the performance of plug thieves is only superior when sampling from poor flowing compressible blends. The sampling error can significantly increase if a plug-thief is selected over side-sampling thieves to collect samples from powders with excellent flow and poor compressibility [24]. Therefore, the type of thief probe should be carefully selected in a case-by-case fashion, addressing the main limitations of both the formulation system and the sampling technique overall.

Perhaps the major limitation of traditional thief sampling is the disturbance of the blend during probe handling. In the process of inserting the thief, the powder bed is disrupted, leading to potential migration of powder particles from the top to inner locations in the bulk. Collecting material from perturbed blend locations results in samples that are a biased representation of the actual state of the blend. The local segregation induced by the thief probe is therefore a significant source of error during powder sampling [6, 7, 44, 45]. Rigorously designed thief probes have been developed over the years to minimize powder bed disruption and facilitate sample collection from undisturbed blend zones [44, 46, 58]. However, independently from the design of the probe, thief sampling is invasive in nature and violates the two Golden Rules of Sampling: (i) samples should be collected while the powder is in motion and (ii) the entire material stream should be sampled in several short time intervals in lieu of only one part being collected the entire time [2]. In violation to these rules, thieves grab samples from stationary powder beds. Also, only a limited number of samples can be collected from carefully selected locations to minimize the disturbance of the blend structure. Given the invasive nature of the technique, it is not feasible to obtain samples representative of all regions of the powder blend [19, 45].

6.8.1 Dynamic Sampling

As an alternative to thief sampling, dynamic sampling methods have been developed to collect representative samples from moving powder blends. Dynamic sampling (or full-stream sampling) occurs while the material is in motion during blender discharge. A major advantage of dynamic sampling is that it follows the golden rules of sampling and reduces errors associated to invasive sampling methods. However, it is important to handle the collected samples carefully to avoid introducing sampling errors. Since the samples collected are relatively large, the technique typically involves the assistance of a secondary sample splitting method. Secondary splitting techniques divide large samples uniformly into multiple smaller sub-samples to facilitate further analysis. *Spinning or rotary rifflers* are by far the best option for secondary splitting of dry powders [39]. For their application, it is required that the powders can be brought to a fairly even flow. The powder is brought into a hopper, from which it is smoothly fed, via a vibrating feeder and a rotating divider, into sample containers. Assuming a large number of rotations of the divider and smooth particle flow, each particle is given the same probability for each of the containers. Various designs exist, which handle different quantities of material. Maximum content of the hopper is about 1 kg, minimum sample about 100 mg. This process should be performed cautiously to ensure that the sub-samples are representative of the larger samples.

Dynamic sampling has significant limitations that must be considered. Contrary to thief sampling, if dynamic samples indicate that additional blending is required, there are no options to continue the blending process given that the blend has been already discharged. As a result, dynamic sampling is not a practical approach to characterize the progression of a blending process. In addition, this approach is unable to intentionally collect samples from targeted regions of the bulk where poor uniformity is suspected [47]. This makes it very challenging to understand, identify and mitigate potential risks to blending quality.

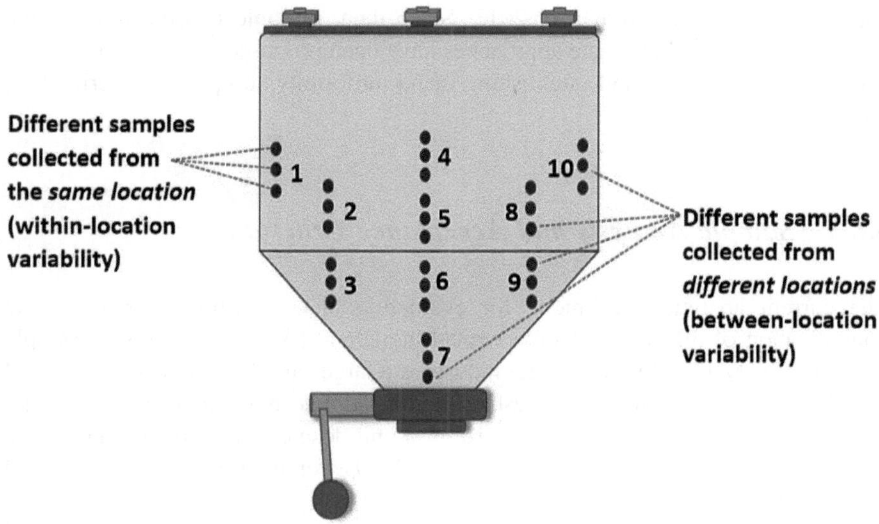

Fig. 6.3 Schematic representation of stratified sampling in a bin-blender

6.8.2 *Stratified Sampling*

The stratified sampling approach focuses on sampling multiple units strategically from high-risk locations or at specific process intervals in the blender or during compression or filling operation [8]. Stratified sampling suggests collecting three replicate samples from at least 10 target locations in the blender where the risks for incomplete mixing or segregation are high. Additionally, it recommends collecting these samples from at least two different depths within the blender. This sampling plan facilitates effective statistical analysis, including the evaluation of within-location variability and between-location variability by variance component analysis (VCA). Within-location variability indicates the variation between the samples collected from the same location, while between-location variability is the variation between samples from multiple locations (Fig. 6.3). Results from VCA can indicate potential root causes of poor uniformity that cannot be determined solely by calculating the overall batch relative standard deviation [5]. High within-location variability is commonly associated to issues at the scale of a unit dose (micro-scale) and can be the result of sampling bias, analytical errors or incorrect sample handling and preparation. High between-location variability indicates macro-scale issues such as inadequate blending or highly influential sampling errors.

The stratified sampling approach for the assessment of blend uniformity was introduced in a draft recommendation submitted by the Product Quality Research Institute (PQRI) in 2002 [8]. While these recommendations were included in 2003 FDA Draft Guidance for Industry [22] and used for many years by pharmaceutical companies, it was withdrawn in 2013. Since then, multiple modifications to the revised guidance and alternate approaches have been proposed for evaluating blend and content uniformity and establishing blend uniformity acceptance criteria [4, 9, 19, 23].

6.8.3 Sample Analysis and Acceptance Criteria

The current analytical technique for evaluating blend uniformity involves the quantification of the active pharmaceutical ingredient (API) in the powder sample by HPLC analysis. While in several industries a large sample is collected, followed by sub-sampling steps for further testing, in the pharmaceutical industry the powder samples are limited to 1–3 times the size of a unit dose collected by a sample thief [51]. The minimum number of locations is 10–15 for tumbling blenders and 20 locations for convective blenders.

Blend uniformity results are typically reported in terms of relative standard deviation (*RSD*) and percent of label claim. *RSD* is the standard deviation (s) of the drug content in the samples expressed as the percentage of the mean ($\bar{X}$):

$$RSD = 100s/\bar{X} \tag{6.1}$$

Percent of label claim is the drug substance content in each sample expressed as a percentage of the amount of drug claimed on the label. The common specification is that samples must have a measured *RSD* smaller than a specified value and a drug content within a narrow range around the label claim.

The typical practice in the pharmaceutical industry has been to implement the acceptance criteria as recommended in 2003 FDA Draft Guidance for Industry. The specification requires that samples report an *RSD* of less than or equal to 5.0% and all individual samples fall within ±10% of the mean of the results. Endo Pharmaceuticals reported blend uniformity results for manufactured batches of *Product A* according to the sampling procedures and acceptance criteria established in 2003 Draft Guidance [30]. Batches of *Product A*, a solid oral tablet of multiple strengths (3.0–18.0% API), were manufactured using dry blending followed by compression. Blend sampling was performed in replicates using traditional thief sampling. Results for blend quality assessment are shown in Fig. 6.4. A sampling error due to inadequate thief probe manipulation biased the results for batch 22 and was corrected for the subsequent batches. Once the sampling issue was resolved, batches met the acceptance criteria of falling within ±10% of the mean drug concentration. The *RSD* values reported for each batch were in the range of 1.0–2.8% *RSD*, which meets blend uniformity specifications as established in 2003 Draft Guidance.

It is not unusual that corporations opt to establish internal specifications for a product depending on multiple factors, such as drug properties. Studies have indicated that the pharmacologic profile of drugs plays an important role in

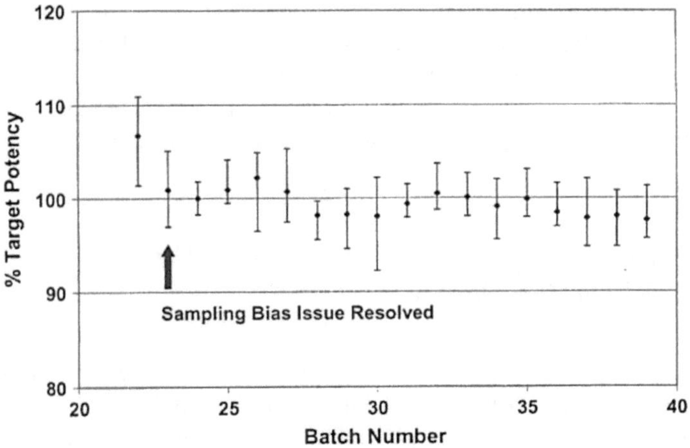

Fig. 6.4 Blend uniformity results (mean and range) for batches 22–39 of Product A [30]. Copyright Parenteral Drug Ass., Inc. (PDA); reproduced with permission

establishing suitable specification limits. Simulated pharmacokinetic profiles for levothyroxine, a narrow therapeutic index drug, showed that dosing units *RSD* of 6.5% would exceed the lower concentration limit. Results indicated that 1 or 2% *RSD* may be acceptable specification limits [5]. Another example involves tablets of warfarin, a narrow therapeutic index medication originally marketed as Coumadine by DuPont Pharma. Given the narrow range between warfarin therapeutic and toxic doses, DuPont Pharma established stricter internal limits for content uniformity of *RSD* no more than 3.0 and 92.5–107.5% of label claim [43].

Internal corporate specifications are also established based on historical data accumulated from prior batch analyses of the product. Table 6.1 shows the relevance of historical batch records to establish meaningful specifications [37].

Besides being a guide for the development of specifications, the historical data for a given product is valuable to identify the presence of irregularities. Out of trend blending results can be indicative of changes in the process that may require attention. For example, a *RSD* of 4.8% typically lies within specifications, but could

Table 6.1 Implementation of historical batch data to guide the development of specification limits for potency

Active ingredient	Antihypertension
Proposed potency limits	93.0–107.0%
Analytical method	Chloroform extraction followed by UV spectrophotometry
Reproducibility	±2%
Batch record (%)	99.3, 98.4, 103.4, 97.9, 101.3, 98.9, 101.8, 99.7, 98.6, 100.4
Acceptable limits	95.0–105%
"House" limits	96.0–104%

Adapted from Lachman et al. [37]

be a red flag if the historical trend for the product has been consistently less than 2.0% *RSD*. Similarly, according to the data shown in Table 6.1, a result of 105% is within the acceptable limits, but out of the historical batch records of the product. While out of trend blend uniformity results might just be indicative that more blending time is required, the source of error or variation should be investigated.

6.8.4 Quantitative Estimation of Sampling Errors and Blending Quality

Errors in sampling of particulate materials consist of two types, viz. ***fundamental errors*** and ***segregation errors***. Fundamental errors are random errors in relation to the constitutional heterogeneity of particulate materials and products, which is due to the discrete nature of particles in relation to differences in particle size, shape and density. They can be expressed by the variance (*Var*), its square root, the standard deviation (s), or the relative standard deviation (*RSD*). The fundamental error represents the lower limit for the sampling error and, thus, indicates the minimum amount of sample for a stated precision.

The theory for binomial distributions provides a way for calculating this fundamental error for (quasi-) two-component mixtures. It states that the variance *Var* of the measured proportion by number of one of the components (p) as [39]:

$$Var(p) = s_p^2 = \frac{p \cdot (1 - p)}{n} \tag{6.2}$$

where p and $(1 - p)$ are the measured proportions by number of the two components, s_p is the standard deviation of p, and n is the number of particles in the sample, which is much smaller than the number of particles in the batch from which the sample is taken. This equation can be used for calculating variance and standard deviation of p—and subsequently, if the particle size distribution (PSD) is known, for the related particle size—for any point of the cumulative number-based size distribution.

For small numbers of particles in a critical size class of interest, usually Poisson statistical theory can be assumed to be valid as limit case of binomial distributions. Here, the variance is equal to the average number $\bar{n}$ of particles and, thus, the standard deviation s_n equals to the square root of that number and the relative standard deviation $RSD(n)$ (%) inversely to the square root of that number:

$$Var(n) = s_n^2 = \bar{n} \tag{6.3}$$

and

$$s_n = \sqrt{\bar{n}} \tag{6.4}$$

and

$$RSD(n) = 100/\sqrt{\bar{n}} \tag{6.5}$$

For example, if 10 particles are measured in a size class of a distribution, the fundamental error (standard deviation) in this number amounts $\sqrt{10} = 3.16$, or 32% relative. This error is especially important at the upper side of a particle size distribution, where few particles contribute a significant volume, since particle volume relates to the third power of its size. Following the above example, 10 particles of 100 μm have the same volume as 10,000 particles of 10 μm and, thus in a mixture, would represent 50% of the volume-based distribution. But when measuring this size distribution by counting 10,000 particles (e.g. by microscopy), the statistical uncertainty (*RSD*) of the 10 large particles and, thus, also of the 50% by volume is still about 32%.

Note: Since the above theories have a statistical nature, they are based on numbers of particles. There are also several equations to calculate the fundamental error based on volume or mass proportions, but they are all approximating equations [39]. A reasonable estimate of the fundamental error of a PSD parameter of a measured volume-based size distribution of spherical and equant particles may be obtained through application of the above equations after conversion of this distribution to the related number-based distribution. Such conversions can be done using the equivalent sizes D of the particles and their density ρ through $M = 1/6\pi\rho D^3$, where M is particle mass. Some examples, which show the great influence of mass to size, are presented in Table 6.2.

The ***segregation error*** is due to distribution heterogeneity as indicated in the previous paragraph. For a fully segregated mixture of 2 components, the variance is given by:

$$Var(p) = s_p^2 = p \cdot (1 - p) \tag{6.6}$$

Note, that this Eq. (6.6) differs from Eq. (6.2) in that the number of particles n is no longer present. Thus, the variance is independent of sample size. The sample

Table 6.2 Conversion of particle size to mass (ρ taken as 1200 kg/m^3)

Particle size (nm/μm)	Particle mass (μg)
1 nm	6.3×10^{-16}
10 nm	6.3×10^{-13}
100 nm	6.3×10^{-10}
1 μm	6.3×10^{-7}
10 μm	6.3×10^{-4}
100 μm	6.3×10^{-1}
1000 μm	$6.3 \times 10^{+2}$

behaves as if only one particle was sampled. This variance represents the maximum value that can be found for a (segregated) mixture. In practice, values will be found in between the values coming from Eqs. (6.2) and (6.6) and they cannot be calculated from theory. The degree of segregation can only be estimated by analyzing a reasonable number of samples (N) from different locations in the powder lot or at different times from the process line and analyzing them separately. **This assessment should always be a step in the characterization of a new product**. It should be noted that such quantitative assessment of the degree of segregation requires that the other parts of the procedure (sample splitting, powder dispersion and measurement) have a relatively small contribution to the overall standard deviation. In the following we will assume that this is the case. Thus, we will use both words 'samples' and 'assays' for the total assessment procedure and use their number similarly for assessing the overall standard deviation of the total of sampling, dispersion and measurement.

An estimate of the variance (s^2) of a measured property y (e.g. concentration, median size or amount in a certain size class) can be obtained from the results of N measurements:

$$s^2 = \frac{\sum (y_i - \bar{y})^2}{N - 1} \tag{6.7}$$

with:

$$\bar{y} = \frac{\sum y_i}{N} \tag{6.8}$$

and

$$RSD^2 = s^2/(\bar{y})^2 \tag{6.9}$$

If this absolute or relative standard deviation is significantly larger than the fundamental error, then there is product segregation, provided that the rest of the procedure is adequate (i.e. has much smaller standard deviation).

The true value μ for the property of interest y of the mixture can be stated with a certain confidence to be within the interval:

$$\mu = \bar{y} \pm \frac{t \cdot s}{\sqrt{N}} \tag{6.10}$$

Here, the t-factor comes from the Student t-test for statistical significance. The t-value depends on the confidence level required and the number of degrees of freedom (number of assays minus 1). It can be taken from statistical tables. For example, for $N = 30$ there is 95% probability ($t = 2.04$) that the true mean critical value to be established for the mixture lies in the range

$$\bar{y} - 0.373 \cdot s < \mu < \bar{y} + 0.373 \cdot s$$

Equation (6.10) can now be used to calculate the required number of samples N_r to reach a stated maximum allowable difference E between estimated and actual value for the characteristic parameter of interest:

$$N_r = (t \cdot s / E)^2 \tag{6.11}$$

It may be clear that for heavily segregated mixtures a large number of primary samples is required for reasonable precision; as a rule of thumb, often a minimum of 30 primary samples is taken in relation to the fact that the t-factor approaches 2 for 95% probability [39].

A measurement procedure for any property of interest in any mixture typically consists of several steps, including primary sampling, secondary sampling, sample dispersion and measurement. Most often, differences in concentration in a batch are included as part of the variance of the primary sampling. For chemical analysis, the quality of secondary sampling can be improved through reduction of particle size by milling. In other cases, e.g. powder flow measurements, dispersion is not necessary. Each step is subject to statistical variations, which contribute to the overall error. In the normal case that the steps are independent of one another and their errors are random, these errors can be combined as variances, provided that the sample amounts are the same:

$$Var_{\text{total}} = Var_{\text{primsampl}} + Var_{\text{secsampl}} + Var_{\text{disp}} + Var_{\text{meas}} \tag{6.12}$$

or

$$s^2_{\text{total}} = s^2_{\text{primsampl}} + s^2_{\text{secsampl}} + s^2_{\text{disp}} + s^2_{\text{meas}} \tag{6.13}$$

For particulate mixtures, the sampling error usually gives the greatest contribution to the overall error. Typically, errors during primary sampling are dominant due to both heterogeneity of the batch and quality of the sampling procedure. Errors due to secondary sampling can be decreased through application of rotary rifflers. In this approach, batch heterogeneity is taken for granted and the overall error in the measured concentration of a relevant component decreased to an acceptable level by increasing the number of primary sample increments.

For assessing the quality of a blending process or the degree of segregation of a powder through analyzing different samples of its contents, the variance of the blend uniformity (Var_{blend}) should be included explicitly in the equations [60]:

$$Var_{\text{total}} = Var_{\text{blend}} + Var_{\text{primsampl}} + Var_{\text{secsampl}} + Var_{\text{disp}} + Var_{\text{meas}} \tag{6.14}$$

or

$$s_{\text{total}}^2 = s_{\text{blend}}^2 + s_{\text{primsampl}}^2 + s_{\text{secsampl}}^2 + s_{\text{disp}}^2 + s_{\text{meas}}^2 \tag{6.15}$$

In 2003, the FDA launched a draft document for assessment procedures of blend [22]. However, this document was withdrawn in 2013 due to changed views, and a new document is not yet available. Recently, Cholayudth proposed to add the variance of sample mass $\left(s_{\text{sammass}}^2\right)$ in the above equations, for situations where a significant degree of variation of sampling masses is involved [9]. Of course, if the size of the different samples collected is about the same, i.e. the variance of their mass is small, then this variance can be neglected.

The above equations mean that best assessment of the variance of the blend uniformity is only possible if the variances of sampling, dispersion and measurement are smaller. If this is not the case and any of these variances is the same or greater than that of blend uniformity, then the assessed variance of blend uniformity is limited by the greater total variance. This would mean less well control of blend uniformity.

Based on the total overall standard deviation, a confidence interval can be calculated for quality aspects in specifications, the limits of which give guidance to blending control and to acceptance or rejection of product lots. Statistical background can be found in e.g., [40].

6.9 On-line Blend Monitoring

Blend monitoring during the process (on-line blend monitoring) offers several advantages for producing quality mixtures. A key feature of on-line blend monitoring is that it allows active control of the blend. That is, the blend can be stopped at the point that it is complete, avoiding the risk of over blending and possibly driving the mixture towards segregation. This approach to designing, measuring and controlling a unit operation in the pharmaceutical industry is referred to as process analytical technology (PAT). Recent years have seen a substantial increase in the use of this approach. It is widely thought to offer advantages for improving both the quality and efficiency of manufacturing drug substances (APIs) and drug products.

On-line monitoring of blending allows for the blend time of a mixture to be adjusted, based on its state of homogeneity. Raw materials are subject to changes in their chemical and physical properties. Using an analytical system to observe the progress of the blend allows the operators to compensate for these changes by modification of the blend time. For example, raw materials blended in a more humid environment may have an increased cohesivity. The blend time for such a mixture might be longer compared to the same materials in a dry environment. An appropriate blend monitoring system would detect that the blend performed in humid conditions was approaching the desired end point more slowly and allow the blend to proceed for the time required for it to be complete.

6.9.1 Analytical Methods

Methods for the on-line monitoring of blends have two major components; the analytical technique and the algorithm that is used to determine the end-point of the blend process. Analytical techniques that are currently described include near-infrared (NIR) spectroscopy, light-induced fluorescence (or laser-induced fluorescence, LIF), Raman spectroscopy and spectroscopy-based chemical imaging techniques. The common feature of these techniques is the ability to make measurements on a timescale that is substantially faster than the blending process. Each has its own measurement properties offering different advantages and disadvantages. The second major component of a blend monitoring method is the algorithm, or criteria for stopping the blend. Substantial data is typically generated during blend monitoring; the ultimate use of this data is to decide when to stop the blend. The choice of mathematical algorithm used for this purpose is important. For example, a typical criterion for a blend to be complete by thief sampling is a 5% *RSD* amongst the samples. An on-line blend might use similar approach. For example, the on-line method might prescribe that the blend is complete when the predicted quantity of API for the last 30 consecutive measurements demonstrates a RSD of less than 5%. Additional details that include the frequency of measurement, the criteria for stopping, volume of interrogation of the analytical technique and the location of sampling all must be considered as the blend end point criteria are developed.

The choice of analytical technique used for on-line blend monitoring must consider flexibility to be on-line, measurement time, interface and volume sampled. The first criterion is that the analytical technique must be adaptable to use on a moving blend container. That is, the analytical system must be mounted to the blender itself. Given that the blender is moving, the analytical system must move with it. Thus, the analytical system must supply its own power and transmit (wirelessly) the data. While there have been reports of stationary spectrometers/ imagers used to monitor a moving blender through a window from a fixed location, these do not represent the typical approach to on-line blend monitoring. Blend times are often on the order of an hour, so a system must be able to operate without external power for that amount of time. There are several options for how the data is transmitted from these systems. The least demanding for wireless transmission of data is that the analytical system stores and processes the data, only sending a signal out when the blend criteria are met. In contrast, a blend monitor may be designed such that the analytical system sends all raw data out to an external data processing system. The external system then executes all of the necessary algorithms on the raw data and ultimately indicates when the endpoint criteria have been met.

In order to make timely decisions about the state of the blend and stop it at an appropriate time, the calculations must be made in a timeframe faster than the rate at which a blend is changing. Reporting of a result requires that data is collected, analyzed and any blend endpoint algorithms applied. For some techniques, such as light-induced fluorescence, this can be very fast; for other multivariate analyses, the data collection and computational time may become significant. Most reports of blend monitoring use a data frequency equivalent to the rotational frequency of the blender; one measurement is made per rotation. There are two reasons for this. First, the measurement is made in a single point in the blend. Until the material at that particular point has had an opportunity to be exchanged, via mechanisms discussed earlier, it is essentially stagnant. Second, measurements of the blend can only be made while material is covering the interface of the analytical instrument. Except in the cases where an interface can be created at a location that is continually covered by material, the frequency that material covers the interface is synchronized with the rotation of the blender (for tumble type blenders). Given that the blend will not change significantly within a rotation, one measurement is sufficient; less frequent measurements are often justified, based on the rate at which the blend changes. For other blenders, such as high-shear or screw-type, it is more likely that a location for the interface can be found that is in continuous contact with the blend mixture. In these cases, the frequency of measurement must be fast enough to make decisions about the blending process relative to the rate of change of the mixture at the interface.

Analytical techniques used for blend monitoring require an interface with the powders in the blender. An example of a typical interface might be a window cut into the shell of the blender, through which the analytical system monitors the blending process. Another example would be an interface for fiber optics terminating on the inside of the blender shell. The interface must meet the same regulatory scrutiny applied to other pharmaceutical processing equipment. Materials which come in direct contact with pharmaceutical products must meet strict guidelines such that they do not alter the quality of the ingredients of a blend [34]. Thus, there are substantial limitations as to the materials from which process interfaces may be constructed. Further, the location of the interface is of substantial importance. It must be located such that there is physical space for the analytical system and that the addition of the analytical system does not disrupt the functionality and safety of the blender. While this seems an obvious criterion, bear in mind that many of the analytical systems are added to blenders that were designed without such systems in mind. Therefore, it becomes an important design consideration when adding such equipment to a blender. Physical requirements notwithstanding, the sampling location is critical for obtaining adequately representative samples from the blend mixture. As previously described, different blenders have characteristic dead zones; that is, areas through which material is exchanged less frequently compared to other regions of the blender. The choice to locate a sensor in such a zone, or away from such a zone is an important choice. Often this is based on the strategy employed for blending. If the analytical technique is adequately sensitive to detect a change in concentration due to material held in a dead zone, then it is sensible to mount the sensor in a location where material turnover is high. A different strategy is to mount the analytical system in the last location in the blender that is expected to reach homogeneity (i.e., sensor monitors the dead zone). The interface and its location must be an integral part of the design and validation of an analytical system used to monitor blending.

The scale of scrutiny concept also applies to on-line measurements [36]. A simple estimate of the volume of material interrogated can be taken from the depth of penetration of a given analytical technique multiplied by the surface area interrogated (i.e., spot size). However, understanding the volume of material interrogated is complicated by the fact that a simple depth of penetration does not accurately describe the sampling of a powder. A more accurate picture is obtained by considering the relative contribution to the measurement as a function of depth. The closer to the surface a material is, the more influence it will have on the resulting measurement. This is further complicated by the dependence of this phenomenon on wavelength. In particular for NIR measurements, information depth (and indeed depth of penetration) are a function of the wavelength at which the data is collected. Estimations for each of the analytical techniques will follow. In general, most of the on-line monitoring techniques tend towards sampling smaller volumes of material compared to a typical pharmaceutical unit dose (e.g., tablet).

6.9.2 *Sampling Considerations for On-line Monitoring*

The challenge with on-line blend monitoring, as with traditional monitoring, is appropriate sampling of the blend. Golden rules for sampling indicate that it is ideal to sample a system in motion and capture a complete cross section of the process [39]. Further, the concept of probabilistic sampling supports the idea that every volume in the entire body of the blend should have the potential to be sampled [27, 28]. The idea that the sample should be in motion is consistent with on-line blend monitoring. There is no need to stop the blend while an on-line measurement is made. However, the idea of a full cross section of the process and the similar idea of probabilistic sampling are not as well founded by on-line blend monitoring. Analytical techniques that are typically discussed for on-line blend monitoring are mounted to a fixed location on the outside of the blender. Thus, they are restricted to a single location on the surface of the blender (the surface at which the mixture and the blender meet). Additional sensors can be added to a blender, but the restriction of being located on the outside of the blend and looking inwards, remains.

Restrictions for the interface between analytical technologies used for on-line blend monitoring present a different set of considerations compared to the traditional thief sampling. Whereas traditional thief sampling represents a number of sampling locations (both near the surface and in the middle of the blend) typically at the end of the blending process, on-line blend monitoring represents a single location in the blend analyzed throughout the entire blending process. Monitoring the blend continuously from a single location supposes that the interface through which the measurement is made is refreshed by new material throughout the blending process. Further, the representative nature of this sample of material at the interface must be questioned. Appropriate blend monitoring methods must undergo rigorous analytical method validation to demonstrate not only the accuracy of the measurement, but also the suitability of the sampling for the specific blender and formulation. A typical approach to the use of on-line blend monitoring data is to consider the standard deviation of consecutive measurements as an indicator of the state of homogeneity of the blend. When the standard deviation is at a minima, the blend may be considered to be complete.

6.9.3 Near-Infrared Spectroscopy

Near infrared spectroscopy is the most commonly reported technique for blend monitoring. The NIR region of the electromagnetic spectrum is approximately 800–2500 nm (12,500–4000 cm^{-1}). In this region, overtones and combination bands from infrared active vibrations are measured. The strongest of the signals in the NIR region find their source in the strongest dipoles in analytes. Thus, in pharmaceutical powders O–H and N–H bonds tend to produce the most absorption per function group. Bonds from C–H are typically observed due to the large number present in a typical powder sample. A significant reason for the utility of NIR spectroscopy for measurement of pharmaceutical powders is that the absorptivity of these powders in the NIR region is quite low. As a result of the poor extinction coefficient, measurements are typically made directly on powders without the need for sample preparation. Further, the scattering coefficients in both reflectance and transmission for pharmaceutical materials in the NIR region are quite large. This facilitates longer path lengths and results in a depth of penetration on the order of millimeters [11].

Analytical NIR instruments have been on the market for a number of years and are quite mature. There are a number of purpose-built NIR spectrometers. Specifically, they have wireless capabilities and a small form factor. Many vendors make interface kits available with their spectrometers to facilitate the secure mounting and appropriate optics for making measurements on industrial scale blenders. Optical requirements for NIR spectroscopy lend themselves to the use of quartz or sapphire optics. These materials are typically regarded as safe in the interface with pharmaceutical powders. However, care must be taken to assure that any anti-reflective coatings are compatible with the materials and process. Typical instruments used for blending include Fourier-transform, diode-array, rotating filter, linear variable filter (LVF), acousto-optical tunable filter (AOTF) and micro-electro-mechanical systems (MEMS). The specific capabilities of each type of spectrometer are described elsewhere [56].

Additional considerations of an appropriate spectrometer include the reliability of the instrument, instrument to instrument similarity, susceptibility to vibration and temperature, internal diagnostics capabilities, and health and safety hazards. Instrument reliability is often a function of the NIR source and external optics, such as fiber optic probes. In considering the reliability, the availability of the instrument to change the source and the ramifications for the methods associated with a given instrument for that change are important. The similarity of replicates of the same model of an instrument is a very important feature for blend monitoring. Method development for blending can be a resource intensive process and the ability to transfer data between instruments depends on the similarity of the data produced by different instruments. Substantial changes to baselines or to spectral response between instruments can necessitate intricate procedures for transferring methods between different spectrometers (and hence different manufacturing sites). Many instruments offer a broad range of internal diagnostics. These can vary from the ability to collect a reference or dark scan internally to full performance qualification testing. Internal testing abilities may require that the instrument be taken off-line to perform these tests. At a minimum instruments are typically not available for on-line measurements during diagnostic tests. The availability of internal diagnostics can become a substantial issue when access to the spectrometer is difficult, such as high-containment environments. In such situations, even a relatively simple operation, such as collecting a reference scan can be quite complicated if it requires an external reference source. Environmental considerations are very important in a pharmaceutical environment. The susceptibility of the instrument to vibration, temperature changes, humidity changes and other environmental intrusions must be well understood. The specific capabilities of a spectrometer to maintain its performance in the face of these changes must be part of the documentation associated with method development and validation. In addition to the instrument response to external environmental changes, the potential for the instrument to affect the health and safety in its environment must be considered. The primary hazards associated with NIR spectrometers are as a heat source and a potential ignition source (sparks). Many vendors offer NIR systems with high levels of containment that address these issues.

Data from NIR measurements requires substantial mathematical manipulations for appropriate interpretation. Chemometrics is typically required as part of an on-line measurement utilizing NIR spectroscopy. There are two general approaches to data interpretation for monitoring blending, qualitative and quantitative. Qualitative NIR spectroscopy analysis monitors spectral deviations over time. Examples include calculation of a moving window standard deviation of an optical parameter (reflectance, absorbance) at one or multiple wavelengths of several subsequent spectra, followed by the determination of an overall standard deviation plotted against time [25, 52]. Other approaches are based on dissimilarity indices, principal component analysis, soft independent modeling of class analogies (SIMCA) [16, 25], image analysis [38], and principal component modified bootstrap error-adjusted single-sample technique [17]. The development of these qualitative methods is often efficient, utilizing production runs only. However, suitable thresholds for stopping the blend must be determined and justified.

In contrast, quantitative modeling is based on monitoring concentration of constituents over the course of the blending process. Quantitative predictions from regression models relate NIR powder spectra to the concentration of the mixture components [3, 32, 33, 55]. The blend end-point is determined by an algorithm utilizing the predicted content of the blend, number of samples and target quantity for constituents. The blend process is considered complete when the variance for a

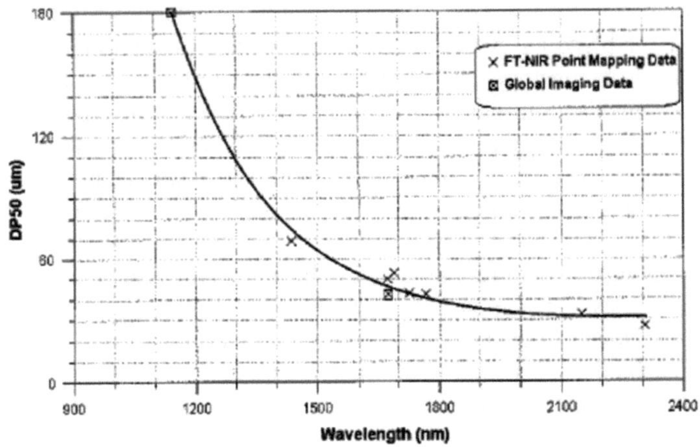

Fig. 6.5 Wavelength dependence of the DP_{50} [10]. Copyright ©2002 SAGE Publications, Ltd. Reprinted with permission

specified time window is below a pre-specified limit. The information about each ingredient provided by quantitative NIR methods represent a potential advantage over qualitative approaches. These methods may be more helpful in root cause analysis and in real-time release strategies. The information ensures that the active pharmaceutical ingredient (API) and excipients are adequately distributed and within content specifications.

The requirement of analyst time and significant quantities of material for quantitative model development is a potential pitfall of this approach. Large calibration designs, spanning relevant variables (e.g., component concentration and scale), is required. Typically, the focus is on the prediction of API concentration, however other blend constituents may be considered as well. In addition, potential interferences such as blend scale must be incorporated into the calibration design to ensure robustness.

When employing NIR spectroscopy, the depth of penetration of light and the spot size must be considered when determining the volume of interrogation for a given probe. There are a number of ways in which this volume has been quantified. Unfortunately, most of these studies have been conducted on consolidated materials. An early effort to understand this critical parameter used the depth of penetration into a series of cellulose sheets. The depth of penetration to which the radiation penetrates into the sample giving a signal corresponding to an intensity of 50% of the pure substrate (DP_{50}) is expressed as a function of wavelength (Fig. 6.5). The x-axis illustrates the wavelength range for a given measurement, and the y-axis represents the DP_{50} of the light [10]. The depth of penetration of light in a tablet was characterized by the fraction of information reflected back to the surface as a function of depth (Fig. 6.6). Here the totality of information was interrogated in the first 2 mm of the tablet surface, at a wavelength of 1200 nm [54]. Recently, milk powder was used to study the depth of penetration at which a

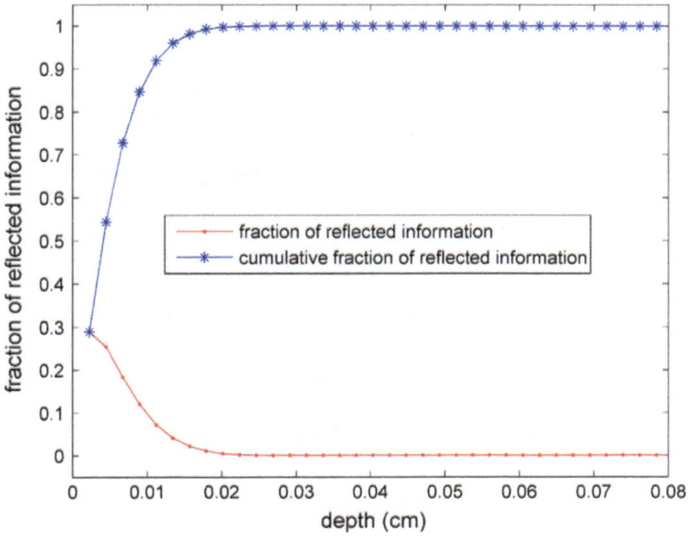

Fig. 6.6 Depth profile for 1200 nm light penetrating into a tablet. The red curve indicates the fraction of information returned from a given depth; the blue curve is the cumulative quantity of information returned at a given depth [53]. Reprinted with permission from Elsevier

melamine contaminant could be detected. The results of this study indicated that the ability to detect melamine began decreasing when 2 mm of milk powder was placed on top of it. When that depth was increased to 5 mm the ability to detect and the melamine was substantially impeded [31]. The depth of penetration of light into a given sample is a complex phenomenon and is not easily summarized, however it is often assumed to be in the 1–3 mm range. An example of the application of this information to a measurement for a spectrometer with a 10 mm diameter spot size is a volume of interrogation of 157 mm^3 (depth $\times$ πr^2). Given a powder density of 1.3 g/cm^3, the mass sampled in this example would be approximately 200 mg. Thus, with an appropriate spot size a powder sample can be obtained with a mass similar to that of the final tablet.

As an example, NIR was used to quantitatively monitor blending of acetaminophen (APAP), microcrystalline cellulose (MCC) and lactose. The concentration of the three constituents was varied over eight blends, each run in triplicate. A quantitative partial least squares model using pre-processed spectra was developed for each of the three constituents. Figure 6.7 illustrates the predictions for the mixture during a blend run.

In this work, the algorithm used to determine the end point of the blend was the root mean square from nominal value (*RMSNV*) of the predicted concentrations of the blend [55].

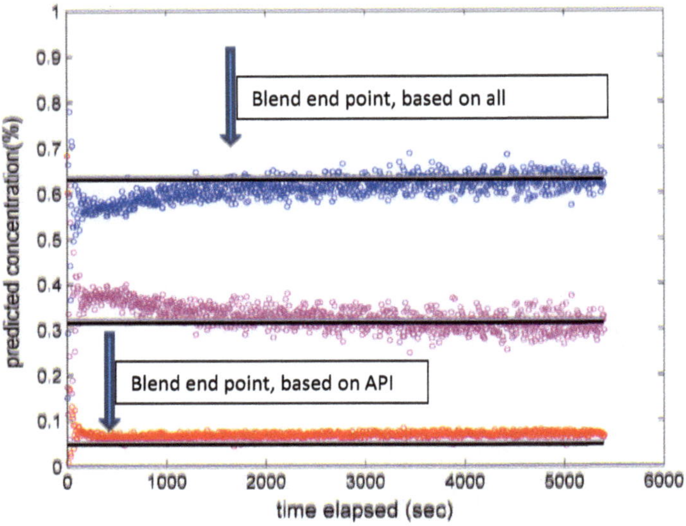

Fig. 6.7 Predicted concentration plot for a powder blend. The nominal value for each component (w/w) was: MCC (63.3%, top/blue), lactose (31.7%, middle/magenta) and APAP (5%, bottom/red). The solid lines represent the nominal values for each component [55]. Reprinted with permission from Elsevier

$$RMSNV = \sqrt{\frac{(P_{APAP} - N_{APAP})^2 + (P_{LAC} - N_{LAC})^2 + (P_{MCC} - N_{MCC})^2}{3}}$$

where P_{APAP}, P_{LAC} and P_{MCC} represent the NIR prediction of the acetaminophen, lactose and microcrystalline cellulose concentrations, respectively; and N_{APAP}, N_{LAC} and N_{MCC} represent the label claim, or target concentrations of acetaminophen, lactose and microcrystalline cellulose. This particular example of an end point criterion incorporates both the idea of an absolute concentration target, the difference between the predicted and target concentration; and the familiar blending concept of a lack of change, or minimization of variance. Thus, the blend end point is verified by both an agreement with an absolute concentration and an acceptable level of variance between rotations. Specifically, a Student's t-test is used to compare the *RMSNV* value for a moving block of 12 data points (here, one minute of data) and zero, using a 95% confidence interval. Note that the blend end point based on *RMSNV* reflects the fact that the excipients (lactose and MCC) take longer to blend than the API. Thus, an end point based on API would have ended the blend too early.

Despite the fact that near-infrared is a widely applied on-line measurement tool for blend monitoring, it has limitations. First, it is a secondary technique. An NIR method must be calibrated using a series of samples with known composition. This often requires a large volume of materials to create mixtures in the appropriate scale blender on which to calibrate. Further, those mixtures are typically analyzed by a primary method to determine the concentration of constituents. There is often a degree of uncertainty in the exact composition of the blend that is immediately in front of the sensor. Recent studies have demonstrated a method for calibrating blends using small quantities of materials suitable for use across a broad range of blender scales [41, 42]. This method has not yet been widely adopted. Another limitation is the relative concentration for which NIR is suitable to measure [13]. Like the other spectroscopic techniques, NIR is limited to interrogating a volume element at the interface between the powder mixture and the blender shell, if the internal workings of the blender are to remain undisturbed. Location for the sensor is critical and has the potential to induce sampling based errors. Finally, NIR typically requires multivariate mathematical treatments (chemometrics) to treat the data. While this criterion is not unique to NIR, the proper application of chemometrics to complex analytical systems requires a specific skill set.

6.9.4 Raman Spectroscopy

Raman spectroscopy is useful for measuring both organic and inorganic molecules and is responsive to changes in the dipole moment of bonds. Thus, Raman is responsive to a wide range of organic and inorganic bonds. It is considered complementary to infrared, and therefore NIR spectroscopy. The bands measured in Raman spectroscopy are much narrower than those in NIR spectroscopy and are often not overlapping. They are typically assignable to specific functional groups from which the bands originate. This greatly enhances the specificity of Raman measurements.

While Raman spectroscopy for blend monitoring is less mature than NIR spectroscopy, it remains a viable option for specific applications. Raman spectra can typically be collected on the order of one spectrum per 10 s. The rate of collection is highly dependent upon the system and the spectrometer used to collect the data. There are a few portable, battery powered Raman spectrometers available. These spectrometers are designed for material identification methods and do not have many of the features required for on-line measurements. Reports of on-line use of Raman spectrometers are primarily those in which a fixed Raman probe is interfaced with a blender. This results in the placement of the Raman probe being invasive with respect to the blender operation. However, in cases where material is flowing past the probe, such as continuous blending, Raman becomes a viable option.

When considering the volume of material interrogated by a Raman spectrometer, we must consider both the diameter of the illumination and detection optics, and the depth of penetration. Similar to NIR, the depth of penetration for Raman was reported as 1.7–2.0 mm [1, 13]. However, there are specially designed Raman probes capable of probing well beyond the surface of a sample, specifically Spatially Offset Raman Spectroscopy, or SORS.

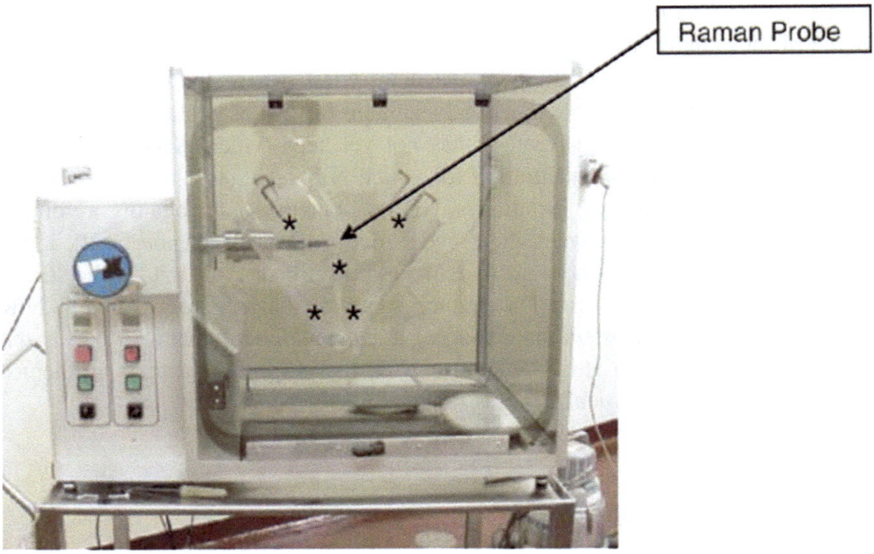

Fig. 6.8 Raman spectroscopy used to monitor blending. Asterisks indicate locations for sample thief [29]. Reprinted with permission from Elsevier

Similar to near infrared spectroscopy, methods based on Raman are typically multivariate. Thus, appropriate systems for maintaining multivariate models must be in place. The data collected by Raman spectroscopy consists of much more distinct peaks than does NIR; however, chemometric methods are typically employed to create appropriate calibrations. Methods that involve selection of specific regions of Raman spectra are often employed to facilitate calibration [15].

Common sources of interference with Raman spectroscopy include temperature and fluorescence. The effect of temperature is to shift the location of peaks as a function of the sample temperature, including any heating associated with the illumination of the sample. Fluorescence is a common interference when Raman spectroscopy is used. Several common pharmaceutical excipients have the potential to fluoresce. An example is microcrystalline cellulose. Typically, MCC contains trace amounts of lignin, which fluoresces strongly. As the lignin content is variable, the amount of fluorescence from microcrystalline cellulose can vary substantially. This complicates method development using Raman spectroscopy for blends containing MCC.

An example of a blending process monitored by Raman spectroscopy utilizes a fixed probe inside a V-blender (Fig. 6.8) [29]. This study utilized a low dose drug (1% w/w) formulation to demonstrate the capabilities of Raman spectroscopy for blend monitoring. The Raman results were compared to thief samples taken at specific intervals and to tablets produced from the blend. The authors report significant correlation (p-values <0.03, determined from correlation analysis) between the Raman measurement, the standard deviation of the HPLC results from thief samples and the HPLC results from tablets made from the blend.

6.9.5 *Light-Induced Fluorescence*

Light-induced fluorescence, or laser induced fluorescence (LIF), is another promising (but not well developed) technique for monitoring blending. This technique is highly sensitive and is often very specific due to the requirements that the analyte must fluoresce [14, 26, 35]. Further, the sensitivity of technique is enhanced by the fact that few pharmaceutical excipients typically fluoresce strongly, compared to the fluorescence of APIs. The specificity of an LIF method is a function of the wavelength used for excitation and detection. In the currently available LIF instruments, these wavelengths are set using optical filters. Interferences with the measurement come from other agents that fluoresce at similar excitation/emission wavelengths and quenching of the fluorescence of the analyte by agents such as water or oxygen. Optics for interfacing LIF with a blender require only that they are transparent to the visible frequencies of light used for excitation and emission.

A feature of measurements by LIF is that they are inherently univariate. The measured signal is the intensity of fluorescence from the sample. Calibration requires regression of the measured fluorescence against the known sample concentration to determine a slope and bias for each type of mixture. The predicted concentration of the analyte is tracked over time and blend end point is determined by a criterion specifying the minimum acceptable standard deviation. The volume of interrogation for LIF is similar to that of NIR, on the order of millimeters. Another advantageous feature of LIF is the frequency with which measurements are made, on the order of hundreds of measurements/second; this offers the potential to monitor very rapidly changing blends (e.g., continuous blenders).

6.10 Summary

While there are a variety of techniques available to monitor on-line blending, NIR remains the dominant choice in the pharmaceutical industry. Other methods including thermal effusivity and imaging are under development, but have not achieved the popularity of NIR. There are a number of potentially promising other rapid analytical techniques that have yet to be reported for use with blending. The primary obstacle continues to be requirements related to interfacing an analytical system with a moving blender shell and the potential errors associated with sampling. As these obstacles are addressed, it is anticipated that the current challenges, such as very low dose drugs ($<0.1\%$), may be addressed.

6.11 Definitions, Abbreviations and Symbols

Active pharmaceutical ingredient (API)	Any component that provides pharmacological activity or other direct effect in the diagnosis, cure, mitigation, treatment, or prevention of disease, or to affect the structure or any function of the body of man or animals [21]
Aspect ratio	Ratio of minimum to maximum dimension of a particle (in some areas of work, the inverse value is more commonly used)
Binder	An excipient used in tablet manufacturing to impart adhesion between particles [12]
Boundary lubricant	An excipient used in tablet manufacturing to prevent adhesion of the powder mixtures to the surface of the tablet press (punches and dies) [12]
Blend uniformity	Relative standard deviation of a series of samples removed from a powder blender [30]
Chemometrics	Mathematical and statistical methods used to extract information from multivariate systems
Cohesivity	Stickiness of a powder, caused by strong attractive forces between individual particles
Compression aid	An excipient used in tablet manufacturing to facilitate the compression of a powder mixture into a tablet
Confidence interval	Interval, in which measured data or their mean value fall at an assigned probability (typically 95 or 99%)
Content uniformity (tablet)	The variation and content of API in tablets [59]
Critical quality attribute	A physical, chemical, biological, or microbiological property or characteristic that should be within an appropriate limit, range, or distribution to ensure the desired product quality [34]
Compressibility	The reduction in volume as a result of an applied force

Direct compression process	A pharmaceutical manufacturing process where powders are blended and compressed, without additional manufacturing steps
Disintegrant	An excipient used in tablet manufacturing to facilitate the mechanical disruption of a tablet when it comes in contact with gastrointestinal fluids
Drug product	A finished dosage form (e.g., tablet) [21]
Drug substance	See "active pharmaceutical ingredient"
Equant	Particle shape with similar particle dimensions (aspect ratio near 1)
Equivalent particle size	Diameter of a sphere that has the same property as the observed particle in relation to a given measurement principle
Free flowing	Condition of a powder, the primary particles of which easily flow or can be made to flow, e.g. under influence of gravity
Segregation	Separation of particles in a mixture (e.g. in a powder) due to differences in size, shape and/or density
Label claim	The quantity of drug intended to be found in a dose of a drug product
Primary particle	Basic particle, which cannot be separated unless by breakage
Powder	A collection of (primary) particles
Therapeutic dose	The minimum dose to obtain the desired therapeutic effect of a drug
Toxic dose	A dose of drug that exceeds toxic concentrations
Unit operation	A single step in a batch-wise manufacturing process (e.g., blending)
API	Active pharmaceutical ingredient
BU	Blend uniformity
CRM	Completely random mixture
CU	Content uniformity
FDA	United States Food and Drug Administration
HPLC	High-performance (or pressure) liquid chromatography
VCA	Variance component analysis
NIR	Near-infrared
PQRI	Product Quality Research Institute

LIF	Light (or laser) induced fluorescence
PAT	Process analytical technology
PCA	Principal component analysis
PLS	Partial least squares
SIMCA	Soft independent modeling of class analogy
VCA	Variance component analysis
D	Equivalent particle size
s	Measured standard deviation, coming from a series of measurements
Var	Variance
RSD	Relative standard deviation, standard deviation divided by mean value (usually in %)

References

1. Allan P, Bellamy LJ, Nordon A, Littlejohn D, Andrews J, Dallin P (2013) In situ monitoring of powder blending by non-invasive Raman spectrometry with wide area illumination. J Pharm Biomed Anal 76:28–35
2. Allen T (1981) Particle size measurement, 3rd edn. Chapman and Hall, London, UK
3. Anderson CA, Drennen JK, Ciurczak EW (2008) Pharmaceutical applications of near-infrared spectroscopy. Pract Spectrosc 35:585–611 Handbook of Near-Infrared Analysis (3rd ed)
4. Bergum J, Parks T, Prescott J, Tejwani R, Clark J, Brown W, Muzzio F, Patel S, Hoiberg C (2015) Assessment of blend and content uniformity. Technical discussion of sampling plans and application of ASTM E2709/E2810. J Pharm Innov 10(1):84–97
5. Bergum JS, Prescott JK, Tejwani RW, Garcia TP, Clark J, Brown W (2014) Current events in blend and content uniformity. Pharm Eng 34(2):28–39
6. Berman J, Planchard JA (1995) Blend uniformity and unit dose sampling. Drug Dev Ind Pharm 21(11):1257–1283
7. Berman J, Schoeneman A, Shelton JT (1996) Unit dose sampling: a tale of two thieves. Drug Dev Ind Pharm 22(11):1121–1132
8. Boehm G, Clark J, Dietrick J, Foust L, Garcia T, Gavini M, Gelber L, Geoffroy JM, Hoblitzell J, Jimenez P, Mergen G, Muzzio F, Planchard J, Prescott J, Timmermans J, Takiar N (2003) The use of stratefied sampling of blend and dosage units to demonstrate adequacy of mix for powder blends. PDA J Pharm Sci Technol 57(2):64–74
9. Cholayudth P (2017) Establishing blend uniformity acceptance criteria for oral solid-dosage forms. Pharm Technol 41(2):42–52
10. Clarke FC, Hammond SV, Jee RD, Moffat A (2002) Determination of the information depth and sample size for the analysis of pharmaceutical materials using reflectance near-infrared microscopy. Appl Spectr 56(11):1475–1483
11. Dahm DJ, Dahm KD (2007) Interpreting diffuse reflectance and transmittance: a theoretical introduction to absorption spectroscopy of scattering materials. NIR Publications, Chichester
12. Troy BT, Remington PB (2006) The science and practice of pharmacy, 21st edn. Lippincott Williams & Wilkins, Baltimore, MD

13. De Beer T, Burggraeve A, Fonteyne M, Saerens L, Remon JP, Vervaet C (2011) Near infrared and Raman spectroscopy for the in-process monitoring of pharmaceutical production processes. Int J Pharm 417(1–2):32–47
14. Durao P, Fauteux-Lefebvre C, Guay J-M, Abatzoglou N, Gosselin R (2017) Using multiple process analytical technology probes to monitor multivitamin blends in a tableting feed frame. Talanta 164:7–15
15. Dyrby M, Petersen RV, Larsen J, Rudolf B, Norgaard L, Engelsen SB (2004) Towards on-line monitoring of the composition of commercial carrageenan powders. Carbohydr Polym 57(3):337–348
16. El-Hagrasy AS, D'Amico F, Drennen JK III (2006) A process analytical technology approach to near-infrared process control of pharmaceutical powder blending. Part I: D-optimal design for characterization of powder mixing and preliminary spectral data evaluation. J Pharm Sci 95(2):392–406
17. El-Hagrasy AS, Delgado-Lopez M, Drennen JK III (2005) A process analytical technology approach to near-infrared process control of pharmaceutical powder blending: part II: qualitative near-infrared models for prediction of blend homogeneity. J Pharm Sci 95(2):407–421
18. Emady H, Hapgood K, Smith R (2016) Production, handling and characterization of particulate materials. (Particle technology series). Springer, Berlin
19. Esbensen KH, Román-Ospino AD, Sanchez A, Romañach RJ (2016) Adequacy and verifiability of pharmaceutical mixtures and dose units by variographic analysis (theory of sampling)—a call for a regulatory paradigm shift. Int J Pharm 499(1–2):156–174
20. Fan LT, Chen YM, Lai FS (1990) Recent developments in solids mixing. Powder Technol 61 (3):255–287
21. FDA (November 2017), Drugs@FDA glossary of terms. Available: https://www.fda.gov/drugs/informationondrugs/ucm079436.htm
22. FDA (2003) Guidance for industry: powder blends and finished dosage units: stratified in-process dosage unit sampling and assessment
23. Garcia T, Bergum J, Prescott J, Tejwani R, Parks T, Clark J, Brown W, Muzzio F, Patel S, Hoiberg C (2015) Recommendations for the assessment of blend and content uniformity: modifications to withdrawn FDA draft stratified sampling guidance. J Pharm Innov 10(1):76–83
24. Garcia TP, Taylor MK, Pande GS (1998) Comparison of the performance of two sample thieves for the determination of the content uniformity of a powder blend. Pharm Dev Technol 3(1):7–12
25. Gonzalez F, Pous R (1995) Quality control in manufacturing process by near-infrared spectroscopy. J Pharm Biomed Anal 13(4/5):419–423
26. Guay J-M, Lapointe-Garant P-P, Gosselin R, Simard J-S, Abatzoglou N (2014) Development of a multivariate light-induced fluorescence (LIF) PAT tool for in-line quantitative analysis of pharmaceutical granules in a V-blender. Eur J Pharm Biopharm 86(3):524–531
27. Gy P (2004) Sampling of discrete materials. Chemom Intell Lab Syst 74(1):39–47
28. Gy P (2004) Sampling of discrete materials-a new introduction to the theory of sampling. Chemom Intell Lab Syst 74(1):7–24
29. Hausman DS, Cambron RT, Sakr A (2005) Application of Raman spectroscopy for on-line monitoring of low dose blend uniformity. Int J Pharm 298(1):80–90
30. Howard-Sparks M, Gawlikowski A (2004) Evaluation of blend uniformity and content uniformity based on 2003 stratified sampling guidance and 1999 blend uniformity analysis guidance: product A. PDA J Pharm Sci Technol 58(4):222–230
31. Huang M, Zhu Q, Huang M, Kim MS, Chao K, Qin J, Mo C, Esquerre C, Delwiche S (2016) Penetration depth measurement of near-infrared hyperspectral imaging light for milk powder. Sensors (Basel) 16(4):441
32. Igne B, Talwar S, Drennen JK, Anderson CA (2013) Online monitoring of pharmaceutical materials using multiple NIR sensors—part II: blend end-point determination. J Pharm Innov 8(1):45–55

33. Igne B, Zacour B, Shi Z, Talwar S, Anderson CA, Drennen JK (2011) On-line monitoring of pharmaceutical materials using multiple NIR sensors—Part I: blend homogeneity. J Pharm Innov 6(1):10–23
34. International Conference for Harmonisation (2009) Pharmaceutical development Q8(R2)
35. Karumanchi V, Taylor MK, Ely KJ, Stagner WC (2011) Monitoring powder blend homogeneity using light-induced fluorescence. AAPS Pharm Sci Tech 12(4):1031–1037
36. Lacey PMC (1954) Theory of particle mixing. J Appl Chem 4:257–268
37. Lachman L, Lieberman HA, Kanig JL (1986) The Theory and practice of industrial pharmacy. Lea & Febiger, Philadelphia
38. Ma H, Anderson CA (2007) Optimisation of magnification levels for near infrared chemical imaging of blending of pharmaceutical powders. J Near Infrared Spectrosc 15(3):137–151
39. Merkus HG (2009) Sampling of particulate material. Particle size measurements: fundamentals, practice, quality. Springer, Dordrecht, Netherlands, pp 73–116
40. Merkus HG (2009) Statistical background. Particle size measurements: fundamentals, practice, quality. Springer, Dordrecht, Netherlands, pp 469–512
41. Mohan S, Momose W, Katz JM, Hossain MN, Velez N, Drennen JK III, Anderson CA (2018) A robust quantitative near infrared modeling approach for blend monitoring. J Pharm Biomed Anal 148:51–57
42. Momose W, Katz JM, Drennen JK III, Anderson CA (2015) Development of NIR methods for blend analysis using small quantities of materials. J Pharm Biomed Anal 10(1):36–46
43. Muselik J, Franc A, Dolezel P, Gonec R, Krondlova A, Lukasova I (2014) Influence of process parameters on content uniformity of a low dose active pharmaceutical ingredient in a tablet formulation according to GMP. Acta Pharm 64(3):355–367
44. Muzzio FJ, Goodridge CL, Alexander A, Arratia P, Yang H, Sudah O, Mergen G (2003) Sampling and characterization of pharmaceutical powders and granular blends. Int J Pharm 250(1):51–64
45. Muzzio FJ, Robinson P, Wightman C, Dean B (1997) Sampling practices in powder blending. Int J Pharm 155(2):153–178
46. Muzzio FJ, Roddy M, Brone D, Alexander AW, Sudah O (1999) An improved powder-sampling tool. Pharm Technol 23(4):92–110
47. Popo M, Romero-Torres S, Conde C, Romañach RJ (2002) Blend uniformity analysis using stream sampling and near infrared spectroscopy. AAPS Pharm Sci Tech 3(3):61–71
48. Poux M, Fayolle P, Bertrand J, Bridoux D, Bousquet J (1991) Powder mixing: some practical rules applied to agitated systems. Powder Technol 68(3):213–234
49. Prescott JK, Garcia TP (2001) A solid dosage and blend content uniformity troubleshooting diagram. Pharm Technol North Am 25(3):68 70, 72, 74, 76, 78, 80, 82, 84, 86, 88
50. Rippie EG, Faiman MD, Pramoda MK (1967) Segregation kinetics of particulate solids systems. IV. Effect of particle shape on energy requirements. J Pharm Sci 56(11):1523–1525
51. Romañach RJ (2015) Sampling and determination of adequacy of mixing. In: Pharmaceutical blending and mixing, Wiley, pp 57–78
52. Sekulic SS, Ward HW, Brannegan DR, Stanley ED, Evans CL, Sciavolino ST, Hailey PA, Aldridge PK (1996) On-line monitoring of powder blend homogeneity by near-infrared spectroscopy. Anal Chem 68(3):509–513
53. Shi Z, Anderson CA (2010) Application of Monte Carlo simulation-based photon migration for enhanced understanding of near-infrared (NIR) diffuse reflectance. Part I: depth of penetration in pharmaceutical materials. J Pharm Sci 99(5):2399–2412
54. Shi Z, Anderson CA (2010) Pharmaceutical applications of separation of absorption and scattering in near-infrared spectroscopy (NIRS). J Pharm Sci 99(12):4766–4783
55. Shi Z, Cogdill RP, Short SM, Anderson CA (2008) Process characterization of powder blending by near-infrared spectroscopy: blend end-points and beyond. J Pharm Biomed Anal 47(4–5):738–745
56. Simpson MB (2007) Near-infrared spectroscopy for process analytical technology: theory, technology and implementation. In: Bakeev K (ed) Process analytical technology, Wiley
57. Sommer K (1986) Sampling of powders and bulk materials. Springer, Berlin

58. Susana L, Canu P, Santomaso AC (2011) Development and characterization of a new thief sampling device for cohesive powders. Int J Pharm 416(1):260–267
59. The United States Pharmacopia (USP) (2015) Uniformity of dosage units ⟨905⟩
60. Weinekötter R (2016) Mixing of solid materials. In: Merkus HG, Meesters GMH (eds) Production, handling and characterization of particulate materials. Springer International Publishing, Cham, pp 291–326
61. Williams JC (1986) Mixing of particulate solids. In: Uhl V, Gray JB (eds) Mixing, theory and practice, vol 3. Academic Press, Orlando, FL
62. Yeung CC, Hersey JA (1979) Ordered powder mixing of coarse and fine particulate systems. Powder Technol 22(1):127–131

Carl A. Anderson, Ph.D. is an associate professor, Head of the Division of Pharmaceutical and Social Sciences, and the co-founder and director of the Duquesne University Center for Pharmaceutical Technology. He is also the co-founder of Strategic Process Control Technologies, LLC. He received his B.S. degree in chemistry from California State University—Fullerton; and his Ph.D. from The University of Texas at Austin, also in chemistry. He has worked at Duquesne University since 2002. Dr. Anderson's research is heavily influenced by the seven years he spent working for Aventis Pharmaceuticals as an analytical development scientist. During that time, he was directly responsible for implementing process analytical technology solutions. He continues to collaborate and consult with pharmaceutical manufacturers and derives a significant portion of his funding from these projects. The focus of his research has been to develop and implement methods to effectively characterize statistically relevant numbers of samples of pharmaceutical products during manufacturing to better characterize their quality. Dr. Anderson has published papers describing the development, implementation, validation and chemometric analysis for on-line spectroscopic methods. He has advised a number of students through their Ph.D. studies with dissertations focused on improvement of analytical capabilities and characterization of quality as related to pharmaceutical manufacturing. Dr. Anderson was a founding member of NIPTE and has led several project teams related to pharmaceutical manufacturing and quality.

Natasha L. Velez is a Ph.D. student and head teaching assistant at the Duquesne University's Graduate School of Pharmaceutical Sciences. She obtained her B.S. degree in chemistry from the University of Puerto Rico—Mayagüez Campus (UPRM). Her interest for the pharmaceutical sciences was highly influenced by the two years she dedicated to undergraduate research in the Analytical and Pharmaceutical Lab at UPRM, followed by an undergraduate internship in the Pharmaceutical Engineering Program at Rutgers University. Natasha has assumed a leading role in a NIPTE QbD-based project, a collaboration between Duquesne University, Purdue University and the University of Connecticut, funded by the U.S. Food and Drug Administration (FDA). Her research has been focused on the implementation of process analytical technology (PAT), quality by design (QbD) strategies and relevant statistical analyses for facilitating effective formulation development and enhancing pharmaceutical blending efficiency.

Chapter 7
Guidance on Drug Substance Particle Size Controls for Solid Oral Dose Forms

Jon Hilden, Christopher L. Burcham, Stephen D. Stamatis, Jim Miesle and Carrie A. Coutant

Abstract Drug substance particle size is a critical property affecting drug product performance. Smaller particles dissolve faster and may improve bioavailability of the drug as a result. Smaller particles are typically dispersed more uniformly, leading to lower inter-tablet potency variation. Unfortunately, smaller particles can also result in poor powder handling characteristics or other processing issues. For example, powders can fail to flow through hoppers or stick to tooling surfaces, leading to poor tablet weight uniformity or tablet appearance issues. During crystallization, smaller particles can also be difficult to filter from the crystallization media. This can lead to higher levels of residual solvents and other impurities. Particle design and size selection are therefore critical to achieving a balance between manufacturability, bioavailability, and content uniformity. In this chapter, the impact of particle size on bioavailability is introduced and the impact on content uniformity is considered in depth. Control of the variability of tablet potency is discussed in relation to the overall particle size distribution. The risk-based selection of a positive control such as a screen to limit the maximum allowable particle size is discussed in relation to the occurrence of rare, but highly super-potent tablets.

7.1 Introduction

Developing a new pharmaceutical product requires that both the new chemical entity and formulated drug product (e.g. tablet or capsule) be safe and efficacious. Physical properties of the drug substance, such as particle size, particle size distribution, crystal form, and crystal habit can affect the safety, efficacy, and bioavailability of the drug product. For example, a change in crystal form can reduce the solubility and dissolution rate, impacting the bioavailability of the drug. The physical properties of the drug substance can also affect content uniformity

J. Hilden (✉) · C. L. Burcham · S. D. Stamatis · J. Miesle · C. A. Coutant
Eli Lilly and Company, Lilly Corporate Center, Indianapolis, IN 46225, USA
e-mail: hildenjl@lilly.com

H. G. Merkus et al. (eds.), *Particles and Nanoparticles in Pharmaceutical Products*, AAPS Advances in the Pharmaceutical Sciences Series 29,
https://doi.org/10.1007/978-3-319-94174-5_7

(CU, also referred to as uniformity of dosage units, UDU), especially for low dose products (e.g. ≤ 1 mg). Adequate dose uniformity cannot be achieved in cases where individual drug substance particles or agglomerates are so large that they individually account for an appreciable fraction of the intended unit dose. Therefore, drug substance particle size must be controlled to ensure safety and efficacy.

Establishing a physical property control strategy for a drug substance starts prior to Phase I clinical trials and continues through product registration and commercialization of a new drug product. The physical property control strategy includes the identification of the physical properties to be controlled, the acceptable ranges to control the properties within, and the quality attributes that are impacted by the physical property [7]. The physical property control strategy also includes the documented rationale, analytical methods, production process, and process controls used to ensure that the desired materials are routinely produced during normal operation.

Pharmaceutical product development should follow *Quality by Design* principals as introduced by the ICH Q8 through Q10 standards [9–11]. The concept of Critical Quality Attributes (CQAs,) such as content uniformity, is introduced in these guidance documents. The physical property control strategy must align with the CQAs identified for the drug product and should consist of the following elements:

1. Understanding factors that impact a CQA and developing the rationale that links the control of an identified physical property to the CQA.
2. Identification of allowable physical property ranges, within which drug product critical quality attributes will be achieved.
3. Established analytical methods that ensure the accurate characterization of the physical property. Defined rationale for the selection of the test and applicability of the method to the measurement of the physical property.
4. A defined process (i.e., unit operations) demonstrated to ensure the physical property is achieved.
5. Necessary process controls to ensure that the process operates as intended.
6. Appropriate packaging and storage conditions used to store the drug substance and drug product that do not adversely impact the physical property.

The physical property control strategy should evolve throughout the course of product development. Until the final clinical dose is determined, it may not be possible to finalize particle size specifications since the dose and particle size may affect content uniformity and bioavailability. Once the dose is fixed, the robustness and capability of both the physical property control step (e.g. milling or crystallization) and the drug product manufacturing process should be considered, along with analytical variability in the specification test, before control ranges are established.

The particle size distribution (PSD) of the drug substance may impact content uniformity and bioavailability, as well as manufacturability. Of these, the assurance of patient safety and bioavailability are the most significant factors and are the subject of this chapter. The impact of PSD to bioavailability is briefly considered in Sect. 7.2 before turning our attention to the impact of PSD on content uniformity for the remainder of this chapter.

7.2 Particle Size Impact on Bioavailability

Bioavailability, defined as the fraction of drug that enters systemic circulation, is a function of the permeability, solubility, and dissolution rate of a drug product. These properties are the basis for the biopharmaceutics classification system (BCS), which categorizes drugs into four classes based on their solubility and permeability [21]. Permeability and solubility are properties of a drug substance related to the molecular structure. As a result, these properties are inherent to the molecule and typically not affected by processing conditions, though they may be modified by functional excipients in the formulation [1, 14]. In order for absorption from the gastrointestinal tract to occur, drug must first be in solution. Therefore, dissolution is a CQA for most solid oral drug products.

The dissolution rate is a function of the solubility and particle size of the drug substance, as well as the dosage form design (e.g. controlled release versus immediate release). The relationship between drug substance particle size and dissolution rate is given by Eq. 7.1, which is a modified form of the Noyes-Whitney equation [16].

$$\frac{dX_s}{dt} = \frac{-3D_{diff}X_0^{1/3}X_s^{2/3}}{\rho h r_0}\left(C_s - \frac{X_d}{V}\right) \tag{7.1}$$

where dX_s/dt is the dissolution rate, D_{diff} is the diffusion coefficient, X_0 is the initial mass of drug, X_s is the mass of drug at time t, ρ is the true density of the drug substance, h is the diffusion layer thickness, r_0 is the particle radius, C_s is the equilibrium solubility, X_d is the mass of drug in solution, and V is the media volume. The dissolution rate increases with decreasing particle size, making particle size a potential key control element to ensure appropriate drug release and subsequent bioavailability.

The key assumption related to the Noyes-Whitney equation is that the concentration at the surface of the dissolving particle is equal to the solubility of the drug in given media. This implies that the rate of dissolution is controlled not by dissolution at the solid liquid interface, but rather by diffusion of the dissolved drug through an unstirred solvent surrounding the particle. For many drugs, this is the case; further discussion is included in Chap. 8, and the interested reader is referred to the following review article [3]. When applied to drug product dissolution, this equation also assumes that the particles are well dispersed in the medium (i.e. no hydrodynamic effects such as sedimentation), and that there are no formulation or process effects on dissolution.

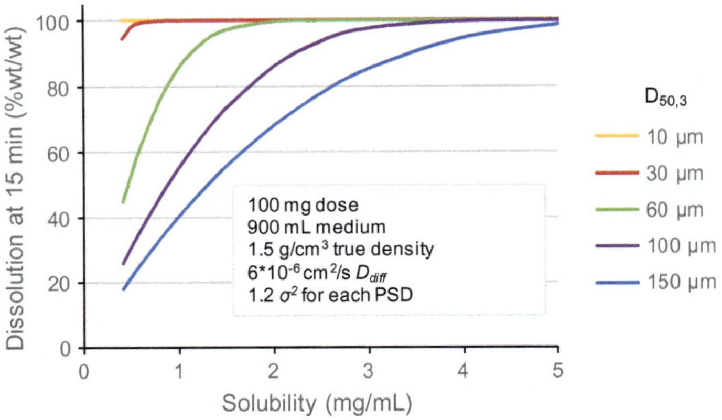

Fig. 7.1 Release at 15 min as a function of solubility and particle size

While reduced particle size is frequently considered to enhance dissolution of poorly soluble drug substances, it should also be considered in the design of highly soluble drugs. In order to meet the requirements for BCS I/III classification of highly soluble drugs, the drug product must meet the definition of rapidly (>85% in 30 min) or very rapidly dissolving (>85% in 15 min) for BCS I and III respectively. Figure 7.1 shows the relationship between solubility and the amount of drug released at 15 min in 900 mL of media for a theoretical drug product containing 100 mg of drug substance with a range of mean particle size (assuming a log-normal distribution with standard deviation, σ). At lower solubility values, smaller particles are required to achieve very rapid dissolution, even though the entire range shown would meet the BCS definition of highly soluble (e.g. 100 mg of drug is soluble in 250 mL of media at a solubility of 0.4 mg/mL).

The presence of agglomerates in the drug substance can have a negative impact on dissolution rate and bioavailability. If agglomerated particles do not readily disintegrate and disperse and the voids between the fused particles are smaller than the diffusional path length, then the observed dissolution rate will be similar to that of a monolithic particle of the same size as the agglomerate [2]. However, a loosely agglomerated particle may have an effective surface area comparable to the primary particles, and have little impact on dissolution. Figure 7.2 shows example dissolution profiles for a theoretical highly soluble, rapidly dissolving drug product with varying levels of large agglomerate particles having a density equivalent to the primary particles. The example assumes a dose of 100 mg, 900 mL of medium, 0.4 mg/mL solubility, and primary particle size of 10 μm. The small particles dissolve rapidly, while the agglomerates dissolve very slowly, resulting in a very shallow slope after the first 5 min. This effect is due to low effective surface area of the large agglomerates, and in some cases may limit bioavailability of the drug.

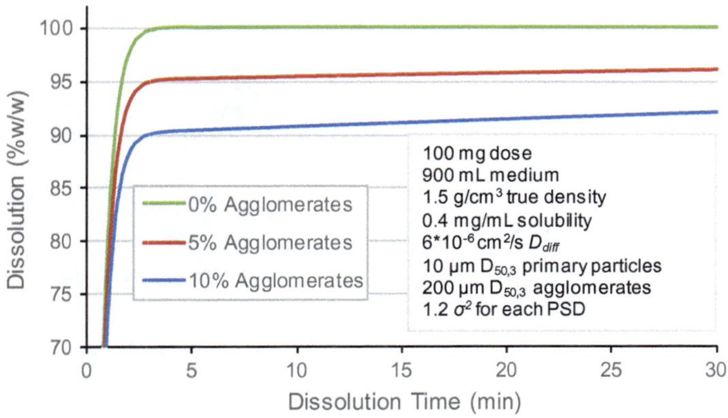

Fig. 7.2 Effect of agglomerates on dissolution

Commonly used particle size measurement techniques, such as laser diffraction, have some limitations when using the particle size results from these instruments to predict dissolution and bioavailability. Laser diffraction may not differentiate between various agglomerated particles, and may underestimate the number of large particles present in a distribution [13]. The sample preparation technique may also introduce errors if particles are broken during suspension. If laser diffraction is used to quantify the particle size distribution, confirmation using other techniques such as optical microscopy or scanning electron microscopy is recommended.

7.3 USP Requirements for Uniformity of Dosage Units

The United States Pharmacopeia (USP) prescribes control for the content uniformity such as tablets and capsules as shown in Fig. 7.3. Unit dose uniformity is described in the USP in terms of drug substance content, which is defined as the mass of drug substance in an individual dosage unit [20]. The content uniformity is assured by achieving two limiting values:

A. The coefficient of variation, C_v, of the measured drug substance content is controlled. For a content distribution centered at 100%, such as that illustrated in Fig. 7.3, a USP-defined 'Acceptance Value' effectively limits the measured content C_v to <6.25%.[1]
B. The maximum and minimum observed content of a sample is controlled. It indicates that no observed dosage units may exceed 125% or fall below 75% of

[1]The language of the USP Chapter <905> indicates an L1 acceptance value of 15%, which when divided by the acceptability constant of 2.4 (associated with tier 1 testing of n = 10 dosage units) gives an effective C_v of 6.25%.

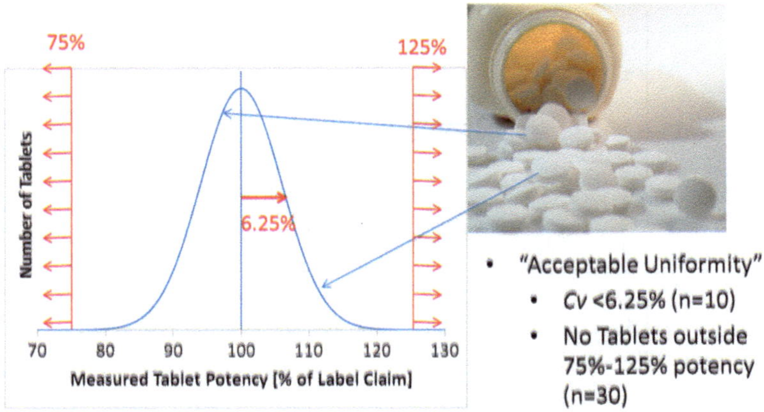

Fig. 7.3 Content distribution centered at 100% label claim and with C_v = 6.25%

the target content.[2] In this chapter, we only address the problem of dosage units with contents exceeding 125%, and not of dosage units containing less than 75%. The related physics of sub-potency are different from those of super-potency because the presence of a single large agglomerate can lead to a significant dose increase. The impact of a large agglomerate on the sub-potency of the remainder of the lot is minimal. As a result, sub-potency is controlled by meeting the C_v target.

As particle size impacts unit dose uniformity, the two levels of uniformity control translate into two levels of particle size control. One applies to content C_v. The other applies to the maximum observable dose. These two levels of particle size control are described in the following two sections.

7.3.1 *Control of Content* C_v

Unit Dose Uniformity (UDU) or content uniformity is concerned with ensuring that each dosage unit contains the prescribed amount of drug linking directly to the "strength" element of SISPQ. This ensures the potential for sub-potent or super-potent doses is minimized. Testing for content uniformity is defined by the USP [20] using a tiered approach with limits on the measured C_v of unit dose potency. Testing consists of measuring the potency of a specified number of dosage units. The minimum, maximum, mean, and the relative standard deviation of measured potency must fall within certain defined bounds in order to meet the

[2]The language of the USP Chapter <905> indicates an L2 acceptance value 25, which when applied to a batch with mean individual contents of 100% (M = 1) provides an acceptable range of 75–125%.

acceptance criteria. Tier 1 testing is conducted with 10 dosage units. If the mean potency of the set is within 1.5% of the target, then the C_v must be no more than 6.25%. If the Tier 1 conditions are not met, then Tier 2 testing is done with a total of 30 samples (the initial 10 samples plus 20 additional samples). If the mean potency of the set is within 1.5% of the target and no individual tablets have a potency less than 75% of target or greater than 125% of target, then the C_v must be no more than 12.5%. If the criteria are met, then uniformity has been confirmed.

A number of factors can affect content uniformity. Two common factors are inadequate mixing of excipients with the drug substance and segregation of excipient or drug substance during drug product manufacturing (e.g. through powder transfers or blending operations). Content uniformity can also be affected by the particle size distribution of the drug substance. If the unit dose is low, a single, large drug substance particle could contain a significant portion of the dose. To better understand the effect the particle size distribution can have on content uniformity, a number of models have been developed predicting the relative standard deviation for each unit dose. Johnson [15] developed one of the first models assuming:

1. The drug substance is composed of spherical particles.
2. A random and dilute mixture that follows a Poisson distribution of drug substance in the drug product.
3. A constant number of particles in the dosage unit.
4. A representative particle size distribution of drug substance in the dosage unit
5. A randomly mixed system of drug substance in the drug product (i.e., no segregation of the drug substance from excipients and a fully randomized blend of drug substance with excipient).

With these assumptions, Johnson's model predicts the coefficient of variance (as a percentage) for the content uniformity for drug loadings greater than 1% and less than 10% as:

$$C_v = 100y\left(\frac{\pi\rho}{6G}\right)^{1/2}\left(\sum_{i=1}^{N} f_i D_i^3\right)^{1/2} \tag{7.2}$$

where ρ is the density of the drug substance, G is the target dose (mass) of drug in each tablet and f_i is the mass fraction of drug of particle size D_i in the tablet, and y the excipient mass fraction. For excipient mass fractions lower than 1% this expression can be simplified to:

$$C_v = 100\left(\frac{\pi\rho}{6G}\right)^{1/2}\left(\sum_{i=1}^{N} f_i D_i^3\right)^{1/2} \tag{7.3}$$

Johnson's model was extended by Yalkowsky and Bolton [22] who considered the problem based on the probability of meeting the (then current) USP requirements for content uniformity. They assumed the particle size distribution follows a log-normal distribution parametrized by the arithmetic mean particle size on a number basis, $D_{m,0}$, and the arithmetic coefficient of variation of the particle size distribution, C. They maintain the rest of the same assumptions as Johnson and derive the following expression for the coefficient of variation in dosage content uniformity:

$$C_v = 100\left(\frac{\pi\rho}{6G}\right)^{1/2}\left(D_{m,0}^3\left(1+C^2\right)^{12}\right)^{1/2} \tag{7.4}$$

It is important to note that the parameters on a *number basis* are not the same as the more commonly measured geometric mean diameter and geometric coefficient of variation on a *mass or volume basis*. The advantage of Yalkowsky and Bolton's approach is that the particle size distribution is described with two parameters, the arithmetic mean particle size and the coefficient of variation of the number based distribution. This has advantages from a design or planning perspective, but for a retrospective analysis (i.e. will this particle size distribution be expected to give good content uniformity) resolution is lost. For a retrospective analysis, using a model, which utilizes the full particle size distribution, is much preferred for the estimation of the variance in content uniformity attributed to particle size effects.

Most particle size analyzers in use do not report a number distribution by default, but instead report a volume based distribution. As a result, Eq. 7.4 cannot be used directly. For this reason, Rohrs et al. [19] expanded on the models proposed by Johnson and Yalkowsky and Bolton for a volume based distribution. The derivation also assumes that the particle size distribution is a log-normal distribution. The log normal distribution is specified with a geometric mean particle size, $D_{50,3}$, in place of the arithmetic mean, and the coefficient of variation to a geometric standard deviation, σ_g. The derivation for coefficient of variation assuming a log-normal distribution on a volume basis is:

$$C_v = 100\left(\frac{\pi\rho}{6G}\right)^{1/2}\left(D_{50,3}^3\exp\left(4.5\ln^2\sigma_g\right)\right)^{1/2} \tag{7.5}$$

This expression can be rearranged to calculate the maximum geometric mean particle allowable for a specified geometric standard deviation of the particle size distribution and specified coefficient of variation for content uniformity:

$$D_{50,3} = \sqrt[3]{\left(\frac{6G}{\pi\rho}\right)\exp\left(-4.5\ln^2\sigma_g\right)\left(\frac{C_v}{100}\right)^2} \tag{7.6}$$

If σ_g has not been regressed, and the distribution is approximately log normal, σ_g can be estimated using the following expression:

$$\sigma_g = \left(\frac{D_a}{D_b}\right)^{\frac{1}{z_a - z_b}} \tag{7.7}$$

where D_a and D_b are the particle size, and z_a and z_b are the z-values of the standard normal distribution at the ath and bth percentiles. The $D_{90,3}/D_{50,3}$ ratio for a log-normal distribution is equivalent to $(\sigma_g)^{1.28}$.

An alternative but equivalent approach to Johnson's model is presented by Hilden et al. [5]. The authors take a similar approach to Johnson and cast it in terms of a less traditional moment of the particle size distribution, the $D[6,3]$. Hilden et al. do not require the assumption of a constant number of particles in the dosage unit and results apply only to low drug loads, a worst-case condition for content uniformity concerns. They derive an expression that is different, but mathematically equivalent to Eq. 7.2 for the coefficient of variation:

$$C_v = \sqrt{\frac{\pi\rho}{6G} D[6,3]^3} \tag{7.8}$$

where the $D[6,3]$ is defined as:

$$D[6,3] = \left(\frac{\sum_{i=1}^{N} n_i D_i^6}{\sum_{i=1}^{N} n_i D_i^3}\right)^{\frac{1}{3}} \tag{7.9}$$

where n_i is the number fraction in the ith bin.[3] When rewritten in terms of a mass fraction for each particle size instead of a number fraction, this expression is equivalent to Johnson's expression. The summation applies to a frequency (as opposed to cumulative) distribution, and the subscript i indicates the particle size bins over which the summation is calculated.

In comparison with the Rohrs model and the Yalkowsky and Boltan model, the Hilden (Eq. 7.8) and Johnson (Eq. 7.2) models hold for any drug substance particle size distribution, regardless of the distribution shape (e.g. bimodal, log-normal, multimodal, etc.). The further advantage of the Hilden model versus the Johnson model is a matter of convenience. The D[6,3] can be reported directly by particle size analysis software as a single scalar statistic similar to the more common D[4,3] (volume weighted mean diameter) or D[3,2] (surface area weighted mean diameter) statistics.

It is important to note that the D[6,3] is generally independent of other statistics such as the D_{90} (90th percentile of the cumulative undersize volume distribution), especially in cases where the shape of the particle size distribution changes between batches. This is especially true for mixtures where the upper 10% w/w of the

[3]The number fraction is **not** equivalent to the mass fraction in the ith bin. To convert from a number fraction to a volume (or mass) fraction: $f_i = \frac{n_i D_i^3}{\sum_{i=1}^{N} n_i D_i^3}$.

particles are much larger than the D_{90}. Thus, control of a D_{90} does not constitute control of the $D[6,3]$. Moreover, as shown later, limitations of the analysis technique to measure small percentages of very large particles may contribute to inaccuracy of the $D[6,3]$ and as a result in the prediction of content uniformity.

All of the above content uniformity models predict only the particle size contribution to unit dose uniformity. However, when content uniformity is determined experimentally, other sources of variability may affect the results. These include assay variability, non-random mixing, particle segregation during processing and other non-idealities. In order to ensure these effects are accounted for, the C_v associated with particle size effects should be limited to much less than the USP limits (e.g. $C_v < 1\%$).

It is important to note that the content uniformity models predict an ideal coefficient of variation. It is possible that the measured coefficient of variation could be much less than the predicted value. If this is observed, it may suggest that the drug substance is undergoing attrition in the drug product manufacturing process (i.e. the measured particle size does not reflect the actual particle size in the drug product). The maximum upper bound on particle size should be no greater than that predicted by a content uniformity model, unless there is strong experimental evidence that the drug substance is undergoing attrition in the drug product manufacturing process in a way that is expected to be scalable.

Inherent in this discussion, the PSD that is measured by analytical means must be representative of the PSD in the final dosage unit. For example, if agglomerates are present and broken in the analytical method, but not broken in the formulation process, the C_v would be under-predicted leading to a false sense of security.

7.3.2 Control of Maximum Observable Dose

One issue that can arise with low dose compounds is that of a super-potent tablet due to the presence of a large agglomerate or particle of drug substance representing a *substantial* portion of the intended dose. The effect of the size of a single spherical particle (with density 1.3 g/cm^3) on a 1 mg dose is presented in Fig. 7.4 and Table 7.1.

From this data it can be concluded that, if one requires e.g. a maximum of 1 tablet per million at 125% of dose, it means that at most only one 716 μm particle (representing 0.25 mg) may be present in 1 kg of API. Similarly, if one requires e.g. a maximum of 1 tablet per million at 200% of dose, it means that at most only one 1137 μm particle may be present in 1 kg of API. Thus, it will be clear that adequate dose uniformity cannot be achieved in cases where individual drug substance particles or agglomerates are so large that they individually account for an appreciable fraction of the intended unit dose.

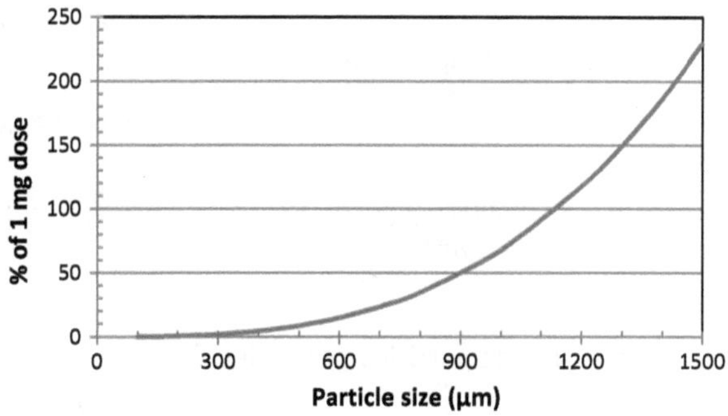

Fig. 7.4 Contribution of a single particle to a 1 mg dose

Table 7.1 Impact of single particle mass to 1 mg dose

Particle size (μm)	% of 1 mg dose
200	0.54
400	4.4
569	12.5
716	25.0
902	50
1000	68
1137	100
1300	150
1431	200

For discussion purposes, f is defined as the probability that a unit dose will exceed a large multiple (e.g. 125%) of the intended dose. Two questions must then be answered:

1 How does f relate to the drug substance particle size distribution?
2 Since f will never be 0, what is an acceptable upper limit for f?

To answer question 1, it is useful to consider the hypothetical example in Fig. 7.5. This illustrative example has the following characteristics:

(1) The tablets contain a nominal dose of 1 mg
(2) The drug substance particle size distribution is bimodal as follows:

a. The majority (99.9 wt%) of the drug substance particles are small (85 μm) and provide only a small contribution to the content C_v.
b. A very small fraction of the drug substance particles (0.1% by weight) are large (0.13 mg, or ~576 μm).

With the above conditions set forth in this example, the average number of large particles in each tablet, n_{Ave}, can be determined as follows:

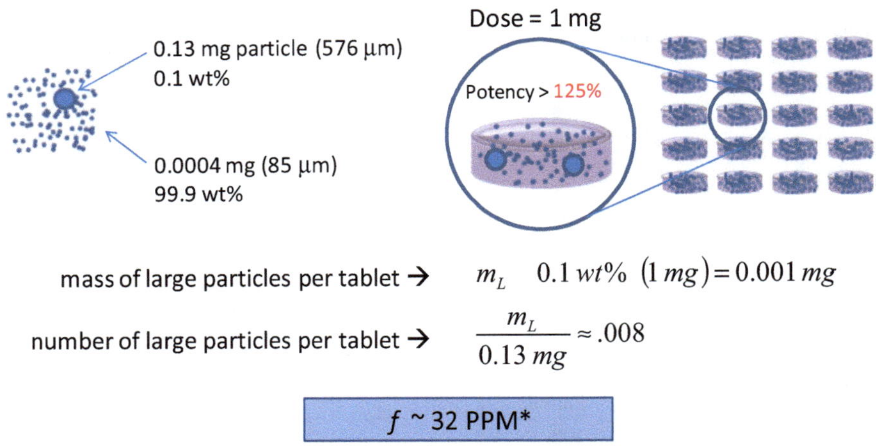

Fig. 7.5 Pictorial description of a worst-case particle size distribution

$$n_{Ave} = \frac{(wt\,fraction\ of\ large\ particles) \cdot G}{(weight\ of\ a\ large\ particle)} = \frac{(0.1\%) \cdot 1\,\text{mg}}{0.13\,\text{mg}} = 0.008 \tag{7.10}$$

Thus, on average, 8 large particles exist for every 1000 tablets. However, these 8 large particles are not evenly distributed among the 1000 tablets. The vast majority of tablets will certainly contain no large particles and it is highly probable that 8 tablets from a random sampling of 1000 tablets would each be found to contain one large particle. However, a small but finite probability exists for any given tablet to contain two large particles. An even smaller, but also finite probability exists for a single tablet to contain 3 large particles. The probability for a randomly selected tablet to contain 1, 2, or 3 large particles can be determined from the Poisson distribution, which is defined as follows:

$$f = \exp(-\gamma)\frac{\gamma^k}{k!} \tag{7.11}$$

where γ denotes the average number of large particles per tablet, f denotes the probability that a randomly selected tablet will be found to contain k large particles, and the! indicates the factorial function.

Equation 7.11 can be applied to the example in Fig. 7.5 to determine the probability that a tablet will exceed the 125% potency limit. Since each large particle in this example individually weighs 0.13 mg, then the presence of two or more large particles is required to cause the tablet to exceed 125% of the intended dose of 1 mg. The probability that a tablet will contain two large particles can be estimated using Eq. 7.11 as follows:

$$f = \exp(-0.008)\frac{0.008^2}{2 \cdot 1} = 3.2 \times 10^{-5} \quad (7.12)$$

In this case, 32 tablets per million are expected to contain two large particles. Similarly, tablets can exceed the 125% limit by containing 3 large particles (an event that occurs with a frequency of 8.5×10^{-8}), 4 large particles (a frequency of 1.7×10^{-10}), and so on. Thus, it is necessary to use the cumulative Poisson distribution function to determine the number of tablets that will exceed the 125% limit by containing 2 *or more* large particles. This calculation is easily performed using standard spreadsheet functions (e.g. one can perform this calculation as $f = 1 - \mathrm{POISSON}(\lambda - 1, k, \mathrm{TRUE})$ in Microsoft Excel).

The calculation of f illustrated in Fig. 7.5 can be repeated for other bi-disperse particle size distributions where large particles are present in different sizes and in different amounts. An example is shown in Fig. 7.5 where the log of the expected tablet frequency (f) is plotted against the tablet content, expressed as a percentage of the intended dose. The figure corresponds to a bi-disperse distribution in which large particles are present in the amount of 5% by weight of the drug substance. Each curve corresponds to large particles of a different size, expressed as a percentage of the intended dose. The smaller particles for each distribution are considered negligibly small. The yellow curve, for example, corresponds to large-particles of sufficient size to individually account for 20% of the intended Dose. The curve shows that tablets are expected to exceed 200% of the intended dose (x-axis) with a frequency of $\sim 10^{-5}$ (y-axis). In other words, 10 in 1 million tablets would be expected to exceed 200% of the intended dose.

Figure 7.6 also shows that the probability for a tablet to contain a single large particle decreases when the particles are larger. For example, the left-most curve (corresponding to large particles that weight 1% of the intended dose) approaches zero on the y-axis. This indicates a high likelihood (near 100%) for tablets to contain at least one large particle. This simply reflects the high prevalence of 1% particles in order to achieve a weight fraction of 5% of the drug substance. For the right-most curve (corresponding to large particles that weigh 800% of the intended dose), less than 1 in 100 tablets are expected to contain even a single large particle. However, each of those tablets would have an unreasonably high potency. Thus adequate dose uniformity cannot be achieved in cases where individual drug substance particles or agglomerates are so large that they account for an appreciable fraction of the intended unit dose.

The examples provided above in relation to Fig. 7.6 are based on an idealized bi-disperse particle size distribution where the large particles are present in the amount of 5 wt%. Additional figures are provided in Appendix to consider the impact of bi-disperse PSD's containing different weight fractions of large particles. These additional figures show that smaller amounts of large particles within the PSD result in lower frequencies of super-potent tablets. For example, a PSD containing 1% of large particles (Fig. 7.9), each weighing 20% of the intended dose would result in tablets exceeding 200% potency with a frequency of $\sim 10^{-8.5}$.

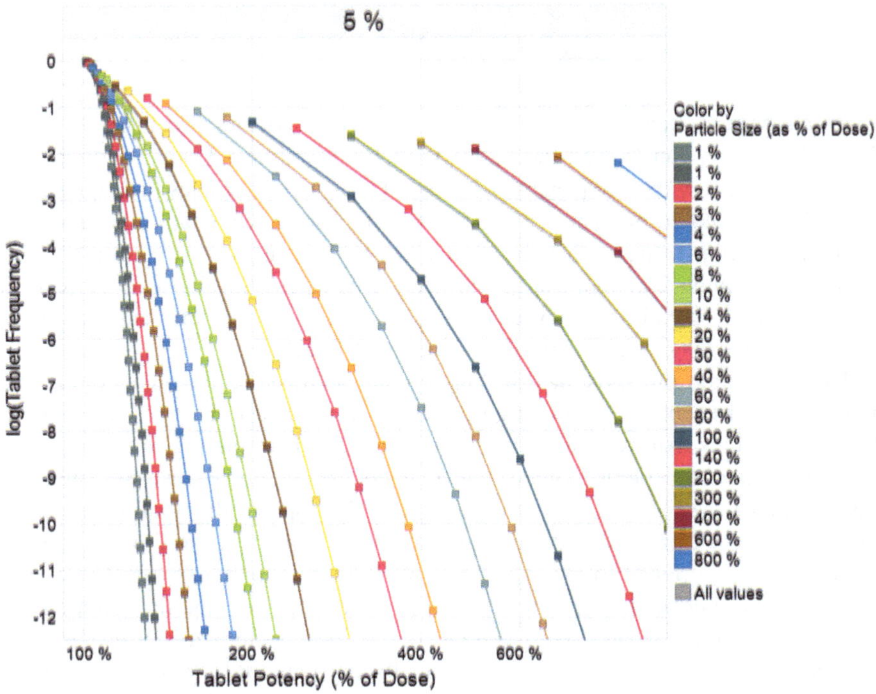

Fig. 7.6 Expected frequency of super-potent tablets for drug substance distributions with 5% w/w of large particles

Compare this to the example described above where the frequency was expected to be $\sim 10^{-5}$. Similarly a PSD containing 10% of large particles (Fig. 7.11), each weighing 20% of the intended dose would result in tablets exceeding 200% potency with a frequency of $>10^{-4}$.

7.4 PSD Controls for Super-Potent Tablets

Ideally, the model described above may be used to inform the particle size control strategy. Specifically, the model establishes guidelines for the maximum size of any large particles and for the maximum fraction of any large particles within the PSD.

The maximum size of large particles can be limited with a positive control such as a screen on a conical mill. The conical mill reduces the size of large agglomerates by forcing them through the screen. The screen size then defines the maximum particle size in the milled drug substance. There are practical limitations in the screen size below which the screen will blind or otherwise foul, preventing passage of material. The formulation process may also be augmented with high-intensity mixing equipment such as intensifier bars to disrupt agglomerates. However, agglomerate breakage in such processes has been shown to follow exponential decay kinetics [6] and it can be difficult to demonstrate complete particle breakage.

The PSD measurement controls the maximum fraction of any large particles (below the positive control limit). However, the analytical method limit of detection (LOD) for large particles must be considered. It can be difficult to quantify or even detect the low-level presence of large particles or agglomerates within a batch of drug substance using traditional measurement techniques. For example, the LOD for large particles using laser diffraction can range between 3 and 5% [17, 18]. This can especially occur when the analytical method applies aggressive dispersion conditions to the particles. For example, capillary action with liquid methods (especially with the addition of ultrasound) can disintegrate agglomerates that may otherwise survive the drug product formulation process. Similarly, dry methods may apply a high-pressure air jet to disperse particles and consequently disrupt agglomerates. In these cases, the agglomerates may not be detected by the method, but may be expected to exist in the drug product. In cases of great concern, an analytical screening method may be necessary to achieve a smaller LOD for large particles.

7.5 Acceptable Risk

The model described by Fig. 7.6 predicts only the expected rate of occurrence of super-potent tablets. It does not determine whether a particular rate is acceptable. A risk assessment is required to assess acceptability.

Protection of the patient by managing the risk to drug product quality is of paramount importance. Therefore, preventing harm to the patient is the primary focus of any risk assessment activity. Several risk assessment tools exist for identification and quantification of risk, with the failure modes and effects analysis (FMEA) or failure modes, effects, and criticality analysis (FMECA) being the most prevalent. ICH Q9 [11] defines risk as the combination of the probability of occurrence of harm and the severity of that harm, expressed as the product of these two rankings. While the FMEA and FMECA approaches generally employ separate rankings of severity, occurrence, and detection, it may be more convenient to combine the occurrence and detection into a single parameter that reflects the overall probability of harm and aligns with the ICH risk definition.

Given the semi-quantitative nature of the FMEA or FMECA risk assessment, it is necessary to define both the number and meaning of the levels employed in the analysis. Various levels of resolution are routinely used when performing an FMEA, ranging from a three point to a ten point scale. One example of these ranking scales for the quantification of severity, occurrence, and detection within the framework of the pharmaceutical industry is given in the ISPE Product Quality Lifecycle Implementation Guide [12]. Another is given in a briefing pack to the ICH Q9 guidance [4].

An example has been adapted from the ICH Q9 Briefing pack and is shown in Table 7.2. The table includes a five point scale for severity, occurrence, and detection. It should be noted that the probability of occurrence rankings in Table 7.2 is extremely conservative. For example, a ranking of "Probable" corresponds roughly to the lifetime odds of drowning in a swimming pool (Insurance Information Institute, n.d.). An occurrence ranking of 1 corresponds to an extraordinarily low defect rate of <1 in 10^{12}. This defect rate is effectively zero in comparison with the number of tablets that could possibly be produced for a particular drug product over the lifetime of the drug product.

The occurrence and detection aspects in Table 7.2 can be combined into one metric to align with ICH guidelines—the probability of harm, as shown in Fig. 7.7. The resulting probability of harm is then used in conjunction with the severity of harm to determine an overall risk level, as shown in Fig. 7.8. For example, if the probability of detection value for a particular event is 3 and the corresponding probability of occurrence value is 3, then Fig. 7.7 indicates a probability of harm value of 3. If the corresponding severity of harm ranking for this event is 3, then Fig. 7.8 indicates a medium risk level (yellow).

The risk may vary on a molecule-to-molecule basis and depends upon the dose, mechanism of action, and potential side effects related to a super-potent dose (e.g. >125% of label claim). As such, the severity rankings are generally tied to the properties of the specific molecule being assessed. For application of the risk grids shown in Figs. 7.7 and 7.8 to the example of super-potent tablets (e.g. Fig. 7.5), the

Table 7.2 Example FMEA rankings for severity, occurrence, and detection

Rank	Severity	Occurrence (failure or defect rate)	Detection
5	**Dangerously High**—A failure that results in death	**High** (Probability > 10^{-3})	**Non-detectable**—highly unlikely that the defect would be detected
4	**Very High**—A failure that causes injury	**Probable** (10^{-4} < Probability $\leq 10^{-3}$)	**Low**—Small likelihood that the defect would be detected
3	**Moderate**—A failure that causes customer complaints and dissatisfaction	**Occasional** (10^{-6} < Probability $\leq 10^{-4}$)	**Moderate**—Possible but not certain that the defect would be detected
2	**Minor**—The patient may notice discomfort but is not significant enough to warrant a complaint	**Rare** (10^{-12} < Probability $\leq 10^{-6}$)	**High**—Likely the defect would be detected
1	**None**—Failure is not noticed by the patient	**Improbable**—not expected to occur in the lifetime of the product (Probability $\leq 10^{-12}$)	**Almost Certain**—The defect would almost certainly be detected

PROBABILITY OF HARM						
		Proability of Occurrence				
Detection Rank		1	2	3	4	5
	5	1	3	4	5	5
	4	1	3	3	4	5
	3	1	2	3	3	4
	2	1	2	2	3	3
	1	1	1	2	2	3

Fig. 7.7 Probability of Harm matrix approach

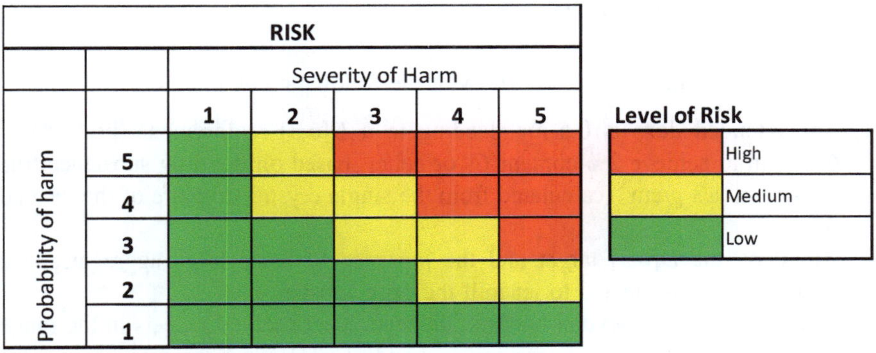

Fig. 7.8 Combined grid for risk level determination

probability of detection can be assumed to be "Non-Detectable"—a ranking of 5 on the detection scale in Table 7.2. This is assumed because analytical testing of 10–25 tablets is unlikely to detect rare occurrences of super-potent dosage units. The probability of harm can thus take values of 1, 3, 4, or 5 (top row in Fig. 7.7) depending on the probability of occurrence. For any occurrence level of non-detectable super-potent tablets above the "Improbable" ranking of 1, the probability of harm takes a value of 3 or above (Fig. 7.7), resulting in an overall medium or high risk level at severity rankings above 2 (Fig. 7.8). In order to mitigate this risk, two approaches may be taken. The first is to improve the probability of detection of the super-potent dosage unit, which likely increases the overall sampling and testing required. This approach is unlikely to be effective without testing a substantial portion of tablets. The second approach is to decrease the probability of occurrence by utilizing a positive control in the form of a terminal screening operation, thereby dictating the maximum potential particle size of the drug substance and reducing the potential defect rate to a value of "Improbable".

7.6 Typical Scenarios and Example Calculations

In this section, an example scenario typically encountered by development teams will be discussed along with several key questions and solution procedures. Specifically, the following are considered:

1. Determination of the best screen size
2. The impact of the limit of detection for small percentages of large particles in PSD
3. The impact of dose on the selection of the best screen size
4. The impact of screen size.

7.6.1 *General Scenario*

Consider a molecule in development with the following conditions:

(1) The intended dose is 0.5 mg, leading to a $D[6,3]$ requirement (Eq. 7.8) of 66.5 μm to achieve 2% content C_v or better, based on the drug substance true density of 1.3 g/cm^3 (calculated from the single crystal structure of the desired form).
(2) Based on the $D[6,3]$ target and the presence of some large aggregates, the project team has elected to jet mill the drug substance.
(3) Over the course of several batches, the team has observed $D_{90,3}$'s in the range of 15–20 μm. On two occasions, a handful of large aggregates were also observed lying on top of formulated powders in a tumble bin blender. On both occasions, the aggregates were collected and found to be rigidly bound (i.e. not formed by static charge). On another occasion, a single tablet was measured to contain 150% of the intended dose.
(4) The team plans to use a positive control (e.g. a screen) to remove large aggregates from the drug substance and has determined that a 280 μm mesh screen blinds during screening but the 460 μm screen is acceptable.
(5) The particle size method is laser diffraction-based and spiking studies suggest that the method can only detect the presence of large drug substance aggregates if they occur at levels above 5% by weight. In other words, there is a 5% limit of detection for large aggregates.
(6) Medical associates have indicated that patients can tolerate a single dose of up to 1 mg (i.e. 200% of the intended dose) without any measureable side effects, but a single dose beyond 2 mg (400%) can be life threatening.

7.6.2 Example Calculation 1: Determine the Best Screen Size

Given the decision of the team to employ a positive particle size control, the key question to the team is what is the best screen size to employ?

A potential solution to the above problem statement can be reasoned as follows:

(1) Items 2 and 3 above suggest that the particle size can be bimodal. Most of the particles are very small and do not significantly impact content uniformity, but a small percentage of particles (aggregates) is very large. The aggregates are large enough to significantly impact tablet content uniformity and they are strong enough to survive into the drug product.

(2) Item 4 above indicates that the proposed screen would limit the largest possible particles to 460 μm. Assuming spherical particles (reasonable for aggregates) particles passing through the screens could have masses of up to 0.0663 mg, or 13.3 percent of the intended 0.5 mg dose.

(3) Item 5 above indicates an analytical method limit of detection of 5%. Given this limit of detection, Fig. 7.6 may be used to estimate the frequency of high dose tablets. The frequency of high dose tablets is then used to determine a risk value as follows:

   a. **Risk of a 200% of dose tablet**: In Fig. 7.4, one interpolates between the 10% (green) and 14% (brown) curves to determine that a 200% of dose tablet (x-axis) would occur at a frequency, $f < 10^{-7}$ (y-axis). This probability of occurrence translates into a rating of 2 on the occurrence scale (Table 7.2). With a detection rank of 5, the probability of harm value is 3 (Fig. 7.4). Given that medical has indicated that a dose of 200% of target would result in no measurable side effect, the severity of harm is 1 (Table 7.2), the risk associated with this scenario is low (Fig. 7.4).

   b. **Risk of a 400% of dose tablet**: The probability of this is negligibly small. Following the 14% (brown) curve in Fig. 7.4, $f < 10^{-12}$ even for 250% potency tablets. As stated previously, an f-value below 10^{-12} provides acceptable "low" risk for any severity of harm.

(4) The team concludes that use of the 460 μm screen presents an acceptable low risk on the risk grid and thus elects to use this screen in manufacturing of the drug substance.

7.6.3 Example Calculation 2: Impact of LOD

A second question the team may be asked relates to the LOD. For the example described above, consider the risk of a 200% of dose tablet assuming an analytical method limit of detection of 10% instead of 5%.

Solution:
Since the method limit of detection is 10%, we use Fig. 7.4. Following the 14% (brown) curve, a 200% of dose tablet is expected to occur with a frequency of up to $\sim 10^{-5}$. This probability of occurrence translates into a rating of 3 on the occurrence scale (Table 7.2). With a detection rank of 5, the probability of harm value is 4 (Fig. 7.4). Given that medical has indicated that a dose of 200% of target would result in no measurable side effect, the severity of harm is 1 (Table 7.2) the risk associated with this scenario is low (Fig. 7.4).

7.6.3.1 Example Calculation 3: Impact of Dose

For the same conditions as Example Calculation 2, what is the risk of a 400% of dose tablet assuming a dose of 0.05 mg instead of 0.5 mg. What is the required *D* [6,3] to allow a 2% content uniformity C_v?
Solution:
Since the dose is 0.05 mg the *D*[6,3] must fall below 30.9 μm in order to meet the C_v target of 2% (Eq. 7.8). Particles passing through the 460 μm screen now individually account for 133% of the intended dose. Fig. 7.4 indicates that tablets are expected to exceed 400% potency in this case with a frequency of $\sim 10^{-4}$, which corresponds to a probability of occurrence level of 3–4 (Table 7.2). The probability of harm is thus 4–5 (Fig. 7.4). The corresponding risk is high (Fig. 7.4). It would be ill-advised to make tablets under these conditions, even if the *D*[6,3] target is met. In this case, the team may incorporate a smaller screen size in the drug product process to remove any large particles from the API.

7.6.3.2 Example Calculation 4: Impact of Screen Size

For the same conditions as Example Calculation 1, what is the risk of a 200% of dose tablet and a 400% of dose tablet using a 600 μm screen size (versus 460 μm screen).
Solution:
Particles passing through the 600 μm screen can individually weigh up to 0.147 mg, or 29% of the intended 0.5 mg dose assuming spherical particles. Interpolating between the 20% (yellow) and 30% (pink) curves in Fig. 7.4, the expected frequency of 200% of dose tablets is slightly above 10^{-4}. This probability of occurrence translates into a rating of 4 on the occurrence scale (Table 7.2). With a probability of detection of 5, the probability of harm value is 5 (Fig. 7.4). Given

that medical has indicated that a dose of 200% of target would result in no measurable side effect, the severity of harm is 1 (Table 7.2) the risk associated with this scenario is low (Fig. 7.4). In this case, the occurrence of a 400% tablet is below 10^{-12}, and is not expected to ever occur. Thus, the risk is low.

7.7 Summary

The above discussion provides guidance on the selection of appropriate particle size target and controls to ensure a safe level of content uniformity and consideration of bioavailability. A two part control strategy consisting of a $D[6,3]$ measurement and a positive control. Guidance is provided for the selection of an appropriate $D[6,3]$ target and screen size. Since risk can never be fully eliminated in any process, a risk-based approach is provided in the selection of screen size.

7.8 Definitions, Abbreviations and Symbols

Tablet potency	API content of a tablet as a percentage of its intended dose
API	Active pharmaceutical ingredient
BCS	Biopharmaceutics classification system
CQA	Critical quality attribute
CU	Content uniformity
FMEA	Failure modes and effects analysis
FMECA	Failure modes, effects and criticality analysis
ICH	International Conference of Harmonisation
ISPE	International Society for Pharmaceutical Engineering
LOD	Limit of detection
NCE	New Commercial Entity
ppm	Parts per million
PSD	Particle size distribution
SISPQ	Safety, Integrity, Strength, Purity and Quality
UDU	Unit dose uniformity
USP	United States Pharmacopoeia
C_s	Equilibrium solubility
C_v	Coefficient of variation
D	Particle size (spherical diameter)

Di	ith percentile of diameter size distribution
Di,j	ith percentile of diameter size distribution type j, where j = 0 corresponds to number distribution, j = 1 corresponds to length, j = 2 corresponds to area, and j = 3 corresponds to volume (or mass under assumption of constant density)
$D[6,3]$	Mean particle size of a PSD according to Eq. (7.2)
D_{diff}	Diffusion coefficient
f	Probability that a unit dose exceeds a large multiple of the intended dose
h	Diffusion layer thickness
m_L	Mass of a large particle
m_s	Mass of a small particle
N_L	Number of large particles in a tablet
r_0	Particle radius
t	Time
X_0	The initial amount of drug
X_d	Mass of drug in solution
X_s	The amount of drug
V	Media volume
ρ	Density of the drug substance
σ_g	Geometric standard deviation of a lognormal distribution

Acknowledgements The authors would like to acknowledge Dale Greenwood, and Susan Reutzel-Edens for extensive editorial review as well as Tim Kramer, Rick Berglund, Chad Wolf, Rich Meury, Sal Garcia-Munoz, and Paul Stroud for excellent technical feedback and discussions.

Appendix: Expected Frequencies of Super-Potent Tablets for Bi-Disperse PSD's with Differing Weight Fractions of Large Particles

See Figs. 7.9, 7.10 and 7.11

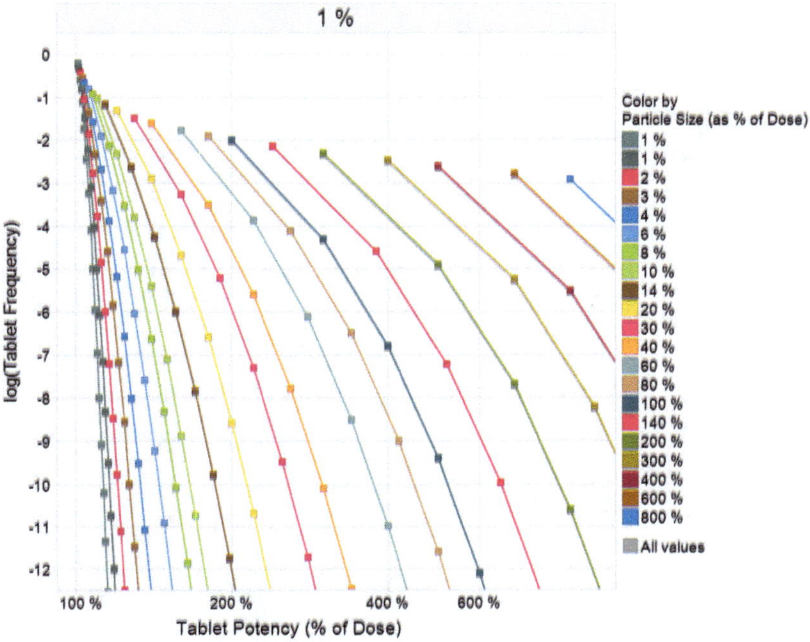

Fig. 7.9 Expected frequencies of super-potent tablets given a bi-disperse PSD with 1% w/w of large particles

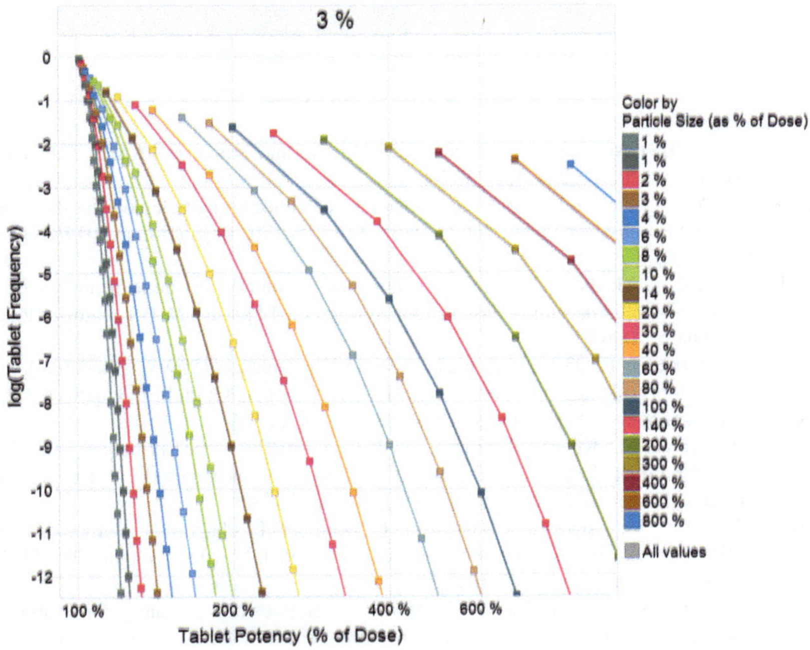

Fig. 7.10 Expected frequencies of super-potent tablets given a bi-disperse PSD with 3% w/w of large particles

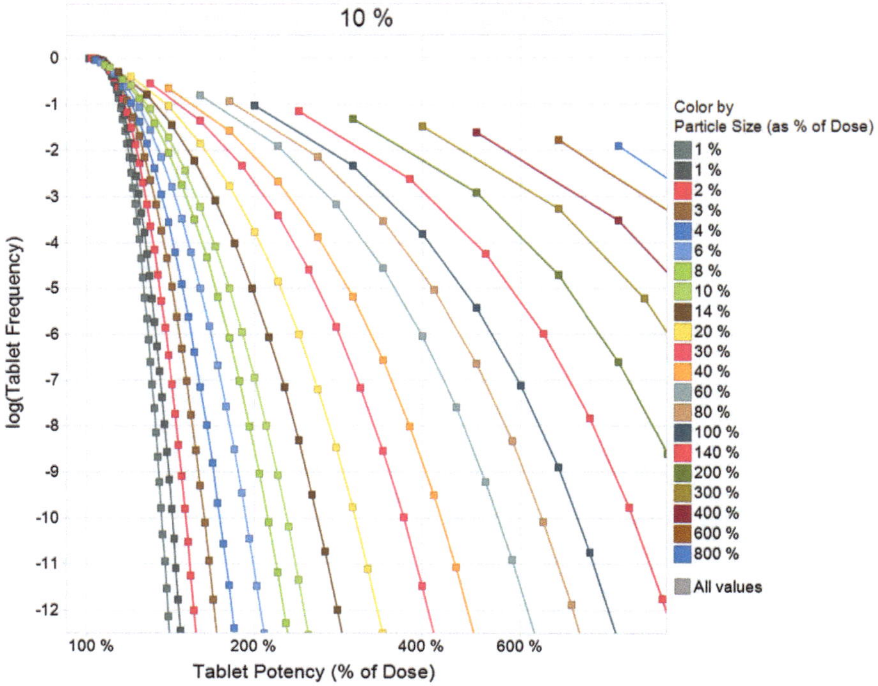

Fig. 7.11 Expected frequencies of super-potent tablets given a bi-disperse PSD with 10% w/w of large particles

References

1. Aungst BJ (2012) Absorption enhancers: applications and advances. AAPS J 14:10–18. https://doi.org/10.1208/s12248-011-9307-4
2. de Villiers MM (1996) Influence of agglomeration of cohesive particles on the dissolution behaviour of furosemide powder. Int J Pharm 136:175–179. https://doi.org/10.1016/0378-5173(95)04380-2
3. Dokoumetzidis A, Macheras P (2006) A century of dissolution research: from Noyes and Whitney to the biopharmaceutics classification system. Int J Pharm 321:1–11. https://doi.org/10.1016/j.ijpharm.2006.07.011
4. Harbour G (2006) ICH Q9 Briefing pack II, Online Presentation [WWW Document]. URL http://www.ich.org/fileadmin/Public_Web_Site/ICH_Products/Guidelines/Quality/Q9/Q9_Briefing_Pack/Tools_-_Applications.pdf. Accessed 29 Aug 17
5. Hilden J, Schrad M, Kuehne-Willmore J, Sloan J (2012) A first-principles model for prediction of product dose uniformity based on drug substance particle size distribution. J Pharm Sci 101:2364–2371. https://doi.org/10.1002/jps.23130
6. Hilden J, Schrad M, Reynolds J, Hartmann C, Sommer T, Pletcher T, Zettler A (2017) Investigation of an intensifier-bar tumble bin scale-up model. Powder Technol 305:723–738. https://doi.org/10.1016/j.powtec.2016.09.064
7. Iacocca RG, Burcham CL, Hilden LR (2010) Particle engineering: a strategy for establishing drug substance physical property specifications during small molecule development. J Pharm Sci 99:51–75. https://doi.org/10.1002/jps.21801

8. Insurance Information Institute (n.d.) Mortality Risk [WWW Document]. URL http://www.iii.org/fact-statistic/mortality-risk. Accessed 29 Aug 17
9. International Conference on Harmonisation (2009) Pharmaceutical Development (Q8)R2
10. International Conference on Harmonisation (2008) Pharmaceutical Quality System (Q10)
11. International Conference on Harmonisation (2005) Quality Risk Management (Q9)
12. International Society for Pharmaceutical Engineering (2011) Product Quality Lifecycle Implementation (PQLI) from Concept to Continual Improvement, Part 1—Product Realization using Quality by Design (QbD): Concepts and Principles
13. ISO 13320, Particle Size Analysis—Laser Diffraction Methods. International Organization for Standardization
14. Jain S, Patel N, Lin S (2015) Solubility and dissolution enhancement strategies: current understanding and recent trends. Drug Dev Ind Pharm 41:875–887. https://doi.org/10.3109/03639045.2014.971027
15. Johnson M (1972) Particle size distribution of the active ingredient for solid dosage forms of low dosage. Pharm. Acta Helv 546–559
16. Kesisoglou F, Wu Y (2008) Understanding the effect of API properties on bioavailability through absorption modeling. AAPS J 10:516–525. https://doi.org/10.1208/s12248-008-9061-4
17. Ma Z, Merkus HG, de Smet JGAE, Verheijen PJT, Scarlett B (1999) Improving the sensitivity of forward light scattering technique to large particles. Part Part Syst Charact Banner 16:71–76. https://doi.org/10.1002/(sici)1521-4117(199906)16:2<71::aid-ppsc71>3.0.co;2-2
18. Merkus HG (2009) Particle size measurements: fundamentals, practice, quality. In: Particle Technology Series. Springer, The Netherlands
19. Rohrs BR, Amidon GE, Meury RH, Secreast PJ, King HM, Skoug CJ (2006) Particle size limits to meet USP content uniformity criteria for tablets and capsules. J Pharm Sci 95:1049–1059. https://doi.org/10.1002/jps.20587
20. United States Pharmacopeial Convention (2015). Chapter <905> Uniformity of Dosage Units, in: USP 38-NF 33 S1. United States Pharmacopeial
21. U.S. Department of Health and Human Services, Food and Drug Administration, Center for Drug Evaluation and Research (CDER) (2015) Draft Guidance for Industry: Waiver of In Vivo Bioavailability and Bioequivalence Studies for Immediate-Release Solid Oral Dosage Forms Based on a Biopharmaceutics Classification System Revision 1
22. Yalkowsky SH, Bolton S (1990) Particle size and content uniformity. Pharm Res 7:962–966. https://doi.org/10.1023/A:1015958209643

Jon Hilden holds a bachelor's degree in Materials Engineering from San Jose State University, and Masters and Ph.D. degrees in Materials Science and Engineering from Purdue University. Jon spent 1 year performing post-doctoral research split between the Industrial and Physical Pharmacy department and the Materials Science and Engineering department at Purdue, followed by 5 years in the Materials Characterization group at Bristol Myers Squibb, and 10 years in the Formulation and Process development area of Eli Lilly and Co. Jon has been an active contributor in the mathematical modeling and data sciences aspects of pharmaceutical development.

Christopher L. Burcham is a Senior Engineering Advisor at Eli Lilly and Company, in the Small Molecule Design and Development department within Product Research and Development. He earned his BS in Chemical Engineering from the University of Illinois and Ph.D. also in Chemical Engineering from Princeton University. His career started with The Dow Chemical Company. Since joining Lilly he was responsible for the formation of the Particle Design Lab and has led many other technology initiatives. He currently is the Commercialization Team Leader responsible for the commercialization of a late phase development project.

Stephen D. Stamatis completed his Ph.D. in Chemical Engineering from Purdue University in 2010 focusing on heterogeneous catalysis and Bayesian statistical methods. After completing a postdoctoral fellowship at the University of Iowa in Pharmaceutics, he joined Eli Lilly and Company in 2013. His current work focuses on oral absorption modeling under uncertainty and other drug product performance concerns.

Jim Miesle holds a bachelor's degree in chemical engineering and a master's degree in pharmaceutical engineering. He has worked in the area of solid oral dosage form process development at Lilly for the past 9 years, focused mainly on granulation, tablet compaction, and coating unit operations. Prior to Lilly, he worked at McNeil Consumer Healthcare in dosage form development. Jim is a registered professional engineer in the state of Indiana.

Carrie A. Coutant received a doctorate in analytical chemistry from University of Michigan, studying programmable selectivity for high-speed gas chromatography with Professor Richard Sacks. She joined Lilly as a senior analytical chemist, working as the lead analyst for a number of late stage development projects. Later, Dr. Coutant transitioned to the drug product performance group, where her responsibilities included the development of in vitro methods for characterizing and controlling drug product performance, and the application of these methods from clinical development through registration and commercialization of the drug product. She currently works in the product performance prediction group, with a focus on leveraging both in-vitro and in-silico models for drug product performance.

Chapter 8
Effects of Particle Size, Surface Nature and Crystal Type on Dissolution Rate

Giuseppina Sandri, Maria Cristina Bonferoni, Silvia Rossi, Carla M. Caramella and Franca Ferrari

Abstract Solid drug delivery systems are crucial formulations for the oral route. In such systems, particle size and polymorphism have a strong impact on drug dissolution and on drug absorption. Starting from the role of particle size in dissolution rate, the Noyes-Whitney equation, the modified form by Nernst-Brunner and the cube root equation are here described. According to these equations diffusion of a solute through a boundary layer around the particles is the rate limiting step for both drug dissolution and absorption and, thus, depends on the specific (external) surface area of the particles, the diffusion coefficient of the dissolved drug, the thickness of the boundary layer and the drug solubility. In relation to this, good wetting of the particle surface by the surrounding liquid and adequate particle dispersion play an essential role. Information from dissolution rates suggests that the thickness of the boundary layer is constant for larger particle sizes, but dependent upon size for smaller particles. Given the larger surface area of smaller particles, the attention has been directed to nanosystems and on their relevance to the bioavailability of poorly soluble drugs. A second advantage of such drug systems is that solubility increases on decreasing particle size, according to the Freundlich–Ostwald equation. The impact of polymorphism, pseudopolymorphism and amorphous form on drug dissolution and bioavailability is also described. Since dissolution and absorption are closely related, the effect of particle size and polymorphism on drug absorption is described. Moreover, regulatory implications of particle size and polymorphism are reviewed.

8.1 Introduction

Drug molecules with limited aqueous solubility are becoming increasingly prominent in the research and development of new chemical entities. This type of molecules provides many challenges in pharmaceutical development, since a slow

G. Sandri (✉) · M. C. Bonferoni · S. Rossi · C. M. Caramella · F. Ferrari
Department of Drug Sciences, University of Pavia, viale Taramelli 12, 27100 Pavia, Italy
e-mail: giuseppina.sandri@unipv.it

H. G. Merkus et al. (eds.), *Particles and Nanoparticles in Pharmaceutical Products*, AAPS Advances in the Pharmaceutical Sciences Series 29,
https://doi.org/10.1007/978-3-319-94174-5_8

dissolution in biological fluids and a consequently poor bioavailability determine a negative effect in efficacy, for drugs administered via the oral route. In fact, drug bioavailability from solid dosage forms due to gastro-intestinal absorption, often reflects in vitro dissolution rate when drug dissolution in the gastro-intestinal fluids is the rate-determining step for absorption rather than drug diffusion across the intestine wall.

Advances in pharmaceutical sciences have led to several approaches aimed to face and solve this challenge. Among the strategies for improving and maximizing dissolution rate, a key role is assumed by particle size reduction that increases the surface area available for dissolution.

The granulometric properties of pharmaceutical powders (both active ingredients and excipients) play an important role on their behavior in technological as well as biopharmaceutically relevant processes. Furthermore, such properties are fundamental in the design and development of drug delivery systems, where the drug is loaded as a solid, since there is a close influence of particle size on technological processes and on biopharmaceutical properties of formulations.

However, the increase in dissolution rate due to the increase of surface area does not improve the equilibrium solubility and is not enough to provide adequate bioavailability enhancement especially for drugs with a very poor aqueous solubility. Crystal engineering and, in particular, habit modification, polymorphism and pseudopolymorphism offer an alternative and a potential fruitful method for ameliorating aqueous solubility, dissolution rate and consequent bioavailability of a poorly soluble drug. It is well known that the different polymorphs can exhibit different solubility and different dissolution rate, leading to non-equivalent bioavailability of the different forms.

In addition, formulation strategies could influence the available/apparent surface area thanks to the improvement of formulation wettability.

The present chapter is focused on the role played by particle size and solid state (polymorphism or pseudopolymorphism) on aqueous solubility and dissolution properties. Moreover, the surface properties of formulations and the formulative strategies aiming at improving drug solubility and dissolution rate are discussed. In addition, the effect of nanonization technology on bioavailability of poorly soluble drugs is reviewed.

8.2 Dissolution Models for Particles

Particle size plays a key role in the dissolution rate of poorly soluble drugs in oral dosage forms, which is mainly dependent on interfacial surface area: in fact, the dispersion of particles in aqueous dissolution media is a prerequisite to trigger the dissolution process.

In 1897 Noyes and Whitney first proposed the diffusion controlled model to explain the dissolution phenomenon [44]. They suggested that keeping surface area

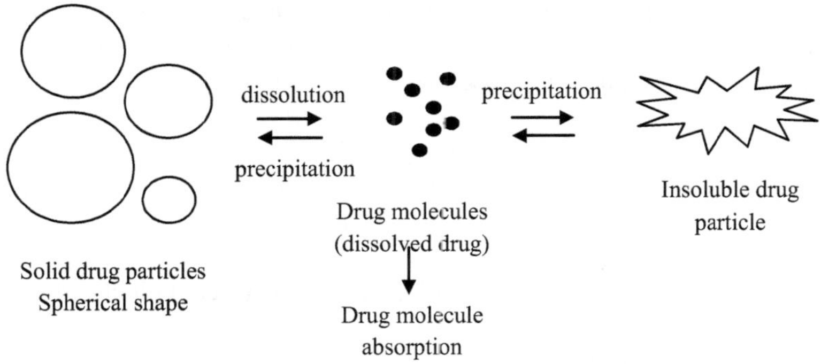

Fig. 8.1 Scheme of the dissolution model: solid spherical particles of different diameter dissolved in gastrointestinal fluids [51], copyright Springer, reproduced with permission. Precipitation can occur leading to particles similar in shape to the original ones or to insoluble particles. Drug molecules can be absorbed

constant, the dissolution rate is directly related to the differences between solubility and bulk solution concentration:

$$\frac{dm}{dt} = k \cdot (C_s - C_t) \tag{8.1}$$

where

dm/dt dissolution rate
C_s equilibrium solubility of a substance, i.e. the concentration of its saturated solution
C_t concentration of a substance in the bulk medium at time t
K constant (mass transfer coefficient)

The Noyes-Whitney Eq. (8.1) introduced the concept of the relationship between particle size and dissolution rate, based on the assumption that in the dissolution process the solubilization step is fast, whereas the subsequent diffusion of drug molecules through a diffusion boundary layer surrounding the particle is the rate limiting step. Figure 8.1 illustrates the dissolution model, based on the Noyes–Whitney equation for spherical particles with a given particle size distribution.

In 1904 Nernst and Brunner [7, 40] modified Noyes-Whitney equation suggesting that a rapid equilibrium (i.e. saturation) is achieved at the solid-liquid interface during dissolution: consequently a diffusion occurs through a thin layer of static (unstirred) solution called diffusion layer, to the bulk solution. In most cases, diffusion across this diffusion layer is rate-controlling, which effectively converts the heterogeneous dissolution phenomenon into an homogeneous, liquid phase diffusion process. The authors explained the constant k of Eq. (8.1) in terms of surface area, diffusion coefficient and diffusion layer thickness; the model is called film model and the relevant equation is Eq. (8.2) and it became one of the most used equations to describe drug dissolution:

$$\frac{dm}{dt} = \frac{A \cdot \mathbb{D}}{\delta}(C_s - C_t) \tag{8.2}$$

where

- A surface area of the undissolved solid drug in contact with the dissolution medium
- $\mathbb{D}$ diffusion coefficient of the solute
- δ thickness of the diffusion boundary layer.

On the basis of Nernst and Brunner's interpretation of the dissolution phenomenon, the diffusion coefficient $\mathbb{D}$ is influenced by medium viscosity (it decreases on increasing the viscosity) whereas the thickness of the diffusion layer δ is influenced by both medium viscosity and stirring rate: it increases at increasing viscosity and δ decreases on increasing stirring rate.

Consequently, particle dissolution is not only dependent on their surface area, but also on δ [24]. Different theories on the relation between δ and particle size have been developed. In particular, Hixson assumes δ to be constant; consequently the increase in surface area is the unique driving force for dissolution rate enhancement. However, other models do not assume that the thickness of the boundary layer remains constant on changing particle size.

When C_t does not exceed 10% of C_s, the so called sink conditions are satisfied, and C_t can be considered negligible in comparison to C_s. Under these conditions, the dissolution rate can be considered directly proportional to drug solubility and to particle surface area, which in turn depends on its particle size and shape.

It must be noticed that Nernst and Brunner's hypothesis of an unstirred liquid diffusion layer adherent to solid surface is hydrodynamically unrealistic, although it allows the complex dissolution process to be analyzed due to its simplicity. The diffusion layer model is commonly applied to a wide range of dissolution phenomena although the presence of an unstirred layer is not supported by fluid dynamics and liquid motion can occur at distances from the solid surface much closer than the thickness of a typical diffusion layer. Therefore, the diffusion layer is actually a boundary layer, characterized by a concentration gradient. Another limitation of Nernst and Brunner's model is that diffusion layer cannot be calculated independently based on hydrodynamics, but is obtained only by fitting of the equation to experimental data.

To overcome the limitations of the diffusion layer model (Nernst and Brunner equation), a further examination of the dissolution process proposed two combined mechanisms for the transport of solute molecules from a solid to a stirred liquid: a molecular diffusion, due to concentration gradient, and a convectional diffusion of molecules along the stirred solvent.

In this perspective, the dissolution process is described by the general Eq. (8.3) for convective diffusion in three dimensions (convective dissolution model) [55]:

$$\frac{\partial C}{\partial t} = \mathbb{D}\left(\frac{\partial^2 C}{\partial x^2} + \frac{\partial^2 C}{\partial y^2} + \frac{\partial^2 C}{\partial z^2}\right) - v_x\frac{\partial C}{\partial x} - v_y\frac{\partial C}{\partial y} - \frac{\partial C}{\partial z}. \tag{8.3}$$

where v_x, v_y and v_z are the liquid stirring rates in the x, y, and z direction of Cartesian coordinates.

According to this model, solvent stream movement depends on its distance from solid surface: in the bulk solution convection dominates over diffusion, while close to the solid surface, where solvent stream movement is slow and concentration gradient is large, the contribution of the diffusion is comparable or greater than convection. This latter layer is called convective diffusion layer [46].

The differences between Nernst's layer and convective diffusion layer can be summarized in the following terms:

(1) in Nernst's layer, solvent is considered as unstirred, whereas in convective diffusion model it represents a high rate gradient;
(2) in Nernst's model, mass transport tangential to solid surface is not considered, whereas in convective diffusion model, convection and diffusion both across and along the diffusion layer are taken into account;
(3) Nernst's layer is considered to be constant, whereas the thickness of convective diffusion based diffusion layer can vary along the solid surface depending on position and solvent stream.

This model is more realistic from a hydrodynamic point of view.

Another equation has been derived from the Noyes-Whitney equation by Hixson and Crowell [22]. Such equation was proposed to evidence key role played by the initial radius of the particles that undergo dissolution is evident in the Hixson-Crowell Eq. (8.4), and it is also known as Cube root equation:

$$\sqrt[3]{m_t} = \sqrt[3]{m_0} - K\,t \tag{8.4}$$

where

K "cube root" dissolution constant
m_t drug amount undissolved at time t
m_0 initial drug amount
t time

The amount of undissolved drug, m_t, is related to the remaining volume of solid powder and, consequently, to the cube of the mean volume particle size, $D_{3,0}$. In fact, the Hixson-Crowell equation is rigorously applicable to systems constituted by monodispersed spherical particles under sink conditions, although it is quite commonly used to describe in an approximate way the dissolution behavior of non-spherical particles.

To describe the dissolution of very small particles Higuchi and Hiestand [19, 20] proposed the two- thirds root Eq. (8.5), where drug amount undissolved at time t has a two third root dependency on particle weight:

$$\sqrt[3]{(m_t)^2} = \sqrt[3]{(m_0)^2} - K\,t \tag{8.5}$$

As illustrated in Eq. (8.2), dissolution rate is directly related to specific surface area of the solid. Dissolution rates can, therefore, be measured either maintaining the surface area constant (intrinsic dissolution rate measurements), or taking into account the variation of particle surface area during dissolution. Nystrom et al. [45] proposed to measure particle size with Coulter Counter apparatus to calculate the surface area of a solid undergoing dissolution. This approach involved to determine weight of particle population dissolved as a function of time. Under the assumption that the shape of the particles remains constant [9], the authors introduced a new parameter, the surface specific dissolution rate (G) (8.6):

$$G = \frac{m_0 - m_t}{t((A_0 - A_t)/2)} \tag{8.6}$$

where

G	surface specific dissolution rate
m_0	initial weight of solid particles
m_t	weight of solid particles at time t
A_0, A_t	surface area of solid particles at time zero and time t,
t	time

The proposed parameter G relates to the weight loss divided by the loss of surface and, thus, to the inverse of the specific surface area (surface area per unit weight).

In fact, among the different mean diameters used for particle size characterization, the area-weighted Sauter mean diameter, $D_{3,2}$, gives information directly related to specific surface area and, for this reason, it is profitably used in the pharmaceutical field.

It was however noticed that the increase in dissolution rate due to particle size decrease could be higher than expected on the basis of surface area increase. This finding can be explained taking into account that a decrease in particle size is accompanied by a decreased thickness of the boundary layer δ [5]. This effect is especially pronounced for materials with mean particle size less than 5 µm [3, 5] and was confirmed by Mosharraf and Nystrom [36] in a study involving sparingly soluble materials such as griseofulvin, glibenclamide, barium sulphate and oxazepam.

Higuchi et al. [18] assumed that the diffusion boundary layer thickness decreases proportionally with the particle size. According to this assumption, about 10 fold increase in surface area would cause a theoretical increase in dissolution rate by a 100 factor compared with the initial dissolution rate of micronized material.

Niebergall et al. [42] proposed that δ correlates with the square root of the particle radius while Hintz and Johnson proposed the concept of a transitional particle size of 30 μm, implying that δ is constantly equal to 30 μm for particles larger than 30 μm, whereas the diffusion layer is equal to particle radius for particles smaller than 30 μm [21]. Sheng et al. [53] found a dependency of δ on medium agitation, which seems logical if one assumes also convection.

Anyhow, all the classical dissolution models were developed for spherical particle shape, whereas pharmaceutical powders are mainly characterized by irregular shapes.

Pedersen and Brown [48], on the contrary, investigated for ten common crystal shapes how much the shapes contribute to the dissolution profile in comparison to spheres.

Later Dali and Carstensen [10] added a shape factor in the cube-root law and fitted dissolution profiles of crystals characterized by different shapes. They found that the shape factor is not constant, but changes as particles dissolve.

Another aspect to be considered is polydispersity of particles present in a powder or resulting from solid form disintegration. In fact, the classical particle dissolution models and in particular the cube root law are intended for single monodisperse particle dissolution, whereas pharmaceutical powders are composed from particles having different size.

Given these considerations, it can be argued that dissolution rate calculations are strongly dependent on theoretical model considered.

Most modern software packages used for plasma level simulation (e.g. GastroPlus®) consider a constant δ value for simulating drug dissolution. This causes only an approximate predictability of the mathematical models and suggests that dissolution rate should be experimentally determined to obtain reliable results.

8.3 Drug Factors Affecting Dissolution

Drug factors that affect dissolution are particle size, wettability, solubility and the form of the drug: whether it is a salt or a free form, crystalline or amorphous.

8.3.1 Surface Area and Particle Size

According to Eq. (8.2), an increase in total surface area of drug in contact with the dissolution medium causes an increase in dissolution rate. Considering that each drug particle is wetted, the effective surface area will be directly proportional to drug particle size of the drug. Hence the smaller the particle size the greater the effective surface area exhibited by a given mass of drug and consequently the dissolution rate. Therefore, particle size reduction is likely to result in increased bioavailability, provided drug absorption is limited by dissolution rate.

The first results dealing with the relation of particle size with dissolution rate and therefore with absorption, involve sulfamidics such as sulfadiazine, sulfaisossazole, sulfathiazole and sulfamethizole [25, 26, 56].

The solubility of the drug dramatically influence the relationship between particle size and dissolution rate; this appeared evident especially for poorly soluble drugs, whereas no influence could be found for freely soluble drugs. In the case of sulfadimetossine a great influence of particle size on rabbit gastro-intestinal absorption was demonstrated: the increase in particle size corresponded to a decrease in C_{max} (maximum plasma concentration) and an increase in t_{max} (time to reach maximum concentration), due to slower gastro-intestinal absorption [26].

Similar results were found in man for other drugs. In particular [41] put clear evidenced that particle size, among the various biopharmaceutical factors, plays an important role on the oral absorption of phenitoin, and, in turn, can result in serious clinical consequences due to the dose dependent metabolism and narrow therapeutic range of the drug [41]. A positive effect of particle size reduction on bioavailability was observed also for digoxin [23, 52] and, although with some discordant results, for griseofulvin [4, 14].

Simoes et al. [54] evaluated the influence of particle size and related properties, in particular specific surface area and solubility, on dissolution rate of a sparingly soluble drug, indomethacin. As expected, a strong influence of particle size on the dissolution rate was found. A correlation was established between mean dissolution time (MDT) and mean particle size of various indomethacin fractions: the dissolution rate increased with a reduction in particle size, confirming the correlation between MDT and mean particle size.

Micronization, by dry milling (jet milling, ball milling, pin milling) to 2–5 μm sizes was proposed as a routinely method to reduce particle size and, therefore, as a useful method to improve bioavailability of poorly soluble drugs. This strategy was proven successful also in more recent studies, such as the one by Ning et al. [43], where drug micronization was shown to produce a solid dispersion of glimepiride and to improve its bioavailability in beagle dogs, as evidenced by AUC (area under blood plasma concentration-time curve), C_{max} and t_{max}.

Although micronization represents a simple and attractive technology to enhance dissolution rate by increasing surface area, it has often limited success. For some drugs, in particular for the hydrophobic ones, micronization and other dry particle size reduction techniques can determine aggregation and consequent decrease in drug effective surface area in contact with dissolution medium. Wet milling in presence of a stabilizer has been proposed to overcome aggregation.

A wetting agent, such as polysorbate 80, could be also added to the formulation of a hydrophobic drug to increase dissolution thanks to the improvement of particle wetting and solvent penetration into the particles, and consequent reduction of particle aggregation and maintenance of a large effective surface area. It must however be underlined that an increase in effective surface area does not determine a corresponding increase in absorption rate, if dissolution is not the rate limiting step. Furthermore, poorly soluble drugs even though micronized, tend to be eliminated from the gastro-intestinal tract before absorption due to an insufficient solubility in the gastro-intestinal fluids.

8.3.2 Nanonization

Nanonization of a drug substance is the logical subsequent step to micronization to achieve surface area increase and dissolution rate enhancement. Nanonization means the reduction of drug particle size to the sub-micrometer range and could be obtained by two different approaches, "top down" and "bottom up". The "top down" approach basically relies on mechanical attrition to reduce large crystalline particles into nanoparticles [28]. Examples of the "top down" approach include Elan's NanoCrystal® wet-milling technology [35] and SkyePharma's Dissocubes® high-pressure homogenization technology [28, 38]. The "bottom up" approach is based on controlled precipitation/crystallization [49]. The process is based on the dissolution of the drug in a proper solvent and its controlled precipitation to form nanoparticles, through addition of an anti-solvent (usually water). This technology is available from DowPharma (Midland, MI, USA) and BASF Pharma Solutions (Florham Park, NJ, USA). NANOEDGE® technology (Baxter) represents a hybrid approach that employs both "bottom up" and "top down" methods through microprecipitation and homogenization [28]. The resulting nanosuspensions are usually stabilized by the addition of surfactants or polymers to avoid agglomeration. These technologies are usually taken into account to increase dissolution rates and to enhance bioavailability of poorly soluble drugs. The success of the nanonization has been confirmed by several commercial available products containing nanosized drugs (Table 8.1).

The underlying basis for dissolution-limited bioavailability and its improvement by nanonization is illustrated in Fig. 8.2 [13, 51].

Nanonization increases saturation solubility, according to the Freundlich–Ostwald equation (8.7):

$$C_S = C_\infty \exp\left(\frac{2\lambda M}{r\rho RT}\right) \tag{8.7}$$

where

C_s saturation solubility of the nanonized drug
C_∞ saturation solubility of an infinitely large drug crystal
λ crystal/medium interfacial tension
M drug molecular weight
R particle radius
ρ particle density
R gas constant
T absolute temperature

Table 8.1 Marketed products containing drug nanocrystals [51], copyright Springer, reproduced with permission

Trade name	INN name	FDA approval	Company	Adminstration route	Indications	Therapeutic benefit
Casamet®	Nabilone	2005	Lilly	Oral	Antiemetic	High bioavailability
Rapamune®	Sirolimus	2000	Pfizer (Wyeth)	Oral	Innumosuppresant	High patient compliance, tablet formulation instead of solution
Emend®	Aprepitant	2003	Merck	Oral	Antiemetic	High bioavailability, no food effects
Tricor® Lyphantyl®	Fenofibrate	2004	Fournier Pharma, Abbott Laboratories	Oral	Hypocolesterolemic	No food effects
Triglide®	Fenofibrate	2005	Sciele, Shionogi Pharma Inc.	Oral	Hypocolesterolemic	No food effects
Megace® ES	Megestrole acetate	2005	PAR Pharmaceuticals	Oral	Appetite stimulant	No food effects, high patient compliance
Invega® Sustenna® Xeplion®	Paliperidone palmitate	2009	Janssen	Parenteral, Intramuscular	Antidepressant	High bioavailability

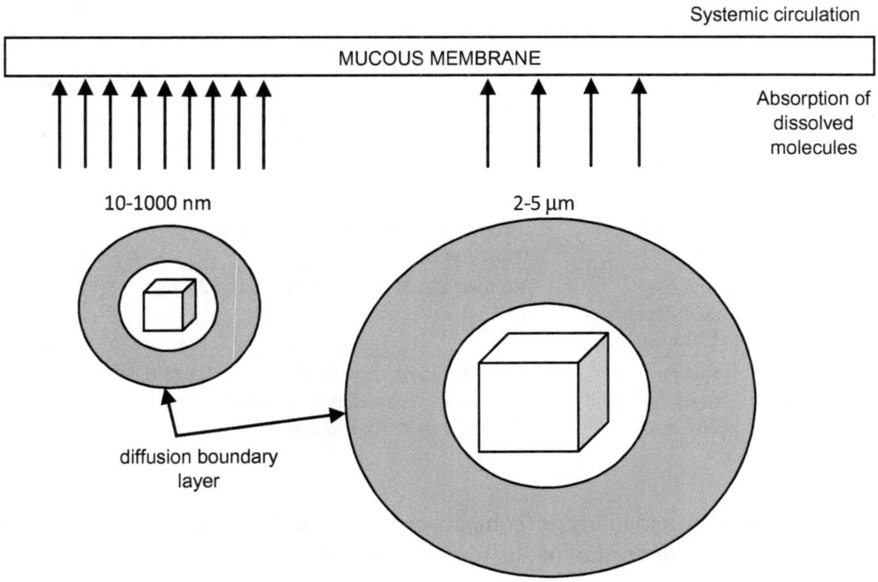

Fig. 8.2 Thickness of boundary layers of nanocrystals (10–1000 nm) and micronized particles (2–5 μm) [51], copyright Springer, reproduced with permission

Equation (8.7) shows that drug solubility of a drug at given conditions depends on particle size and molecular weight. Thus, during dissolution, Ostwald ripening may occur: large particles can grow at expense of dissolving small particles.

According to Eq. (8.7), a 10–15% increase in solubility could be expected for a drug having M = 500, $\rho = 1$ g/ml, particle size of 100 nm and crystal-intestinal fluid interfacial tension $\lambda = 15\text{–}20$ mN m^{-1} in comparison to crystalline drug [29]. A more pronounced increase in solubility was experimentally determined by Muller and Peters [39]: an insoluble antimicrobial drug increased 50% in solubility when the particle size was reduced from 2.4 μm to both 800 or 300 nm. Proportionally, dissolution rate increased and a significantly higher bioavailability was observed. Moreover, the presence of surfactants as stabilizers of nanosuspensions, further enhanced dissolution rate in comparison with micronized suspensions, thanks to an increase in wetted surface.

Drug nanocrystals are generally reported as safe and well tolerated systems intended for many administration routes, such as oral and parenteral ones as well as for ophthalmic and pulmonary administration [36]. In the case of poorly soluble drugs, nanosuspension safety via oral route is sometimes better than that of coarse materials due to an uniform distribution in gastrointestinal fluids and lack of high and prolonged local drug concentration [33]. For the same reason, local tolerability is increased in case of mucosal administration. Moreover, changes of pharmacokinetic parameters, including increased C_{max}, reduced t_{max}, enhanced AUC and

Table 8.2 Changes of pharmacokinetic properties of oral drug nanocrystal formulations compared with conventional formulations [51], copyright Springer, reproduced with permission

	Dosage form		Reference
Fenofibrate	Aqueous nanosuspension (194–356 nm)	1.8–12.5 fold increase in C_{max}; 1.7–17 fold increase in bioavailability; 1.3–2.3 fold reduction in t_{max}	[31]
Fenofibrate	Aqueous nanosuspension (340 nm)	1.67 fold increase in C_{max}; 1.3 fold increase in bioavailability; 4.9 fold reduction in t_{max}	[17]
Aprepitant	Aqueous nanosuspension (120 nm)	No food effect at a dose of 2, 80, 125 mg/kg	[57]
Itraconazole	Aqueous nanosuspension (267 nm)	1.2–1.8 fold increase in C_{max}; 1.2–1.8 fold increase in bioavailability; fasted/fed ratio of AUC markedly reduced	[37]

reduced fasted/fed variability in comparison to conventional formulations, occur after nanocrystals administration, as reported in Table 8.2.

In particular, when drugs are administered as nanocrystals, a high drug concentration gradient takes place between GIT and blood vessel, markedly improving absorption and resulting in a high bioavailability. This behavior is conceivably determined by the increased saturation solubility and dissolution rate of drug nanocrystals in digestive juice. One classic example is danazol, a poorly soluble gonadotropin inhibitor. The absolute bioavailability of marketed danazol conventional microsuspensions (200 mg, 10 μm) was only 5.2% in beagle dogs. When administered as an aqueous nanosuspension (200 mg, 169 nm), an absolute bioavailability of 82.3% could be achieved, with a 15 fold t_{max} reduction and C_{max} increase [34].

8.3.3 *Crystal Forms and Drug Solubility*

Since the middle of the 18th century it has been known that many substances can be obtained in more than one crystalline form. This property is referred as polymorphism and each crystalline form is known as polymorph. Crystal form is crucial since the structure of a given compound can have a deep impact on its solid state properties. For instance, it has been demonstrated that various polymorphs could exhibit different solubility and dissolution rates that sometimes lead to non-equivalent bioavailability of the different forms [50]. Another variation in crystalline form is pseudopolymorphism, that refers to a crystalline form in which solvent molecules represent a part of the structure. If the solvent is water, the pseudopolymorph is called hydrate. The entrapment of solvent molecules within the crystal lattice occurs in an exact molar ratio with the crystallizing substance.

Since different lattice energies (and entropies) are associated with different polymorphic forms, they determine differences in physical properties but also exhibit different solubility and dissolution rate.

A solid having higher lattice free energy is a less stable polymorph (metastable/ unstable) and tends to dissolve faster since the release of a higher amount of stored lattice free energy increases solubility and hence represents the driving force for dissolution. As a consequence, a supersaturated solution not thermodynamically stable is obtained, that moves back to true solubility causing the excess solute to precipitate. In contrast, a stable polymorphic form is characterized by a slower dissolution rate. As for hydrophobic drugs, having a very limited aqueous solubility, the use of a metastable form (enough stable to be considered in a drug product) represents an option to improve dissolution and bioavailability. However, the risk associated with the use of a metastable form is its possible conversion to the stable form during product life (processing and storage), with consequent changes in properties [6].

The use of metastable polymorphs with improved dissolution properties improves in vivo performance through increasing oral bioavailability. In fact, it is believed that the equilibrium solubility is not the important factor for the enhancement of oral absorption.

Pseudopolymorphs, in particular hydrates are often characterized by different properties in comparison with the anhydrous form; in particular they can have faster or more frequently slower dissolution rate than the anhydrous form. The hydrate form usually reaches the true equilibrium solubility whereas the anhydrous one initially forms a supersaturated solution.

The solubility of crystalline materials increases when particle size is reduced to submicron levels: this is described in Gibbs-Thomson Eq. (8.8) [16]:

$$\ln \frac{c(r)}{c*} = \frac{2M\gamma}{\upsilon RT\rho r} \tag{8.8}$$

where

- $c(r)$ solubility of particles of radius r
- $c*$ normal equilibrium solubility of the substance
- R gas constant
- T absolute temperature
- ρ density of the solid
- M molar mass
- γ interfacial tension
- ϑ number of separated species in the solution phase formed from one mole of solute; it is 1 for non electrolytes
- r particle radius

This equation was firstly proposed to describe the formation of liquid droplets from the vapor phase and considers that species having smaller curvature radius and consequently smaller size would be expected to possess higher vapor pressures due an excess of surface free energy that characterize small drops possess.

The thermodynamic similarity between vapor pressure of liquids and solution pressure (interfacial tension) of solution-dispersed solids suggests an analogous behavior of small solid particles in solution. Equation (8.9) considers curved particles although this assumption is not realistic due to the facetted nature of most crystalline particles. Moreover the interfacial tension varies both with solvent and with crystal surface chemistry and hence it relates to crystal habit. Hammond et al. [15] evaluated the effect of morphological changes on the solubility of small particles of different aspirin crystals: the interfacial tension γ was calculated by using the following equation [16]:

$$\gamma = \frac{zE_{att}d}{2V_{cell}N_A} \tag{8.9}$$

where

Z numbers of molecules in the unit cell of volume V_{cell}
E_{att} attachment energy
d interplanar spacing
V_{cell} unit cell volume
N_A Avogadro's number

In relation to crystal morphology of aspirin, the higher the surface tension the higher the solubility enhancement; this is markedly influenced by morphology rather than by particle size. Solubility of aspirin in water (in which it is poorly soluble) was slightly increased for larger particles (2 μm diameter) whereas reducing particle size from 0.2 to 0.02 μm, the enhancement effect was more significant compared to that observed in a good solvent, that can efficiently wet particle surface [15].

Carbamazepine is an anticonvulsant antiepileptic drug having at least four polymorphic forms and a dihydrate one (pseudopolymorph) [30]. In Fig. (8.3a, b) the dissolution and plasma profiles of carbamazepine are reported. Form III (mean diameter 19.5 μm) exhibited a rapid increase in drug concentration and a maximum value at beginning of the dissolution process; then, the concentration gradually decreased due to precipitation of dihydrate crystals (Fig. 8.3a). In contrast, Form I (mean diameter 13.9 μm) maintained high constant concentration over 4 h. These results suggest that the conversion was faster from Form III to dihydrate, than that from Form I. Dihydrate (mean diameter 13.4 μm) presented a lower drug concentration at the initial stage. These differences in dissolution profiles caused marked differences in bioavailability of the two polymorphs and the dihydrate form in comparison with drug solution when a high dosage (200 mg) was administered in dogs (Fig. 8.3b). Form I showed the highest plasma concentration levels, whereas Form III and the dihydrate were similar and Form I showed highest AUC and relative bioavailability (rBA %), with respect to drug solution in PEG equal to rBA of 68.7%, Form III presents intermediate AUC and rBA equal to 47.8%, while the dihydrate is characterized by the lowest AUC and 33.1% values.

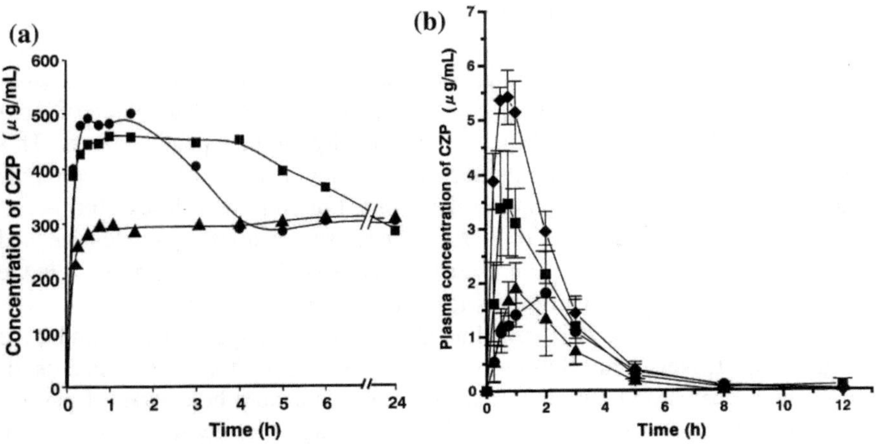

Fig. 8.3 **a** Dissolution and **b** plasma profiles of carbamazepine Form I (square), Form III (circle) and dihydrate (triangle) [30]; copyright Elsevier, reproduced with permission

8.4 Biopharmaceutical Concerns

Dissolution and particle size have become key points in pharmaceutical technology for the assessment of drug delivery system performance, due to their impact on biopharmaceutical behavior, since drug absorption requires the drug in solution at the absorption site [1, 12]. When the dissolution is the limiting step for absorption, all the factors that affect dissolution are potentially determinant for drug bioavailability.

In 1993, a biopharmaceutics drug classification scheme was proposed by Oh et al. [46] to correlate in vitro drug dissolution with in vivo bioavailability thus recognizing that dissolution and gastrointestinal permeability are fundamental parameters controlling rate and extent of drug absorption. In particular the scheme was based on the consideration that the dose fraction absorbed from suspensions is a function of four dimensionless parameters: absorption number (*An*), dissolution number (*Dn*), dose number (*Do*) and initial saturation (*Is*). *An* is the ratio of radial absorption rate to axial convection rate; *Dn* is the ratio of residence time in intestine to the dissolution time. *Dn* depends on drug properties such as solubility, diffusivity, density and, in particular, it is inversely proportional to the square of the initial particle size. Thus, a way to increase *Dn* is to decrease particle size [47]. *Do* is the ratio of dose concentration to solubility and *Is* is the ratio of drug luminal concentration at the beginning of the intestine to drug solubility.

When a drug is administered via oral route as a suspension, particle size decreases down the tube so that the dissolution rate from particles increases. Drugs with low *Do* and low *Dn* can be completely absorbed by reducing their particle size, while the absorption of drugs with high *Do* and low *Dn* is solubility limited and requires an enhancement of solubility, in addition to micronization, to increase the fraction absorbed.

Good knowledge of bioavailability of a drug is also important for regulatory requirements, in view of assuring constant quality and reproducibility of drug products and, consequently, adequate predictability of therapeutic effects. This concern has led during the last decades to the identification of all the possible critical parameters of the different dosage forms. For solid oral dosage forms and liquids containing undissolved drug substances (suspensions), particle size and polymorphism are considered parameters to be necessarily taken into account in the definition and control of quality specifications.

As for particle size the ICH guideline Q6A on product specifications under the paragraph 3.3.1, New Drug Substances, states that "testing for particle size distribution should be carried out, and acceptance criteria should be decided if particle size affects dissolution rate, bioavailability or stability of a new drug substance in solid or suspension drug product", while for polymorphism (polymorphs, pseudopolymorphs and amorphous forms) the same guideline reports that "Differences in these forms could, in some cases, affect the quality or performance of the new drug products. In cases where differences exist which have been shown to affect drug product performance, bioavailability or stability, then the appropriate solid state should be specified."

Indications relevant to the possible role of particle size and polymorphism on drug product properties, in particular on bioavailability, are detailed in decision tree # 3 and decision tree # 4 of the same guideline (Figs. 8.4 and 8.5), respectively.

In ICH Guideline Q6A the role of particle size and polymorphism in processability, content uniformity and stability of a drug product is clearly recognized. The first question, however, regards the possible effect of particle size and polymorphism on solubility, dissolution and bioavailability parameters strictly related to each other. In case of possible effects of particle size and polymorphism on product performance, definitions of specifications will be required (ICH GL).

The observations reported in early studies [8] on the relevance of particle size on dissolution rate, that is particularly evident when poorly soluble drugs are concerned, have been more recently revised and reconsidered, in a regulatory perspective, in the frame of the Biopharmaceutical Classification System (BCS) [2].

BCS was developed to avoid in vivo bioequivalence studies for drug products characterized by rapid in vitro dissolution, being based on drugs that are very soluble and have good permeability (class I). In these cases, in vivo studies can be substituted by in vitro studies based on dissolution tests: it is accepted that differences observed in the rate and extent of in vivo absorption may be due to a difference in the in vivo dissolution rates. BCS based biowaiver approach can be extended also to class III drugs (with good solubility and poor permeability). Drugs with low solubility and good permeation (class II) and with low solubility and poor permeation (class IV) pose much more concerns from a regulatory point of view, since small changes in physical-chemical properties can significantly affect dissolution and absorption and require more extensive in vivo characterization.

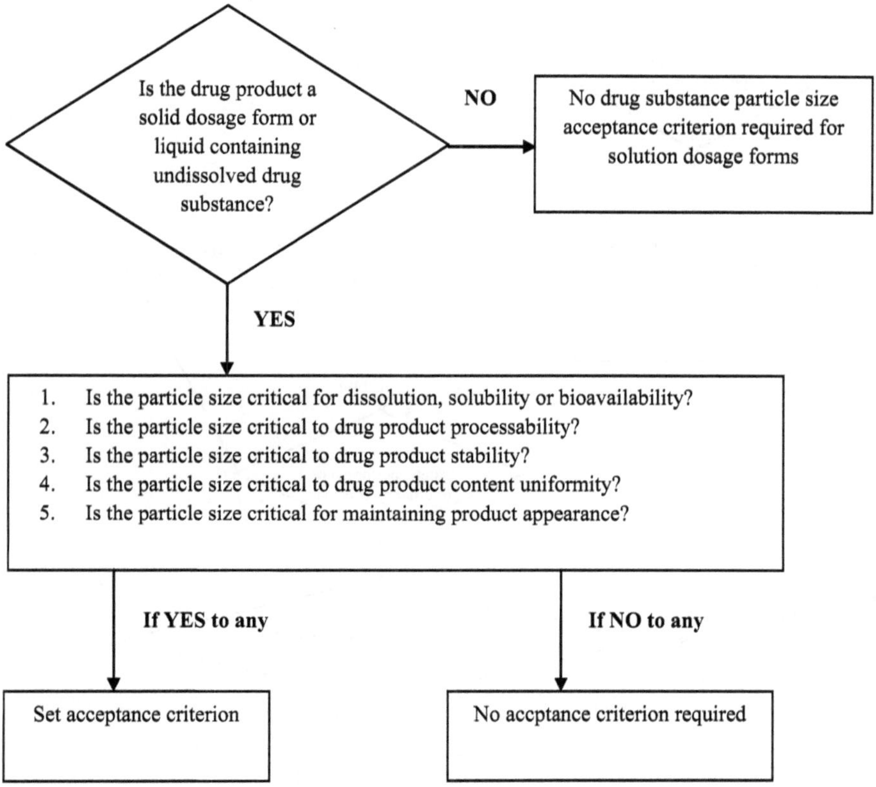

Fig. 8.4 Decision tree on particle size reported in Q6A ICH guideline relevant to product specifications, paragraph 3.3.1, New Drug Substances

The biowaiver is a significant step started by FDA in reducing the number of human clinical studies and burdens on the pharmaceutical companies during the approval process, besides the rationales are based on purely scientific principles. To develop dissolution tests that better predict the in vivo performance of drug products a biologically relevant dissolution medium is often suggested by regulatory agencies.

To this aim, Dressman et al. [11] have investigated the physiological condition of the human gastrointestinal tract, taking into account the composition, the volume, the flow rate and the mixing pattern of the gastrointestinal fluids, to set up appropriate dissolution media. Given these the composition of dissolution media, the hydrodynamics and the duration of the tests are critical parameters of a predictive dissolution test. In particular the fed and the fasted states and consequently the dissolution media that simulate those conditions are crucial to consider the influence of food on drug absorption.

The increasing attention for this topic depends on the fact that it is estimated that about 40% to 60% of the new drug entities exhibit poor solubility, with related uncertainty on bioavailability and clinical use [32].

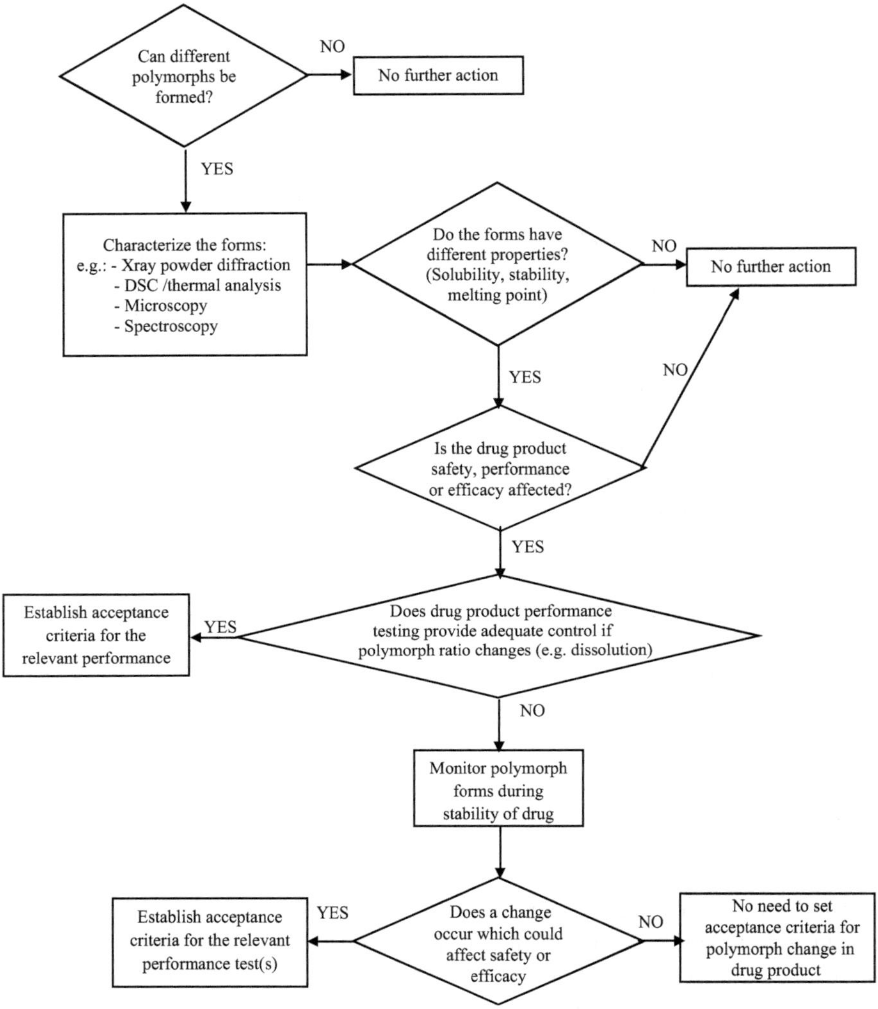

Fig. 8.5 Decision tree on polymorphism reported in Q6A ICH guideline relevant to product specifications, paragraph 3.3.1, new drug substances

BCS represents therefore a reference in formulation development, since poorly soluble drugs are the major challenge for the design of drug delivery systems. As evidenced by Kawabata et al. [27], the physical-chemical properties of a drug and in particular polymorphism and amorphous forms, and particle size among them, are key factors for reproducibility, bioavailability and therapeutic effect of a given formulation (Fig. 8.6). In particular, particle size reduction has been therefore suggested as technological strategy for class II and IV drug formulation, both as micronization and as reduction to nanocrystals.

Class I **High solubility/high permeability** 1) IR dosage forms	**Class II** **Low solubility/high permeability** 1) Crystal modifications - metastable polymorphs - salt formation Cocrystal formation 2) IR solid oral dosage forms with surfactants 3) Particle size reduction - Micronization - Nanocrystals 4) Amorphization 5) Cyclodextrin complexation 6) Lipid formulations - self emulsification systems - liquid filled capsules 7) pH modification
Class III **High solubility/low permeability** 1) IR solid dosage forms with absorption enhancers 2) IR solid dosage forms	**Class IV** **Low solubility/low permeability** 1) Combinations of approaches for BCS class II and absorption enhancers 2) Same approaches as BCS class II

PERMEABILITY

SOLUBILITY

Fig. 8.6 Formulation design of poorly water soluble drugs based on BCS: basic approaches and practical applications (modified from [51])

The regulatory concerns and the recognition of particle size relevance for bioavailability led in more recent years to the development of an intensive research about possible prediction of bioavailability in physiologically based models simulating gastrointestinal transit and absorption. In some studies samples containing drug of different drug particles, characterized by different dissolution profiles according to Noyes Whitney equation, have been evaluated through this modeling approach. Among the commercial softwares available for pharmacokinetic simulation are GastroPlus™, PK-Sim® and Stella® softwares.

8.5 Final Remarks

The interest in particle size and solid state characterization, in particular polymorphism, and on their influence on bioavailability is renewed by regulatory requirements. As for particle size, the introduction of the new nanonization technology to obtain nanocrystals is an approach well-adaptable to drugs having different chemical-physical properties in order to solve bioavailability criticism. Several kinds of nanosized active ingredients are nowadays commercially available and others are undergoing development studies. The employment of excipients to stabilize nanocrystals, and in particular the use of surfactants for nanocrystal surface modification should further enhance drug bioavailability and could achieve prolonged release and targeted (site-specific) drug delivery.

In addition, nanotoxicological investigations of drug nanocrystals should be extensively carried out to better understand the fate of nanocrystals at a cellular level and to find new potential for nanocrystal applications for innovative treatment approaches.

8.6 Definitions, Abbreviations and Symbols

Absorption number	Ratio of the mean residence time to the absorption time
Dissolution number	Ratio of mean residence time to mean dissolution time
Dose number	Mass divided by an uptake volume of 250 ml and the drug solubility
Fasted condition	Without eating food
Fed condition	After eating food
Food effect	Effect of the presence of food on drug fate in gastrointestinal tract
Initial saturation	Ratio of drug luminal concentration at the beginning of the intestine to drug solubility
Intrinsic dissolution	Dissolution rate of pure substances under the condition of constant surface area
Sink conditions	Conditions where C_t is negligible in comparison to C_s
AUC	Area under the blood plasma concentration-time curve
BCS	Biopharmaceutical Classification System
GIT	Gastrointestinal tract
GL	Guideline
ICH	International conference on harmonization
IR	Immediate release
A	Surface area
An	Absorption number
C_{max}	Maximum plasma concentration
C_s	Solubility of the substance
C_t	Concentration at time t
C_∞	Saturation solubility of an infinitely large drug crystal
$c(r)$	Solubility of particles of radius r
$c*$	Normal equilibrium solubility of the substance
D	Particle size (diameter of an equivalent sphere)
$D_{3,2}$	Area-weighted Sauter mean diameter
$D_{4,3}$	Volume-weighted mean size

Do	Dose number
Dn	Dissolution number
$\mathbb{D}$	Diffusion coefficient of a solute
d	Interplanar spacing
dm/dt	Dissolution rate
E_{att}	Attachment energy
G	Surface specific dissolution rate
Is	Initial saturation
M	Drug molecular weight,
m	Drug amount
N_A	Avogadro's number
R	Gas constant
r	Particle radius
T	Absolute temperature
t	Time
t_{max}	Time to reach maximum plasma concentration
y	Number of separated species in the solution phase formed from one mole of solute
v_x, v_y, v_z	Liquid stirring rates in the *x*, *y*, and *z* direction
Z	Numbers of molecules in the unit cell of volume V_{cell}
δ	Thickness of the diffusion boundary layer around a particle
γ	Interfacial tension
λ	Crystal/medium interfacial tension
ρ	Solid particle density

References

1. Abdou HM (1989) Dissolution, Bioavailability and Bioequivalence. Easton, Mack
2. Amidon GL, Lennernas H, Shah VP, Crison JR (1995) A theoretical basis for a biopharmaceutic drug classification: the correlation of in vitro drug product dissolution and in vivo bioavailability. Pharm Res 12:413–420
3. Anderberg EK, Nyström C, Bisrat M (1988) Physico-chemical aspects of drug release. VII. The effect of surfactant concentration and drug particle size on solubility and dissolution rate of Felodipine, a sparingly soluble drug. Int J Pharm 47:67–77
4. Atkinson RM, Bedford C, Child KJ, Tomich EG (1962) Effect of particle size on blood griseofulvin levels in man. Nature 193:588–589
5. Bisrat M, Nyström C (1988) Physicochemical aspects of drug release. VIII. The relation between particle size and surface specific dissolution rate in agitated suspensions. Int J Pharm 47:223–231
6. Blagden N, de Matas M, Gavan PT, York P (2007) Crystal engineering of active pharmaceutical ingredients to improve solubility and dissolution rates. Adv Drug Deliv Rev 59:617–630
7. Brunner E (1904) Reaktionsgeschwindigkeit in heterogenen Systemen. Z Phys Chem 43:56–102

8. Caramella C (1991) Utilizzazione dei parametri granulometrici nella progettazione di forme farmaceutiche. Boll Chim Farm 130:43–51
9. Carstensen JT (1980) Solid pharmaceutics: mechanical properties and rate phenomena. Academic Press Inc., New York
10. Dali MV, Carstensen JT (1996) Effect of change in shape factor of a single crystal on its dissolution behavior. Pharm Res 13:155–162
11. Dressman JB, Amidon GL, Reppas C, Shah V (2008) Dissolution testing as a prognostic tool for oral drug absorption: immediate release dosage forms. Pharm Res 15:11–22
12. Dressman JB, Reppas C (2000) In vitro–in vivo correlations for lipophilic, poorly water-soluble drugs. Eur J Pharm Sci 11:73–80
13. Gao L, Liu G, Ma J, Wang X, Zhou L, Li X (2012) Drug nanocrystals: in vivo performances. J Control Rel 160:418–430
14. Gibaldi M (1984) Biopharmaceutics and clinical pharmacokinetics, 3rd edn. Lea and Febiger, Philadelphia, pp 44–63
15. Hammond RB, Jeck S, Ma CY, Pencheva K, Roberts KJ, Auffret T (2009) An examination of binding motifs associated with inter-particle interactions between facetted nanocrystals of acetylsalicylic acid and ascorbic acid through the application of molecular grid-based search methods. J Pharm Sci 98:4589–4602
16. Hammond RB, Pencheva K, Roberts KJ, Auffret T (2007) Quantifying solubility enhancement due to particle size reduction and crystal habit modification: case study of acetyl salicylic acid. J Pharm Sci 96:1967–1973
17. Hanafy A, Spahn-Langguth H, Vergnault G, Grenier P, Tubic Grozdanis M, Lenhardt T, Langguth P (2007) Pharmacokinetic evaluation of oral feno! brate nanosuspensions and SLN in comparison to conventional suspensions of micronized drug. Adv Drug Deliv Rev 59:419–426
18. Higuchi T (1963) Mechanism of sustained-action medication. Theoretical analysis of rate of release of solid drugs dispersed in solid matrices. J Pharm Sci 52:1145–1149
19. Higuchi WI, Hiestand EN (1963) Dissolution rates of finely divided drug powders I: effect of a distribution of particle sizes in a diffusion-controlled process. J Pharm Sci 52:67–71
20. Higuchi WI, Rowe EL, Hiestand EN (1963) Dissolution rates of finely divided drug powders II: micronized methylprednisolone. J Pharm Sci 52:162–164
21. Hintz RJ, Johnson KC (1989) The effect of particle-size distribution on dissolution rate and oral absorption. Int J Pharm 51:9–17
22. Hixson AW, Crowell JH (1931) Dependence of reaction velocity upon surface and agitation. Ind Eng Chem 23:923–931
23. Jounela AJ, Pentikainen PJ, Sothmann A (1975) Effect of particle size on the bioavailability of digoxin. Eur J Clin Pharmacol 8:365–370
24. Junemann D, Dressman JB (2012) Analytical methods for dissolution testing of nanosized drugs. J Pharm Pharmacol 64:931–943
25. Kaneniwa N, Watari N (1974) Dissolution of slightly soluble drugs. I. Influence of particle size on dissolution behavior. Chem Pharm Bull 22:1699–1705
26. Kaneniwa N, Watari N, Iijima H (1978) Dissolution of slightly soluble drugs. V. Effect of particle size on gastrointestinal drug absorption and its relation to solubility. Chem Pharm Bull 26:2603–2614
27. Kawabata Y, Wada K, Nakatani M, Yamada S, Onoue S (2011) Formulation design for poorly water-soluble drugs based on biopharmaceutics classification system: basic approaches and practical applications. Int J Pharm 420:1–10
28. Keck CM, Muller RH (2006) Drug nanocrystals of poorly soluble drugs produced by high pressure homogenization. Eur J Pharm Biopharm 62:3–16
29. Kesisoglou F, Panmai S, Wu Y (2007) Nanosizing—oral formulation development and biopharmaceutical evaluation. Adv Drug Deliv Rev 59:631–644
30. Kobayashi Y, Ito S, Itai S, Yamamoto K (2000) Physicochemical properties and bioavailability of carbamazepine polymorphs and dihydrate. Int J Pharm 193:137–146

31. Li X, Gu L, Xu Y, Wang Y (2009) Preparation of fenofibrate nanosuspension and study of its pharmacokinetic behavior in rats. Drug Dev Ind Pharm 35:827–833
32. Lipinski C (2002) Poor aqueous solubility—an industry wide problem in drug discovery. Am Pharm Rev 5:82–85
33. Liversidge GG, Conzentino P (1995) Drug particle size reduction for decreasing gastric irritancy and enhancing absorption of naproxen in rats. Int J Pharm 125:309–313
34. Liversidge GG, Cundy KC (1995) Particle size reduction for improvement of oral bioavailability of hydrophobic drugs: I. Absolute oral bioavailability of nanocrystal- line danazol in beagle dogs. Int J Pharm 125:91–97
35. Merisko-Liversidge E, Liversidge GG, Cooper ER (2003) Nanosizing: a formulation approach for poorly-water-soluble compounds. Eur J Pharm Sci 18:113–120
36. Mosharraf M, Nystrom C (1995) The effect of particle size and shape on the surface specific dissolution rate of microsized practically insoluble drugs. Int J Pharm 122:35–47
37. Mou D, Chen H, Wan J, Xu H, Yang X (2011) Potent dried drug nanosuspensions for oral bioavailability enhancement of poorly soluble drugs with pH-dependent solubility. Int J Pharm 413:237–244
38. Muller RH, Jacobs C, Kayser O (2001) Nanosuspensions as particulate drug formulations in therapy. Rationale for development and what we can expect for the future. Adv Drug Deliv Rev 47:3–19
39. Muller RH, Peters K (1998) Nanosuspensions for the formulation of poorly soluble drugs I. Preparation by a size-reduction technique. Int J Pharm 160:229–237
40. Nernst W (1904) Theorie der Reaktionsgeschwindigkeit in heterogenen Systemen. Z Phys Chem 47:52–55
41. Neuvonen PJ (1979) Bioavailability of phenytoin: clinical pharmacokinetic and therapeutic implications. Clin Pharmacokinet 4:91–103
42. Niebergall PJ, Milosovich G, Goyan JE (1963) Dissolution rate studies. 2. Dissolution of particles under conditions of rapid agitation. J Pharm Sci 52:236–241
43. Ning X, Sun J, Han X, Wu Y, Yan Z, Han J, He Z (2011) Strategies to improve dissolution and oral absorption of glimepiride tablets: solid dispersion versus micronization techniques. Drug Dev Ind Pharm 37:727–736
44. Noyes AA, Whitney WR (1897) The rate of solution of solid substances in their own solutions. J Am Chem Soc 19:930–934
45. Nyström C, Mazur J, Barnett MI, Glazer M (1985) Dissolution rate measurements of sparingly soluble compounds with the Coulter counter model TAII. J Pharm Pharmacol 37:217–221
46. Oh DM, Curl RL, Amidon GL (1993) Estimating the fraction dose absorbed from suspensions of poorly soluble compounds in humans: a mathematical model. Pharm Res 10:264–270
47. Oh DM, Curl RL, Yong CS, Amidon GL (1995) Effect of micronization on the extent of drug absorption from suspensions in humans. Arch Pharm Res 18:427–433
48. Pedersen PV, Brown KF (1976) Theoretical isotropic dissolution of non spherical particles. J Pharm Sci 65:1437–1442
49. Rabinow BE (2004) Nanosuspensions in drug delivery. Nat Rev Drug Discov 3:785–796
50. Saifee M, Inamdar N, Dhamecha DL, Rathi AA (2009) Drug polymorphism: a review. Int J Health Res 2:292–306
51. Sandri G, Bonferoni MC, Ferrari F, Rossi S, Caramella C (2014) The role of particle size in drug release and absorption. In: Merkus HG, Meesters GMH (eds) Particulate products—tailoring properties for optimal performance, vol 19. Particle technology series, chap 11. Springer, Switzerland, pp 323–341
52. Shaw TRD, Carless JE (1974) The effect of particle size on the absorption of digoxin. Eur J Clin Pharmacol 7:269–273
53. Sheng JJ, Sirois PJ, Dressman JB, Amidon GL (2008) Particle diffusional layer thickness in a USP dissolution apparatus II: a combined function of particle size and paddle speed. J Pharm Sci 97:4815–4829

54. Simoes S, Sousa A, Figueiredo M (1996) Dissolution rate studies of pharmaceutical multisized powders—a practical approach using the Coulter method. Int J Pharm 127 (283):291
55. Wang J, Flanagan DR (2009) Fundamentals of dissolution. In: Qiu Y, Chen Y, Zhang G, Liu L, Porter W (eds) Developing solid oral dosage forms, pharmaceutical theory and practice, chap 13. Elsevier, USA, pp 309–318
56. Watari N, Hanano M, Kaneniwa N (1980) Dissolution of slightly soluble drugs. VI. Effect of particle size of sulfadimethoxine on the oral bioavailability. Chem Pharm Bull 28:2221–2225
57. Wu Y, Loper A, Landis E, Hettrick L, Novak L, Lynn K, Chen C, Thompson K, Higgins R, Batra U, Shelukar S, Kwei G, Storey D (2004) The role of biopharmaceutics in the development of a clinical nanoparticle formulation of MK-0869: a beagle dog model predicts improved bioavailability and diminished food effect on absorption in human. Int J Pharm 285:135–146

Giuseppina Sandri obtained her Ph.D. in Chemistry and Pharmaceutical Technology in 2003. She became assistant professor in 2008 and Associate Professor in 2015 at Drug Science Department (Faculty of Pharmacy) at University of Pavia.

She is Member of board of teachers of Ph.D. on "Chemical and Pharmaceutical Sciences and Industrial Innovation" Univ. Pavia, Coordinator of Master Course on "cGMP complinace and Validation in Pharmaceutical Industry and Member of board of teachers of Master Course on "Preformulation, Pharmaceutical Development and Control of Medicines" Dept. Drug Sciences, Univ Pavia, Erasmus Delegate for Pharmacy. Reviewer for "National Research Foundation" Sud Africa, member of reviewers for MIUR, member of Ph.D. committee at University of Granada (Spain), member of Editorial Advisory Board of Journal of Pharmaceutical Sciences, member of Editorial Panel of EC Ophthalmology.

Current research concerns the development of semisolid and solid formulations (sponge-like dressings, scaffolds, particulate systems) for the treatment of mucosal, corneal and skin lesions. Attention focuses on the choice and the characterization of bioactive polymers to be combined with hemoderivatives and eventually with antimicrobials to promote wound healing. In this area of interest, medical devices and antimicrobial based formulations based on electrospun nanofibers are currently in study for heath regeneration.

She received awards and was funded for her research activities.

She takes part in several joint research projects with foreign universities and pharmaceutical companies. She serves as reviewer for several scientific journals. She is author and co-author of 82 contributions published on scientific journals, 12 book chapters, 3 patents [Hindex: 24; number of citations: 1670 (Scopus)].

Maria Cristina Bonferoni graduated in 1984 in Pharmaceutical Chemistry and Technology (Ph. D. in 1991). She is associate professor since 2001, at the Department of Drug Sciences, University of Pavia. She teaches Applied Pharmaceutical Chemistry and Design and analysis of experiments in Pharmacy and Ph.D. courses.

The research interests concern pharmaceutical development of conventional and controlled release formulations. Early work involved the study of hydrophilic polymers for oral matrix tablets. Swelling/erosion and diffusion/dissolution processes occurring in hydrophilic matrices have been studied and related to polymer rheological behaviour (viscosity and viscoelasticity) under different conditions of pH and ionic strength. Other research interest involved mucoadhesion mechanisms, development of mucoadhesive dosage forms and of methods to measure mucoadhesion properties.

Complexes between anionic polymers and basic drugs have been studied and employed in oral prolonged release formulations. Drug-polymer and polymer-polymer ionic interactions have been assessed to obtain self assembling microparticulate systems intended for mucosal delivery and for ophthalmic administration of decongestant drugs and antibiotics. Colloidal systems (polymeric nanoparticles and micelles, SLN) have been designed and characterized for the delivery of poorly soluble drugs (cyclosporine A, clarithromycin, clotrimazole, Ag sulfadiazine) and peptidic drugs (insulin).

Current topics involve the development of therapeutic systems aimed to wound healing, based on bioactive polymers such as chitosan and its derivatives or glucosaminoglycans (chondroitin sulphate and hyaluronic acid) loaded with antioxidant agents, antiinfective drugs and platelet hemoderivatives rich in growth factors. Attention is also paid to methods for the biopharmaceutical characterization of these systems by means of cell cultures. Amphipilic chitosans for the stabilization of nanoemulsions are considered to develop chitosan coated nanoparticles loaded with antioxidants and polyphenols.

Dr. Bonferoni is author of 148 articles, 12 patents, 12 book chapters.

Silvia Rossi has received her degree in Pharmaceutical Chemistry and Technology at the School of Pharmacy, University of Pavia in 1988, and her Ph.D. in 1995. She became assistant professor in 1998 and associate professor in 2006 at the above mentioned School of Pharmacy. She is at present active both in undergraduate and graduate education at the Department of Drug Sciences of Pavia University.

The scientific expertise of Silvia Rossi is in the field of dosage form technology and biopharmaceutical implications and includes: tablet disintegrants, characterization of hydrophilic polymers (swelling and water uptake phenomena), characterization of semisolids by means of viscosity and viscoelastic measurements, mucoadhesion evaluation and study of the mechanisms, study of absorption enhancement mechanisms.

Current research projects are in the field of the design and evaluation of drug delivery systems and include buccal and vaginal semisolid (gels, in situ gelling systems) and solid (films, matrices) formulations, microparticulate systems for ophthalmic delivery, sponge-like dressings and particulate systems for cutaneous application, nanofibers for wound healing. The research work resulted in 140 research papers on scientific journals, 16 book chapters, 11 patents and more than 280 contributions to scientific meetings.

Carla M. Caramella, full professor (1986–2015) of Pharmaceutical Technology and Biopharmacy at the University of Pavia, is presently Contract professor at the same University, Dept. of Drug Sciences. She has recently been nominated Professor Emeritus of the University of Pavia. She holds teaching responsibilities at both undergraduate, graduate and post-graduate levels and is the Director of the Master course in Pharmaceutical Technologies and Regulatory Affairs and the Italian coordinator of the European Master course in Nanomedicine for Drug delivery. She has served as Dean of the Faculty of Pharmacy and as Director of the Interuniversity consortium TEFARCO INNOVA and was nominated AAPS Fellow in 2001.

Her scientific and technical expertise includes: control of APIs, powder solid state characterization, excipient functionality studies, tablet disintegration, dissolution, in vitro diffusion studies, in vivo-in vitro correlation. She was the recipient of the "Ralph Shangraw Memorial Prize", IPEC America Foundation, 2008, for innovative research in the fields of pharmaceutical excipients. She has pioneered the use of rheology in understanding mucoadhesion phenomenon. More recently she has fostered the use of cell cultures as bioassays in formulation development of advanced dressings for wound healing. She has published 223 published papers, including original papers, reviews, published abstracts and miscellaneous publications, 16 book chapters, one text-book on Pharmaceutical Technology and is co-inventor of 15 patents (H index: 39 (Google Scholar), citations over 4000).

She has supervised the research work of about 40 people including Ph.D. students, post-doc and grant holders in collaboration with about 20 european and extraeuropean Universities and has coordinated a number of research projects funded by different Institutions and Pharma companies.

She is also member of Regulatory Bodies appointed by the Italian Agency for Medicines (AIFA) and Ministry of Public Health. She is Expert EMA and EDQM for the Quality assessment of medicinals.

Franca Ferrari has received her Pharm. Sci. degree in Pharmaceutical Chemistry and Technology at the University of Pavia in 1981 and Ph.D. in Pharmaceutical Analysis and Technology, at the same University, in 1984. She became assistant professor in 1984 at the School of Pharmacy, University of Pavia, where she subsequently served as associate professor (1998–2015). From 2016 she is full professor at the same Faculty, where she is presently active both in undergraduate and graduate education.

From 2002 to 2010 and from 2014 to 2016 she has been the coordinator of the II level Masters Course in Preformulation, Pharmaceutical Development and Control of Medicines.

The scientific expertise of Franca Ferrari is in the field of dosage form technology and biopharmaceutical implications and includes: study of tablet disintegrants and of disintegration process, characterization of hydrophilic polymers(swelling and water uptake phenomena), characterization of semisolids by means of viscosity and viscoelastic measurements, evaluation and study of mucoadhesion mechanisms, evaluation and study of absorption enhancement mechanisms.

Prof. Ferrari's current research projects are in the field of the design and evaluation of drug delivery systems and include buccal and vaginal semisolid (gels, in situ gelling systems) and solid (films, matrices) formulations, micro- and nanoparticulate systems and sponge-like dressings for cutaneous and mucosal application.

Her research work resulted in 153 referred contributions published on scientific journals (127 papers and 26 abstracts), 18 book chapters, 7 patents and more than 300 contributions to scientific meetings.

Chapter 9
Amorphous APIs: Improved Release, Preparation, Characterization

Sheila Khodadadi and Gabriel M. H. Meesters

Abstract Producing Amorphous Active Pharmaceutical Ingredients offers an enhanced drug release that is caused by the increase in its dissolution rate. This improvement enables higher bioavailability and bioactivity of such solid APIs. Possibilities to control the drug release and its bioactivity offer new opportunities to design more effective and stable medications. This chapter is a general introduction to Amorphous APIs and most commonly discussed production and characterization methods.

9.1 Introduction

A large number of newly discovered Active Pharmaceutical Ingredients (API) exhibit poor water solubility caused by their lipophilic properties and/or crystalline structure. For an API to become bioactive, it first needs to become bioavailable mainly by entering the cell membrane through molecular diffusion after dissolution. Therefore increasing the solubility and dissolution rate of solid particles will enhance bioavailability and, in relation, its bioactivity [20]. Transforming APIs crystalline structure into its functional amorphous counterpart, increasing the solvent exposure surface and accessibility of hydrophilic groups will promise higher solubility, improvement in dissolution rate and thus will increase bioavailability through faster API release [7, 13, 28]. Amorphous APIs lack the short and long range molecular order of their crystalline counterparts, which results in higher molecular mobility enabling faster drug release. Before going further into details of technologies to produce amorphous APIs, one needs to understand what are amorphous materials and their relevant characteristics.

S. Khodadadi · G. M. H. Meesters (✉)
Delft University of Technology, Delft, Netherlands
e-mail: G.M.H.Meesters@tudelft.nl

S. Khodadadi
e-mail: s.khodadadi@tudelft.nl

H. G. Merkus et al. (eds.), *Particles and Nanoparticles in Pharmaceutical Products*,
AAPS Advances in the Pharmaceutical Sciences Series 29,
https://doi.org/10.1007/978-3-319-94174-5_9

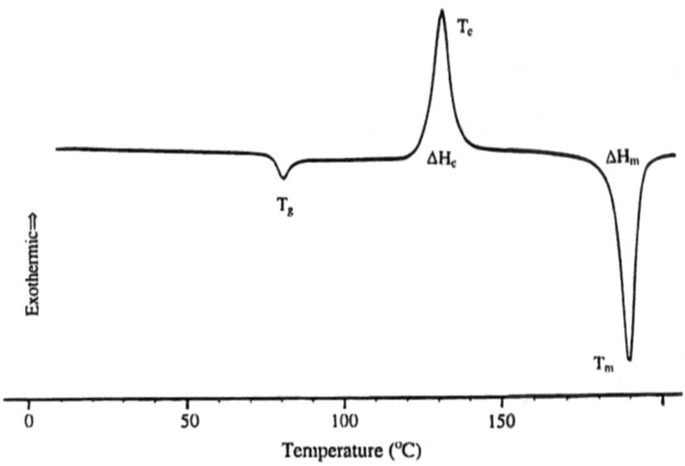

Fig. 9.1 DSC scan of amorphous sucrose. The figure illustrates the appearance of glass transition (T_g), the crystallization temperature exotherm (T_c) and the melting temperature-endotherm (T_m). [15]. Copyright Elsevier. Figure reprinted with permission

One of the first physical characteristics of amorphous materials is the glass transition temperature (T_g), the temperature at which the transition from liquid to glassy state occurs. Parameters such as molecular size, shape, chemical structure, strength of intermolecular interactions and molecular packing influence the glass transition temperature. The glass transition is not a first order phase transition similar to freezing with a distinct transitional temperature but rather a discontinuity in its thermodynamic, physical and (molecular) dynamics properties. T_g differs by the path (rate) of cooling/heating and thermal history of the compound. Therefore usually a range of T_g is introduced [3, 4, 16, 18, 22, 31, 32]. T_g commonly is measured by Differential Scanning Calorimetry (DSC), a thermoanalytical method measuring the change in heat capacity as temperature is changed. In these measurements, an endothermal peak means energy uptake by the substance and an exothermal peak means energy release. Thus, melting appears as an endothermal peak at the melting temperature (T_m) and crystallization of an amorphous material as an exothermal peak at the crystallization temperature (T_c) (see Fig. 9.1). The area in a peak describes the amount of energy change.

Pharmaceutical products are complex systems, containing many ingredients with different physical properties. In such complex systems, the glass transition temperature might be influenced not only by additives but also by the preparation method and thermal history or aging [4, 15, 16, 22, 31, 32]. The disordered structure and higher molecular mobility of Amorphous APIs increases their solubility and dissolution rate which results in their faster bioavailabity (quicker release). However at the same time it can also enable undesired recrystallization, exposure of active sites, quicker water and oxygen absorption upon exposure to humidity and air, resulting in a less chemically and physically stable component.

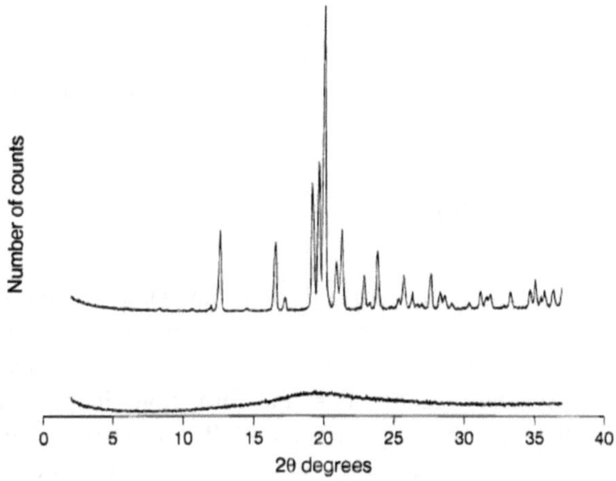

Fig. 9.2 PXRD pattern of crystalline (top) and amorphous (bottom) lactose [15]. Copyright Elsevier. Figure reprinted with permission

Enthalpy relaxation in glassy materials, caused by aging and annealing, also increases molecular mobility and decreases physical and chemical reactivity [2, 14, 15, 31].

In addition to DSC, Powder X-ray Diffraction (PXRD) measurements are also often used to confirm and quantify the transition from crystalline or liquid to amorphous state. In PXRD spectra, the transformation appears as a broad peak in the PXRD scattering spectrum instead of clear peaks indicating the lattice parameters of a crystalline structure (Fig. 9.2).

9.2 Production of Amorphous APIs

To produce amorphous API, usually the crystallization of the API is avoided by a thermodynamic shock during the solidification process (e.g. melt quenched) or the crystalline structure is physically disturbed (e.g. grinding). Different methods have been developed to produce amorphous active pharmaceutical ingredients including: vapor condensation, supercooling of liquid-melt, precipitation from solution, milling, (wet) granulation, compacting crystals, and polymer film coating [15, 21, 35, 38]. However, not all the techniques can be scaled up easily. For small-scale production, grinding, solvent methods such as fast evaporation, rotary evaporation, freeze drying, spray drying, rapid precipitation, or thermal methods such as melting, supercritical fluid, ultra-rapid freezing are often used. For large-scale production freeze drying, spray drying and melt extrusion are most feasible and common [16]. To choose the feasible production method, physical and thermal properties of individual initial components should be considered. As an example, melt-quench methods are applicable for heat-stable components, while grinding is the best suitable for physically stable compounds. Precipitation and spray drying are used for organic solvent soluble compounds, while for freeze drying methods, components need to be water soluble [35]. In the following, some of the widely used and promising methods such as melt quench, grinding, supercritical fluid technologies, spray drying and melt extrusion are summarized.

9.2.1 Melt-Quench

In this method the ordered molecular lattice of the crystalline component is distorted by melting and restored in disordered molecular structure by rapidly reducing the temperature (vitrification) to form an amorphous material. PXRD can be used to identify and quantify this transformation. In glass forming materials and upon vitrification a broad peak in the PXRD scattering appears instead of clear peaks that indicate the lattice parameters of a crystalline structure (see Fig. 9.2) [11, 12, 29].

9.2.2 Grinding

In this methods the lattice (ordered) structure is physically distorted to provide a disordered molecular structure. Among different grinding method used in industry, Cryo-grinding has the advantage of dispersing the generated heat during the grinding process and thus avoiding thermal degradation. It is important to note that not all crystalline compounds fully transform to amorphous material by this method and that the efficiency of this transformation depends on initial crystalline material, processing time and conditions [17]. As can be seen from Fig. 9.3, cimetidine transforms to an amorphous structure after 180 min of grinding, as evidenced by its broad band in Powder X-ray Diffraction spectra compared to unmilled sample, while naproxen does not completely lose its ordered structure at similar conditions.

9.2.3 Supercritical Fluids

Technologies based on supercritical fluids (SCF) have been developed to obtain pharmaceutical materials in a tailored physical state under controlled processing conditions. The high compressibility and diffusivity of supercritical fluids along with the possibility of fine-tuning pressure and solvent evaporation rate differentiate SCF technologies from other common methods. In addition, supercritical fluids have become known as a "green" alternative that reduces residual contents of volatile organic compounds in the drug product to satisfy the regulatory restrictions. However, despite its high potential in drug delivery, the high cost of the equipment and the still limited knowledge of this area form a barrier for large scale application of SCF in pharmaceutical industry. Supercritical fluids exist in an intermediate

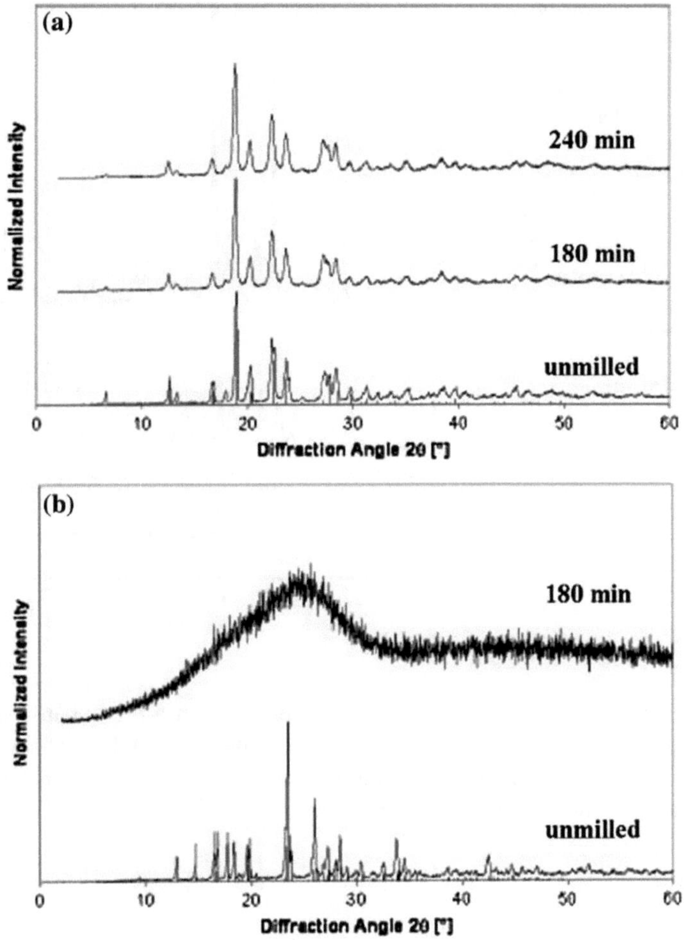

Fig. 9.3 PXRD patterns of milled (at different processing times) and unmilled samples of **a** naproxen and **b** cimetidine [17]. Copyright Elsevier. Figure reprinted with permission

phase between liquid and gas phase. Different approaches have been developed to produce pure solid state (crystal, polymorph and amorphous) components from supercritical fluids, where SCF is used as a solvent or anti-solvent. Rapid Expansion of a Supercritical Solution (RESS) is an example where a supercritical fluid is used as solvent. In this process, solid forms by rapid solution expansion caused by sudden decompression of the supercritical solution. In processes where a supercritical fluid is used as an anti-solvent, solid formation is due to the sudden decrease in solubility power of solvent upon addition of SCF (precipitation). Usually, polymorph and crystals form by these processes [23]. Ibuprofen was processed by Rapid Expansion of a Supercritical Solution into a Liquid Solvent (RESOLV) at

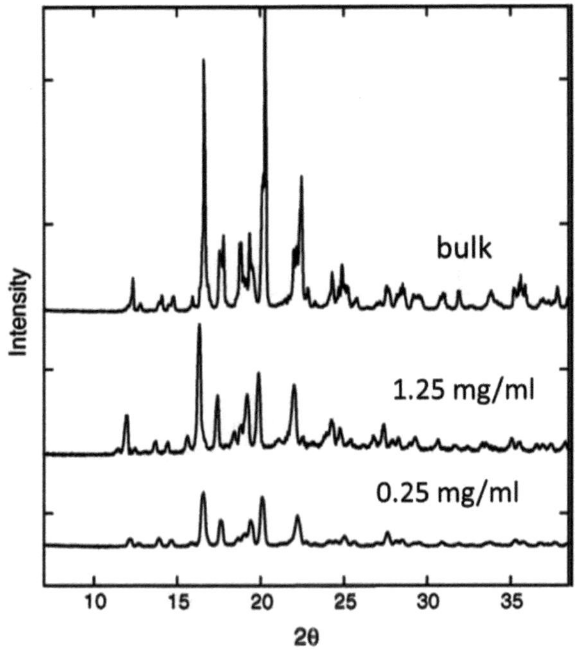

Fig. 9.4 PXRD patterns of bulk isoprofen and nanoparticles produced by RESOLV method at 1.25 and 0.25 mg/ml isoprofen concentrations [25]. The reduction in peak intensities indicates less crystalline structure in the produced nanoparticles. Lower concentration of isoprofen results in higher degree of disordered material. Copyright by Elsevier. Figure reprinted with permission

20 MPa (using supercritical CO_2 fluid as a solvent) with a pre-expansion temperature of 40 °C, using PVP, PEG and PVA polymers as stabilizing agents in an aqueous receiving solution. The resulting product showed reduction in crystallinity (Fig. 9.4) and an increase of solubility [25].

9.2.4 Freeze Drying

Freeze drying or lyophilisation has numbers of application in food and pharmaceuticals, among them preservation and stabilization of living materials which can be damaged by simple freezing due to ice formation, freeze concentration effect or recovery process. Many biological products including various vaccines (e.g. Measles Virus Vaccine Live) or even anti-blood clot medicine Streptokinase and Wasp Venom Allergenic Extract are lyophilized to increase their stability and shelf life.

Freeze drying is a process that can be described in three main steps: (a) an initial freezing process, (b) a primary drying (sublimation) phase and (c) a secondary drying to eliminate the final traces of water. The freeze drying process need to be designed and controlled based on the initial material and desired characteristics of final product including its T_g.

The prepared aqueous solution or suspension has a freezing temperature that is mainly lower than its individual components (water). Size of the formed crystals is affected by freezing rate and degree of supercooling. Fast freezing results in smaller ice crystals formation that reduces the chances on restructuring and damaging the active material. However, it makes the sublimation process slower and, if not completed, will result in higher amount of water residues. The residual water (usually 5–10% in this stage) should be removed in the secondary drying process. Slow freezing results in bigger ice crystals, the formation of which may disturb the structure of active materials, but yields less restrictive channels in the sublimation process. By application of reduced pressure instead of heat during the sublimation process, frozen solvent transforms to the gaseous state without experiencing liquid state. If water pockets (unfrozen state) are present in this stage, they can expand and compromise the stability of the resulting solid material. The rate of ice sublimation is a critical factor to reach a stable product. This rate depends on the difference in vapor pressure of the component to the pressure of the ice condenser to avoid molecular migration from higher pressure to lower pressure area. To reach an optimum drying, it is also important that the temperature at which a product is freeze dried is balanced between the temperature that the component remains frozen and the temperature that maximizes its vapor pressure. Additional steps might be introduced to enhance the properties of the final product and its T_g [27].

This process is widely applied for both small and large scale production. However, one of the main challenges, when producing full capacity in large scale, is the differences in heat and mass transfer that depends on the load condition, dryer design, and the container closure system. Therefore to have the desired amorphous product, careful setup considerations during process development and scale-up are critical [24].

9.2.5 *Spray Drying*

This method has been used in both small and large scale production and when the components are compatible with organic solvent. Organic solvents are mainly used in the spray drying technique to enhance the solubility of low water-soluble APIs dispersed in polymers. Aqueous solvents are used when API is water soluble and the drying conditions are changed accordingly. After dissolving API, the solution is sprayed in the drying chamber for quick solvent removal (~1 s). The spray drying process (as simplified and shown in Fig. 9.5) includes the following steps: (a) atomization of the liquid stream, (b) interaction of the feed droplets with heated drying gas resulting in evaporation of solvent and solid particle formation, (c) isolation and collection of solid product, (d) secondary drying if needed. Nitrogen atmosphere is used to provide inert atmosphere when organic solvents are present. As the atomized droplets contact drying gas, solvent evaporates and solid particles are formed during the particle residence in the drying chamber. The solid particles are usually separated from the gas stream by using a cyclone [9, 19, 33].

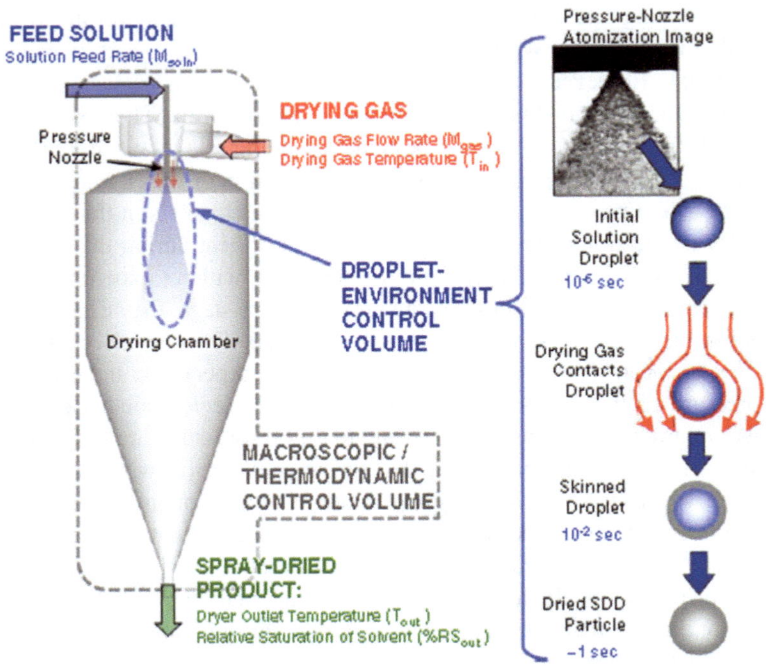

Fig. 9.5 Process flow chart for spray drying process [9]. Figure reprinted from [9], open source

In some cases, especially in large-scale batches, solvents need to be removed completely from the spray dried product. Then, secondary drying by tray drying, fluid bed drying, etc. is added to the process. The spray dried material can be used later as one of the ingredients for different granulation processes such as blending, compression and capsule filling [9, 19, 33]. Solvent thermal properties, solid concentration, solution feed rate, nozzle size, atomization pressure, inlet temperature, drying gas temperature, drying gas flow rate and drying outlet temperature are the parameters that directly influence the process and final product. Pressure nozzles are offering simplicity and scalability in tuning droplet size for pharmaceutical application [9, 36]. To avoid crystal formation and mixed products (crystals/amorphous) usually fast drying is preferred.

The morphology of the formed solid particle is also influenced by drying condition. When hot and fast drying condition is applied, as droplet skin forms, the droplet's temperature is near or above the boiling point of the solvent. Therefore, the vapor pressure inside the droplet keeps it inflated as it dries out. In such conditions, particles with hollow sphere morphology are formed. However in cold and slow drying conditions, droplet's temperature drops below the boiling point of the solvent as droplet skin forms which causes the collapsed particle to form "raisin" morphology (Fig. 9.6) [9].

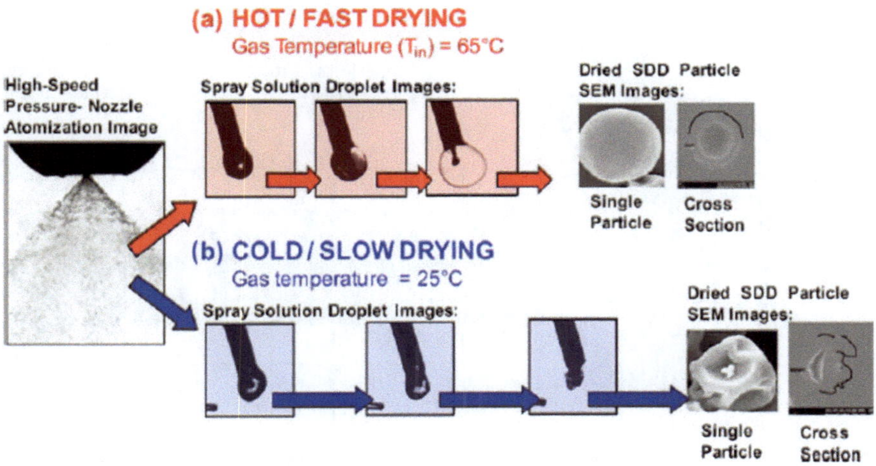

Fig. 9.6 Influence of drying condition on morphology of produced particle. Figure reprinted from [9], open source

9.2.6 *Melt Extrusion*

Melt extrusion has been widely used in polymer processing and found its application in pharmaceutical and drug delivery field. This method has many advantages compared to traditional pharmaceutical processing techniques that were discussed before, including absence of solvents, fewer processing steps and continuous operation. In melt extrusion, API and polymer are heated up to the polymer's melting temperature. Therefore, one critical consideration is to know how heat and shear stress affect different APIs. Exceeding the thermal and stress tolerance of APIs will impose the risk of reduced API's desirable bioactivity. To reduce processing time, processing temperature and pressure, pharmaceutical graded polymers with low processing temperature and thermally stable ingredients are more favorable [5, 8, 26, 30, 35].

This method has been adopted to mix drugs with different polymeric carriers for various solid dosage of active pharmaceutical ingredients. In general, the extrusion channel is composed of three main sections: feeding, transition and compression. The solid is compressed, melted and plasticized in the compression section and, by passing through the metering section, a homogeneous melt is extruded at a uniform delivery rate from the die section [5, 8, 30, 35] (Fig. 9.7).

One of the main issues in this method is heat degradation of the API. Therefore, monitoring the temperature of the extruder head and die as well as the pressure in extruder and die is crucial. Also, excluding oxygen and reducing moisture exposure will reduce the chances of unwanted chemical degradation during the processing [5, 26, 30, 35].

In melt extrusion, ingredients need not only to improve processability but also to retain (and not inhibit) the API bioactivity. Parameters to consider are: molecular

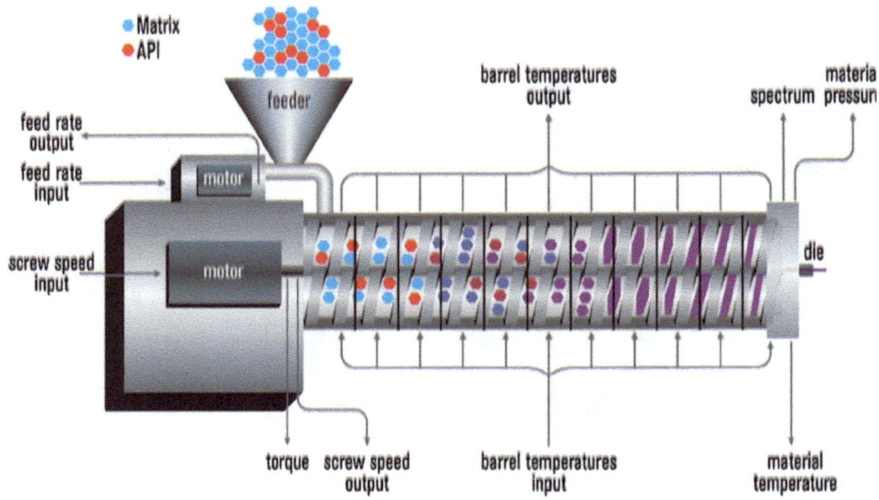

Fig. 9.7 Schematic of melt extrusion system for APIs [26]. Copyright Springer. Figure reprinted with permission

weight of the matrix polymer, glass transition and melting temperature of the amorphous or semi-crystalline polymer (and degradation of API and other components), sensitivity of the matrix or drug towards heat and shear force, miscibility of drug and polymer and possibility of using a compatible plasticizer to lower the processing temperature [5, 8, 26].

A variety of polymers and excipients with a wide range of processing temperatures have been used as drug carriers in thermal processing methods. Polymers with low melt viscosity and high thermal conductivity are favorable. In addition, interactions between drug and polymers and effects of miscibility on processing conditions need to be taking into consideration. H-bonding or ion-dipole interactions might cause recrystallization or formation of undesirable bonding and, hence, change the composition melting temperature [5, 8, 30, 35].

A variety of pharmaceutical products from transdermal patches and ophthalmic inserts pellets and films with various shapes and sizes to solid API dispersion formulated in tablets and capsules is produced by this method. A comprehensive list of these polymers and examples of drug substances can be found in [8].

9.3 Active Amorphous Pharmaceutical Ingredient Characterization

Different methods are used to characterize and screen the degree of crystallization, bond formation, miscibility, dispersion and morphology of produced amorphous pharmaceutical ingredients, including powder diffraction, small-angle X-ray and neutron scattering, thermal analysis (DSC, dynamic mechanical analysis), dielectric spectroscopy, IR, Raman, NMR spectroscopy, solution calorimetry, optical microscopy, scanning electron microscopy (SEM), atomic force microscopy (AFM). Scattering and diffraction methods provide information on degree of crystallization, amorphous/crystal ratio, structural coordinates and bond formations at molecular level. Molecular mobility can also be monitored and measured by different dynamic scattering and spectroscopic methods [17, 29, 37].

In preparation of quick release APIs, one of the initial steps in formulation design is to identify the compatible components, screening their interaction and mixing properties. Amorphous material prepared with different methods can demonstrate different physical properties and stabilities. In mixing the ingredients, miscible components have better stability than physically dispersed and mixed products. Phase separation can be avoided in miscible components with a strong molecular interaction leading to higher physical and chemical stabilities in Amorphous APIs [17, 29, 37].

To illustrate the role of molecular interactions in miscibility, molecular mobility and stability, amorphous solid dispersions of indomethacin (IMC) and sodium indomethacin (NaIMC) over a range of compositions are prepared by two methods of [1] physically mixing of amorphous IMC and amorphous NaIMC and [2] co-precipitation from methanol solution [34]. In this study, DSC measurements indicated a phase separation by exhibiting two T_g values for materials prepared by physical mixing while the sample prepared by co-precipitation exhibited one T_g value that was higher than predicted T_g for the ideal miscible composition. The higher T_g was attributed to the stronger interaction in the amorphous state prepared by co-precipitation, which was confirmed by FTIR analysis. It was suggested that such interactions inhibit the isothermal crystallization of IMC by NaIMC. PXRD measurements (Fig. 9.8) indicate the higher physical stability of amorphous material formed by co-precipitation method after different storage time and temperature as compared to the composites prepared by physical mixing [34]. It has to be noted that in this comparison, shown in Fig. 9.8, the samples are kept at lower temperature (4 °C vs. 60 °C) for longer period (14 months vs. 2 weeks). It appears that the component prepared by physical mixing shows more tendency to crystallize and the intensity of the crystalline peaks increases by mole fraction of NaMI (see Fig. 9.8a vs. b), while the crystalline peak does not appear in the prepared samples by co-precipitation due to better miscibility that is apparent by showing one T_g. However, when samples are stored at 60 °C for 2 weeks, relatively lower stability was observed indicated by formation of crystalline peak that shows the same trend as in previous storage condition (see Fig. 9.8c vs. d). The higher intensity of the crystalline peaks (that even appears in samples prepared by co-precipitation) despite their shorter storage time is due to the increase in molecular mobility at higher storage temperate.

Dispersion screening is done to single out the best dispersion condition which provides faster dissolution and higher bioavailability. Performance screening is the initial step to investigate dissolution, stability, bioavailability of a simple prototype formulations (powder in capsule) for further in vitro-in vivo correlation (IVIVC) evaluations [10, 30].

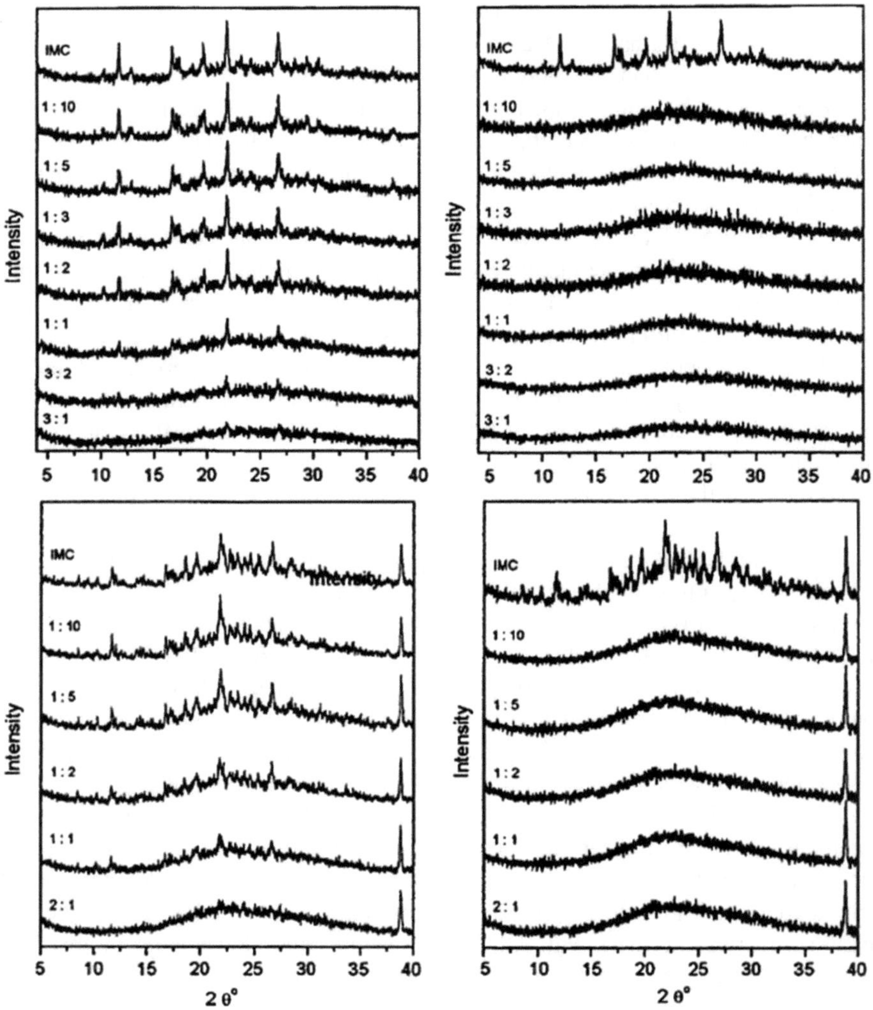

Fig. 9.8 PXRD patterns for NaIMC and IMC when prepared by physical mixing and stored at **a** at 4 °C for 14 months and **c** at 60 °C for 2 weeks. Mixtures prepared by coprecipitation methods and stored at **b** 4 °C for 14 months and **d** 60 °C for 2 weeks [34]. Copyright Elsevier. Figure reprinted with permission

The drying method has a significant impact on numerous properties of the dried powders such as particle morphology, surface composition and undesired exposure of active sites resulting in physical and chemical processing and storage instability. Different drying conditions and formulation design result in different particle morphology, packing and molecular interaction, which influence the dissolution rate as well as stability of the product. A study on Met-hGH powder, produced by different drying methods demonstrated that spray dried (and freeze dried) powders had higher protein surface coverage and specific surface area than other methods.

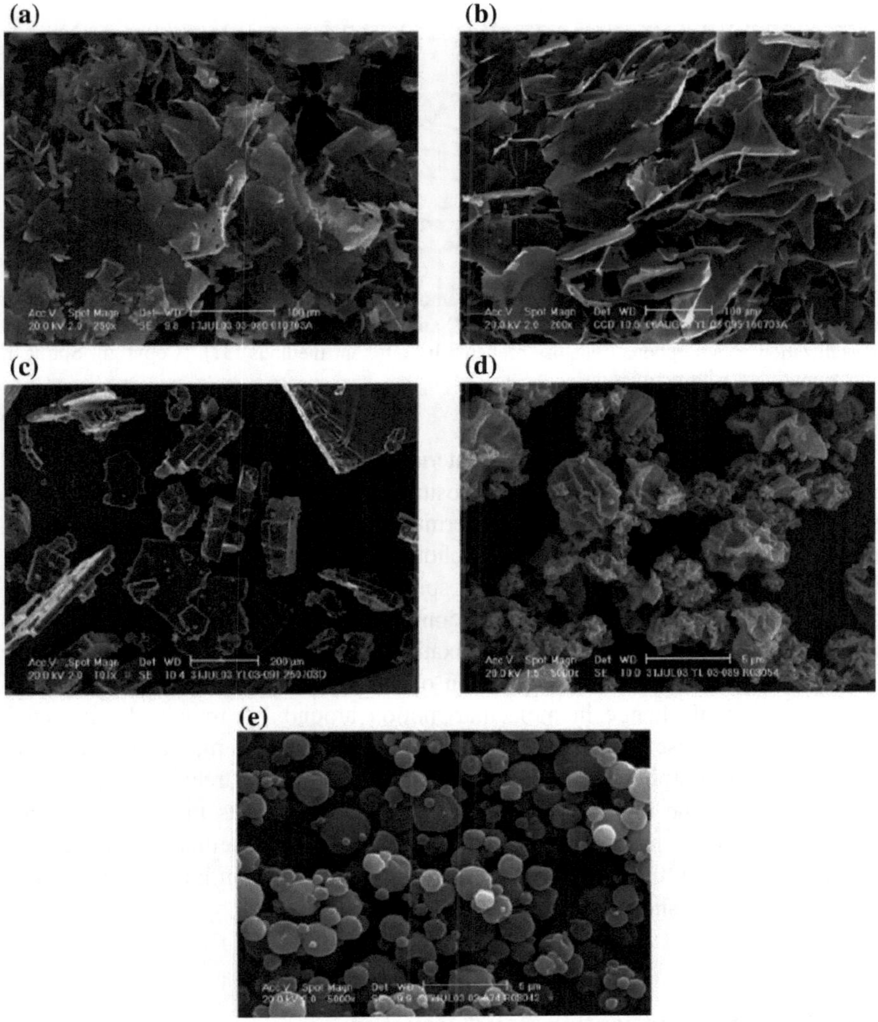

Fig. 9.9 SEM images of dried Met-hGH illustrate morphology differences of dried powders produced by different drying methods and different formulations: **a** Freeze dried Met-hGH/trehalose with Tween-20; **b** Freeze dried Met-hGH/trehalose with Tween-20 annealed prior to primary drying; **c** Film dried Met-hGH/Ficoll 70 with Tween-20; **d** Spray dried Met-hGH/trehalose without Tween-20; **e** Spray dried Met-hGH/trehalose with Tween-20 [1]. Copyright Elsevier. Figure reprinted with permission

The lowest protein surface coverage, observed in powders produced by film drying, might contribute to their improved stability. However, adding excipients such as trehalose has demonstrated to influence stability of Met-hGH more effectively which can be related to the bond formation and changing the molecular packing and therefore reducing molecular mobility and relatively influencing fragility and T_g of the component [1] (Fig. 9.9).

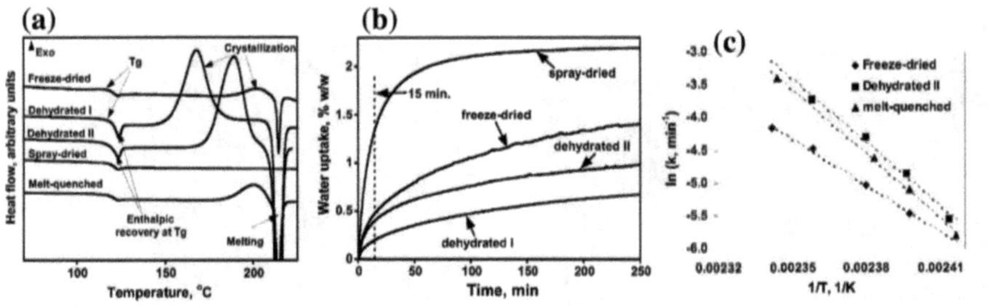

Fig. 9.10 Influence of preparation method of amorphous trehalose on properties **a** DSC heating scans; **b** water vapor sorption behavior at 25 °C and 10% RH; **c** Arrhenius plot of the isothermal crystallization of amorphous trehalose prepared by different methods [31]. Copyright Springer. Figure reprinted with permission

Molecular mobility of amorphous materials is believed not only to be influenced by the formulation and chemical composition but also by the production method and pre-process treatments causing thermal history or ex-process treatments by aging-annealing. In a study where amorphous anhydrous trehalose was prepared by four different methods—freeze-drying, spray-drying, melt quenching and dehydration at different temperature, duration and atmospheric conditions—it was observed that, while the enthalpic relaxation and crystallization behaviors were influenced by the method of preparation of amorphous trehalose, the T_g and fragility (the rate of change in molecular motion around T_g) remained unaffected. Amorphous trehalose samples prepared by different methods had large differences in particle size, morphology, and surface area. In this study, trehalose prepared by dehydration had the highest tendency to crystallize, whereas there was no crystallization in melt quenched amorphous trehalose. The method of preparation influenced not only the rate and percentage of water sorption but also the recrystallized kinetics as shown in Fig. 9.10 [31].

9.4 Specific Applications and Benefits

Amorphous APIs and dispersions are metastable forms of materials. Kinetic conditions usually are applied to trap their amorphous metastable state. While different methods are employed to produce AAPIs, large scale production remains limited mainly due to physical and chemical sensitivity of final product to processing variables. Freeze-drying, spray drying and melt extrusion are the most common methods to produce amorphous solid dispersions. Amorphous APIs are later formulated in tablets and capsules.

Controlled release drugs are among the emerging topics in the drug delivery. However, their advancement is limited by their low solubility, which not only inhibits their bioavailability but also asks for more innovations in their production methods. Identifying and developing approaches to increase APIs solubility are of high interest to remove this obstacle into introducing newly developed medications to the market.

Combining bioavailability enhancing technology with controlled release techniques enables extended delivery of drugs with low solubility. Amorphous dispersion of a formulation using hot melt extrusion has been done already for a number of applications and in combination with other methods. Layered tablets are one of the examples that gives the advantage of control release. *Intelence (Etravirine)* is an approved medication for HIV treatment. Spray drying technology is used to prepare the amorphous phase of API with HPMC dispersion. The amorphous dispersion is formulated into tablets with microcrystalline cellulose derivatives, colloidal silicon dioxide, croscarmellose sodium, magnesium stearate and lactose monohydrate. Tablets are stirred in water until the water becomes cloudy and to reach the timely desired bioactivity, the medication must be taken with food.[1] Usually food will reduce the negative side effect of medication on digestive system and provide the desired pH in the body (blood or stomach) to enable release and absorption of the active ingredient.

Kaletra (Ritonavir/lopinavir) is another example of HIV medication that prevents human immunodeficiency. *Kaletra* is produced as soft gelatin capsule and solution and marketed in 2000.[2] Both products initially had to be refrigerated for storage. In 2005, a new tablet was formulated and produced by an improved melt extrusion technology. In this marketed medication Copovidone (cross-linked PVP) was used to enhance the stability of the produced amorphous dispersion product [6].

To conclude, one of the effective approaches to enhance drug release and increase bioavailability of APIs is to produce these products in amorphous forms. Such solid amorphous structure enables faster solvent accessibility for a rapid dissolution. However, in comparison with crystalline structure, amorphous materials might exhibit lower chemical stability and be more sensitive to light and humidity due to the increased exposure of their active sites. Depending on the thermal and solvent tolerance range of the APIs and their relevant additives, different production methods can be chosen. Melt-extrusion, freeze drying and spray drying process are the most commonly used methods at large scale. Different characterization techniques are usually used to determine the effectiveness of the process in terms of producing amorphous material. DCS and PXRD are among the most used methods that indicate the existence (change in) T_g (and disappearance of T_m) and change in crystalline structure of the component, respectively.

[1]http://www.intelence-info.com/about-intelence/about-intelence; http://www.natap.org/2008/Pharm/Pharm_07.htm. Last accessed 16/2/2017.

[2]http://www.thebody.com/confs/ias2005/pdfs/WeOa0206.pdf. Last accessed 16/2/2017.

9.5 Abbreviations and Symbols

AAPI	Amorphous Active Pharmaceutical Ingredients
API	Active Pharmaceutical Ingredients
DSC	Differential Scanning Calorimetry
FTIR	Fourier-Transform Infrared Spectroscopy
IMC	Indomethacin
IVIVC	In-Vitro-In Vivo Correlation
Met-hGH	Methionine (N-terminal) human growth hormon
NaIMC	Sodium Indomethacin
PEG	Polyethylene glycol
PVA	Poly(vinyl alcohol)
PVP	Polyvinylpyrrolidone
PXRD	Powder Xray Diffraction
RESOLV	Rapid Expansion of a Supercritical Solution into a Liquid Solvent
RESS	Rapid Expansion of a Supercritical Solution
SCF	Supercritical Fluid
T_g	Glass Transition Temperature
T_m	Melting Temperature

References

1. Abdul-Fattah AM, Lechuga-Ballesteros D, Kalonia DS, Pikal MJ (2008) The impact of drying method and formulation on the physical properties and stability of methionyl human growth hormone in the amorphous solid state. J Pharm Sci 97(1):163–184
2. Andronis V, Yoshioka M, Zografi G (1997) Effects of sorbed water on the crystallization of indomethacin from the amorphous state. J Pharm Sci 86(3):346–351
3. Angell CA (1995) Formation of glasses from liquids and biopolymers. Science 267(5206): 1924–1935
4. Bhugra C, Pikal MJ (2008) Role of thermodynamic, molecular, and kinetic factors in crystallization from the amorphous state. J Pharm Sci 97(4):1329–1349
5. Breitenbach J (2002) Melt extrusion: from process to drug delivery technology. Eur J Pharm Biopharm 54(2):107–117
6. Breitenbach J (2006) Melt extrusion can bring new benefits to HIV therapy. Am J Drug Deliv 4(2):61–64
7. Brouwers J, Brewster ME, Augustijns P (2009) Supersaturating drug delivery systems: the answer to solubility-limited oral bioavailability? J Pharm Sci 98(8):2549–2572
8. Crowley MM, Zhang F, Repka MA, Thumma S, Upadhye SB, Kumar Battu S, McGinity JW, Martin C (2007) Pharmaceutical applications of hot-melt extrusion: Part I. Drug Dev Ind Pharm 33(9):909–926
9. Dobry DE, Settell DM, Baumann JM, Ray RJ, Graham LJ, Beyerinck RA (2009) A model-based methodology for spray-drying process development. J Pharm Innov 4(3): 133–142
10. Dong Z, Chatterji A, Sandhu H, Choi DS, Chokshi H, Shah N (2008) Evaluation of solid state properties of solid dispersions prepared by hot-melt extrusion and solvent co-precipitation. Int J Pharm 355(1–2):141–149

11. Engers D, Teng J, Jimenez-Novoa J, Gent P, Hossack S, Campbell C, Thomson J, Ivanisevic I, Templeton A, Byrn S, Newman A (2010) A solid-state approach to enable early development compounds: Selection and animal bioavailability studies of an itraconazole amorphous solid dispersion. J Pharm Sci 99(9):3901–3922
12. Guinot S, Leveiller F (1999) The use of MTDSC to assess the amorphous phase content of a micronised drug substance. Int J Pharm 192(1):63–75
13. Hancock BC, Parks M (2000) What is the true solubility advantage for amorphous pharmaceuticals? Pharm Res 17(4):397–404
14. Hancock BC, Shamblin SL, Zografi G (1995) Molecular mobility of amorphous pharmaceutical solids below their glass transition temperatures. Pharm Res 12(6):799–806
15. Hancock BC, Zografi G (1997) Characteristics and significance of the amorphous state in pharmaceutical systems. J Pharm Sci 86(1):1–12
16. Karthik N, Janan J (2008) Amorphous active pharmaceutical ingredients in preclinical studies: preparation, characterization, and formulation. Curr Bioact Compd 4(4):213–224
17. Lin Y, Cogdill RP, Wildfong PLD (2009) Informatic calibration of a materials properties database for predictive assessment of mechanically activated disordering potential for small molecule organic solids. J Pharm Sci 98(8):2696–2708
18. Luthra SA, Hodge IM, Pikal MJ (2008) Investigation of the impact of annealing on global molecular mobility in glasses: optimization for stabilization of amorphous pharmaceuticals. J Pharm Sci 97(9):3865–3882
19. Masters K (1991) Spray Drying Handbook. Burnt Mill, Harlow, Essex, England; New York, Longman Scientific & Technical; Wiley
20. Neervannan S (2006) Preclinical formulations for discovery and toxicology: physicochemical challenges. Expert Opin Drug Metabol Toxicol 2(5):715–731
21. Newman A (2015) Pharmaceutical amorphous solid dispersions. John Wiley & Sons, Hoboken, NJ
22. Ngai KL (2011) Glass-forming substances and systems. Relaxation and Diffusion in Complex Systems. Springer, New York, pp 49–638
23. Pasquali I, Bettini R, Giordano F (2008) Supercritical fluid technologies: an innovative approach for manipulating the solid-state of pharmaceuticals. Adv Drug Deliv Rev 60(3): 399–410
24. Patel SM, Pikal MJ (2011) Emerging freeze-drying process development and scale-up issues. AAPS PharmSciTech 12(1):372–378
25. Pathak P, Meziani MJ, Desai T, Sun Y-P (2006) Formation and stabilization of ibuprofen nanoparticles in supercritical fluid processing. J Supercrit Fluids 37(3):279–286
26. Patil H, Tiwari RV, Repka MA (2016) Hot-melt extrusion: from theory to application in pharmaceutical formulation. AAPS PharmSciTech 17(1):20–42
27. Rey L, May JC (2004) Freeze-drying/Lyophilization of Pharmaceutical and Biological Products. Informa Healthcare, New York, NY
28. Serajuddin ATM (1999) Solid dispersion of poorly water-soluble drugs: early promises, subsequent problems, and recent breakthroughs. J Pharm Sci 88(10):1058–1066
29. Shah B, Kakumanu VK, Bansal AK (2006) Analytical techniques for quantification of amorphous/crystalline phases in pharmaceutical solids. J Pharm Sci 95(8):1641–1665
30. Shanbhag A, Rabel S, Nauka E, Casadevall G, Shivanand P, Eichenbaum G, Mansky P (2008) Method for screening of solid dispersion formulations of low-solubility compounds—Miniaturization and automation of solvent casting and dissolution testing. Int J Pharm 351(1–2): 209–218
31. Surana R, Pyne A, Suryanarayanan R (2004) Effect of aging on the physical properties of amorphous trehalose. Pharm Res 21(5):867–874
32. Surana R, Pyne A, Suryanarayanan R (2004) Effect of preparation method on physical properties of amorphous trehalose. Pharm Res 21(7):1167–1176
33. Thybo P, Hovgaard L, Lindeløv JS, Brask A, Andersen SK (2008) Scaling up the spray drying process from pilot to production scale using an atomized droplet size criterion. Pharm Res 25(7):1610–1620

34. Tong P, Zografi G (2001) A study of amorphous molecular dispersions of indomethacin and its sodium salt. J Pharm Sci 90(12):1991–2004
35. Vasconcelos T, Marques S, das Neves J, Sarmento B (2016) Amorphous solid dispersions: rational selection of a manufacturing process. Adv Drug Deliv Rev 100:85–101
36. Vehring R (2008) pharmaceutical particle engineering via spray drying. Pharm Res 25(5): 999–1022
37. Yu L (2001) Amorphous pharmaceutical solids: preparation, characterization and stabilization. Adv Drug Deliv Rev 48(1):27–42
38. Zografi G, Newman A (2010). Introduction to Amorphous Solid Dispersions. Pharmaceutical Sciences Encyclopedia. Wiley, Hoboken, NJ

Sheila Khodadadi received her Ph.D. in Polymer Science in 2009 from the University of Akron, USA. In 2010, she was awarded two years NRC Postdoctoral Research Associateship to conduct her research at NIST Center for Neutron Research, National Institute of Standards and Technology, USA. Her Ph.D. and postdoctoral research experience was focused on understanding the influence of solvent and temperature on biological systems dynamics at molecular level. After moving to the Netherlands in 2012, she started to work at Delft University of Technology (TU Delft) as a researcher for three years. During that time, she extended her research to product and process development with the focus on producing tailored particles. Currently, she is a guest researcher at TUDelft and an (E)MBA student at Rotterdam School of Management, Erasmus University, NL. Her main research interests are understanding protein dynamics and interactions in pharmaceutically relevant conditions and innovative approaches in bringing fundamental sciences to practical applications.

Gabriel M. H. Meesters has a B.Sc. and M.Sc. in Chemical Engineering with a major in BioProcessTechnology from the Delft University of Technology. He has a Ph.D. in Particle Technology also from the Delft University of Technology. He worked at biotechnology companies like Gist-Brocades in The Netherlands, as well as for Genencor International and currently at DSM in research and development in The Netherlands. In all these functions he was working on formulation and product development. Since 1996 he holds a part time position at the Delft University of Technology, as assistant professor at the faculty of Applied Sciences, first in the Particle Technology group, later the Nano Structured Materials Group and currently in the Product and Process Engineering group. He supervised over 15 Ph.D. students and more than 50 M.Sc. students. He published over 70 refereed papers, holds around 15 patents and patent applications and is co-author and co-editor of the books 'Particulate Products—Tailoring Properties for Optimal Performance' (2014; Springer) and 'Production, Handling and Characterization of Particulate Materials' (2016, Springer). He (co-) organized several international conferences in the field of particle technology and was president of the World Congress on Particle Technology in 2010.

Chapter 10
Particle Properties: Impact on the Processing and Performance of Oral Extended-Release Hydrophilic Matrix Tablets

Peter Timmins and Carl Allenspach

Abstract The role of the particulate properties of the drug substance and the release rate controlling polymer in hydrophilic matrix extended release tablets, principally those based on hydroxypropyl methylcellulose, are reviewed. The effect on drug release of drug particle size and how this interacts with drug solubility properties are considered and also how control of particle size and amount of polymer together are key to producing robust products with a desired, reliable rate of drug release. The utility of the recently introduced direct compression grade hydroxypropyl methylcellulose is described and observations as to the triboelectric properties of the cellulose ether are given.

10.1 Introduction

Extended release formulations are dosage forms fabricated to allow the gradual release of drug from a dosage over a period of time, from just a few to up to 24 h, in a defined manner or rate, to allow for slower absorption and more prolonged levels of drug in the bloodstream compared with a conventional or immediate-release dosage form. They are also referred to as prolonged release, sustained release, and controlled release dosage forms. They can offer the benefit of reducing dosing frequency for patients, affording better adherence and compliance which can also improve efficacy. Additionally, the extended release of drug leading to extended

P. Timmins (✉)
Bristol-Myers Squibb, Reeds Lane, Moreton, Merseyside CH46 1QW, UK
e-mail: p.timmins@hud.ac.uk

Present Address:
P. Timmins
Department of Pharmacy, University of Huddersfield, Huddersfield HD1 3DH, UK

C. Allenspach
Bristol-Myers Squibb, 1 Squibb Drive, New Brunswick, NJ 08940, USA
e-mail: carl.allenspach@bms.com

H. G. Merkus et al. (eds.), *Particles and Nanoparticles in Pharmaceutical Products*,
AAPS Advances in the Pharmaceutical Sciences Series 29,
https://doi.org/10.1007/978-3-319-94174-5_10

drug absorption can reduce the C_{max} (maximum concentration of the drug in the bloodstream) minimising risk of peak concentration-associated side effects.

Since their first description in the patent and pharmaceutical science literature over 50 years ago [11, 40], hydrophilic matrix tablets have continued to be a popular option for extended-release oral dosage forms. This probably relates to their use of widely accepted and relatively inexpensive materials as the basis of the extended-release mechanism (along with other conventional excipients such as compression aids, fillers, lubricants, binders, etc.), that the method of manufacture usually involves conventional pharmaceutical unit operations equipment and that a significant understanding of the factors affecting drug release from these dosage forms exists, thus better enabling a quality-by-design (QbD) approach to their creation.

In considering the impact of particle properties on the processing and performance of hydrophilic matrix tablets it will be of value to understand the mechanism of drug release from these systems and also to consider the commonly-used manufacturing options. The mechanism of drug release from these dosage forms involves the development of a surface gel layer of hydrated, swelled polymer on immersion of the dosage form into aqueous fluid. This hydrated polymer layer is dynamic, modulating the ongoing ingress of fluid into the dosage form with associated further swelling, and is also susceptible to erosion under shear, so providing the platform for drug dissolution and diffusion with erosion of the hydrated polymer offering an additional mechanism for drug release [3, 12, 14, 37, 49, 51, 53, 65, 73, 77].

The kinetics of drug release can be explored through consideration of the movement of different fronts inside the evolving matrix tablet over time after its immersion in an aqueous environment. The first of these fronts is the boundary between the remaining dry core (also described as being or containing glassy polymer) and the hydrated polymer (also described as being or containing rubbery polymer), and defined as the swelling front. A second front is the boundary between the discernible outer edge of the hydrated polymer and the bulk medium, defined as the erosion front. The concentration of polymer in the hydrated gel decreases sharply close to the interface between hydrated matrix and bulk fluid which may enable the erosion mechanism, especially for low viscosity polymer and for matrices based on formulations with low concentration of polymer. The other boundary is within the hydrated polymer gel layer between where solid drug exists in the presence of dissolved drug and where there is only drug in solution in the hydrated polymer gel, defined as the diffusion front (Fig. 10.1) [13, 15].

It is important to understand that the fronts are dynamic. The swelling front migrates into the matrix from the outside over the duration of drug release, as the dosage form continues to take up fluid. Towards the end of drug release the matrix may be completely hydrated and the swelling front no longer exists. For very water soluble drugs, the diffusion front may have very close proximity to the swelling front, especially for low initial content of drug in the dosage form (i.e. with low dose drugs) and as drug release progresses may possibly disappear (i.e. become coincident with the swelling front). Drug concentration in the matrix at the very

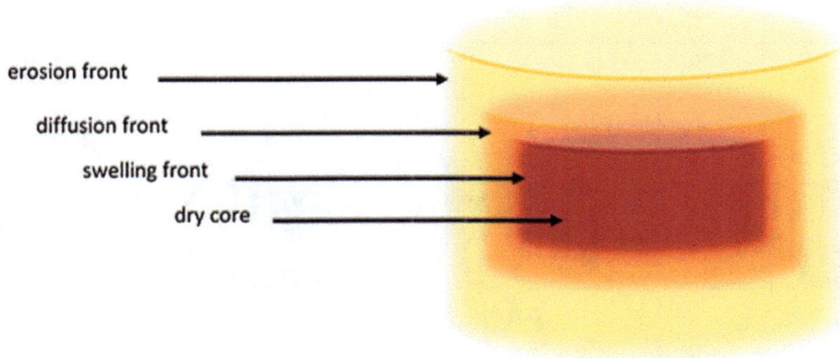

Fig. 10.1 The evolving fronts within a hydrating hydrophilic matrix tablet a little time after immersion in an aqueous environment when the fronts are properly established. Coloured for emphasis and shown as if able to "see into" the tablet

beginning of drug release will be uniform, but once the process of drug release gets underway then a gradient may establish, dependent on drug solubility, where drug concentration in the hydrated matrix declines from the diffusion front towards the erosion front, due to diffusional mechanism of drug release. The depth of hydrated matrix from erosion front to swelling front is driven by the balance between the liquid penetration rate into the matrix and erosion of the hydrated polymer gel at the erosion front. It is the balance of diffusional release, fluid penetration (swelling) and hydrated polymer gel erosion that contribute to defining the overall drug release rate from these matrix systems.

The typical manufacturing operations for these dosage forms, whether dry granulation, wet granulation or, where possible, direct compression, centre around the blending of powders, and the flow of these blends during processing (especially with dry granulation and tableting where blending and flow will be affected by the particle size/shape of the components) (Fig. 10.2). Particle properties of the excipient and drug materials used to make the product play a role on processabilty of matrix formulations as with all solid dosage forms [38, 68]. The cohesivity/adhesivity of all the formulation components will be impactful in these unit operations, along with triboelectrics, which can be inherent or generated during processing. Intrinsic material properties (plastic, elastic or brittle nature) along with particle properties (particle size and shape, hence specific surface area) will impact compaction properties during dry granulation and final tableting steps, as well as possibly affecting the milling properties.

Multiple, and interacting, formulation and process factors can affect drug release from hydrophilic matrix tablets [73]. However, as drug release processes involve the wetting and swelling of individual polymer particles to yield the hydrated gel layer and this aspect, along with the dissolution of drug from solid particles within the gel layer (at the outer boundary of the diffusion front), indicate the specific importance of particle properties to the manufacturing processes for these dosage forms and the extended-release performance of the dosage form.

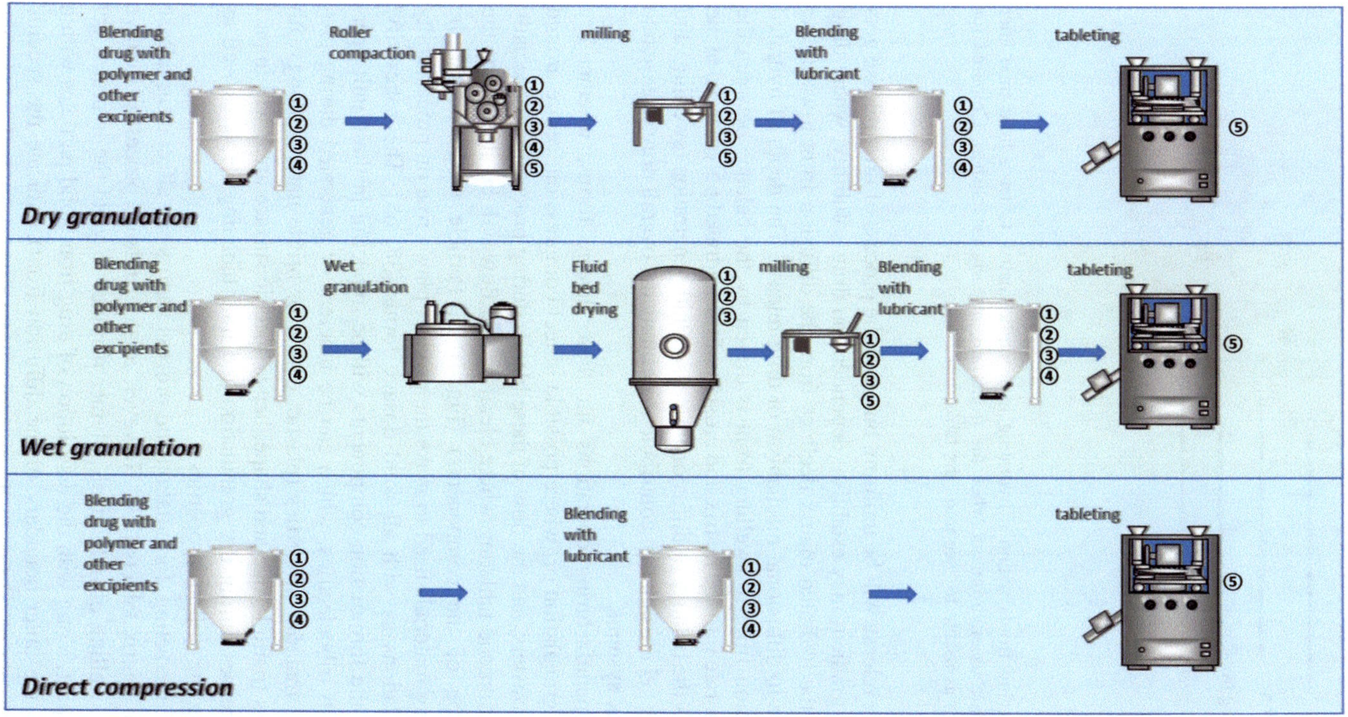

Fig. 10.2 Commonly-used processes for manufacture of hydrophilic matrix tablets with the drug and formulation property impacts noted. Particle cohesion ① and adhesion ② properties, as well as triboelectric ③ properties, will be important at blending and milling stages. Starting material particle size/shape ④ will be important for blending and flow, being critical for direct compression where no particle enlargement processes like dry granulation or wet granulation are employed. The compaction properties ⑤ which the formulation components bring to the blend, which will be influenced by their particle properties as well as their intrinsic properties, and will affect roller compaction, potentially milling (plastic deformation versus brittle fracture compaction mechanisms) and are particularly critical during final tableting steps

This chapter will discuss and review the research literature describing the influence of particle properties of active pharmaceutical ingredient, hydrophilic polymer and other dosage form excipients on the processing and performance of hydrophilic matrix extended-release tablets. As the available published work considered in this chapter has almost exclusively focused on how the particle properties of hydroxypropyl methylcellulose (hypromellose, HPMC) impact the performance of hydrophilic matrix extended release tablets employing this polymer, it is worth at this point just noting aspects of the types of this polymer utilized in their formulation. HPMC is employed in three substitution grades, described in the United States Pharmacopoeia as 2208, 2910 and 2906, and which have, respectively, 19–24% methoxyl and 7–12% hydroxypropoxyl; 28–30% methoxyl and 7–12% hydroxypropoxyl; 27–30% methoxyl and 4–7% hydroxypropoxyl group substitution on the methyl cellulose backbone. The 2208 and 2910 grades are the most commonly employed in extended release drug products. They are available in a range of viscosity grades (based on viscosity of a 2% w/v concentration of polymer in water) from as low as 3 mPa·s up to 100,000 mPa·s. The full range of viscosity options are not available in all substitution grades, but matrix tablets tend to employ the higher viscosity grades.

The following sections describe the influence of both API and polymer particle properties on extended release.

10.2 Active Pharmaceutical Ingredient (API)

As there appear to be no reports in the scientific literature of the impact of API particle properties on the processing of hydrophilic matrix extended-release tablets (although the impact would be similar to immediate release dosage forms and relate to the drug load in the formulation), the rest of this section will focus on the influence of the API particle properties (particle size, specific surface area, and solubility) on the drug release performance of hydrophilic matrix extended-release tablets.

We can consider theoretically that high content of poorly flowing API in a powder blend for compression might result in processing problems such as feed into dry granulation roller compactor and consequent output ribbon quality, as well as tableting issues and weight uniformity impacting the drug content uniformity. Similarly, high levels of API with poor compression properties in a dosage form might dominate the compression properties of the blend with polymer and other excipients, to manifest as processing problems. The inherent compaction properties of the cellulose ethers commonly employed in extended release formulations are such that, when used at recommended levels of around 30% w/w in a dosage form along with high levels of poorly compressible drug (say 50–60% w/w of the dosage form), dry granulation or direct compression processes can be problematic.

Therefore, wet granulation may be preferred in these cases. As with all solid dosage forms, consideration needs to be given to the particle size of the API in relation to the other materials in the formulation to avoid the potential for segregation resulting in variability of individual tablet drug content across a tableting run, which can be especially problematic with low dose drug direct compression processes.

10.2.1 API Particle Size

One property of the API that can be changed by processing (e.g. milling) that may affect drug release rate from a hydrophilic matrix tablet is particle size. There is, however, a dependency of the magnitude of the effect on the water solubility of the drug and the drug : polymer ratio [75]. Promethazine hydrochloride, aminophylline, and propranolol hydrochloride, freely water soluble drugs, show a very limited effect of increasing drug particle size on release rate from the matrix. Increasing particle size by a factor of three or more (63–90 μm sieve cut to 180–250 μm and 250–500 μm sieve cut) led to a very small increase in release rate except in the case where a low ratio of polymer to drug was employed. Here, there was a more marked increase in release rate only at the largest size, 250–500 μm, sieve cut, which was likely due to an increase in hydrated matrix fragility engendered by the presence of coarse, undissolved particles in the matrix [20, 21]. Although rate of release at steady state for a water-soluble drug was overall not sensitive to API particle size, it was observed that the time to establishing steady state, and hence the magnitude of an initial burst effect was affected [66]. For coarse potassium chloride particles (315–400 μm sieve fraction as opposed to 63–100 μm sieve fraction) included in a hydrophilic matrix, there was a more marked initial burst associated with the increased time to the establishment of the coherent gel layer around the hydrating tablet. This modified the overall shape of the full drug release profile and reduced the time to 50% of API being released. The effect was explained as being due to limited presentation of polymer at the tablet surface due the more significant presentation of API from the coarse particles, which at the initiation of polymer hydration delayed the evolution of the complete gel layer around the tablet [66].

For more poorly water-soluble drugs there was a more marked effect of drug particle size. For indomethacin at a constant drug: HPMC polymer ratio, as drug particle size was increased the drug release rate decreased [22]. The effect of moving from a 63–90 μm sieve cut to a 125–180 μm sieve cut resulted in a near halving of the dissolution rate at low polymer to drug ratio. At a higher polymer drug ratio, there was a slightly more marked reduction in dissolution rate [19]. In the case of diclofenac sodium, when drug release studies were conducted in water where the drug is soluble, an effect on the rate of release and the mechanism of release was noted with changes in drug particle size [76]. Rate of release was found to increase with decrease in particle size and the mechanism of release was shifted closer to an erosion-dominated drug release mechanism as opposed to a diffusion-dominated mechanism with an increase in drug particle size. Smaller particle size was likely to allow drug to dissolve more readily in the hydrated matrix, enabling a greater role for diffusion. Larger particles were expected to dissolve more slowly, and be more prone to erosion at the hydrated matrix surface, i.e. the erosion front and the diffusion front effectively coincide [76].

With poorly soluble drugs, that there are drug particles in equilibrium with the drug in solution within the hydrated gel layer can result in some interesting dynamics of the position of this layer due to the swelling of the matrix. As the matrix hydrates and swells, the boundary defining the edge of the diffusion front is transported outwards with the swelling hydrated gel, effectively "pushing" the undissolved drug particles through the hydrated gel [7]; such translocation of insoluble particles in a hydrating hydrophilic matrix has been separately described for a model system [1]. The presence of insoluble drug particles in the hydrated gel layer may reduce its resistance to erosion, indeed generating cracks within the hydrated gel, and so these insoluble particles may affect drug release rate as the diffusion front moves out in the gel layer during swelling and may cause some acceleration of the erosive delivery contribution to drug release mechanisms, particularly in the later part of the overall drug release process when most of the matrix has become hydrated [7]. As smaller particle size of API means more particles per unit mass, reducing particle size may translate to greater impact on polymer swelling and erodibility, although no specific research to explore this appears to have been undertaken.

Altering API particle size can influence the percolation threshold for the rate controlling polymer [54]. Percolation theory was first applied to solid pharmaceutical dosage forms to better characterize the role and influence of the individual ingredients to the properties of the finished dosage form [45, 46].

Percolation theory can consider the hydrophilic matrix tablet as a binary system (e.g. of API and hydrophilic polymer, or of polymer and all the other components regarded as a single component). Percolating clusters are formed when particles of the same component are in contact with each other throughout the system and essentially form a continuous phase. This can only occur if the component is present at a concentration that exceeds its percolation threshold. For HPMC it is the concentration at which its hydrated swollen particles are in contact with each other to create a cluster that percolates the entire wetted part of the matrix and yields the coherent strong gel layer. Below the percolation threshold the HPMC will not form the required percolating cluster and although it may apparently provide for a coherent gel layer, the gel layer may be fragile and not offer for good extended release properties [54].

In the case of hydrophilic matrix tablets, it is usually applied to consideration of the amount of polymer (more accurately the volume occupied in the dosage form by the polymer) required to offer robust drug release performance. Once above the percolation threshold for the polymer, further increase in the amount of polymer in the dosage form has no significant effect on drug release rate. Reduction of the level of polymer to below the percolation threshold can lead to increasing susceptibility of the hydrated dosage form to erosion from shear forces (in vitro and in vivo) and hence an effect on drug release (rate increases below the percolation threshold with a reducing level of polymer and the variability in release rate from tablet to tablet may increase). Typically for HPMC the percolation threshold is around 20–30% polymer by weight in the formulation and appears to hold for a wide range of drugs [2, 24, 25, 29–31, 54]. Increasing the particle size of the model drug potassium chloride from a 50–100 μm sieve fraction to a 200–250 μm sieve fraction, at a fixed level of drug and polymer and with controlled particle size of the polymer, led to a small increase in the percolation threshold by around 2% (i.e. more polymer is needed to assure a robust product with increasing particle size of the API); however, as will be described in a subsequent section, the effect of increasing the polymer particle size is more impactful [54]. The effect of API particle shape (aspect ratio), for example the impact of acicular/needle-shaped crystals compared with prism-shaped crystals might be important but, based on published literature, appears not to have been investigated.

10.2.2 API Solubility

API solubility influences drug release rate. When incorporated into the same hydrophilic matrix tablet formulation drugs with a lower aqueous solubility, such as theophylline (1 part is soluble in 120 parts of water) show slower release rates compared to a more freely soluble drug such as chlorpheniramine (1 part is soluble in 4 parts water), tested under the same conditions [48]. There is also an interaction of API solubility with API particle size on drug release rate. For freely soluble drugs (promethazine hydrochloride, propranolol hydrochloride and aminophylline) there was only a small effect, where increasing API particle size resulted in a small increase in drug release rate. This was most marked in formulations containing low levels of polymer (drug: polymer ratio 2.8:1 compared with 0.6:1) and a very large drug particle size fraction (250–500 μm). It was observed that at this extreme the hydrated matrix was fragile and tended to disintegrate [20, 21].

In summary, in order to ensure robust performance of hydrophilic matrix tablets it is important to take into account the API powder properties such as particle size, solubility, compression properties, impact to powder flow as well as the drug load and relationship to the percolation threshold when establishing a viable design space. Determination of the API properties which impact the Critical Quality Attributes (CQAs) of the hydrophilic matrix extended release tablet, and enable setting product specifications supports the QbD approach to product development and helps ensure product performance will be robust.

10.3 Polymer

10.3.1 Effect of Polymer Particle Size on Drug Release

Polymer particle size has been shown to influence the rate of drug release from a hydrophilic matrix tablet formulation and also to affect the mechanism (erosion or diffusion mechanism dominance) by which drug is released. The control of the initial burst release of drug prior to establishing a coherent gel layer was sensitive to polymer particle size.

Alderman [3] explored the effect of using a coarse sieve fraction (material retained on a 150 μm mesh aperture sieve) of hydroxypropyl methylcellulose 2208–4000 MPa s grade compared with unsieved material and material retained on sieves of different mesh aperture (Table 10.1). Material passing through the 150 μm aperture screen was 75% by weight of the original polymer, so properties of the unsieved material might be expected to be dominated by the behavior of the finer fraction component.

It was shown that product formulated with this coarse material yielded tablets that exhibited essentially immediate release of drug (Fig. 10.3), due to failure to establish a coherent gel layer and disintegration of the tablet, although it should be noted the formulation studied had a low level of polymer (10%) [3].

This level of polymer is below the anticipated percolation threshold of 20–30% polymer and this challenge to forming a coherent gel layer may be exacerbated by the large particle size. It has been suggested that the polymer particles need to be close enough together in the dosage form so as they are wetted, hydrate rapidly (smaller particles might be expected to completely hydrate more rapidly than coarse particles) and swell, so that they rapidly interact to establish the gel layer [3, 9]. For a given mass of polymer of coarse particle size, there will be fewer particles in the tablet compared with the same mass of fine particle size material, and so the time to swell and coalesce will be greater, which may allow rapid fluid penetration into the dosage form leading to its disintegration instead of the desired gradual drug release. Increasing polymer level to beyond 20% can mitigate this particle size effect [9].

Indeed, it has been proposed that there is a threshold level for polymer particle size, and reducing polymer particle size below that has no effect on drug release [9, 16, 35, 56, 57] although the effect may be drug dependent [57]. By considering

Table 10.1 Particle size distribution (nested sieve data) of a batch of commercial HPMC 2208–4000 MPa s grade

Fraction	Screen size (mesh aperture, μm)	% by weight retained
A	425	1
B	212	14
C	150	10
D	75	43
E	Through 75	32

Adapted from Alderman [3]

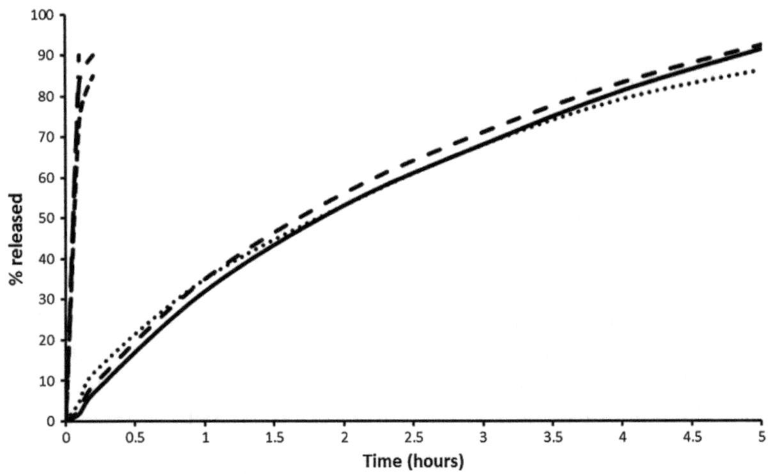

Fig. 10.3 Drug release from HPMC 2208 hydrophilic matrix tablets made with different sieve fractions of polymer. Size fractions as in Table 10.1: A— · —, B······, C— —, D········, E– – –, unsieved polymer——. Redrawn from data in reference [3]

drug (aspirin) release from matrix tablets prepared using a series of sieve fractions of HPMC 2208–15,000 MPa s grade it was determined that a mean particle size of 113 μm (as measured by laser diffraction for the 70–80 μm sieve fraction[1]) was the threshold value below which reducing the polymer mean particle size further (20–30 and 40–50 μm sieve fractions) had no effect on release rate and mechanism [35]. Above that mean size (100–120 and 160–200 μm sieve fractions), the rate of release was increased, and the role of erosion/disintegration in the release rate mechanism was enhanced.

In addition, the effect was influenced by polymer amount employed in the formulation, with rate of release being most affected by use of low levels of greatest mean particle size polymer [9, 35, 56]. Also, the initial rate of release (contributing to a burst effect) was increased by increasing the polymer mean particle size and by reducing the amount of polymer in the formulation. Where mean polymer particle size was 113 μm or greater, then a rapid initial burst release of drug was not observed at greater than 5% polymer level; when the mean particle size was increased to 180 μm a burst effect was still observed at 20% polymer.

In another study, a threshold value of 177 μm, based on the upper edge of the sieve fraction studied, was identified as above which little control of drug release was offered by HPMC 2208–4000 MPa s grade, and where disintegration of tablets mostly occurred. This observation and this threshold value applied to slightly soluble drugs (theophylline), soluble drugs (promethazine hydrochloride) and very

[1]Given the elongated shape of the particles, a significant size difference can be expected between the results coming from sieve analysis and laser diffraction sizing, since these techniques use different principles for size measurement.

soluble drugs (metoprolol tartrate) [16]. It was suggested that very soluble drugs might actually be the more sensitive to the impact of polymer particle size, as the spread between the drug release profiles of tablets made with the different sieve fractions appeared greater for metoprolol tartrate compared to the other two less water soluble drugs [16]. With diclofenac sodium as a model drug it was observed that increasing the particle size of HPMC 2208–15,000 MPa s grade polymer in the matrix tablet from a 45–75 μm, to a 75–150 μm and to a 150–250 μm sieve fraction resulted in a decrease in lag time for equilibrium of the hydration of the matrix before erosion and movement of the hydration front into the matrix was established with increasing particle size. This effect manifested as a marked burst effect for the largest particle size fraction of polymer [76].

There are, however, further factors at work to be considered when exploring particle size fractions prepared by sieving to try to assure consistency of drug release rate. By preparing sieve fractions it is possible that inadvertent creation of material with distinct properties relative to the bulk material may occur, leading to unexpected drug release performance of dosage forms prepared from sieve fractions. The viscosity (2% w/v polymer in water) of sieve fractions of different viscosity grades of HPMC 2208 was found not to be consistent with that of the unsieved polymer, with the finest sieve fractions usually showing markedly lower viscosity and the coarsest fractions sometimes possessing higher viscosity, compared to the unsieved material [57]. Where polymer viscosity is a characteristic impacting drug release, e.g. where an erosion-dominated mechanism is involved, preparing sieve fractions of polymer may confound such efforts to contribute to drug product robustness. Commercially-available hydroxypropyl methylcellulose material of controlled particle size and specifically designated for extended release dosage form applications is offered (Methocel™ CR, Dow Chemical Company; Benecel™, Ashland) and can avoid having to screen material with its associated risks [4, 17].

Material of equivalent chemistry from different vendors may have different particle properties, potentially resulting in non-equivalent performance. Although it was demonstrated that there was an effect of polymer mean particle size on in vitro drug release, and indeed in vivo drug pharmacokinetics for propylthiouracil formulated into matrix tablets containing 30% w/w HPMC 2910 from two different vendors (Methocel E4 M and Metolose™ 60SH), the effects of particle size modification of the polymer, either by sieving or by micronization, was different for the material from the two vendors [42]. For one polymer (E4 M) material, using a coarse fraction obtained by sieving (sieve fraction >125 μm) resulted in very rapid (all drug released within 1 h) in vitro drug release compared with the product made with the <125 μm sieve fraction (60% drug release at 8 h), whereas for the second material (60SH) an increase in drug release rate was noted for the coarser polymer fraction but the effect was much less marked (coarse fraction showed approximately 75% drug release at 8 h versus about 60% released at 8 h for the finer fraction). Micronizing the polymers only really affected the performance of the E4 M polymer, slowing the amount of drug released at 8 h to 40%, whereas for the 60SH polymer the amount of drug released at 8 h for the formulation made with the

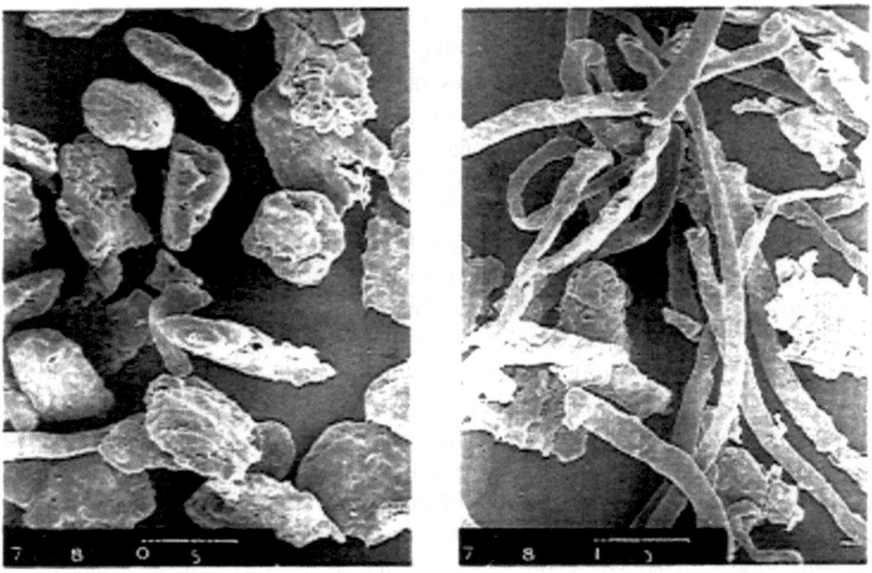

Fig. 10.4 Electron microscopy images of sieved (90–125 μm fraction) hydroxypropyl methylcellulose 2910, E4 M (left image) and 60SH (right image); 150× magnification [42]; copyright Springer; reproduced with permission

<125 μm sieve fraction and that made with polymer subjected to micronization were more or less the same. The shape of the particles may have contributed to this difference in behavior, with the E4 M particles appearing as somewhat irregular, oval-shaped particles, whereas the 60SH polymer showed mostly fibrous material (Fig. 10.4). Not only will this distinct particle shape difference affect sieving and milling, it also might impact hydrated matrix forming behavior [42]. Caution must be exerted therefore when substituting polymer from different vendors into an established product, and in considering a substitution, careful material characterization, including proper examination of particle properties, including microscopic form, will be important.

Interestingly, comparing drug release from HPMC formulations made from polymer of the same measured viscosity (4000 MPa s) but of different substitution chemistry (2208, 2910 and 2906) suggested an effect of particle size and particle shape on drug release for the 2910 grade. Sieve fractions allowed the separation of different size and shape fractions of this grade, the shape fractions being defined as having a more fibrous particle or a more granular particle appearance under optical microscopy. For equivalent chemistry and particle size fractions the shape fractions exhibited different drug release profiles. This substitution grade also showed higher lot-to-lot variability compared to the other two types, in terms of packing properties, particle shape and drug release from matrices [8].

The percolation threshold for the polymer is influenced by its particle size. Percolation theory would imply that in a binary system, the effect of increasing the polymer particle size is similar to a reduction in API particle size—the relative particle size of each component, not its absolute particle size impact the overall characteristics of the system. For a water soluble drug, potassium chloride, it was shown that the smaller the excipient/drug particle size ratio then the smaller the polymer percolation threshold. Large excipient particles are less effective in percolating the system and result in the increase in the percolation threshold [55].

Initial tablet porosity is an important factor in defining the percolation threshold [55]. The effect of making very small matrix tablets (3 mm diameter) for a given formulation composition with defined particle properties of drug and polymer was found to shift the percolation threshold higher compared with regular size tablets (7 mm diameter) from the same blend [58]. This was related to differences in tortuosity (having a convoluted, difficult pathway) between the very small and regular size matrix tablets. The very small tablets had a higher porosity. A pore of given size in a small tablet would occupy greater overall volume than a pore of the same size in a larger tablet and so permit a greater extent of water ingress before the polymer can hydrate and influence drug release. Hence an increase of polymer level sufficient to achieve rapid gel layer formation more effectively is required, which translates as a higher percolation threshold in small size tablets. It should be noted that this was only really manifest with a coarse sieve fraction (125–355 μm) of polymer [58], so not only having the appropriate amount polymer but controlling its particle size is perhaps even more important when formulating very small size hydrophilic matrix tablets.

The hydration and swelling of HPMC can be impaired by the presence of sugars in the release medium or added into the dosage form [79] and it has been demonstrated that sieve fractions 63–90 μm and <63 μm showed some resistance to these effects, hence selecting a fine particle fraction may provide for the preparation of more robust HPMC matrix formulations [78].

As noted earlier in this chapter, virtually all published research on particle properties and hydrophilic matrix tablets has been undertaken on HPMC, but the effect of polymer particle size on drug release from hydrophilic matrix tablets in that body of work was confirmed as translating to another increasingly used polymer, polyethylene oxide. Increasing the polyethylene oxide sieve fraction particle size used in preparing tablets was found to result in an increase in drug (theophylline) release rate and a change in mechanism of drug release, and these effects were more marked at low polymer content in the tablets, particularly when high molecular weight polymer was used [71].

In summary, the polymer properties are of critical importance to the performance of hydrophilic matrix tablets. Polymer particle size and shape as well as the level of polymer in the formulation drive the drug release characteristics. For this reason, suppliers provide many different grades of polymer with well controlled particle properties specifically designed for extended release formulations. Also for this reason material from different suppliers should not necessarily be considered interchangeable, even between similar grades, and studies should be performed prior to switching sources. Consideration needs to be given to the drug solubility, the drug: polymer particle size ratio and porosity impact to the percolation threshold and even to other materials in the formulation which could impact the hydration and swelling of the polymer.

10.3.2 Effect of Polymer Particle Properties on Dosage Form Processing—Flow and Compaction

As Fig. 10.2 illustrates, it is important to take the powder properties into account in the manufacture of hydrophilic matrix dosage forms. Flow, compressibility and compactability of the hydrophilic polymer, as well as its triboelectric properties, may have influence on the ability to produce extended release tablets employing these polymers. Particle size and particle size distribution can affect flow, and due to presentation of surfaces for particle-particle bonding, can impact compaction properties. This may be especially true of HPMC 2208 given it exhibits a predominantly plastic deformation mechanism for compaction, and this is affected by plasticization of the polymer induced by increasing its moisture content or by deliberately introducing hydrophilic plasticizers such as propylene glycol [33, 59, 60].

The range of factors, mechanical properties, and polymer chemistry aspects, influencing the powder compaction properties of the cellulose ethers, notably hydroxypropyl methylcellulose for extended release products, have been recently reviewed [27].

Improvement in compactability of hydroxypropyl methylcellulose (2208, 2910 and 2906 grades) with decrease in particle size was found. Data generated at different humidity conditions suggested that the effect was driven by changes in moisture distribution in the polymer that was mostly related to the particle size changes, although there was a lesser effect associated with the methoxy/hydroxypropoxy substitution ratio for the polymers studied [52]. The facilitation of compression with particle size was suggested as being due to improved plastic deformation at the early stages of compression associated with moisture induced plasticisation, and enhanced flow/particle movement due to a lubricating action of moisture at the polymer particle surfaces. Elastic recovery of compacted polymer is related to particle size, with greater elastic recovery being seen with coarser sieve fraction of HPMC [52] and consequently generates challenges in making robust (high crushing strength) tablets with very coarse sieve fraction polymer [56]. Particle size of the polymer was further confirmed as a significant factor affecting the tensile strength of compacts of HPMC 2208, with the finest particle fraction (<45 μm) of the lowest viscosity grade (100 MPa s) investigated having the best compaction properties [61]. However, this low viscosity grade may not necessarily allow for the desired drug release profile and so the final formulation developed, and the manufacturing process employed, may have to accommodate the compaction properties of the polymer.

That a finer particle grade fraction of polymer is thus key to formulating robust dosage forms in terms of the drug release profile, and that polymer—which may have poor inherent compaction properties—can be a significant part by weight of the formulation may lead to poor powder blend flow and means that a direct compression manufacturing approach is unlikely to be successful. Whilst roller compaction for dry granulation may be feasible [69] feed into the compactor could be problematic unless coarse grade polymer is selected, which had the disadvantage that it tends to result in tablets of lower tensile strength [34]. Wet granulation is an option for enabling tableting of hydrophilic matrix tablets through particle enlargement of powder blends, but as the (likely high viscosity grade) hydrophilic polymer is a significant component of the blend this process option can be challenging. Granulation water amount, water addition rate, and in the case of HPMC substitution chemistry, can influence granule formation, granule size and density, and compactability [5, 36, 70].

Hence, two manufacturers of HPMC, Dow and Ashland, have introduced direct compression grades of this polymer to help the efficient manufacture of hydrophilic matrix tablets. The HPMC direct compression (DC) versions of the standard (controlled release, CR) materials have been developed to overcome the challenges presented by CR grade polymer, associated with its fibre-like small particles, prone to surface charging and exhibiting cohesive nature and so provide for improved flow with the goal of enabling direct compression processes for hydrophilic matrix tablets. This provides a manufacturing benefit, avoiding additional processing steps and possible control challenges related to dry granulation (additional steps of roller compaction and milling, with associated reduced compactability) and wet granulation (additional steps of wet granulation, drying, and milling, difficult granulation end point control, risk to moisture sensitive products).

Comparative particle size distribution data for the polymers from the two suppliers is given in Tables 10.2 and 10.3 and scanning electron micrographs are given in Figs. 10.5 and 10.6. The Dow DC grade material exhibits larger particles relative to the regular CR grades, which will better assure powder flow. In contrast the sieve analysis suggests that the Ashland DC material has similar or slightly smaller particles relative to the standard CR grade. It must be noted that these reported

Table 10.2 Particle size analysis of CR and DC grade HPMC 2208 from Ashland (sieve analysis)

HPMC grade	Geometric mean particle size (μm)	Median particle diameter (d_{50}, μm)	d_{10} (μm)	d_{90} (μm)	Distribution width (d_{90}–d_{10})/d_{50} (%)
4000 MPa s CR	60.1	74.8	33.8	183.3	199.9
4000 MPa s DC	51.4	71.5	28.2	142.3	159.6
100,000 MPa s CR	55.5	72.8	28.7	199.7	234.9
100,000 MPa s DC	48.5	69.2	27.1	138.7	161.3

Table 10.3 Particle size analysis (sieve analysis) of CR and DC grade HPMC 2208 from Dow adapted from [32]

HPMC grade	Median particle diameter ($d_{50,}$ µm)	d_{16} (µm)	d_{84} (µm)	Distribution width (d_{84}–d_{16})/d_{50} (%)
4000 MPa s CR	80.7	52.6	108.8	69.6
4000 MPa s DC	96.9	63.8	129.9	68.2
100,000 MPa s CR	66.7	31.6	101.8	105.2
100,000 MPa s DC	91.1	61.5	120.6	64.9

particle size distributions we generated using sieve analysis which is subject to influence of particle shape. In addition, agglomerated materials may present a challenge to volume-based particle size methods and small sub-populations of larger agglomerates within a cellulosic excipient can have beneficial effects on flow properties, without necessarily being identified by sieve analysis [26]. This may be the case here in determining a difference or similarity of particle size of CR and DC grade materials by sieve analysis.

DC grade polymer from Ashland appears as agglomerates of fibre-like particles to form larger particles with an elongated aspect, looking like larger versions of the particles in the regular CR grade. Material for direct compression from Dow appears as ovoid particles, looking like agglomerates of smaller fibre-like particles. This direct compression grade material appears to be made of the fibre-like particles that the CR grade shows under the scanning electron microscope. The surface of the particles in the Ashland direct compression material appears smoother than those that make up the Dow direct compression grade polymer.

As might be expected the upward shift in particle size for the DC grades of the material from Dow led to improved flowability as indicated by Carr index [10] moving from 37.9% for 4000 MPa s CR grade to 29.4% for 4000 MPa s DC grade and from 38.3 to 33.7% for 100,000 MPa s CR grade and 100,000 MPa s DC grade respectively [32]. Although these changes suggested improvements in flowability with the DC grade, the flow properties of even the DC grades are still classified as poor to very poor as indicated by the values of the Carr index.

Polymer particle shape and size will influence packing in a powder bed which impacts its bulk density. This may modulate the flow properties of a blend for tabletting containing the polymer. Moving to the modified particles in the case of Ashland DC material does not have an effect on bulk density but DC material from Dow has a lower bulk density relative to the CR material from that supplier [4]. These differences likely relate to particle shape effects on packing and/or the change in distribution width on making particles for the DC grade material.

By exploring the comparative properties of CR grades and DC grades in pilot scale high speed tableting trials on commercial-equivalent tablet formulations demonstrated that the DC grade of the Dow HPMC 2208–100,000 MPa s yielded tablets of a lower standard deviation for weight uniformity but of lower crushing strength [43], associated with the inherently poorer compactability of the DC grade

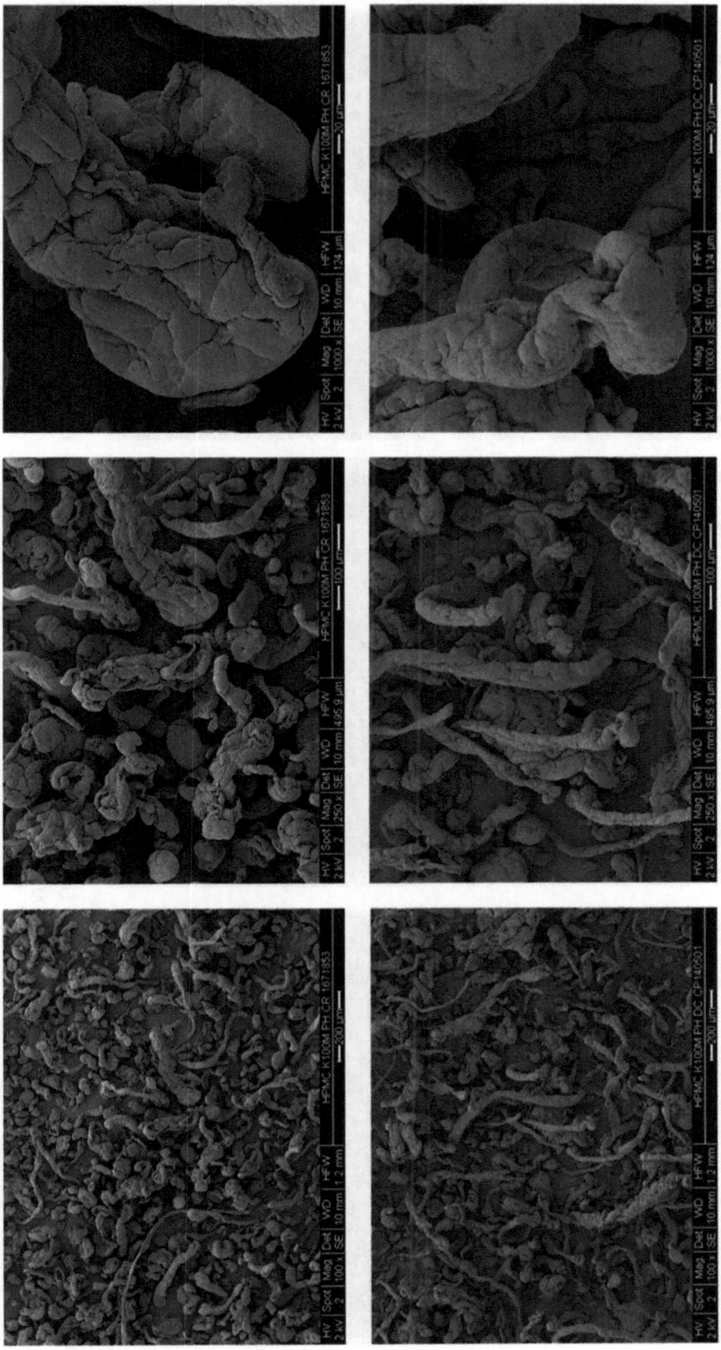

Fig. 10.5 Scanning electron micrographs of CR and DC grades of HPMC 2208–100,000 MPa s grade from Ashland. Top CR grade—left 100×, middle 250×, right 1000×; bottom DC grade—left 100×, middle 250×, right 1000×

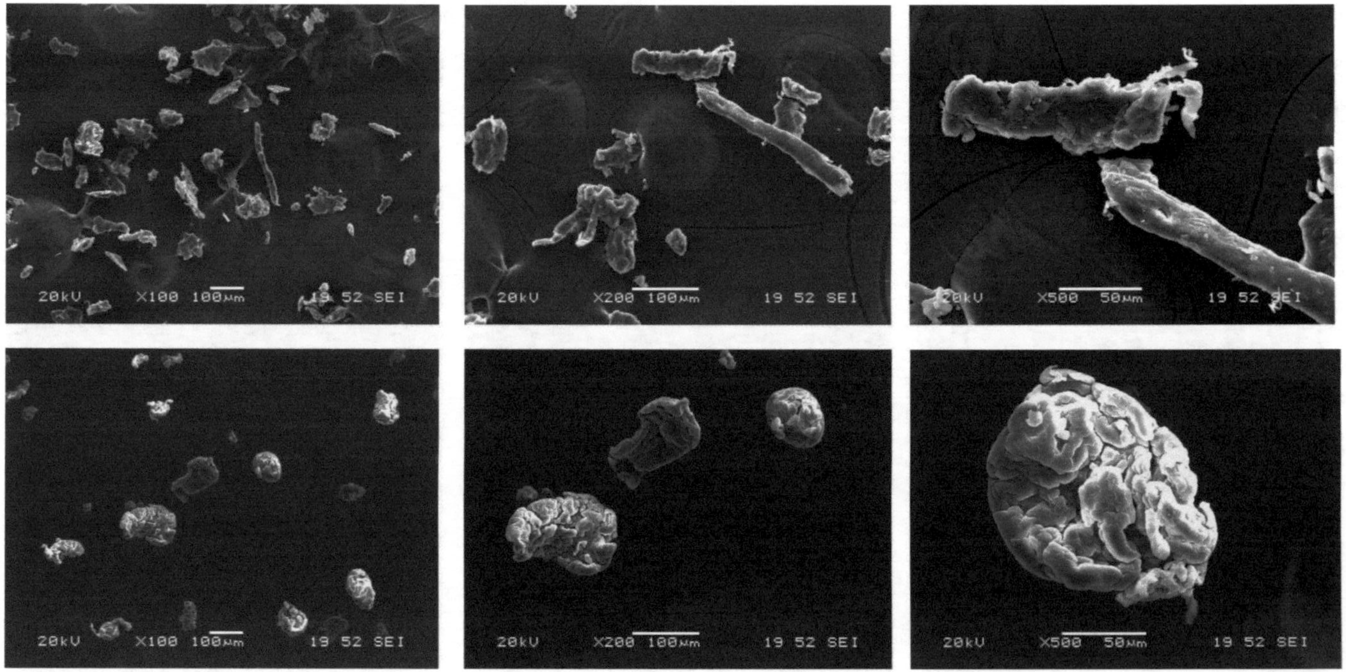

Fig. 10.6 Scanning electron micrographs of CR and DC grades of HPMC 2208–100,000 MPa s grade from Dow. Top CR grade—left 100× middle 200×, right 500×; bottom DC grade—left 100×, middle 200×, right 500×

relative to the CR grade [32]. The latter effect is perhaps due to the increased particle size reducing the available surface for particle-particle contacts key to bonding for plastically deforming material [32] or may be associated with the particle enlargement technique employed in the fabrication of the DC grade that consumes some of the compactability of the polymer. Investigation of Ashland DC and CR grade HPMC 2208–100,000 MPa s in a different extended release tablet formulation showed that again the DC grade of polymer improved flow on a high-speed tablet press as indicated by improvements in weight uniformity, as well as a beneficial increase in tablet crushing strength compared with tablets made with CR grade polymer (suggesting the DC grade has overall better compactability relative to the CR grade). This also was demonstrated for a formulation with a high load (50% w/w) of a poorly flowing API (metformin hydrochloride) [67].

These studies indicate that the poor flow potential indicated by Carr's index do not necessarily translate to a problem with high speed manufacturing, and indeed an alternate dynamic flowability test clearly confirmed the marked improvement on blend flowability when prepared with DC grade polymer, and that this allows a simple mixing and high speed tabletting process [64]. The application of DC grade polymer to high speed continuous matrix tablet manufacture has been demonstrated, where the DC polymer enabled the production of tablets with improved quality relative to the use of CR grade [18, 74]. It was observed that there may be a shift in the percolation threshold in the case of DC material relative to the CR grade, not surprisingly given the shift in mean particle size and the known influence of polymer particle size on percolation threshold, as indicated by a greater sensitivity to mixing conditions of drug release rate for tablets made with DC grade [18].

10.3.3 Tribo-Electric Properties of HPMC

HPMC as used in extended release dosage forms is an insulating material. It has relatively low bulk density and particle size and similar to many pharmaceutical materials can be subject to tribo-electric charging. Powder handling and processing causes particles of a pharmaceutical powder to be subject to impact, sliding and friction between themselves and the walls of the processing equipment resulting in electrical charging of the particles. The static charge on the particles causes them to attract or repel each other or to adhere to grounded surfaces. This can result in problems in blending powders, including loss of material, poor homogeneity, and segregation [72].

HPMC was found to charge positively against a stainless-steel surface and adhere to the surface [28]. Methoxy and hydroxypropoxy substitution and polymer molecular weight was found to impact the magnitude of the acquired charge and the consequent degree of adhesion. HPMC 2910 was found to acquire a greater degree of charge relative to the same molecular weight (4000 MPa s grade) HPMC 2208. For HPMC 2208 increased viscosity grade material acquired a smaller charge (4000 MPa s > 15,000 MPa s > 100,000 MPa s). For each grade reducing the particle size increased the magnitude of the acquired charge as expected. For binary mixtures of polymer with a model drug, flurbiprofen, which charged negatively against stainless steel, showed that the charging effect of the two materials could be partially cancelled, the degree being dependent on the relative proportions of drug and polymer. At 15% by weight polymer, the HPMC:flurbiprofen mixture there was no overall net charge on the mixture and adhesion of the blend to stainless steel was very low [28]. These studies suggest that a knowledge of charging propensity of drug, polymer, and other excipients along with their mixtures may be useful in mitigating risk of adhesion to surfaces during processing. On the positive side, the higher viscosity grades of HPMC 2208 often used in hydrophilic matrix tablets have the lowest propensity to develop charge and adhere.

10.4 Particulate Properties of Added Excipients in Hydrophilic Matrix Tablets

The effect of added excipients (added to the binary mix of API and polymer) in hydrophilic matrix tablets has shown variable effects in increasing or decreasing drug release rate dependent on the nature of the drug (soluble or poorly soluble), the added excipient or excipients (soluble or insoluble) and the amount of drug included or excipient employed, often in complex ways and often generating conflicting data amongst similar studies [23, 39, 41, 44, 47, 50, 63, 80]. Most of these investigations focussed principally on measuring drug release rate changes as a consequence of excipient addition, with some studies aiming to elucidate mechanisms but usually indirectly (e.g. erosion rates). The tools to directly visualise and explore the effect of these variables on the initiation and evolution of the gel layer have been developed [6] and have applied to exploring the effect of the solubility of and the amount of excipient incorporated into a hydrophilic matrix [62], but to date no exploration of the influence of the particle size of the added excipient has been reported.

10.5 Definitions and Abbreviations

Some of the terms used freely in this chapter, including as acronyms, are described in greater detail in this section.

Percolation threshold

Percolation theory can consider the HPMC hydrophilic matrix tablet as a binary system (e.g. of HPMC as one component and all the other components together regarded as the second component). Percolating clusters are formed when particles of the same component are in contact with each other throughout the system and essentially form a continuous phase, a point where the component exceeds its percolation threshold. For HPMC it is the concentration (more accurately the volume occupied in the dosage form by the polymer) at which its hydrated swollen particles are in contact with each other to create a cluster that percolates the entire wetted part of the matrix and yield the coherent strong gel layer. Below the percolation threshold although it may apparently provide for a coherent gel layer, the HPMC will not form the required percolating cluster and so the gel layer may be fragile and not offer for robust extended release properties (e.g. very sensitive to erosion by shear) resulting in high variability in release between tablets and a possible effect on clinical variability.

Extended, controlled, sustained, prolonged, and modified release

The term "controlled release" is widely used to imply that some action has been taken to assure that drug release from a dosage form will occur in a predefined way and distinct from immediate release (on occasion referred to as "prompt release"). The terminology controlled release has been largely superseded, but alternate nomenclature can operate in different territories based on whether the European Pharmacopoeia (EP), the United States Pharmacopoeia (USP) or the International Conference on Harmonization (ICH) is adhered to. The USP has a blanket term to cover all oral dosage forms that not designed to provide for immediate release of their contained drug, and this blanket category is "modified release". Under modified release we have the two subclasses of "delayed release" (DR) and "extended release (ER or, sometimes, XR). Enteric coated or gastro-resistant products that have an acid-insoluble polymer coat that should only dissolve and allow drug release once the dosage from has left the stomach, are a typical example of DR products. ER products release their contained drug over an extended time period, usually several hours, and hydrophilic matrix tablets represent a widely-applied approach to extended release products. The EP categorises modified release products similarly to the USP but uses the term prolonged release instead of extended release. ICH has adopted the USP approach to terminology.

API: Active pharmaceutical ingredient

ICH defines API as "any substance or mixture of substances intended to be used in the manufacture of a drug product and that, when used in the production of a drug, becomes an active ingredient in the drug product. Such substances are intended to furnish pharmacological activity or other direct effect in the diagnosis, cure, mitigation, treatment or prevention of disease or to affect the structure and function of the body" which is broad based and includes biologics as well as small molecule compounds.

HPMC: Hydroxypropyl methylcellulose

HPMC is a modified cellulose, which is a β-1-4 glycosidic-bonded glucose linear polymer, with ether linked methoxyl and hydroxypropoxyl substituents attached to the hydroxyl groups of the cellulose chain. Typically, it is prepared from cellulose by reacting it with methyl chloride and propylene oxide. HPMC has a greater degree of methoxy substitution relative to hydroxypropoxy substitution.

QbD: Quality by Design: A systematic approach to drug product development that begins with predefined objectives and emphasizes product and process understanding plus process control, all based on sound science and quality risk management. Modeling and simulation is an established part of quality by design that is taking on a greater role in product development, from both a manufacturing and process understanding, as is evidenced by its appearance in regulatory guidances and scientific literature.

Acknowledgements The authors would like to thank Dr. Kofi Asare-Addo, Department of Pharmacy, University of Huddersfield for the images in Fig. 10.6.

References

1. Adler J, Jayan A, Melia CD (1999) A method for quantifying differential expansion within hydrating hydrophilic matrixes by tracking embedded fluorescent microspheres. J Pharm Sci 88:371–377
2. Aguilar-de-Leyva A, Cifuentes C, Rajabi-Siahboomi AR, Caraballo I (2012) Study of the critical points and the role of the pores and viscosity in carbamazepine hydrophilic matrix tablets. Eur J Pharm Biopharm 80:136–142
3. Alderman DA (1984) A review of cellulose ethers in hydrophilic matrices for oral controlled release dosage forms. Int J Pharm Tech Prod Manuf 5:1–9
4. Ashland (2013) Benecel™ PH DC HPMC, direct compression HPMC. Ashland Speciality Ingredients, Wilmington DE, USA
5. Bajdik J, Baki G, Kelemen A, Pintye-Hódi K (2008) The effect of wetting of powder mixture on the preparation of hydrophilic matrix granules with high-shear granulator. Chem Eng Res Des 86:1–7
6. Bajwa GS, Hoebler K, Sammon C, Timmins P, Melia CD (2006) Microstructural imaging of early gel layer formation in HPMC matrices. J Pharm Sci 95:2145–2157
7. Bettini R, Catellani PL, Sant P, Massimo G, Peppas N, Colombo P (2001) Translocation of drug particles in HPMC matrix gel layer: effect of drug solubility and influence on release rate. J Controlled Release 70:383–391
8. Bonferoni MC, Rossi S, Ferrari F, Bertoni M, Caramella C (1996) A study of three hydroxypropylmethyl cellulose substitution types: effect of particle size and shape on hydrophilic matrix performances. STP Pharma Sci 6:277–284
9. Campos-Aldrete ME, Villafuerte-Robles L (1997) Influence of the viscosity grade and the particle size of HPMC on metronidazole release form matrix tablets. Eur J Pharm Biopharm 43:173–178
10. Carr RL (1965) Evaluating flow properties of solids. Chem Eng 72:163–168
11. Christensen GL, Dale LB (1962) Sustained release tablet. US Patent US3065143
12. Colombo P (1993) Swelling-controlled release in hydrogel matrices for oral use. Adv Drug Del Rev 11:37–57
13. Colombo P, Bettini R, Massimo G, Catellani PL, Sant P, Peppas N (1995) Drug diffusion front movement is important in drug release control from swellable matrix tablets. J Pharm Sci 84:991–997
14. Colombo P, Bettini R, Santi P, De Ascentis A, Peppas N (1996) Analysis of the swelling and release mechanisms from drug delivery systems with emphasis on drug solubility and water transport. J Controlled Release 39:231–237

15. Conte U, Colombo P, Gazzaniga A, Sangali ME, La Manna A (1988) Swelling-activated drug delivery systems. Biomaterials 9:489–493
16. Dow Chemical Company (2006) Technical bulletin: using Dow excipients for controlled release of drugs in hydrophilic matrix systems. The Dow Chemical Company, Midland, MI, USA, pp 16–17
17. Dow Chemical Company (2013) Technical bulletin: cellulose ethers—a technical review. The Dow Chemical Company, Midland, MI, USA, p 3
18. Ervasti T, Simonaho S, Ketolainen J, Forsberg P, Fransson M, Wikström H, Folestad S, Lakio S, Tajarobi P, Abrahamsén-Alami S (2015) Continuous manufacturing of extended release tablets via powder mixing and direct compression. Int J Pharm 495:290–301
19. Ford JL (2014) Design and evaluation of hydroxypropyl methylcellulose matrix tablets for oral controlled release: a historical perspective. In: Timmins P, Pygall SR, Melia CD (eds) Hydrophilic matrix tablets for oral controlled release. AAPS Press/Springer, New York, NY, USA
20. Ford JL, Rubinstein MH, Hogan JE (1985a) Formulation of sustained release promethazine hydrochloride tablets using hydroxypropylmethylcellulose matrices. Int J Pharm 24:327–338
21. Ford JL, Rubinstein MH, Hogan JE (1985b),Propranolol hydrochloride and aminophylline release from matrix tablets containing hydroxypropylmethylcellulose. *Int. J. Pharm.* 24:339–350
22. Ford JL, Rubinstein MH, Hogan JE (1985c) Dissolution of a poorly water soluble drug, indomethacin, from hydroxypropylmethylcellulose controlled release tablets. J Pharm Pharmac 37, Suppl., 33P
23. Ford JL, Rubinstein MH, McCaul F, Hogan JE, Edgar PJ (1987) Importance of drug type, tablet shape, and added diluents on drug release kinetics from hydroxypropylmethyl-cellulose matrix tablets. Int J Pharm 40:223–234
24. Fuertes I, Caraballo I, Miranda A, Millán M (2010) Study of critical points of drugs with different solubilities in hydrophilic matrices. Int J Pharm 383:138–146
25. Fuertes I, Miranda A, Millán M, Caraballo I (2006) Estimation of the percolation thresholds in acyclovir hydrophilic matrix tablets. Eur J Pharm Biopharm 64:336–342
26. Gamble JF, Chiu W, Tobyn M (2011) Investigation into the impact of sub-populations of agglomerates on the particle size distribution and flow properties of conventional microcrystalline cellulose grades. Pharm Dev Tech 16:542–548
27. Ghori MU, Conway BR (2016) Powder compaction: compression properties of cellulose ethers. Br J Pharm 1:19–29
28. Ghori MU, Šupuk E, Conway BR (2014) Tribo-electric charging and adhesion of cellulose ethers and their mixtures with flurbiprofen. Eur J Pharm Sci 65:1–8
29. Gil EC, Colare AI, Bataille B, Brouillet F, Caraballo I (2009) Estimation of the percolation thresholds in ternary lobenzarit disodium-dextran-HPMC hydrophilic matrix tablets: effects of initial porosity. Eur J Pharm Sci 38:312–319
30. Gonçalves-Araújo T, Rajabi-Siahboomi AR, Caraballo I (2008) Application of percolation theory in the study of an extended release verapamil hydrochloride formulation. Int J Pharm 361:112–117
31. Gonçalves-Araújo T, Rajabi-Siahboomi AR, Caraballo I (2010) Polymer percolation threshold on HPMC extended-release formulation of carbamazepine and verapamil. AAPS PharmSciTech 11:558–562
32. Grdešič P, Vrečer F, Ilič I (2016) Flow and compaction properties of hypromellose: new directly compressible versus the established grades. Drug Dev Ind Pharm 42:1877–1886
33. Hardy IJ, Cook WG, Melia CD (2006) Compression and compaction properties of plasticised high molecular weight hydroxypropylmethylcellulose (HPMC) as a hydrophilic matrix carrier. Int J Pharm 311:26–32
34. Heiman J, Tajarobi F, Gururajan B, Juppo A, Abrahansén-Alami S (2015) Roller compaction of hydrophilic matrix tablets—combined effects of processing variables and drug/matrix former particle size. AAPS PharmSciTech 16:267–277

35. Heng PWS, Chan LW, Easterbrook MG, Li X (2001) Investigation of the influence of mean HPMC particle size and number of polymer particles on the release of aspirin from swellable matrix tablets. J Controlled Release 76:39–49
36. Herder J, Adolfsson Å, Larsson A (2006) Initial studies of water granulation of eight grades of hypromellose (HPMC). Int J Pharm 313:57–65
37. Hogan JE (1989) Hydroxypropylmethylcellulose sustained release technology. Drug Dev Ind Pharm 15:975–999
38. Hou H, Sun CC (2008) Quantifying effects of particulate properties on powder flow properties using a ring shear tester. J Pharm Sci 97:4030–4039
39. Huang YB, Tsai YH, Yang WC, Chang JS, Wu PC (2004) Optimization of sustained-release propranolol dosage form using factorial design and response surface methodology. Biol Pharm Bull 27:1626–1629
40. Huber HE, Dale LB, Christensen GL (1966) Utilization of hydrophilic gums for the control of drug release from tablet formulations I. Disintegration and dissolution behavior. J Pharm Sci 55:974–976
41. Jamzad S, Tutunji L, Fassihi R (2005) Analysis of macromolecular changes and drug release from hydrophilic matrix systems. Int J Pharm 292:75–85
42. Kabanda L, Lefebvre RA, Van Bree HJ, Remon JP (1994) In vitro and in vivo evaluation in dogs and pigs of a hydrophilic matrix containing propylthiouracil. Pharm Res 11:1663–1668
43. LaBella G, Theuerkauf J, Rogers TL (2014) Pilot scale evaluation of Methocel™ DC2 for robust matrix tablet manufacture by direct compression. Poster presented at the controlled release society annual meeting, Chicago, IL, USA
44. Lapidus H, Lordi NG (1968) Drug release from compressed hydrophilic matrices. J Pharm Sci 57:1292–1301
45. Leuenberger H, Rohera BD, Hass C (1987) Percolation theory—a novel approach to solid dosage form design. Int J Pharm 90:213–219
46. Leuenberger H (1999) The application of percolation theory in powder technology. Adv Powder Technol 10:323–352
47. Levina M, Rajabi-Siahboomi AR (2004) The influence of excipients on drug release from hydroxypropyl methylcellulose matrices. J Pharm Sci 93:2746–2754
48. Levina M, Rajabi-Siahboomi AR (2014) An industrial perspective on hydrophilic matrix tablets based on hydroxypropyl methylcellulose (hypromellose). In: Timmins P, Pygall SR, Melia CD (eds) Hydrophilic matrix tablets for oral controlled release. AAPS Press/Springer, New York, NY, USA
49. Li CL, Martini LG, Ford JL, Roberts M (2005) The use of hypromellose in oral drug delivery. J Pharm Pharmacol 57:533–546
50. Loftipour F, Nokhodchi A, Saeedi M, Norouzi-Sani S, Sharbafi J, Siahi-Shadbad MR (2004) The effect of hydrophilic and lipophilic polymers and fillers on the release rate of atenolol from HPMC matrices. Farmaco (Lausanne) 59:819–825
51. Maderuelo C, Zarzuelo A, Lanao JM (2011) Critical factors in the release of drugs from sustained release hydrophilic matrices. J Control Release 154:2–19
52. Malamataris S, Karidas T, Giodas P (1994) Effect of particle size and sorbed moisture on the compression behaviour of some hydroxypropyl methylcellulose (HPMC) polymers. Int J Pharm 103:205–215
53. Melia CD (1991) Hydrophilic matrix sustained release systems based on polysaccharide carriers. Crit Rev Ther Drug 8:395–421
54. Miranda A, Millán M, Caraballo I (2006) Study of the critical points of HPMC hydrophilic matrices for controlled drug delivery. Int J Pharm 311:75–81
55. Miranda A, Millán M, Caraballo I (2007) Investigation of the influence of particle size on the excipient percolation thresholds of HPMC hydrophilic matrix tablets. J Pharm Sci 96:2746–2756
56. Mitchell K, Ford JL, Armstrong DJ, Elliott PNC, Hogan JE, Rostron C (1993) The influence of particle size of hydroxypropyl methylcellulose K15 M on its hydration and performance in matrix tablets. Int J Pharm 100:175–179

57. Mitchell SA, Balwinski KM (2007) Investigation of hypromellose particle size effects on drug release from sustained release hydrophilic matrix tablets. Drug Dev Ind Pharm 33:952–958
58. Mohamed FAA, Roberts M, Seton L, Ford JL, Levina M, Rajabi-Siahboomi AR (2015) The effect of HPMC particle size on the drug release rate and the percolation threshold in extended-release mini-tablets. Drug Dev Ind Pharm 41:70–78
59. Nokhodchi A, Ford JL, Rowe PH, Rubinstein MH (1996) The influence of moisture content on the consolidation properties of hydroxypropylmethylcellulose K4 M (HPMC 2208). J Pharm Pharmacol 48:1116–1121
60. Nokhodchi A, Ford JL, Rowe PH, Rubinstein MH (1996) The effect of moisture on the Heckel and energy analysis of hydroxypropylmethylcellulose 2208 (HPMC K4 M). J Pharm Pharmacol 48:1122–1127
61. Nokhodchi A, Rubinstein MH, Ford JL (1995) The effect of particle size and viscosity grade on the compaction properties of hydroxypropylmethylcellulose 2208. Int J Pharm 126:189–197
62. Pygall SR (2008), Critical processes in drug release from HPMC controlled release matrices. Ph.D. thesis, University of Nottingham
63. Rekhi GS, Nellore RV, Hussain AS, Tillman LG, Malinowski HJ, Augsburger LL (1999) Identification of critical formulation and processing variables for metoprolol tartrate extended release (ER) matrix tablets. J Control Release 59:327–342
64. Rogers TL, Hewlett KO, Theuerkauf J, Balwinski KM (2013) Assessing how the physical properties of enhanced powder flow of HPMC affect process control during direct compression of matrix tablets. Tablets Capsul (October):14–22
65. Roy DS, Rohera BD (2002) Comparative evaluation of rate of hydration and matrix erosion of HEC and HPC and study of drug release from their matrices. Eur J Pharm Sci 16:193–199
66. Salomon JL, Vuagnat P, Doelker E, Buri P (1979) Importance of technology and formulation for the mechanism of release of potassium chloride from hydrophilic matrices. 2: Influence of compression force, drug particle size and tablet thickness. Pharm Acta Helv 54:86–89
67. Schwing JA, Davis M, Tewari D, Dürig T (2014) Evaluation of directly-compressible hypromellose in commercial-scale manufacturing of a modified-release matrix tablet. Poster presented at the AAPS annual meeting. San Diego, CA, USA
68. Shah UV, Karde V, Ghoroi C, Heng JYY (2017) Influence of particle properties on powder bulk behavior and processability. Int J Pharm 518:138–154
69. Sheskey P, Pacholke K, Sackett G, Mather L, Polli J (2000) Effect of process scale-up on robustness of tablets, tablet stability, and predicted in vivo performance. Pharm Tech 24 (11):30–52
70. Sheskey PJ, Williams DM (1995) Comparison of low-shear and high-shear wet granulation techniques and the influence of percent water addition in the preparation of controlled-release matrix tablet containing HPMC and a high-dose, highly water-soluble drug. Pharm Tech 19 (3):80–91
71. Shojaee S, Enami P, Mahmood A, Rowaiye Y, Dukulay A, Kaialy W, Cumming I, Nokhodchi A (2015) An investigation on the effect of polyethylene oxide concentration and particle size in modulating theophylline release from tablet matrices. AAPS PharmSciTech 16:1281–1289
72. Šupuk E, Zarrebini A, Reddy JP, Hughes H, Leane MM, Tobyn MJ, Timmins P, Ghadiri M (2012) Triboelectrification of active pharmaceutical ingredients and excipients. Powder Tech 217:427–434
73. Timmins P, Desai D, Chen W, Wray P, Brown J, Hanley S (2016) Advances in mechanistic understanding of release rate control mechanisms of extended release hydrophilic matrix tablets. Ther Deliv 7:553–572
74. Van Snick B, Holma J, Cunningham C, Kumar A, Vercruysse J, De Beer T, Remon JP, Verbaet C (2017) Continuous direct compression as manufacturing platform for sustained release tablets. Int J Pharm 519:390–407

75. Vázquez MJ, Perez-Marcos B, Gomez-Amora JL, Martinez-Pacheco R, Souto C, Concheiro A (1992) Influence of some technological variables on drug release from hydrophilic matrices. Drug Dev Ind Pharm 18:1355–1375
76. Velasco MV, Ford JL, Rowe P, Rajabi-Siahboomi AR (1999) Influence of drug: hydroxypropyl methylcellulose ratio, drug and polymer particle size and compression force on the release of diclofenac sodium from HPMC tablets. Int J Pharm 57:75–85
77. Wen X, Nokhodchi A, Rajabi-Siaboomi A (2010) Oral extended release hydrophilic matrices: formulation and design. In: Wen H, Park K (eds) Oral controlled release formulation design and drug delivery: theory to practice. Wiley, Hoboken, NJ, USA
78. Williams HD, Ward R, Culy A, Hardy IJ, Melia CD (2010) Designing HPMC matrices with improved resistance to dissolved sugar. Int J Pharm 401:51–59
79. Williams HD, Ward R, Hardy IJ, Melia CD (2009) The extended release properties of HPMC matrices in the presence of dietary sugars. J Control Release 138:251–259
80. Williams RO, Reynolds TD, Cabelka TD, Sykora MA, Mahaguna V (2002) Investigation of excipient type and level on drug release from controlled release tablets containing HPMC. Pharm Dev Tech 7:181–193

Peter Timmins was most recently Executive Director, Drug Product Science and Technology, Research and Development with Bristol-Myers Squibb at Moreton, Merseyside, UK, with 37 years in the pharmaceutical industry. In addition to being head of the Moreton site he led a group based both in the UK and the USA responsible for developing drug delivery solutions for small and large molecule candidates, including device technologies. Since retiring from Bristol-Myers Squibb he has continued to pursue research interests in the oral delivery of small molecule candidates where issues of poor aqueous solubility need to be resolved, or where modified-release technology is required. Peter has a BPharm degree and a Ph.D. from the University of Bradford, and currently has advisory, visiting and honorary academic appointments at the University of Bradford, the University of Central Lancashire, the University of Huddersfield and the University of Nottingham. He was designated a Fellow of the Royal Pharmaceutical Society in 2015 (p.timmins@hud.ac.uk).

Carl Allenspach is currently an Associate Director in Manufacturing Technology at Bristol-Myers Squibb. He has 20 years' experience working in the pharmaceutical industry at Searle/Pharmacia/Pfizer and Novartis. He started his career in analytical development and quickly moved into solid formulations where he developed expertise in excipients and manufacturing processes for both immediate and controlled release formulation. He has supported many initiatives over the years such as continuous manufacturing and compaction simulation, has developed an expertise in many pharmaceutical processes, and has significant scale-up and technology transfer experience. Carl holds a B.S. in Chemistry from Roosevelt University in Chicago and is enrolled part-time in the Pharmaceutical Science Ph.D. program at Rutgers University (carl.allenspach@bms.com).

Chapter 11
The Role of Particulates in Film Coating of Pharmaceutical Tablets

Anneke M. Dijkhuis-Bouwman

Abstract Many pharmaceutical tablets have a coating. This chapter discusses the critical quality attributes of coating liquids, their preparation, and the coating process of the tablets. Moreover, it describes how the requirements can be fulfilled in practice. A coating layer can be applied for many reasons, from cosmetic purposes to advanced drug delivery. A coating liquid generally consists of a polymer, a plasticizer, a colorant or opacifier, and sometimes components to stabilize the coating liquid or to improve wetting of the tablets. In the coating process, the liquid is sprayed onto the tablet cores whilst they are moving. The inlet air ensures drying of the tablets. During this drying phase the polymers coalesce and form a smooth thin layer around the tablet of more or less constant thickness. Good quality of the coating liquid is the basis for well coated tablets. Therefore, the preparation of a coating liquid should ensure a lump free, viscous liquid with a set amount of solids.

11.1 Introduction

Pharmaceutical tablets are coated for many reasons. Next to anti-counterfeiting and market branding, it can also be used to improve mechanical strength, improve ease of handling, improve swallow ability, mask taste, improve appearance, or influence the drug delivery.

Generally speaking, the quality of the film coating is dependent on four factors: the tablet or pellet to be coated, the film coating formulation chosen, the film coating process, and the storage conditions of the drug product. The tablet, where both the composition and the shape of the tablet are important, is chosen to obtain the desired release profile. After development, it is usually not changed anymore. The coating process is determined by formulation of both the tablet and the coating liquid, where the coating liquid generally contains a polymer, a plasticizer, and a colorant. Of course, storage conditions of the product are dictated by the product.

A. M. Dijkhuis-Bouwman (✉)
Meppel, The Netherlands
e-mail: dijkhuisanneke@gmail.com

H. G. Merkus et al. (eds.), *Particles and Nanoparticles in Pharmaceutical Products*, AAPS Advances in the Pharmaceutical Sciences Series 29,
https://doi.org/10.1007/978-3-319-94174-5_11

The production of coating liquids requires a combination of various powder particles and an ultimately viscous liquid, which makes this production process not very easy. Despite the complexity of the composition, in literature preparation of coating liquids is described as 'being performed'. If one wants to know the optimal way of preparing such a liquid, the exact actions on how to prepare a liquid are not described.

This chapter therefore focuses on the requirements for a coating formulation, the preparation of that formulation, and the influence of the powders in the formulation on the preparation. It will deal with the production of aqueous coating liquids for film coating purposes. Starting with the materials necessary, followed by the effects of the use of these materials (the processability), and the properties of the fluid desired for the film coating process. Then follows the description of the process of making a coating liquid, and it will end with PAT, Process Analytical Technology.

11.1.1 Purposes of Applying Coating Layer

Tablets can be coated for cosmetic purposes. It can be done to improve appearance, but also to improve recognizability by the patient. The latter is considered as both trade-marking and protection of intellectual property. Next to that, the appearance of a tablet can be used against counterfeiting by applying a specific logo, or specific color. In Sect. 11.3.6, even the addition of PEG as a tracer has been described for counterfeiting purposes [22].

Furthermore, a coating can be applied to mask taste. Or to increase shelf life, for instance, by protecting a product from moisture or oxygen, or UV radiation. A remark has to be made that moisture sensitive products might not benefit from a coating layer. A recent study shows that the efficacy of the moisture barrier is fairly limited. In this study it is shown that when applying a moisture barrier coating, the moisture related degradation of the model drug is even enhanced over a non-coated tablet. The equilibrium moisture content was higher in the coated tablets; the probable cause is that moisture can enter the tablet despite the coating, but it is harder for the tablet to release the moisture due to the coating layer [32].

A coating can also be used as the way to apply the active pharmaceutical ingredient (API), or to modify the release of the API [15]. In these cases, coating uniformity is extremely important.

Moreover, a coating layer could be used to separate incompatible substances. And finally, in case a tablet surface is very rough or dusty, a coating layer can be applied to reduce friction, and therefore handling (packaging) is easier [14, 16, 38]. The composition of coating liquids is discussed in Sect. 11.4.

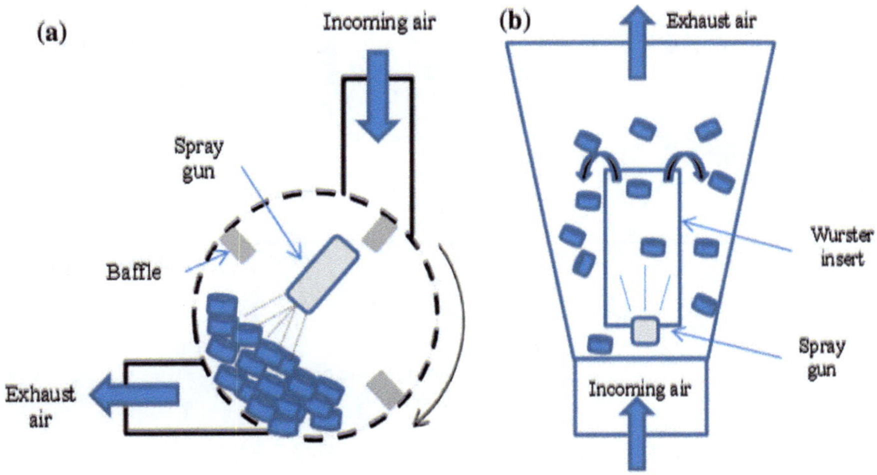

Fig. 11.1 **a** Conventional pan coating. The black arrow shows the rotation of the pan. The pan itself is perforated. **b** Wurster principle. The curved arrows show the direction of the tablets. On the bottom side, they are drawn in the Wurster insert again

11.1.2 Coating Process

Coating is generally performed in a (semi) perforated pan (see Fig. 11.1a). Cores are loaded into the coater, hot air (whether or not conditioned) is led through the tablet bed. Before spraying coating liquid from the top, the cores are dried, to ensure no excess moisture in the cores will be present. During the whole process, the cores are mixed in the pan by rotation of the pan. Baffles inside the pan ensure mixing in all dimensions.

In case of using a Wurster coater, which is a fluid bed coater, air flow is used to move the tablets. The concept is that the tablets are separated by the air flow, and a coating formulation is sprayed from the bottom of the bed onto the particles. Using the Wurster principle, less spray drying is expected. The process is schematically depicted in Fig. 11.1b.

The whole coating process, here written down for conventional pan coating, is described in phases, since every phase has its own important aspects. First phase is production of the coating liquid, the details of which are in Sect. 11.6. The important parameters are viscosity of the suspension, particle size of the particles in the suspension, and lumpiness of those particles. The coating liquid characteristics can influence the subsequent coating process steps.

After loading the cores and drying them whilst rotating, the second step in the coating process is spraying the coating liquid onto the tablet bed. When correct atomization of the coating liquid is achieved, small individual droplets are formed that all arrive on the surface in the same droplet stage. However, upon one revolution of the coating drum, the coating liquid on the tablets need to be dried to prevent overwetting. An equilibrium between spraying and drying needs to be set.

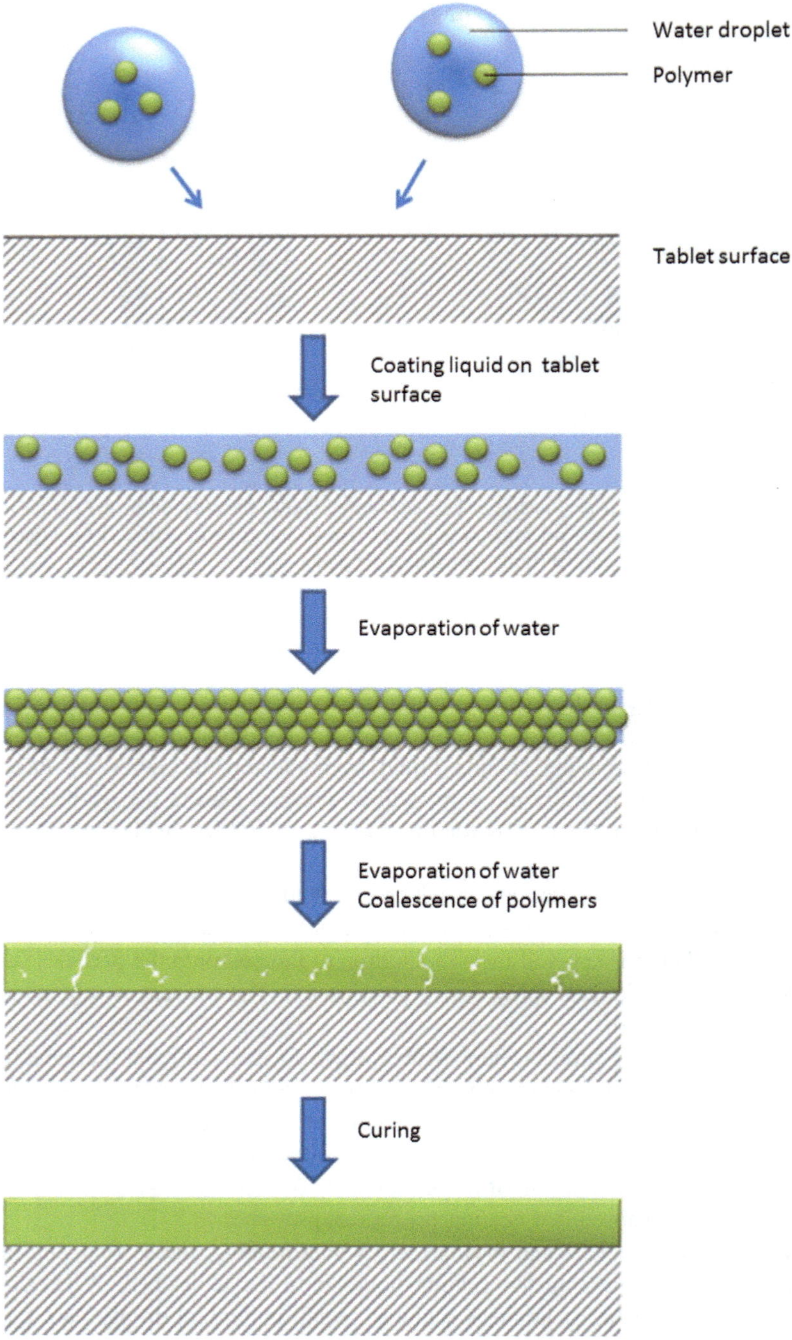

Fig. 11.2 Schematic presentation of the film formation mechanism from aqueous polymer dispersions. Modified from [45]

The spraying phase itself has also several important stages. In these stages, the behavior of the coating liquid is important. First, the droplet is formed, where droplet size is the most important parameter for the process. Not only process parameters influence the droplet size, but also density, surface tension, and viscosity of the coating liquid. After formation of the droplet, the droplet travels from the spray gun to the tablet bed. In this phase, spray drying is undesired. This is of course influenced by the starting droplet size and the process parameters (mainly inlet air temperature, air flow, spray rate). Finally, the droplets reach the surface of the tablet, which is called impingement. There the behavior of the droplets onto the tablet is important. Both wettability, and spreading of the polymers onto the surface determine coating quality [14].

Finally, the drying or curing step is important. It is schematically depicted in Fig. 11.2. During the drying phase, the polymer suspension becomes more concentrated. In case of water-soluble plasticizer particles, the plasticizer easily transports into the polymer, and will be homogenously distributed in the coating layer. This transportation into the polymer phase is of course dependent on the compatibility of the plasticizer with the polymer [5]. In the drying or curing phase, time is important to allow a proper film formation. Curing can be done by using temperature or humidity treatment. However, the best is the combination of the two, which allows the polymers to form a closed layer [44]. Typically, this process will take a few hours [19].

This chapter deals with aqueous coating liquids. However, tablets need to dissolve or disintegrate in aqueous environments as well. Process wise, there is a contradiction. Therefore, one must take care to keep the coating process as dry as possible. To reach that, the dry matter content of the coating liquid has to be as high as possible. Of course, this is also energetically optimal, since the least water needs to be evaporated.

11.2 Critical Process Parameters for Film Coating

In the process, tablets need to stay intact. This requires the starting material (the tablet cores) to have an adequate quality. Tablet hardness must be high enough to withstand the forces in the coating pan. Next to that, friability must be low, so no wear of the tablet is taking place. In case friability is not low enough, and wear of the tablets occurs, powder particles will appear in the coater pan. This can lead to powder (which might be even API) in the coating layer, which might affect the appearance of the coating layer. This rough surface is called 'orange peel'. The final tablet item to be considered is the porosity of the tablets. Preferably, the tablets to be coated should have low porosity, because otherwise part of the water can enter into the tablet during the spraying phase. Due to the fact that water from the coating liquid is entering the pores of the tablet immediately, the time for the polymers to coalesce, and thus forming the film, is reduced drastically. This is influencing the film forming process negatively [25].

Next to that, as discussed in 11.1.2, the coating process needs to be as dry as possible. However, a too dry coating process leads to drying of the droplets before they reach the tablet surface (spray drying), which leads to powder in the coater pan. This powder will affect the appearance of the tablets. Moreover, the amount of coating to be applied is mostly determined by weighing the amount of coating liquid. In case the coating liquid is spray dried in the process, despite the fact that the amount sprayed will be correct, the required coating layer on the tablets (weight increase of the tablets) might not have been reached.

Adequate setting of coating conditions can ensure the correct balance between temperature in the coater, airflow through the coater, air flow used for spraying, and spray flux. Modeling the process could help finding the optimal settings for the process [7, 49]. For instance, for variability of the coating mass on the tablets, the coating process needs to be optimized. Again, this is dependent on the equipment used. Since this optimization is costly, the coating process is often modeled. Boehling et al. did not just model their process, but they also proved their modeling by experiments on the large scale. They found that the number of nozzles and the spray rate were having the highest impact on the intra-tablet coating variation. They also showed that the modeling can be a good prediction, and is very useful to prevent large scale testing [7].

Coating efficiency is affected by many factors, for instance tablet size and shape, since this determines bed density. But also pan load, solid concentration and composition of the coating liquid, pan speed, spray variables, properties of the polymers used [14, 50]. Wang et al. [50] performed multivariate data analysis to optimize coating efficiency. They used 24 batches to prove that the most important variable to improve coating efficiency is the coating liquid delivery. Important items for coating liquid delivery are: spray rate, atomizing air flow rate (together determining droplet size), and spray pattern rate (together with spray rate determining how many droplets will reach the tablets). These findings were in accordance with Toschkoff et al., who stated that this knowledge is especially important for scaling up [49]. Next to that, they found that the fill grade of the coater is also important. A low fill grade has a lower variation of content through the bed, however, a high fill grade has a higher coating efficiency. At the end: increasing the amount of nozzles mounted ensures a decrease in variation. The latter of course also has an optimum: too many nozzles results in spraying the coater.

The coating liquid used can also influence the process. Viscosity of the coating liquid is here the most important parameter. This is governed by the composition of the coating liquid (see Sect. 11.3). Agrawal and Pandey describe the interactions between coating liquid and coating process. They have defined (among other parameters) viscosity of the coating liquid, solids content and surface tension as formulation factors that affect the coating process. Higher solids content is desired to reduce the coating time. However, higher solids content will increase viscosity as well, which impedes the coating process. Also, a higher solids content might lead to a higher surface roughness of the coated materials [1].

Prerequisite of the coating liquid that it is free of air bubbles. These bubbles disturb the process, since they reduce the spray rate of liquid (where there is air, there can't be liquid), and they interrupt the spray cone.

11.2.1 Viscosity

The coating liquid contains polymer particles, which has an effect on viscosity. Polymer type, and polymer chain length, together with the extent and size of the embranchment, determine the viscosity of the aqueous suspension. Often a product has a number in its commercial name, referring to viscosity value that can be obtained using the product at 2% w/w aqueous solution[1] at 20 °C [14].

For proper spraying, the coating liquid should have a maximum viscosity of approx. 500 MPas. This corresponds to a standard HPMC based coating of 15% w/w polymer concentration. In case of the use of other cellulose derivates, the content must be lower. Therefore, HPMC is often used. The high concentration means a lower solvent content, therefore a shorter processing time since less solvent needs to be removed [14].

Simply said, all additives to the simple polymer dispersion lead to an increase in viscosity, although of course the increase depends on the additive used. And, if the coating liquid shows pseudoplastic behavior, the viscosity is dependent on the shear rate used as well [14].

Whenever viscosity is high, avoid heating an HMPC solution. Viscosity will increase enormously at the thermal gelling temperature, which is around 60 °C [33].

Since the coating layer should have a smooth surface, the particle size of the insoluble compounds needs to be very small. A typical layer thickness is 20 to 100 μm. This layer is applied in several steps. As a consequence, the insoluble particles added to the coating liquid should have a maximum size of approximately 1 μm. If particles are larger, they will interfere in the polymer layer, possibly making the polymer layer permeable. Especially when the coating is applied as a barrier layer, this is an issue. Next to that, the larger the insoluble particles, the rougher the surface appears to be. Therefore, the smaller the insoluble particle, the better. However, these small particles will increase viscosity.

Next to that, particle shape is important: the more it deviates from spherical, the more it will increase viscosity. These items have been investigated thoroughly in the last decades, not only for HPMC solutions, but also for other polymer dispersions [14].

---

[1]Typically, HPMC and similar dispersions are named solutions, since they look quite transparent; in fact, however, they are dispersions of hydrated polymer particles.

11.2.2 Tackiness/Stickiness

For a good adhesion between coating layer and substrate, a maximum area of contact (defined by viscosity and surface tension, wettability) and a good intrinsic bonding between chemical groups of the polymer and the tablet are necessary. Tackiness, or stickiness, is the property of a polymer that is also dependent on the viscosity. Due to the viscoelastic properties, the polymers exhibit some of the characteristics of liquids, so they will wet a surface to which they are pressed. Due to their elasticity, they will resist separation when stressed. Therefore, on the one hand, the stickiness helps in adhesion of the polymer to the tablet. On the other hand, it is problematic in the coating process, and it is a limitation for use of some of the polymers. If the polymers are too sticky, in the coating pan twins can be obtained: tablets that are stuck to each other by the sticky coating layer. A smaller, but still undesirable, effect can be picking [3, 14]. Picking means that due to short sticking of tablets to each other, the coating layer is disrupted, and therefore, part of the coating layer is transferred from one tablet to the other.

11.3 Critical Product Quality Attributes

11.3.1 Strength of Coating Layer

Simply said, a film must not crack. Besides the fact that it is not visually attractive, the coating must be closed in case it is a functional coating. In the full process until packaging, a lot of mechanical strengths are applied on the film-coated tablets. The tensile strength of the film, therefore, has to be high enough. Tensile strength means the stress applied at the point which the film breaks. If tensile strength is high, the film will not break easily.

In addition, the film needs to be rigid. This relates to the E-modulus of the film, which is a measure for stiffness. The higher the value for E-modulus, the higher the force necessary for elongation of the film, therefore, the more rigid the film [14].

11.3.2 Thickness of Coating Layer

Generally, the applied coating layer is between 20 and 100 μm thick [14]. However, the thickness distribution of the coating layer is determined by the shape of the tablet. Freireich et al. have performed DEM simulations where this is nicely demonstrated. It is visible from all shapes that the (flat) side of tablet contains a lower layer thickness than the (curved top). For round tablets, the difference is 80 μm (side) to 100 μm (top). For triangular shapes the difference is even larger: 50 μm (side) to 90 μm (top) [17]. For a cosmetic purpose, layer thickness is less important and therefore may vary. However, when API is incorporated in the coating, layer thickness should be constant. Moreover, when using a coating layer to obtain modified release or when having very bitter taste of API, where a coating layer can be used for taste masking [8], layer thickness is important. In this case, the layer thickness determines the effectiveness of the barrier. Unfortunately, very little data is found on layer thickness, layer thickness distribution, and layer thickness variation. Especially for above mentioned items, investigation in thickness (variation) should more often be performed. The new PAT techniques, described in Sect. 11.7, look very promising in this respect.

11.3.3 Modified Release

Many latexes and pseudolatexes can be used for modified release. In the use of aqueous coatings, the layer needed to obtain the desired function is thicker than with comparable organic coatings [47].

11.3.4 Appearance, Recognizability, Counterfeiting

The appearance of a tablet is not only its color. Also the texture of the tablet determines the visual attractiveness. To have the coating layer result in a smooth appearance (therefore a small surface roughness), an undercoat can be used. A first coating layer can overcome all roughness originating from the tableting process. This can be a simple coating layer. On top of that, a second layer can be applied, where colorants can be used. Due to the smooth surface obtained from the first layer, the surface appearance of the final product is improved. The undercoat can also be applied to provide for a moisture barrier. This has been described in several publications [12, 39].

Under close examination, a film coating layer is non-homogeneous. This is caused by two factors, one is caused by the way the coating is built up in layers during the process, and two is the fact that insoluble particles (for instance, pigments) are added to the coating liquid [14].

Logo bridging is also quite often taken as a measure for the coat quality. It has been shown by Pandey et al. that also the composition of the tablet to be coated has an effect on the appearance of the coating layer. An increase in magnesium stearate has a negative effect on the appearance. The same holds for an increase in sodium lauryl sulphate, and they appear to have a synergistic effect [35].

In many cases a colored coating is used, but a transparent coating layer can also be applied. Rotta et al. have described how to analyze color and transparency [40]. They made the films not by coating them onto a substrate, but by preparing them in a Petri dish. The film was then taken out and transmittance was measured in a quartz spectrophotometer cell, at specific wavelengths. The three color components can be measured this way. Measuring transmittance at certain wavelengths can also predict the UV radiation sensitivity. However, since transmittance is measured, this cannot be performed on a tablet, since the UV light will not pass the tablets [40].

Rotta et al. show that single HPMC films (without addition of Chitosan, a biopolymer with antibacterial activity) gives no protection against UV radiation. However, combining HPMC with Chitosan in a 50/50 combination, or the pure use of Chitosan gives a very good protection against UV radiation. Other combinations also show a UV protection [40].

11.3.5 API in Coating Layer

In addition to using the coating layer as modified release system, it can also be used as the layer in which active component is added. Reasoning could be to have an immediate release, combined with a tablet that itself has a sustained release formulation. Or to have two incompatible drugs together in one tablet.

Coating actives requires of course a good uniform distribution of the coating layer on the tablets. To achieve this, an investigation in the critical process variables must be performed, which can be dependent on the equipment used. Also, the efficiency of the spraying process needs to be optimized, since a loss of coating liquid now equals a loss in API [39].

Rege et al. [39] did such an investigation into the critical process parameters, by performing an experimental design. They picked atomization pressure as one of the variables. However, the consistency of the coating liquid was not taken into account, besides performing spray pattern tests. Finally, they found that atomization pressure, pan speed and coating duration were critical to obtain a variation of <6% RSD in drug content. Apparently, only at optimum conditions the drug product has been distributed well, and particle size of the drug was appropriate.

Boehling et al. [7] described ensuring content uniformity, and therefore coating uniformity, as a challenge. In this challenge, both intra-tablet and inter-tablet coating variability are considered as critical quality attributes. The latter is the variation in tablet mass, where variability in the coating mass on the tablets is considered to be intra-tablet coating variation.

Wang et al. [50] describe the film coating with active materials in three challenges: determining end-point of the film coating process accurately, having consistent tablet content uniformity, and finally minimizing the loss of API in the process.

The first challenge: determining the end-point, can be tackled by measuring weight gain. However, in case API is coated, a more accurate measurement could be desired. This can be done by using process analytical technology (PAT) to measure layer thickness or content uniformity during the process. Some of these techniques are shown in Sect. 11.7.

The second challenge, to obtain a constant content uniformity in the tablet bed, was investigated by Chen et al. [13]. They incorporated the term 'cycles' to describe the movement of the tablets from the spray zone to the drying zone and vice versa. The more cycles the tablets undergo, the lower the relative standard deviation of the content, thus the more constant content uniformity. Of course, the number of cycles of importance is also related to other factors, like bed density (since a dense bed impairs drying rate), tablet velocity, the size of the spray zone, etc.

The last challenge as stated by Wang et al. [50] is about minimizing API loss. The loss of API is inevitable, since a surplus of coating liquid needs to be made (more coating liquid needs to be prepared than is necessary solely for coating of the tablets), to be able to fill the lines running to the coater, and to cover for losses inside the coater, since coating efficiency is never 100%. However, it is expensive to lose API due to these reasons. Therefore, two coating liquids could be used. One to spray all API, the second to get to the required weight gain. This way the loss due to filling lines and the loss due to the overage made can be minimized. However, coating efficiency still needs to be as high as possible. Coating efficiency is already described in Sect. 11.2.

11.3.6 Tamperproof

Ilko et al. [22] tried to incorporate tamperproof in the tablets itself. They stated this could be preferred over labeling a secondary package with either holograms, digital watermarks, serialization, or Radio Frequency Identity (RFID) chips. As a marker in the product itself ("Forensic markers") stable isotopes, unique fluorescent patterns, saccharides, or scents have been tested. Unfortunately, these might require materials that are considered not safe or not allowed. However, the use of PEG as a marker has already been tried as in doping controls, where the ingested PEG was found in the body fluid [22]. Despite the fact that this material is not recognizable for the patient, it is very interesting to look into this. Ilko et al. [22] used monodisperse PEG (monodisperse with respect to its molecular weight), as tracer material that was measured in the tablets by LC-MS/MS analysis. By combination of several monodisperse PEGs, a sort of code can be created. These fractions can be spiked into the coating dispersion, only a limited amount of PEG is necessary to show the code. However, not all products can be spiked with this material. For instance, if PEG is already used in the formulation, this tracer material cannot be used, since it will disrupt the coding.

11.4 Coating Liquid Composition

To be able to fulfill all above requirements, both for product and for process, a basic ingredient of the coating liquid is a polymer. There are many types of polymers that can be used, each with its specific properties. But all of them have in common that they will form a network around the tablet. In general, the polymers used in traditional film coating are cellulose derivatives or acrylic polymers.

Of course, application of just a polymer is not possible, a carrier liquid has to be used. To avoid problems related to the use of organic solvents, like environmental and safety concerns, the commonly used carrier liquid is water [46]. Furthermore, to make the polymer more pliable, a plasticizer is added. When in the coating liquid, the polymer is present as discrete particles. During film formation, the polymer particles come into contact with each other, due to the evaporation of water. Subsequently, they deform, coalesce, and finally form a film (see Fig. 11.2). Film formation from aqueous dispersions can be improved through decreasing polymer particle size [47].

Finally, insoluble components, e.g., pigments, talcum, or opacifiers are added to improve the appearance of the tablet [14, 46]. Next to these items, sometimes the coating liquid also contains an API. These insoluble components disrupt the film, and therefore reduce the strength of the film.

11.4.1 Polymers

Different types of polymers can be used. The use of a different type of polymer leads to different effects both in the coating liquid and in the applied film. The polymers used are either amorphous or have a very low crystallinity [14].

A general requirement is that the coating application occurs above the minimum film formation temperature of the polymer, since the polymer needs to be in the rubbery state to coalesce with its neighbour particles. This temperature is often close to their glass transition temperature. Generally stated, the temperature should be 10–20 °C above the glass transition temperature [14].

The viscosity of polymer solution is determined by the degree of polymerization. The higher the degree of polymerization, the higher the viscosity. All cellulose derivatives absorb water and form a gel, which also increases viscosity [43]. Next to that, the higher the degree of polymerization, the stronger the film. So in order to have a strong film (which can be necessary to strengthen a fragile tablet), one has to deal with a high viscosity in the film coating process, resulting in a complicated process. On the other hand, using a low degree of polymerization, the resulting film strength could be too low, possibly leading to peeling [33].

The film-forming mechanism of polymer dispersions starts when the droplet hits the surface of the tablet. Water evaporates, and polymer particles are layered together, forming a dense mass. Evaporation continues, and the particles deform and coalesce, provided that they are mobile. Any water still present is pushed out, and the film is complete, and water-insoluble [14].

11.4.1.1 Cellulose Derivatives

Of the cellulose derivatives, HPMC (hydroxypropylmethylcellulose) is the most commonly used. HPMC is a good aid in coating, since it forms a transparent film, that is both tough and flexible. In this sense it is perfect for its use. Moreover, it dissolves completely in the gastrointestinal tract [33].

The first patent for using HPMC in an aqueous solution in film coating was granted in 1962. However, at that moment the viscosity of the solution at low polymer content was so high, that the coating process took a very long time, and therefore was costly. Due to the development of low viscosity types of HMPC, the possibility of using HPMC for film coating became worldwide acceptable. Also the coating equipment had to be adapted, compared to using an organic solvent, this aqueous coating liquid required improved drying of the tablets during the process. This problem was overcome by using side vented pans [33].

When using HPMC with only a low level of pigments, a plasticizer is not necessary in the formulation. However, to prevent poor adhesion of the film to the substrate, a plasticizer can be added [33].

11.4.1.2 Latexes and Pseudolatexes

True latexes are dispersions of acrylic resins in water [26]. Particle size of the polymer is between 10 nm and 1 μm. The size needs to be this small to have a stable dispersion; this size ensures Brownian motion, to prevent sedimentation [25].

Pseudolatexes for sustained release are mainly ethylcellulose latexes. Both Aquacoat® and Surelease® use ethylcellulose as the film former. Despite the fact that they undergo a complete different production process, the behaviour is similar to the true latexes. Also acrylic acid derivatives are pseudolatexes.

PVAc (poly(vinylacetate)) is a synthetic polymer. This material is sold under the brand name Kollicoat®. It originates from ethylene. In coating liquids, quite often a mixture of soluble and insoluble polymers is used. Also with PVAc there are several blends possible, leading to different characteristics of the coating layer. A blend of PVAc and PVA-PEG (a copolymer) shows advantages: pH-independent drug release, where the drug release is adjustable by the ration of these polymers. Next to that, high resistance to mechanical stress [16].

In the film coating process, PVAc materials are favorable, since high polymer concentrations are possible without enormous viscosity increase. Ennslin et al. [16] performed a three-level composite design to investigate the influence of both the film thickness and the film composition on the drug release of coated pellets. The film composition was altered in ratio PVAc/PVAc-PEG, and in concentration plasticizer. They found that the concentration plasticizer did not affect the drug release, whereas both film thickness and polymer composition did have an effect. The higher the PVAc ratio, and the higher the thickness, the slower the release.

Zein is also a pseudolatex. It is a protein of corn origin, that is soluble in alcohol. It shows hydrophobic properties and low water uptake, and therefore it might be useful for coating tablets with very water-soluble drugs. Li et al. [26] succeeded in making films with zein and PEG 400 as plasticizer. These films were compact and smooth, and resulted in moisture resistant films.

11.4.1.3 Polymethacrylates

Polymethacrylates are synthetic (co)polymers known under the trade name Eudragit®. There are different types of Eudragit®, all having a specific application. Their properties are determined by functional groups. Immediate release, delayed release, and sustained release are all possible, dependent on the type chosen [34].

11.4.1.4 Shellac

Shellac is a resin secreted by the female lac bug. Despite the fact that this might not sound attractive, it is familiar to us all, since shellac coatings ensure candies to 'melt in the mouth and not in the hand' [18]. Shellac is a natural form of a thermoplastic. It is used together with other polymers as a composite film as a barrier for moisture [10]. It can be used for enteric coating material. Next to that, it has potential for moisture protection and taste masking [23]. In a combination with for instance HMPC, it is visible with an optical microscope that the Shellac is distributed over the HMPC.

11.4.2 Plasticizers

The role of plasticizers in coating compositions is lowering the glass transition temperature of the polymers. The small plasticizer molecules crawl between the polymer chains, increasing the spacing and free volume. This allows the polymer chains to move past one another even at lower temperatures, the resistance for deformation of the polymer is reduced. That means, that at lower temperature the polymers can unfold and spread onto the surface of the tablets [14, 47].

Next to lowering the glass transition temperature, also the tensile strength of the polymers is lowered by addition of plasticizers, and the elongation and flexibility of the polymer film is increased. Due to all these effects of addition of plasticizers to the polymers, they can also be used to influence the drug release profile [5].

The time between addition of the plasticizer and starting of the coating process is critical, especially when water-insoluble plasticizers are used. There should be sufficient time for the plasticizer particles to migrate into the polymers. If this time is not sufficient, the distribution of the plasticizer in the coating layer might be inhomogeneous, leading to changes in drug release profiles [5].

As a rule of thumb, the concentration of plasticizer in the formulation is recommended to be 20–30% w/w of the polymer solids content [5].

11.4.3 Other Additives

The coating liquid generally contains also other insoluble components. These are applied as colorant, or opacifier. In addition, inorganic colors like titanium dioxide and talcum, or natural colors like carmine also have a positive effect in lowering the tackiness [14]. However, since these components are insoluble, they may have a negative effect on the strength of the film. They are present as particles, and may therefore result in cracks. Generally, the particle size of the insoluble particles should be below 1 μm to ensure the optimal coating layer. Of course, the lower the concentration of insoluble particles in the fluid, the better the coating layer [37].

11.5 Physical Stability of the Dispersion

The physical stability of the dispersion is dependent on the particle size of the particles in the dispersion, and the application of stabilizers. If the particles are very small, Brownian movement and thermal convection prevent them from sedimentation. If particles are larger than 1 μm, after a few hours already sedimentation can be observed [25].

To prevent agglomeration and coalescence of the polymers during storage, a surfactant can be added. Of course, these surfactants will in the end stay in the coating layer, so they can have an effect on the film structure, and therefore on the release profile of the tablet. Possible surfactants used are sodium lauryl sulphate and cetyl alcohol [6].

The stability of the dispersion can be characterized by the zeta potential, the difference in potential between the dispersion medium and the layer of fluid effectively attached to particles dispersed in it. Particles in aqueous dispersion become charged, depending on the pH of the dispersion liquid. If the charge of the particles is small, resulting in a low value for zeta potential, the particles will agglomerate, and the dispersion is not stable. If the zeta potential is high (greater than about 50 mV), particles will repel each other, making the dispersion stable [31].

Another way of stabilization is steric hindrance. If polymers have loops and tails in the solution, these loops and tails prevent particles from agglomerating. Stability due to steric hindrance is very effective. Steric stabilizers may contain an ionic group as well, with its own contribution to stabilization. With an increasing concentration of polymer, the stability is even further increased [42].

Latexes show the tendency to agglomerate around 20 °C. Therefore, shelf life is limited in the absence of stabilizers. Next to that, stability can be very sensitive to pH changes or addition of electrolytes. Also, if very fine pigment particles are added to the latex, they are able to bind the stabilizer, so that the latex mixture have low levels of stabilizer left, which also will lead to a less stable coating liquid [25].

11.6 Practical Work

Based on all the above information for critical quality attributes, whether this is for the product or for the process, a formulation has been chosen to reach the optimal way to disperse this formulation in a liquid.

11.6.1 Methods for Dispersing

A previously said, the information on how to prepare a coating liquid is fairly limited. Schilde et al. describe several methods for dispersing nanosized particles. In such a process, ultrasonic homogenizers, disc mills, three-roller mills, kneaders, and stirred media mills can be used (some of them shown in Fig. 11.3) [41].

Unfortunately, not all these dispersion techniques are also applicable for production of coating liquids. This can be due to the scale of working, but also to the forces applied on the polymers. For dispersion of nano sized particles, sometimes the dispersion technique is also used to decrease the particle size of the powders by de-agglomeration. For the preparation of a coating liquid, however, this is not always necessary. High shear forces causing turbulence should even be omitted when using latex formulations, since they may coagulate, resulting first in an increase in viscosity, and eventual in phase separation [25].

As Schilde et al. describe [41], both a three-roller mill, kneader, and a disc mill are able to process high-solid concentrations. The stress applied to the particle is a shear gradient in the fluid. The solid concentration where these mills are useful is a suspension above 35% w/w solid fraction. In general, this is overpowering for coating liquids, where the solid fraction is around 20%.

For the solid fractions around 15%, both a dissolver or an Ultraturrax can be used. An Ultraturrax can (dependent on its size) easily disperse materials [41]. However, due to the enormous rotational speed, it is hard not to entrap air. And the finer the air bubbles are entrapped, the harder it is to remove them. Especially of course due to the viscosity of the fluid, air bubbles are quite stable in the dispersion.

For HPMC, it is known that at a concentration below 10%, a clear solution can be obtained in just one day by having the liquid stand. First a gel is formed [43]. In time, the HPMC will fully hydrate, and a clear liquid is obtained. However, long term storage (to reach this state) may result in microbiological growth [33].

At small scale (up to 5 kg), production of a coating liquid is fairly easy. If needed, the liquid can be heated and cooled in a small period of time. For stirring, a magnetic stirrer can be used [8], or a blade stirrer [48], or Ultraturrax [9]. At small scale, there is a relatively high effect of shear of liquid on vessel wall and bottom, which will help the mixing process. Also, the distance between bottom and top is small, so an entrapped air bubble can escape in a short time frame. The power needed to run, for instance, the magnetic stirrer is relatively low.

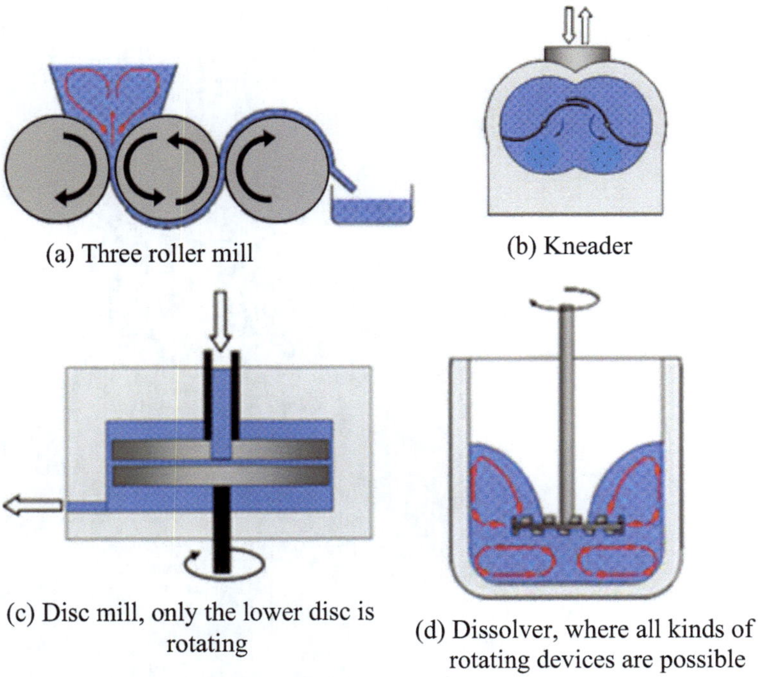

Fig. 11.3 Schematic drawings of equipment. Modified from [41]

11.6.2 Industrial Scale Production

In an industrial setup, the amount of coating liquid required is larger. Due to the increased amount of liquid to be moved, the stirrer blade also needs to be larger. Despite all the sizes of magnets for stirring that can be obtained, a magnet cannot be used in a viscous liquid at a 10 kg scale or larger.

Suppliers of coating excipients show methods how to use their materials on their websites. They all recommend not to use a high shear device (like an Ultraturrax), but a propeller type stirrer (see dissolver in Fig. 11.3). The stirrer blade should approximately cover one third of the diameter of the vessel. Different stirrer blades can be used, in Fig. 11.5 a special type disc blade has been used. If necessary, a second dispersion disc can be mounted on the spindle as well.

The vessel should be high enough to allow a vortex to be made in the liquid. The height of the water level in the vessel should more or less equal the diameter of the vessel. In Fig. 11.4, an example is given of industrial scale production of a coating liquid.

To obtain fully wetted powder particles, the powder should not float onto the surface. Instead, by making a vortex in the fluid, the material is dragged into the water phase. This ensures fully wetted material. Take care that the vortex is not

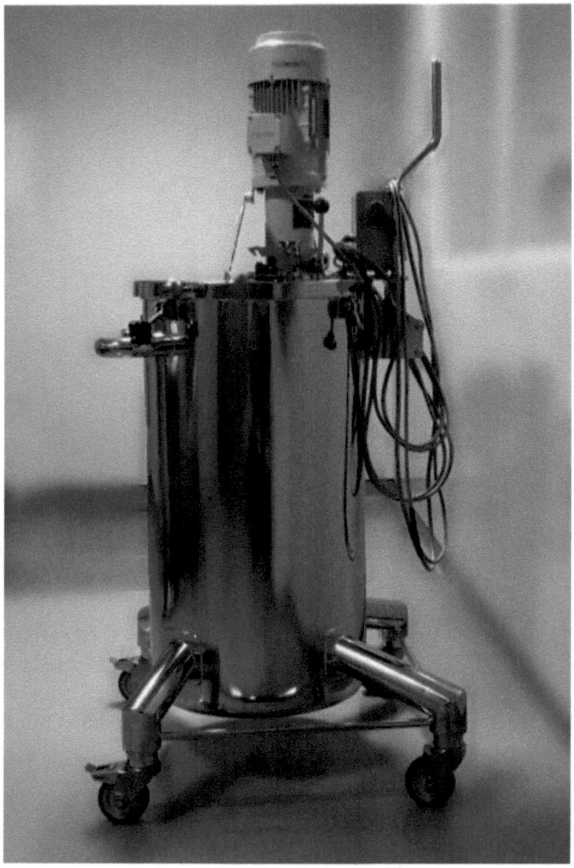

Fig. 11.4 Industrial scale mixer for making coating liquid

reaching the bottom of the vessel, since this would drag the coating material on the bottom. In Fig. 11.5, a laboratory setup with vortex is shown.

To ensure that the material is dragged into the vortex, a funnel can be placed on top of the vessel to direct the powder. Due to the almost immediate viscosity increase by addition of the polymer to the water phase, it is best to add the powder as quickly as possible. When adding slowly, the viscosity increase prevents a proper wettability of the powders.

In all cases, lumps of powder particles have to be avoided. Not only will these lumps block the nozzles of the sprayers, but also not all particles are available in the liquid.

If agglomerated particles of powder enter the coating liquid, they will stay intact as a lump. The powder particles on the outside are wetted first. Upon wetting, swelling takes place on the boundary of powder and water. Due to the swelling and subsequent increase in viscosity, water cannot penetrate further. Therefore, the core of the lump will stay dry. Figure 11.6 shows a lump (created on purpose) of hydroxypropylcellulose (HPC). To show that the inside indeed is dry, the lump has

Fig. 11.5 Laboratory setup for coating liquids. The stirrer makes the vortex, and powder is dragged into the vortex

been cut. The best prevention from lumps is not creating them. A second best alternative is to add a small sieve in the tube towards the nozzles. In case a lump has been created, it will be removed by the sieve and not block the nozzle. Moreover, small lumps can be broken by the shear forces near the sieve surface.

After addition of the powder, the dispersion should be kept in motion by continuous agitation [11]. However, a vortex is only used as long as material is added. At the moment all powder particles are wetted, the rotational speed of the propeller should be lowered to avoid air entrapment.

Water is used for starting the preparation of the coating liquid. So even if air is entrapped at this moment, these air bubbles are removed very quickly due to the low viscosity of water. In case material is added bulk wise in the vortex, experience tells that almost no air is entrapped. After addition, the speed of the propeller is lowered immediately. Working this way leads to an almost foam free liquid. To remove the few air bubble present, just have the fluid rest.

Then it is time to wait. Do not use the coating liquid directly after preparation, since the polymers need to have time to hydrate and form a gel. Time is dependent on the polymer used, but it is advisable to wait at least one hour.

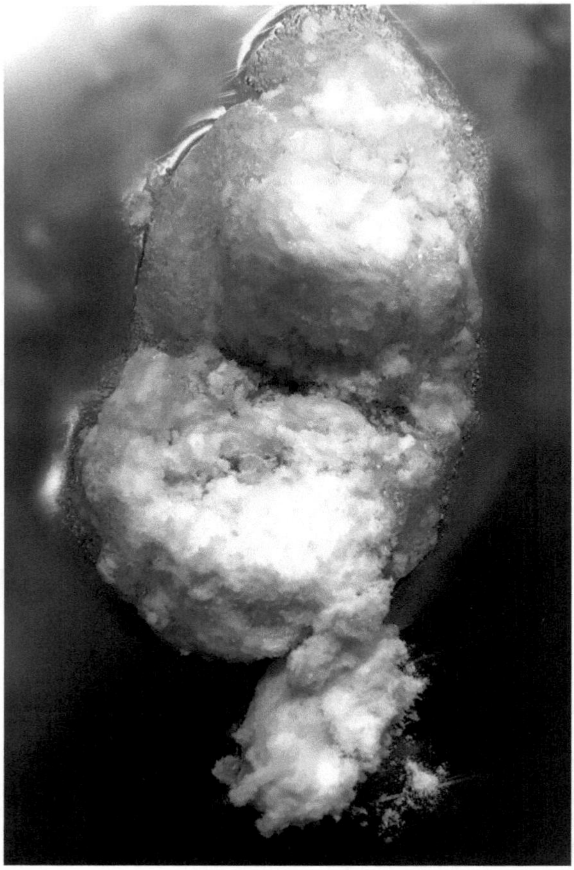

Fig. 11.6 A lump of HPC, cut in two to show the inside. Note the layer of water on the outside, and the dry powder falling out after cutting the lumb

11.6.3 Storage and Microbiological Stability

In this chapter, aqueous formulations are discussed. These are always prone to microbial contamination. Therefore, it is always recommended to test the coating liquid for shelf life with regard to microbial contamination. Of course, in this sense working hygiene, good water quality, and using excipients with a low bio burden are important. If necessary to increase shelf life, for neutral poly(meth)acrylates latexes a possible solution could be the addition of active chlorine, however, a quick use of the liquid is always helpful [25].

11.7 PAT (Process Analytical Technology)

PAT (Process Analytical Technology) usually starts with making a quality by design (QbD) framework. The concept of PAT is to understand the processes, and to measure preferably in-process.

When developing a process by PAT, multivariate data analysis is used to analyse all data obtained. In this way, the critical process parameters are found. Also the operating window—the settings to be used that still will result in good quality product—is set. Finally, by using continuous improvement, the process can be further optimized.

In this framework, the CPP's (Critical Process Parameters) and the CQA's (Critical Quality Attributes) are defined. CQA's are physical or chemical properties that should be within an appropriate limit, range or distribution to ensure the desired product quality. CPP's are process variables whose variability has an impact on a CQA, and therefore require monitoring and control to ensure the desired product quality [21].

In Table 11.1 the relevant CQA's and CPP's are mentioned for the coating of pharmaceutical tablets.

In this approach, novel techniques have been developed to measure the important parameters in-line, on-line, at-line, or off-line. Most of the techniques for PAT are imaging techniques.

In an in-line technique, the measurement takes place in the process, so no material is removed from the process. On-line means that the sample is taken from the process, is analyzed, and can be returned to the process. The difference between

Table 11.1 Relevant CQA's and CPP's for the coating process

Critical quality attributes	Critical process parameters
Preparation of coating liquid	
Viscosity of coating suspension	Duration of powder addition
Lumpiness of the coating suspension	Formation of vortex
Amount of solids in dispersion	Stirring speed
	Waiting time between making and using the coating liquid
Film coating process	
Coating layer thickness	Viscosity of coating suspension
Residual moisture in coated tablet	Lumpiness of the coating suspension
Rigidity/elasticity of the film	Amount of solids in dispersion
Appearance	Inlet/exhaust air flow
Amount of coating defects	Pan rotation speed
Coating efficiency	Spray flux
Release profile	Air flow through the nozzle
Tablet quality	Bed/inlet air/exhaust air temperature

at-line, where the sample is measured close to the process, and off-line, where the sample is taken and measured away from the process, is very small [24].

NIR (Near Infra-Red Spectroscopy) is one of those measurements that, as long as reference data are available, can measure both API concentration in the coating layer, as well as coating layer thickness [4]. The procedures are based on spectral changes originating in varying concentration or layer thickness. Andersson et al. show a measured coating layer thickness of 70 μm [2]. It can be used in-line and on-line. However, it needs a reference method to calibrate the results. This means, that NIR for measuring coating layer thickness is an indirect measuring technique [2, 4, 24, 36].

A direct method for measuring thickness can be measuring the dimensions of the coated tablet. Tablet dimensions are measured by a micrometer gauge. When coating pellets, this can be automated by using laser diffraction, since pellets are more or less spherical, having a size that is measurable using laser diffraction. When measuring tablets, however, measuring dimensions before and after coating cannot be done using laser diffraction, and therefore it is very labor intensive. Many tablets will need to be measured to determine the value of the thickness [24]. Next to that, both the tablets and pellets need to be dry, so spraying should be stopped, drying should be performed to be able to sample.

Another way of measurement is using microscopy; where simply the tablet or pellet is cut into halves, and the thickness is measured visually. However, to be able to use this technique, there must be a clear difference between the core and the coating layer, neither the core can be too brittle that the cutting edges are not sharp [24].

Finally, Terahertz pulsed imaging (TPI) is a non-destructive way to measure coating thickness, both off-line and in-line. Short pulses of terahertz radiation are beamed onto the tablet, and the reflections are measured. It is necessary to know the refractive index of the material, to be able to determine thickness based on the time delay between the reflection pulses [20, 24, 27, 30].

When all above techniques fail, Optical Coherence Tomography could be used in-line to measure coating thickness as well. In this technique, the distance between different interfaces (coating layer-air and coating layer-tablet) is taken as coating layer thickness. This technique is still under investigation, since light scattering is impeding the measurements [28, 29].

11.8 Simulation Studies

It has already been mentioned that coaters can be simulated. In this way, the optimal process parameters can be checked. Not only the coater can be simulated, but also the resulting coating layer onto the tablets can be predicted. Freireich et al. used DEM simulations to predict the variability of the coating layer within a batch [17].

11.9 Final Remarks

This chapter was written not to be complete on the coating process, but to give an idea on how to realize a coating liquid of a good consistency that is ready to be used. In this chapter, all other parameters in the coating process are dependent on that goal. In reality of a manufacturing plant, however, it is the other way around: with the given formulation, the coating liquid must be prepared to be used in the coating process.

11.10 Abbreviations

API	Active Pharmaceutical Ingredient, the active component of a tablet
CPP	Critical Process Parameter, process variables whose variability has an impact on a CQA
CQA	Critical Quality Attribute, physical or chemical properties that should be within an appropriate limit, range or distribution to ensure the desired product quality
DEM	Discrete Element Method, a numerical method to compute the motion and effect of a large number of particles
HPC	Hydroxypropylcellulose
HPMC	Hydroxypropylmethylcellulose
LC-MS	Liquid chromatography-mass spectrometry, an analytical chemistry technique
NIR	Near-infrared spectroscopy
OCT	Optical coherence tomography, a graphical technique to make images
PAT	Process Analytical Technology, a way to design, analyze, and control pharmaceutical manufacturing processes
PEG	Polyethyleneglycol
PVAc	Polyvinylacetate
RFID	Radiofrequence identity
RSD	Relative standard deviation
TPI	Terahertz pulsed imaging, an imaging technique
UV	Ultraviolet light

Acknowledgements I would like to show my gratitude to Astellas Pharma Europe B.V., who supported writing of this chapter by providing access to all scientific literature. Next to that, I would like to present my special thanks to John Lambers, who optimized the pictures in this chapter.

References

1. Agrawal AM, Pandey P (2015) Scale up of pan coating process using quality by design principles. J Pharm Sci 104:3589–3611
2. Andersson M, Josefson M, Langkilde FW, Wahlund K-G (1999) Monitoring of a film coating process for tablets using near infrared reflectance spectrometry. J Pharm Biomed Anal 20:27–37
3. Aulton M, Twitchell A, Hogan J (1997) Physical properties of HPMC solutions and their role in the film coating process and the quality of the coated product. In: Aqueous polymeric coatings for pharmaceutical dosage forms. Marcel Dekker Inc., New York
4. Avalle P, Pollitt M, Bradley K, Cooper B, Pearce G, Djemai A, Fitzpatrick S (2014) Development of process analytical technology (PAT) methods for controlled release pellet coating. Eur J Pharm Biopharm 87:244–251
5. Bodmeier R, Paeratakul O (1997) Plasticizer uptake by aqueous colloidal polymer dispersions used for the coating of solid dosage forms. Int J Pharm 152:17–26
6. Bodmeier R, Guo X, Paeratakul O (1997) Processs and formulation factors affecting the drug release from pellets coated with ethylcellulose-pseudolatex aquacoat. Aqueous polymeric coatings for pharmaceutical dosage forms. Marcel Dekker Inc., New York, pp 81–100
7. Boehling P, Toschkoff G, Knop K, Kleinebudde P, Just S, Funke A, Rhebaum H, Khinast J (2016) Analysis of large-sale tablet coating: modeling, simulation and experiments. Eur J Pharm Sci 90:14–24
8. Brniak W, Maslak E, Jacowicz R (2015) Orodispersible films and tablets with prednisolone microparticles. Eur J Pharm Sci 75:81–90
9. Brock D, Zeitler J, Funke A, Knop K, Kleinebudde P (2014) Evaluation of critical process parameters for inter-tablet coating uniformity of active-coated GITS using terahertz pulsed imaging. Eur J Pharm Sci 88:434–442
10. Byun Y, Ward A, Whiteside S (2012) Formation and characterization of shellac-hydroxypropyl methylcellulose composite films. Food Hydrocoll 27:364–370
11. Cahyadi C, Chan L, Heng P (2015) A comparative study between conventional pan coater and quasi-continuous small batch coater on the stability of tablets containing acetylsalicylic acid. Eur J Pharm Biopharm 90:30–37
12. Castellanos Gil E, Iraizoz Colarte A, Lizardo Lara Sampedro J, Bataille B (2008) Subcoating with Kollidon VA 64 as water barrier in a new combined native dextran/HPMC-cetyl alcohol controlled release tablet. Eur J Pharm Biopharm 69:303–311
13. Chen W, Chang S-Y, Kiang S, Early W, Paruchuri S, Desai D (2008) The measurement of spray quality for pan coating process. J Pharm Innov 3:3–14
14. Cole G, Hogan J, Aulton M (1995) Pharmaceutical coating technology. Taylor & Francis Ltd., London
15. Drechsler M, Garbacz G, Thomann R, Schubert R (2014) Development and evaluation of chitosan and chitosan/Kollicoat Smartseal 30D film-coated tablets for colon targeting. Eur J Pharm Biopharm 88:807–815
16. Ensslin S, Moll K, Metz H, Otz M, Maeder K (2009) Modulating pH-independent release from coated pellets: effect of coating composition on solubilization processes and drug release. Eur J Pharm Biopharm 72:111–118
17. Freireich B, Kumar R, Ketterhagen W, Su K, Wassgren C, Zeitler J (2015) Comparisons of intra-tablet coating variability using DEM simulations, asymptotic limit models, and experiments. Chem Eng Sci 131:197–212
18. Galus S, Kadzinska J (2015) Food applications of emulsion-based edible films and coatings. Trends Food Sci Tech 45:273–283
19. Gendre C, Genty M, da Silva J, Tfayli A, Boiret M, Lecoq O, Baron M, Chaminade P, Pean J (2012) Comprehensive study of dynamic curing effect on tablet coating structure. Eur J Pharm Biopharm 81:657–665

20. Haaser M, Naelapaa K, Gordon K, Pepper M, Rantanen J, Strachan CTP, Zeitler J, Rades T (2013) Evaluating the effect of coating equipment on tablet film quality using terahertz pulsed imaging. Eur J Pharm Biopharm 85:1095–1102
21. "ICH," [Online]. Available: https://www.ich.org/fileadmin/Public_Web_Site/ICH_Products/Guidelines/Quality/Q8_R1/Step4/Q8_R2_Guideline.pdf. Accessed 19 Feb 2017
22. Ilko D, Steiger C, Keller R, Holzgrabe U, Meinel L (2016) Tamper-proof tablets for distinction between counterfeit and originator drugs due to PEG coding. Eur J Pharm Biopharm 99:1–6
23. Joshi S, Petereit H-U (2013) Film coatings for taste masking and moisture protection. Int J Pharm 457:395–406
24. Knop K, Kleinebudde P (2013) PAT-tools for process control in pharmaceutical film coating applications. Int J Pharm 457:527–536
25. Lehmann KO (1997) Chemistry and application properties of polymethacrylate coating systems. In: Aqueous polymeric coatings for pharmaceutical dosage forms. Marcel Dekker Inc., New York
26. Li X, Guo H, Heinamaki J (2010) Aqueous coating dispersion (pseudo latex) of zien improves formulation of sustained-release tablets containing very water-soluble drug. J Colloid Interface Sci 345:46–53
27. Lin H, May R, Evans M, Zhong S, Gladden L, Shen Y (2015) Impact of processing conditions on inter-tablet coating thickness variations measured by terahertz in-line sensing. J Pharm Sci 104:2513–2522
28. Lin H, Dong Y, Shen Y, Zeitler JA (2015) Quantifying pharmaceutical film coating with optical coherence tomography and terahertz pulsed imaging: an evaluation. J Pharm Sci 104:3377–3385
29. Markl D, Hannesschlaeger G, Sacher S, Leitner M, Buchsbaum A, Pescod R, Baele T, Khinast J (2015) In-line monitoring of a pharmaceutical pan coating process by optical coherence tomography. J Pharm Sci 104:3377–3385
30. Maurer L, Leuenberger H (2009) Terahertz pulsed imaging and near infrared imaging to monitor the coating process of pharmaceutical tablets. Int J Pharm 370:8–16
31. Merkus H, Meesters G (2014) Particulate products—tailoring properties for optimal performance; particle technology series, vol 19. Springer, Berlin
32. Mwesigwa E, Basid A (2016) An investigation into moisture barrier film coating efficacy and its relevance to drug stability in solid dosage forms. Int J Pharm 497:70–77
33. Nagai T, Obara S, Kokubo H, Hoshi N (1997) Application of HMPC and HMPCAS to aqueous film coating of pharmaceutical dosage forms. In: Aqueous polymeric coatings for pharmaceutical dosage forms. Marcel Dekker Inc., New York
34. Nollenberger K, Albers J (2013) Poly(meth)acrylate-based coatings. Int J Pharm 457:461–469
35. Pandey P, Bindra D, Gour S, Trinh J, Buckley D, Badawy S (2014) Excipient-process interactions and their impact on tablet compaction and film coating. J Pharm Sci 103:3666–3674
36. Peng T, Huang Y, Mei L, Wu L, Chen L, Pan X, Wu C (2015) Study progression in application of process analytical technology on film coating. Asian J Pharm Sci 10:176–185
37. Plumb A, Rowe R, York P, Doherty C (2003) Effect of varying optimization parameters on optimization by guided evolutionary simulated annealing (GESA) using a tablet film coat as an example formulation. Eur J Pharm Biopharm 18:259–266
38. Porter S, Sackett G, Liu L (2009) Development, optimization, and scale-up of process parameters: pan coating. In: Developing solid oral dosage forms. Elsevier Inc., Burlington, USA
39. Rege B, Gawel J, Kou J (2002) Identifcation of critical process variables for coating actives onto tablets via statistically designed experiments. Int J Pharm 237:87–94
40. Rotta J, Ozorio R, Kehrwald A, Barra GdO, Amboni RdM, Barreto P (2009) Parameters of color, transparency, water solubility, wettability and surface free energy of chitosan/hydroxypropylmethylcellulose (HPMC) films plasticized with sorbitol. Mat Sci Eng C 29:619–623

41. Schilde C, Mages-Sauter C, Kwade A, Schuchmann H (2011) Efficiency of different dispersing devices for dispersing nanosized silica and alumina. Powder Technol 207:353–361
42. Schwarz S, Petzold G, Bhatti Q (2010) Stability for dispersions. Eur Coat J 07:1–10
43. Shokri J, Adibkia K (2013) Application of cellulose and cellulose. In: Cellulose—medical, pharmaceutical and electronic applications. InTech
44. Siepmann F, Siepmann J, Walther M, MacRae R, Bodmeier R (2006) Aqueous HPMCAS coatings: effect of formulation and processing parameters on drug release and mass transport mechanisms. Eur J Pharm Biopharm 63:262–269
45. Siepmann J, Siepmann F (2013) Stability of aqueous polymeric controlled release film coatings. Int J Pharm 457:437–445
46. Suzzi D, Radl S, Khinast J (2010) Local analysis of the tablet coating process: impact of operation conditions on film quality. Chem Eng Sci 65:5699–5715
47. Thoma K, Bechtold K (1999) Influence of aqueous coatings on the stability of enteric coated pellets and tablets. Eur J Pharm Biopharm 47:39–50
48. Tobiska S, Kleinebudde P (2003) Coating uniformity and coating efficiency in a Bohle Lab-Coater using oval tablets. Eur J Pharm Biopharm 56:3–9
49. Toschkoff G, Just S, Knop K, Kleinebudde P, Funke A, Djuric D, Scharrere G, Khinast J (2015) Modeling of an active tablet coating process. J Pharm Sci 104:4082–4092
50. Wang J, Hemenway J, Chen W, Desai D, Early W, Pauchuri S, Chang S-Y, Stamato H, Varia S (2012) An evaluation of process parameters to improve coating efficiency of an active film-coating process. Int J Pharm 422:163–169

Anneke M. Dijkhuis-Bouwman studied both Pharmaceutical Sciences and Pharmaceutical Technology at the University of Groningen. After graduation, she moved to Leiden University to do research in the lipid arrangement in human skin. One year later, she started a Ph.D. study at University of Groningen. The area of research was high shear granulation, a very common process in the pharmaceutical industry. The subject of the thesis was about the process of high shear granulation, and handled both the material used for the equipment, and the material exchange behavior during the equilibrium phase of the high shear granulation process. It resulted in 4 publications in various journals, written as A. M. Bouwman.

After graduation, she started working as a process engineer at Yamanouchi Europe B.V., a company that soon after changed to be Astellas Pharma Europe B.V. In this job, knowledge on more standard pharmaceutical processes was obtained. It is the best combination of both science and practical work! After a couple of years, she changed to be Senior Manager Pharmaceutical Technology.

Chapter 12
Particulates in Semi-Solid Pharmaceutical Products

David Harris

Abstract Particles may be found in a wide range of topical pharmaceutical products, whether as active substances, as functional excipients, or as a constituent part of the drug product, as in the case of emulsion droplets. Formulation of a drug substance as solid particulates introduces a dependence on the dissolution of the drug substance to exert its therapeutic benefit, but at the same time offers the possibility of more sustained drug delivery to the target tissue than may be obtained from a solution formulation. Particles may be employed as emulsifiers (Pickering emulsions) and as thickening agents in gel formulations. Finally, particles may be emulsion droplets in cream or lotion formulations, which usually contain one or more active ingredients dissolved in the dispersed phase. In all of these applications, various attributes of the solid particles may be critical to the functionality of the formulation. These attributes, and the principal techniques used for characterizing particulates for these purposes, are described in this chapter. Also reviewed are the unit operations particularly associated with the manufacture of such products, including mixing, material incorporation and size reduction.

12.1 Introduction

Particulates may be found in a wide range of semi-solid pharmaceutical products, as well as in an even wider range of semi-solid cosmetic and other consumer products. Pharmaceutical examples of semi-solid products include creams, ointments, gels, pastes and lotions. These products are most commonly intended for dermal application, although other applications include ocular, intra-oral or vaginal drug delivery.

Creams are most commonly oil-in-water emulsions in which the drug substance is either dissolved or dispersed, although water-in-oil emulsions are also used in certain applications. Oil-in-water creams are usually water-washable and are

D. Harris (✉)
Merck & Co. Inc., Kenilworth, USA
e-mail: david.harris@merck.com

H. G. Merkus et al. (eds.), *Particles and Nanoparticles in Pharmaceutical Products*,
AAPS Advances in the Pharmaceutical Sciences Series 29,
https://doi.org/10.1007/978-3-319-94174-5_12

relatively aesthetically acceptable. *Ointments* are most often based on oils or polyols, and also contain drug substance which is either dissolved or dispersed; they contain relatively little water, however, and are usually less aesthetically acceptable than creams. *Gels* are generally based on an organic solvent, oil or aqueous vehicle, thickened with a polymer or other viscosity-enhancer; these again can contain drug substance which is either dissolved or dispersed. *Pastes* are generally ointment-like vehicles which contain relatively high levels (typically >50%) of dispersed solid particulates. *Lotions* may be emulsions, suspensions or single-phase systems for dermal application, but are lower in viscosity (generally <50 Pa.s.) and are thus more easily spread on the skin. These definitions are somewhat general, and some semi-solid products defy precise categorization, but this classification is sufficient for the purpose of this chapter. Furthermore, the scientific principles applicable to formulating particulates in semi-solids have many commonalities across these different formulation types, and different routes of administration.

Within semi-solid pharmaceutical products, particulates can be subdivided into three categories:

Firstly, the particulate may be the active ingredient in the formulation, such as for some dermatological cream and ointment products (e.g., betamethasone dipropionate ointment), ophthalmic products (e.g., chloramphenicol eye ointment) and for sunscreens in which the particulates (zinc oxide, titanium dioxide) provide the opacity of the product;

Secondly, the particulate may be a functional excipient, included to provide or enhance a specific attribute of the formulation, most often viscosity or emulsification (e.g., gels, and emulsions based on colloidal silicon dioxide as viscosity-builders or emulsifiers);

Thirdly, the particulate may be an emulsion droplet in a cream or lotion formulation; such emulsions, as noted above, may be of the oil-in-water or (less commonly) water-in-oil varieties, and usually contain one or more active ingredients dissolved in the dispersed phase.

The physical attributes of solid particulate materials for inclusion in semi-solid products can greatly influence the functionality of the product. The most critical attributes of solid particulate materials in this respect are particle size, surface area, shape, polymorphic form and zeta potential. There exists a diverse range of analytical tools which can be used to characterize particulate materials. Some of these tools are relevant to a wide range of particle applications, while others are relevant to very specific applications; the characterization tools that are particularly relevant to semi-solid products will be highlighted in the following sections. Product development should begin, under the Quality-by-Design paradigm, with a target product profile, progress through identification of the critical quality attributes of a product, and ultimately lead to the establishment of critical incoming material attributes and specifications and critical process controls (collectively termed the "control strategy"). Table 12.1 lists the expected Critical Quality Attributes that may be dependent on particle attributes, and the corresponding Critical Material Attributes, for various types of semi-solid product.

The requirements for characterizing the particulates will generally be more stringent for particulates as active substances as compared to particulates as

Table 12.1 Critical quality attributes and critical material attributes for typical semi-solid pharmaceutical products

Formulation type	Critical quality attributes	Critical material attributes
Cream/ointment/gel/lotion formulation—active ingredient in solution	Absence of particulate/ crystalline drug	None applicable
Cream/ointment/gel/lotion formulation—active ingredient in solid state	Drug release rate	Drug crystallinity/ polymorphic form Drug particle size and/or (less-commonly) surface area Shape/morphology
Emulsion (cream/ointment) formulation—particulate as emulsifier	Drug release rate Physical stability/ absence of phase-separation/ Viscosity	Particle size Surface area Shape/morphology
Gel formulation—particulate as viscosity-building agent	Viscosity	Particle size and/or (less-commonly) surface area Shape/morphology Surface energy

excipients, because of the more direct link between drug substance particle attributes and product performance. While excipient particulate attributes are also important to product performance, the level of understanding of that relationship, and the level of control applied to excipient attributes, is generally less rigorous. The following sections discuss the principles of formulation and characterization associated with particulates in the various types of semi-solid products outlined above.

12.2 Particle Attributes Relevant to Semi-Solid Products

It should be evident from the foregoing that the selection of a formulation containing particulate drug substance has significant implications for product performance and drug delivery. It is logical to expect also that the attributes of the drug substance particulates may affect formulation performance. The particulate attributes most likely to be relevant in this way are discussed in the following sections and by several authoritative sources [12, 24, 35].

12.2.1 Particle Size

Particle size is the single most important particle attribute in the context of semi-solid products, and is the particle attribute most likely to be controlled in the incoming raw material and possibly also in in-process and/or in the finished product. Particle size is especially critical where the particulate is the active ingredient in the formulation, in which case its particle size can profoundly affect the clinical effectiveness of the product. It is clear from the Noyes-Whitney equation (see Sect. 12.3.1) that the rate of dissolution of drug particles in a medium will be dependent on their surface area. In practice, the particle size of a solid is generally more easily quantified than the surface area, and so it is more common to measure and control particle size than surface area; the two are obviously geometrically related and so, from a practical standpoint, particle size is generally a suitable surrogate measure.

At this point, however, it is necessary to recognize that, for any system other than perhaps a monodisperse population of spherical particles, there is no absolute measure of particle size. All of the available techniques interpret some size-related attribute of the particle as a measure of the particle size. Different measuring techniques, such as microscopy, sedimentation and laser diffraction, for example, will return different estimates of particle size; furthermore, these estimates of particle size will become increasingly divergent as the particles deviate further from sphericity. Furthermore, it is also necessary to recognize that any population of particles generally possesses a range of sizes, and thus it is necessary to consider the particle size distribution of the material, and not a single particle size. Size distributions may be calculated based on the number of particles, or the volume or mass of particles; these approaches can yield very different estimates of mean or median particle size; number distributions tend to be dominated by the finer particles in a distribution, whereas volume or mass distributions tend to be dominated by the coarser particles in a distribution, which account for most of the volume or mass of the sample. Where a statistically adequate number of particle size data is available, the shape of the size distribution may be calculated; such distributions may approximate a normal distribution or, more typically, a log-normal distribution. Having characterized the particle size distribution of a sample, descriptor(s) must be selected to represent the distribution, and if necessary to control the size of the material by specification. Examples of such descriptors include the mean or the median of the distribution,[1] percentiles of the measured size distribution, and percent of a sample below a certain size. The primary consideration in selecting a size-measuring technique and the descriptor(s) of size to be used is that these should represent the attributes of the particulate material that are most relevant to the quality or performance of the process or finished product. These principles have been described in detail by other workers [22, 35], to which the reader is directed for further information.

An extremely wide range of techniques exists for measuring particle size, with applicability to different material types and size ranges. However, only a relatively

[1]In this book the symbol D is used for particle size. E.g. to number-based distributions a (second) subscript 0 is given; the mean size is expressed as $\langle D_{1;0} \rangle$ and the median size as $D_{50,0}$; to volume-based distributions a (second) subscript 3 is given; the mean size is expressed as $\langle D_{4;3} \rangle$, the median size as $D_{50,3}$.

Table 12.2 Particle size measuring techniques relevant to particulate semi-solid products

Technique	Effective range	Key characteristics
Sieve analysis	25–50 to >1000 μm	Simple and intuitive; commonly used to specify particle size for excipients, but less so for drug substances; >1 g is needed to perform test, even at smallest scale, and yields relatively low-resolution distribution information
Laser diffraction	0.1–2000 μm	Technique of choice for most drug substance particle sizing applications; rapid to perform; reproducible; 1–3 g material generally sufficient to perform test; requires dispersion of particles in air or liquid, but dispersion conditions (particularly for air) require optimization to ensure complete dispersion
Microscopy and image analysis	3–1000 μm using visible light	Material-sparing, and provides additional information on other attributes such as shape, degree of association; tedious and time consuming to develop quantitative size distribution information, unless image analysis is employed
Dynamic light scattering	0.005–1 μm	Applicable only to measurement of dispersions of sub-micron particles in liquid; requires samples with relatively low particulate volume fractions
Particle Tracking Analysis	0.01–0.6 μm	Able to estimate size of nanoparticles in dilute suspension, by visualizing and analyzing the paths of individual particles under Brownian motion
Centrifugal sedimentation	0.02–1 μm	Applicable only to measurement of dispersions of sub-micron particles in liquid; able to measure samples with particulate volume fractions up to 2%
Diffusion Nuclear Magnetic Resonance	1–10 μm	Able to measure droplet size distribution in an emulsion, over a range of volume fractions from about 10–80%
Focused Beam Reflectance Measurement	0.5–3000 μm	Measures chord lengths, from which the particle size distribution is implied; suitable for in-line measurements during processing
Fineness-of-grind (Hegman) gauge	2–250 μm	Fast and easy test to perform; only yields information on the largest particles in a sample; provides no information on the size distribution

small subset of these techniques finds use in the development, manufacture and characterization of semi-solid pharmaceutical products. The most relevant techniques are summarized in Table 12.2 and described briefly in the following sections. These and other particle size measuring techniques have been extensively described elsewhere [22, 35] so the reader is directed to those sources for more detailed information.

Sieve analysis is probably the oldest of these particle sizing techniques. In pharmaceutical applications, woven wire mesh sieves with nominally-square apertures of defined sizes are used. Mesh sizes were historically expressed in terms of the number of holes per inch, although this approach did not account for the width of the wires in the mesh, and so several slightly different mesh size conventions evolved. Although mesh size terminology remain in widespread use today, including in pharmaceuticals, meshes sized in microns provide a universal standard, as codified in ASTM E11 and in other international standards.

Sieve analysis is usually performed using a stack of sieves that are sized so as to pass particles that are approximately $1/\sqrt{2}$ or 1/2 smaller than the sieve above. This technique provides a relatively low-resolution particle size distribution, as generally not more than 5 or so sieves are used in one analysis. However, this technique is extensively used to specify the particle size of pharmaceutical excipients, where precise size control is not deemed critical; common specifications are for not less than 95 or 99% of a material to pass through a sieve of a specified size. Obviously such specifications impose no restrictions on how fine such a material may be, and so are of limited value where product quality may be adversely affected by fine particles. It is also important to note that the passage of particles through a sieve depends on their approaching the apertures with their smallest cross-sections, so this technique is well-suited to particles with aspect ratios close to 1, but may yield less consistent results for particles which are significantly rod-like or plate-like.

Laser diffraction has become the technique of choice for particle size measurement in the pharmaceutical industry, for materials in the applicable size range. This technique is based on the principle that particles scatter light in relation to their size. A laser diffraction instrument measures the scattering pattern of a monochromatic laser light by a particle dispersion. The measured scattering pattern is converted to a particle size distribution using the known relationship between particle size and scattered light intensities at different angles, generally using one of two different models. Mie theory requires knowledge of the optical properties of the sample material. The Fraunhofer approximation offers a simpler approach, which does not require knowledge of the optical properties of the sample, but should be used with caution with samples which might have particles below 50 μm or with relatively transparent materials. Both models assume the particles to be spherical, and so for non-spherical particles, the calculated size distributions are in terms of equivalently-light-scattering spherical diameters.

Adequate dispersion of the sample in the measurement zone is critical in laser diffraction, as incompletely-dispersed agglomerates of particles will be detected as single, large particles. Air dispersion is most commonly used; this is achieved by feeding powder from a vibratory feeder into an air stream, which may be further directed to ensure particle collisions to deagglomerate the sample. It is necessary to optimize the air flow so as to achieve essentially complete deagglomeration with minimal particle breakage. Less commonly, liquid dispersions may be used, particularly for highly cohesive materials, although care is necessary to ensure there is no dissolution of the particles in the test medium.

Laser diffraction is the most commonly-used technique for measuring particle sizes of pharmaceutical actives, in particular, for which specifications are often established for the median particle size (the 50th percentile of the mass distribution), as well a sometimes also the 10th and/or 90th percentiles of the distribution.[2] The most commonly-used laser diffraction instruments in pharmaceutical applications are the Malvern Mastersizer and Sympatec HELOS ranges.

Microscopic techniques, including both light and electron microscopy, are immensely powerful tools for providing qualitative information about a sample. As tools for determining particle size, however, they have significant limitations. Chief among these is the fact that large numbers of particles are needed to provide a statistically accurate estimate of particle size, and relatively small numbers of particles at the large end of a size distribution profoundly affect the overall weight distribution. Automated quasi-static imaging of particles dispersed in an airstream can enable an adequate number of particles to be evaluated to yield a statistically-meaningful result.

Microscopy is not commonly used to measure or control the particle size of pharmaceutical powders, largely due to the above limitations, and given the availability of suitable alternate techniques. However, microscopy may be uniquely suited to measuring and controlling particle size in formulated semi-solid products, such as by confirming the absence of large drug particles or agglomerates, which can arise due to crystal or particle growth, or incomplete dispersion. It must be stressed, however, that measurement of particle size in a semi-solid matrix by microscopy is dependent on the ability of the imaging system to differentiate between the particles of interest and other constituents of the formulation, such as other particles, microcrystalline domains of vehicle components (e.g., petrolatum, hydrocarbon waxes, polyethylene glycol), emulsion droplets, air bubbles, etc.

Dynamic Light Scattering (DLS) or Photon Correlation Spectroscopy is a technique, based on light scattering, which is able to measure the size distribution of sub-micron particles in a liquid sample. The intensity of scattered light at a given angle is measured as a function of time; this signal fluctuates due to Brownian motion of the particles of interest, which causes distances between particles to vary and the scattered light to undergo either constructive or destructive interference. From these temporal fluctuations in scattered light intensity, rates of particle motion, and ultimately particle size, are calculated. DLS is relatively infrequently used in the development or control of semi-solid pharmaceutical products; the principle applications are particle sizing of materials that are significantly smaller than 1 μm such as fumed silicon dioxide and colloidal materials.

Particle tracking analysis (PTA) utilizes light scattered by individual particles in a laser beam, in conjunction with video imaging, to analyze the paths that particles take under Brownian motion. The diffusion coefficients of individual particles are determined from their rates of movement from frame to frame, and the sphere-equivalent hydrodynamic radius of each is calculated using the Stokes-Einstein equation. A size distribution is thus assembled based on these

[2] $D_{10,0}$, $D_{50,0}$ and $D_{90,0}$ are used as symbols for the 10th, 50th and 90th percentile sizes of a number-based distribution, $D_{10,3}$, $D_{50,3}$ and $D_{90,3}$ as the symbols for the 10th, 50th and 90th percentile sizes of a volume-based distribution.

individual particle size estimates [18]. This technique is limited to samples containing relatively low particle concentrations (in the range 10^5–10^{10} per mL).

Sedimentation techniques make use of either gravitation or centrifugation to separate particles suspended in a liquid medium based on their size. Particles settle according to Stokes' Law, with larger particles settling faster than smaller ones. For particles that are significantly smaller than 1 μm in size, and particularly where the difference in density between the particles and the medium is small, the particles are increasingly subject to Brownian motion, and sedimentation rates become very low. This limitation is addressed with centrifugal sedimentation techniques, such as X-ray disc centrifugation and cuvette photocentrifugation. These use X-ray or light absorption to monitor the concentration change over the course of the test and the particle sizes are calculated from the settling times using Stokes' law, resulting in a particle-size-versus-cumulative-mass-undersize distribution. These techniques—and especially X-ray disc centrifugation—are particularly well suited to the measurement of titanium dioxide and zinc oxide, which are used in sunscreens; their high particle densities mean that even very fine (<100 nm) particles sediment under centrifugation. It is notable that X-ray disc centrifugation may also be used to measure the particle sizes of these materials in formulated sunscreen products, which enables the effects of formulation and process variables on particle size to be elucidated.

Diffusion Nuclear Magnetic Resonance (Diffusion NMR) is a non-destructive test that can estimate the droplet size distribution in an emulsion, including at relatively high internal phase volumes. The measurement is based on the restricted diffusion of liquid molecules within a dispersed droplet. A log-normal size distribution is assumed in the calculation [40, 41].

Focused Beam Reflectance Measurement (FBRM) is a technique that measures chord lengths for a population of particles suspended in a liquid medium. A focused laser beam is scanned at high speed, projecting into the sample of interest, and backscattered light, as when the beam encounters a particle in the suspension, is detected by the probe. The chord length of the particle is calculated from the duration of the backscattered signal, knowing the scanning speed. A disadvantage of this technique is that it returns a distribution of chord lengths, not a true particle size distribution. However, the physical configuration of the measuring probe enables it to be inserted into a process vessel which means that it is suitable for in-line process monitoring [31].

The **fineness-of-grind or Hegman gauge** is a simple device for detecting the sizes of the largest particles or agglomerates in a semi-solid product. It consists of a steel plate into which is machined a groove of linearly decreasing depth. A sample of the product is applied at the deep end of the groove and spread with a scraper along the length of the groove. At the point where the diameter of the largest particles or agglomerates is greater than the depth of the groove, the particles begin to leave streaks in the product film. The corresponding particle size is read off the engraved scale by the operator. This test is suitable as a test for the presence or absence of large particles or agglomerates in a sample, but it is limited in that it provides no information as to the particle size distribution.

As noted earlier, each of these techniques may yield different estimates of particle size, relying as they do on different attributes of the particles. This is particularly true for particles with shapes that deviate significantly from spherical (e.g., plate- or needle-like). Also, particles which are agglomerated may yield different estimates of particle size, depending on the degree to which the agglomerates were broken down to primary particles in the course of sample preparation or measurement. However, provided one technique is used consistently, and the approximations inherent in that technique are understood, such techniques may be useful control tools and may be at least directionally predictive of product performance.

The particle size of drug substance particles in semi-solid formulations is most commonly controlled in the incoming drug substance. Measurement of the particle size of solid particles, once formulated into a semi-solid drug product, is considerably more challenging. Of the techniques reviewed above, only microscopy and the fineness-of-grind gauge can be readily applied to determine the particle size of particulates in viscous semi-solid formulations. Particle sizes in less viscous formulations, or in systems which can be diluted without altering the particle characteristics, can also be measured by DLS, centrifugal sedimentation or focused beam reflectance measurement, subject to the limitations of those techniques discussed above. Even then, however, the abilities of any of these techniques to assess particle size in formulated materials may be limited by the presence of other interfering particles or structures.

One of the most critical applications of particle size control in semi-solid formulations is to assure the absence of large drug particles, which can arise due to crystal or particle growth. Two mechanisms can lead to crystal or particle growth in products, particularly where the solid has some degree of solubility in the vehicle. Firstly, exposure of the drug substance to elevated temperatures, or different solvent compositions, during manufacturing, can lead to dissolution of a portion of the drug substance. At the completion of manufacturing, the formulation may be supersaturated with respect to the drug; the excess dissolved drug would be likely to crystallize or precipitate, possibly resulting in the formation of larger particles or crystals than the starting drug substance particle size. Secondly, it is well known that suspension systems are susceptible to the process of Ostwald ripening, whereby the smaller crystals or particles in the system dissolve, and the dissolved material redeposits onto larger crystals or particles. This occurs because molecules on the surface of small particles are energetically less stable than those on the surface of larger particles. To control for such particle growth, microscopic limit tests are frequently applied to semi-solid suspension products.

12.2.2 Surface Area

As noted earlier, surface area is the most important attribute of a solid material with respect to dissolution, and also to other surface-mediated processes such as adsorption. The principal technique for measuring the surface area of a particulate material utilizes the Brunauer-Emmett-Teller (BET) theory for adsorption of gas molecules on a surface. This technique is only applicable to testing of dry samples. Care must be exercised when evaluating the surface area of materials recovered from suspensions; any agglomeration of particles, such as may occur during drying, will result in occlusion of some particle surfaces and underestimation of the surface area. There are no techniques available to directly measure the surface area of a particulate solid once dispersed in a semi-solid formulation.

12.2.3 Crystallinity and Polymorphism

Solid materials can exist as crystalline forms, as amorphous material, or as mixtures of these. Furthermore, many crystalline materials exhibit polymorphism, which is the ability to exist in more than one form or crystal structure. Different polymorphic forms of a solid material possess different crystal energies; this can result in the various forms exhibiting different solubilities in relevant media, dissolution rates, drug delivery profiles and even biological effects. Amorphous solids possess the highest energy of all solid forms, and thus also exhibit the highest solubility; they are generally also metastable, however, meaning that they are at risk of converting to a more energetically-favorable crystalline form. It is thus especially important in

Table 12.3 Summary of most commonly used techniques for characterizing crystallinity and polymorphic forms of particulates

Technique	Characteristics and applications
Single crystal X-ray diffraction	The definitive test for the crystalline form of a solid material; enables the 3-dimensional crystal structure to be elucidated
X-ray powder diffraction (XRPD)	X-ray diffraction applied to a powder sample rather than a single crystal; provides a diffractogram which is generally characteristic of the form; the most widely used technique for form identification in pharmaceutical development
Differential Scanning Calorimetry (DSC) and modulated DSC (mDSC)	Suitable for monitoring crystal forms and form transitions where the different forms exhibit different melting points
Thermo-gravimetric analysis (TGA)	Suitable for detecting weight changes associated with desolvation upon heating of a solvate or hydrate
Solid-state nuclear magnetic resonance spectroscopy (ssNMR)	Able to detect amorphous and different crystalline forms in mixed samples, based on differences in relaxation times; can be used to quantitatively estimate the relative levels of different forms in a sample

pharmaceutical development to be able to quantify amorphous material in mixed crystalline and amorphous systems, and to understand its origin and its fate.

The techniques most commonly used to characterize crystallinity and polymorphism are summarized in Table 12.3.

It should be clear from the above discussion that it is important to be able to characterize the crystallinity and polymorphic forms of particulate drugs substances. It is important to recognize, also that materials can convert between crystal forms during processing (e.g., as a result of exposure to elevated temperature), and that crystalline material can become amorphous (e.g., as a result of mechanical or thermal stress or solvation/desolvation). Also, as discussed in the context of particle size, exposure of the drug substance to elevated temperatures, or different solvent compositions, during manufacturing, can lead to dissolution of a portion of the drug substance, which may crystallize or precipitate in a different form at or after the completion of manufacturing. For these reasons, it may be necessary to confirm the drug substance crystallinity and polymorphic form in in-process materials or in the finished product. As is the case for characterizing particle size, this can be significantly more difficult than characterizing the form of the pure particulate drug substance, due to the presence of multiple interfering materials and signals, and the often low concentration of the drug substance in the formulation. Of the techniques listed in Table 12.3, ssNMR is the technique which is best able to quantify the levels of different solid state forms in formulated products, and can often attain analytical limits of quantitation of <10% amorphous. Other techniques such as DSC/mDSC and XRPD may be able to provide qualitative indications of form changes, but are generally unable to provide quantitative information.

12.2.4 Particle Shape

Particle shape or morphology can vary widely, and is of profound importance to the performance and processing of particulate materials. The International Conference on Harmonization defines some commonly used descriptors of particle shape [13]:

Acicular—slender, needle-like particle of similar width and thickness;

Columnar—long, thin particle with a width and thickness that are greater than those of an acicular particle;

Flake—thin, flat particle of similar length and width;

Plate—flat particles of similar length and width but with greater thickness than flakes;

Lath—long, thin, and blade-like particle;

Equant—particles of similar length, width, and thickness; both cubical and spherical particles are included.

One of the most obvious consequences of these differences in morphology is that the surface area/volume ratios of these different shapes can vary considerably. Just as important, however, is that properties such as material flow, propensity of particles to fracture during processing, and interparticulate interactions can be profoundly affected by particle shape. While particle shape has been extensively studied over the years, there is no universally applicable measure of particle shape. A simple approach to quantification of particle shape is to determine two particle dimensions, such as maximum and minimum diameter, and to calculate their ratio; while not highly informative, this descriptor is often sufficient for practical purposes. The above range of qualitative descriptors of particle shape and appearance have evolved which, although not quantitative, do provide a common language that can be used to describe and compare particles. Shape may best be ascertained by microscopy, and the power of this technique can be enhanced by the addition of computational image analysis. It is essential, though, to recognize the inherent limitations of this technique due to the preferential orientation or alignment of particles that may occur during sample preparation (e.g., needle-like particles align with fluid flow; plate-like particles settle on their faces).

12.2.5 Porosity and Density

Porosity is a measure of the void space in a particle, expressed as a fraction or percentage of the total volume of the particle. It is relevant to the processing of pharmaceutical solids, where it can influence material compressibility and where low particle density can lead to powder blend segregation; it is relevant also to surface adsorption or ion exchange applications. In most applications, however, particle porosity is less critical to semi-solid applications. The porosity of porous or hollow solid particles can be measured by porosimetry, which involves the intrusion of a non-wetting liquid (e.g., mercury) into the material at high pressure; the pore size range and total pore volume can be determined based on the pressure needed to force the liquid into the pores against the opposing force of the surface tension of the liquid. In some tests, what is measured is only the void space which is accessible from the surface of the particle, which does not include fully-enclosed void spaces.

The true density of solid particulates can be determined by helium pycnometry. Most small organic molecules possess true particle densities in the range 1.2–1.4 g/cm^3, although inorganic materials such as titanium dioxide and zinc oxide possess particle densities of 4 g/cm^3 or higher. Most semi-solid vehicles, in contrast, possess densities of around 1 g/cm^3. The wider the discrepancy between the particle and vehicle densities, the greater the risk of particle sedimentation, modulated, of course, by the viscosity of the system.

12.2.6 Association State

Association state refers to whether a solid consists of discrete (primary) particles or aggregates or agglomerates of primary particles. Such aggregates or agglomerates can arise during processing, drying or storage of powders, and can be held together by electrostatic forces or by partial fusion of the primary particles. Aggregates and agglomerates possess larger surface areas than equivalently-sized primary particles, with implications for particle functionality (e.g., dissolution) and processability. Aggregates and agglomerates can also be broken down partially or completely to their constituent primary particles, during measurement of the particle size (e.g., sieve analysis; dispersion in a laser diffraction instrument) and/or during drug product processing. There is thus a risk that measurement of the apparent particle size of an associated material may yield a value which is not reflective of the effective particle size of the material during processing or in the finished product. Microscopy is a useful tool to understand, at least qualitatively, the association state of a particulate material, and to assess such risks.

12.2.7 Zeta Potential

The zeta (ζ) potential is the electric potential at the slipping plane of the double layer at the surface of a dispersed particle, relative to the bulk liquid. The magnitude of the zeta potential indicates the degree of electrostatic repulsion between adjacent particles in a dispersion; it is thus a predictor of the stability of colloidal dispersions, particulate suspensions and emulsions. It is expressed in mV and cannot exceed 100 mV, positive or negative. Zeta potentials of 30 mV or greater are usually sufficient to stabilize dispersions of particles that are greater than 1 μm in size; higher zeta potentials, of 60–70 mV or more, may be needed to stabilize particles that are significantly smaller than 1 μm in size. Dispersions with lower zeta potentials will tend to coagulate or flocculate. Zeta potential cannot be directly measured, but is calculated from the electrophoretic mobility of particles in an electric field.

12.3 Particulates as Active Ingredients

As outlined above, the first broad group of semi-solid products containing particulates is those in which the particulate constitutes the active substance. These include dermatological and ophthalmic products which contain dispersions of pharmacologically active ingredients, and also sunscreen products in which the particulates provides the opacity of the products. This section discusses the key considerations in formulating active ingredients as particulates into semi-solid pharmaceutical products. The physiological characteristics of the skin and cornea, and the general principles of drug delivery to and through these tissues, have been comprehensively reviewed by other workers [25, 26, 34], and will not be recapitulated here. This section focuses rather on those aspects of material characterization, formulation science and drug delivery that are uniquely relevant to particulate systems.

12.3.1 Drug Delivery and Dosage Form Design

The most relevant semi-solid formulation platforms in the context of this review are creams, ointments, gels and pastes, as outlined above. There are many drug-delivery considerations and formulation considerations which may dictate the selection of one formulation type over another, including the target tissue, the nature of the condition being treated, the properties of the drug, commercial factors, etc., but those are largely not relevant to this chapter; the critical aspects of particulate active substances in semi-solid formulations are to a very large degree independent of formulation type.

The nature of the permeation barriers, the pathways of drug permeation, the attributes of the drug that are relevant to permeation, and the thermodynamic considerations governing drug permeation have been reviewed elsewhere for various routes of drug delivery [10, 21, 25, 26]. Crucially, however, the drug is required to be in solution in order to be able to permeate a biological membrane, or to exert a pharmacologic or chemotherapeutic effect in the case of a locally-acting therapy. Drug in particulate form, whether crystalline or amorphous in nature, is unable to permeate through these membranes or exert a biological effect without first dissolving in the vehicle or the surrounding medium. (It should be noted that some workers have reported that nanoparticles—solid particulates in the sub-micron size range—are able to permeate through biological membranes; reviews of the available data suggest there is little evidence that such particles are able to penetrate to the viable layers of the skin [27, 38], although much work is underway probing the mechanisms and the potential implications of such intact particle penetration [1]. The use of solid particulate drugs in semi-solid formulations also imparts some analytical and regulatory challenges and some stability risks, associated with the need to assure the control of particle size from batch to batch, from dosage unit to dosage unit, and over shelf-life. For these reasons, drugs are most commonly formulated in semi-solid products as solutions; the principal reasons why drugs may instead be formulated in these products as solid particulates include the following:

Many drug substances are relatively insoluble, and so it may not be possible to dissolve the requisite amount of drug in the desired mass or volume of vehicle, without resorting to solvent systems which may not be acceptable for aesthetic or safety reasons;

Formulating as a suspension, rather than as a solution, can provide a more sustained release or delivery profile, which may provide a superior clinical response or necessitate less frequent dosing or application.

As discussed above, a necessary first step for a solid drug substance to exert a biological effect is dissolution of the drug into the medium. This process is described by the Noyes-Whitney equation, which applies whether the dissolution process is occurring in a biological system, in a formulation, or in a laboratory setting:

$$dw/dt = k(c_s - c)$$

where $k = \mathcal{D}A/\delta$

In this equation, dw/dt is the rate of increase of the amount of drug in solution; $\mathcal{D}$ is the diffusion coefficient of the drug in the medium; A is the surface area of the drug solid exposed to the medium; δ is the thickness of the diffusion layer around the drug particles, which is dependent on the rate of flow or agitation of the system; c_s is the saturation solubility of the drug in the medium; and c is the concentration of the drug in the bulk of the medium.

Consider a typical semi-solid formulation containing a particulate drug substance upon application to the skin. Some fraction of the drug which is dissolved in the vehicle would partition into the largely lipidic matrix which constitutes the surface of the skin. This partitioning would be dependent on the relative solubility of the drug in the vehicle and in the skin, which would depend in turn upon the physico-chemical properties of the drug and the characteristics of the vehicle and of the skin [25, 26]. This partitioning would furthermore be subject to kinetic considerations, as the rate of partitioning would also be dependent on the diffusivity of the drug in the vehicle and in the skin. Drug permeating into the skin would further diffuse into the underlying tissues from which it may be removed by the vascular circulation. At the same time, diffusion of drug out of the vehicle would cause additional drug to dissolve from the surface of the remaining particulates, due to depletion of the bulk vehicle concentration (c in the equation above). In this way, an equilibrium will be set up between dissolution of the drug substance particulates, partitioning and permeation of the dissolved drug into and through the skin and distribution of drug into, and removal from, the underlying tissues. (It must also be pointed out that any volume or composition changes to the vehicle after application to the skin, such as due to the evaporation of a component or permeation of a component into the skin, could alter the solubility of the drug in the vehicle and/or the partitioning of drug between the vehicle and the skin; this would perturb the equilibration described above, but would not change the fundamental nature of the equilibrium.)

Similar considerations apply to the application of a semi-solid formulation containing a particulate drug substance to sites such as the surface of the eye or the oral cavity. The formulation may or may not become diluted in the local media, depending on its miscibility with aqueous media. Where such dilution occurs, and results in a bulk drug concentration that is lower than the saturation solubility of the drug in the system, additional drug would dissolve from the surface of the remaining particulates. Where the formulation is immiscible with the local media, a similar equilibrium will be set up between dissolution of the drug substance particulates, partitioning of drug out of the vehicle, permeation of the dissolved drug into and through the local epithelia and eventual clearance from the underlying tissues, much as described above for a formulation applied to the skin.

It can thus be seen that the type of semi-solid formulation, and the presence or absence of drug substance particulates in the formulation, can have profound effects

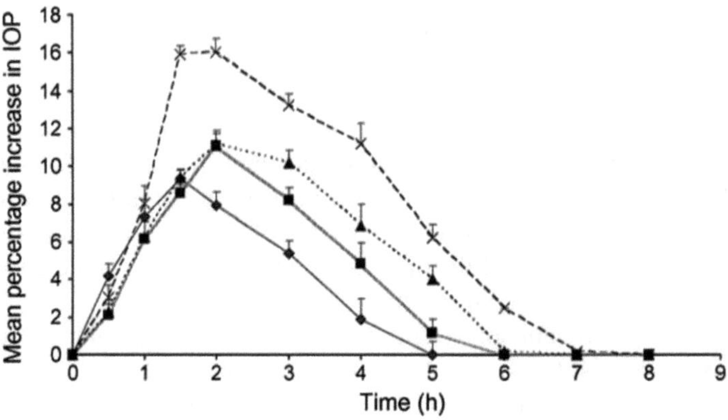

Fig. 12.1 Effects of drug particle size on mean percentage increase in intra-ocular pressure in albino rabbits receiving 50 μL of 0.1% hydrocortisone solution or suspensions of different mean particle diameters: 539 nm (multiplication sign), 1.394 μm (filled triangle), 3.94 μm (filled square), solution (filled diamond) [15]; copyright Elsevier; reproduced with permission

on the rate, duration and extent of drug delivery to the target tissue. Formulations containing the drug in the solution state provide an immediate thermodynamic driving force for drug partitioning into, and permeation through, the local membranes. Formulations containing a substantial fraction of the drug in the solid state provide a lesser thermodynamic driving force for partitioning/permeation, but the possibility of more sustained delivery as the drug particulates dissolve over time, as described in the above examples. However, it is also necessary to consider for how long the drug particulates may be retained at the delivery site, and in the presence of sufficient vehicle to permit continued dissolution and delivery of the drug. In the case of semi-solid formulations of particulates applied to the skin, evaporation or permeation of components of the vehicle may leave drug particulates on the surface of the skin but little remaining formulation to mediate dissolution and permeation of the drug; additionally, residual formulation may be removed from the skin in the course of normal daily activities. In the case of semi-solid formulations of particulates applied to, e.g., the surface of the eye or the oral cavity, normal clearance processes may remove much of the undissolved solid drug more rapidly than the particulates are able to dissolve. Both of these scenarios may result in lower overall extents of drug delivery as a consequence of the particulate presentation of the drug substance.

Kassem and co-workers [15] demonstrated markedly different pharmacodynamic profiles for three different ocular drugs, with solution and suspension formulations of various particle sizes. Their data, reproduced in Figs. 12.1, 12.2 and 12.3, show that for each of these three drugs, the suspension formulations exhibited greater extents and durations of effect as compared to solution formulations. Furthermore, for each of the drugs, there was also a rank order in response as a function of particle size, with the finest particle sizes in each case providing the greatest pharmacodynamic responses.

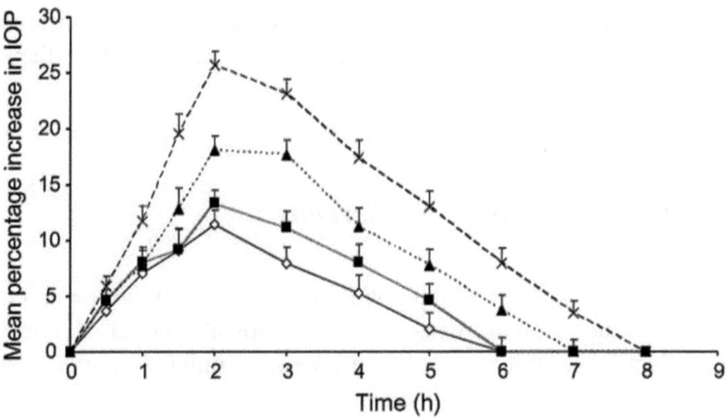

Fig. 12.2 Effects of drug particle size on mean percentage increase in intra-ocular pressure in albino rabbits receiving 50 μL of 0.1% prednisolone solution or suspensions of different mean particle diameters: 211 nm (multiplication sign), 1.626 μm (filled triangle), 4.0 μm (filled square), solution (open diamond) [15]; copyright Elsevier; reproduced with permission

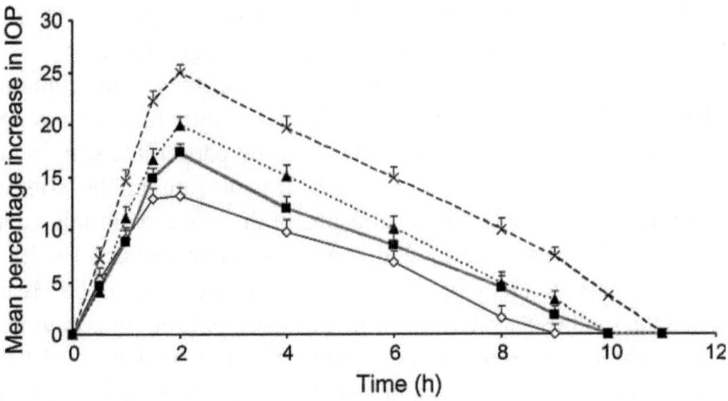

Fig. 12.3 Effects of drug particle size on mean percentage increase in intra-ocular pressure in albino rabbits receiving 50 μL of 0.1% dexamethasone solution or suspensions of different mean particle diameters: 930 nm (multiplication sign), 2.46 μm (filled triangle), 4.87 μm (filled square), solution (open diamond) [15]; copyright Elsevier; reproduced with permission

These results illustrate the complexity of drug delivery from suspension formulations, and the complex influences of particle size on biological performance. The most likely explanation for these observations is that the finest particle sizes (in this case, nanosuspensions) exhibited the longest mean residence times of drug in the pre-ocular space, while at the same time providing the greatest surface areas to facilitate drug dissolution, as explained above.

Considering all of these factors, it is usually desirable for the drug substance particle size in such formulations to be in the micron or even sub-micron range. The chief consideration here is that the dissolution of larger drug particles in the biological medium at the site of action is likely to be slow, and so it is likely that such larger particles will be cleared from the site of action before they are fully dissolved.

12.3.2 Particulate Actives: Sunscreens

A unique class of products, in which solid particulates provide the essential functionality of the product, is sunscreens. These are topical products that are designed to absorb or reflect some of the sun's UV (ultraviolet) radiation and thus protect the skin from sunburn and other UV-mediated damage. The most energetic UV radiation, UVC (λ of 100–280 nm) is absorbed by the ozone layer and thus is not of practical concern at ground level. UVB radiation (λ of 280–315 nm) is mostly absorbed by the ozone layer, but the fraction that does reach ground level is the main cause of sunburn and can also cause direct DNA damage and some forms of skin cancer. UVA (λ of 315–400 nm) is not the main cause of sunburn but can increase the rate of melanoma and photodermatitis; exposure to UVA, and more particularly the lower wavelengths of UVA, can cause indirect DNA damage, mediated by reactive species such as hydroxyl and oxygen radicals.

Sunscreens can function in two different ways: they can contain organic chemicals that absorb UV radiation, and/or they can contain inorganic particulates that scatter, reflect and absorb UV radiation. It is the particulate sunscreens, which utilize titanium dioxide or zinc oxide, which are relevant in the context of this chapter. Titanium dioxide and zinc oxide are both dense materials (densities of $\sim$4.0 and 5.6 g/cm^3, respectively) with high refractive indices ($\sim$2.5 and 2.0, respectively). The particle sizes of titanium dioxide and zinc oxide are critical to their performance in these systems. Solid particles exhibit maximum scattering where their particle size is approximately one half the wavelength of the incident light. Also relevant, however, is the fact that particle sizes below about 100 nm scatter less visible light, whereas larger particles or agglomerates appear white, which can be aesthetically undesirable. The strong dependence of sunscreen performance (as measured by SPF) on particle size is illustrated in Fig. 12.4, which compares two different particle sizes of zinc oxide. The optimal particle sizes of titanium dioxide and zinc oxide depend on their optical properties, and on the desired balance between efficacy and aesthetics; these considerations will not be discussed further here, but the reader is directed to Fairhurst [6] for a much more detailed treatment of this topic. In general, however, titanium dioxide provides more screening of UVB radiation, whereas zinc oxide provides more screening of UVA radiation.

Both titanium dioxide and zinc oxide are formulated in sunscreens as ultra-fine particles, <100 nm in size, at solids loadings in the range 2–8%. The particle sizes of these materials are generally measured by X-ray disc centrifugation; this technique is well-suited to this application as it is able to measure the size distribution of the particulates in the final product. X-ray disc centrifugation is preferred over

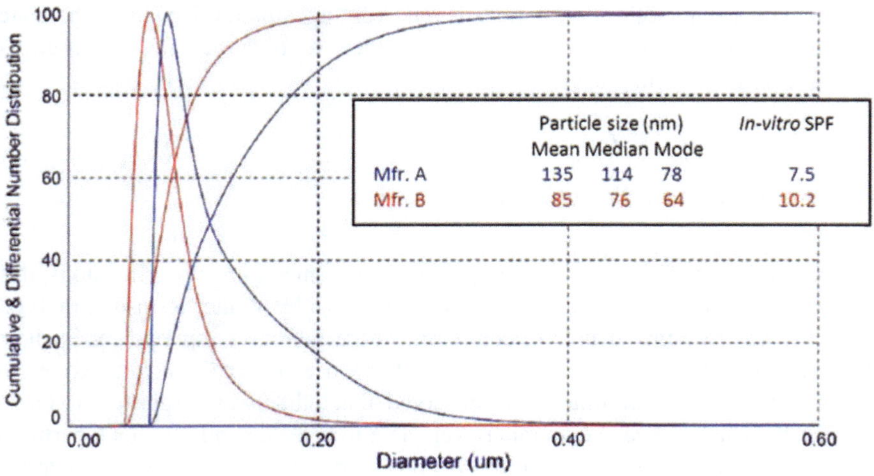

Fig. 12.4 Comparison of particle size distributions and effectiveness as sunscreens for two suspensions of 5% zinc oxide in isopropyl myristate [6]; copyright Springer; reproduced with permission

dynamic light scattering—generally a more widely-used technique for submicron particle sizing—because the latter gives a signal which is skewed towards the larger end of the particle size distribution, whereas the functionality of the solid particulates in a sunscreen is primarily a function of the number-distribution of the particle population.

These materials are considered nanomaterials by virtue of their particle sizes (dimensions less than 1 μm), and thus it is necessary to consider their safety, particularly given their mode of use and their widespread use in the general population. The theoretical risk is that titanium dioxide or zinc oxide particles in sunscreens may be able to penetrate the skin and reach viable cells, and there cause adverse effects. However, based on reviews of the available data, there is no in vivo evidence of harm from the application of these particles in sunscreens. The current evidence indicates that the particles remain on the surface of the skin and in the outer non-viable layers of the skin, and do not penetrate to the viable layers of the skin [27, 38].

12.4 Particulates as Excipients

Excipients in semi-solid formulations are generally present in solution (e.g., antimicrobial preservatives, co-solvents, dissolved polymers), as colloidal dispersions (e.g., hydrated/solvated polymers) or as microcrystalline structures (e.g., amphipathic emulsifiers at emulsion droplet interfaces). The use of solid particulates in semi-solid formulations is relatively uncommon. However, there are two classes of semi-solid formulations in which solid particulates are used as emulsifiers or thickening agents.

12.4.1 Emulsion Stabilizers

Emulsions that are stabilized by solid particles are termed Pickering emulsions; this phenomenon was first recognized over a century ago by Walter Ramsden in 1903 [33], but was recognized independently and more famously reported by Spencer Pickering in 1907 [32]. Pickering described the use of lime (calcium oxide) to emulsify paraffin oil in water, and observed that other solids (plaster of Paris, precipitated chalk, silica and alumina) were able to behave similarly. Such particles associate at the interface between the two phases of an emulsion, and stabilize the emulsion in an analogous way to the more-commonly used amphipathic emulsifiers.

The adsorbed particles at the oil/water interface of a Pickering emulsion can form a rigid shell-like structure, and as such can provide greater stability against coalescence as compared to conventional emulsions stabilized by amphipathic emulsifiers, and can stabilize higher internal phase volumes [8, 9, 20, 37]. The hydrophilicity or hydrophobicity of the surface of the particles is critical to their performance in these systems; this attribute is characterized by the contact angle of the liquid-liquid interface to the surface of the particle measured through the aqueous phase. Contact angles of approximately 90° are optimal because such particles are partially wettable by both liquids and are thus more able to associate at the interface. Furthermore, however, contact angles less than 90° (i.e., more hydrophilic) favor the formation of o/w emulsions while contact angles greater than 90° (i.e., more hydrophobic) favor the formation of w/o emulsions. The size of the particles is also relevant to their performance in these systems; the particle size needs to be significantly smaller than the droplet size of the emulsion to permit tight packing at the droplet interface. Colloidal silicon dioxide, which generally has a median particle size below 50 nm, has been most extensively studied in pharmaceutical applications of Pickering emulsions, although starch (median size of 1–2 μm) has also been used successfully.

Despite their long history, the in vitro skin absorption of drug from a Pickering emulsion was only reported comparatively recently [9], and there are no known examples of Pickering emulsions being used in commercial pharmaceutical products. Researchers continue to explore the utility of such systems, however. Among the potential advantages of these systems over classically-stabilized emulsions are their greater physical stability, their ability to provide different drug release profiles [8, 9], and the potential to avoid the skin irritation that has been linked to conventional emulsifiers. It has also been suggested that the self-assembled particles on the surface of a Pickering emulsion droplet can be fused together forming more durable capsules ("colloidosomes") that may be transferred to a different medium and which may then be used in a variety of applications [4].

12.4.2 Solid-Stabilized Gels

Gels are semi-solid formulations consisting of an organic solvent, oil or aqueous vehicle, thickened by an organic polymer, by small inorganic particles or by a combination of the two. Of interest in this chapter are gels thickened by particles. It is necessary, first, to understand the fundamentals of rheology of particle suspensions. At low particulate concentrations—volume fractions of 0.05 (5%) and below—the viscosity of the system can be described by the following equation:

$$\eta_{rel} = \eta_d/\eta_0 = (1 + 2.5\Phi)$$

in which η_{rel} = relative viscosity, η_d = the viscosity of the particle dispersion, η_0 = the viscosity of the pure liquid and Φ = the volume fraction of the particulates in the system. This relationship was first postulated by Einstein [5]. In this concentration range, and assuming the particle shape to be close to spherical, inter-particle interactions are minimal. At somewhat higher particulate concentrations—volume fractions in the range 0.05–0.30 (5–30%), inter-particle interactions become increasingly-relevant, and the viscosity relationship deviates from linearity thus:

$$\eta_{rel} = (1 - \Phi)^{-2.5}$$

This derivation of the Einstein equation is termed the Brinkman-Roscoe equation, as described elsewhere [7, 23]. Other derivations of the Einstein equation, including the Krieger-Dougherty equation, have been developed for still higher solids fractions, close to maximum packing density, but these are not relevant to the solids fractions that are encountered in semi-solid pharmaceutical systems.

Very few studies have reported topical formulations containing particulate solids alone as the viscosity-building agents [3, 11]. (These compositions are similar to a group of formulations termed "oleogels" in which oils are gelled with organic materials such as sorbitan esters and cholesterol derivatives). These studies of solid-stabilized systems used silica particles under 1 μm in size, and glass microspheres in the size range 20–90 μm. Particulate concentrations greater than 5% were needed in order to achieve viscosities suitable for topical semi-solid formulations. At these higher particulate concentrations, a 3-dimensional network structure is formed through inter-particle interactions. These formulations are thixotropic in nature, displaying an increase in their viscosity at rest due to the formation of these particulate networks, but displaying a temporary loss of viscosity when subjected to mechanical shear, such as by the action of spreading the formulation on the skin. It should be noted, also, that the viscosity of these systems is also

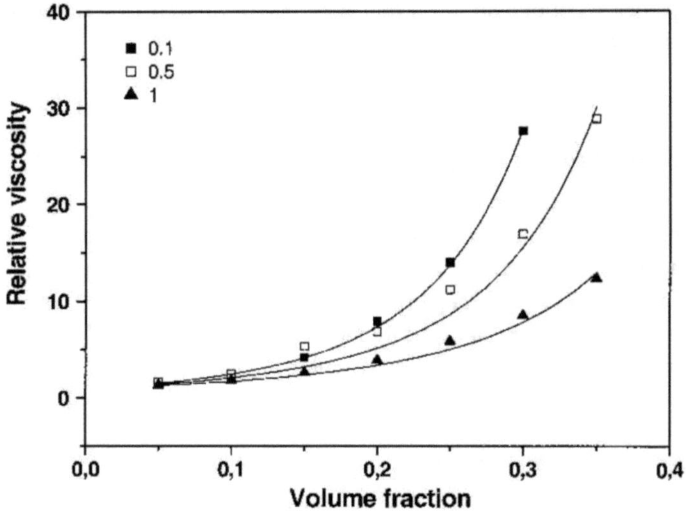

Fig. 12.5 Relative viscosity of particulate suspensions of silica of different particle sizes as a function of particulate volume fraction, measured at constant shear rate: (filled square) 0.1 μm particles; (open square) 0.5 μm particles; (filled triangle) 1 μm particles [3]; copyright Taylor & Francis; reproduced with permission

dependent on particle size, with viscosity increasing more steeply with solids loading with smaller particle sizes. This effect is illustrated in Fig. 12.5, for silica particles of three different particle sizes dispersed in mineral oil.

More commonly, however, solid-stabilized gels contain a small particle size solid, generally colloidal silicon dioxide, although less frequently talc, in combination with an organic polymer, such as carbomer, polyethylene oxide, polyethylene glycol, xanthan gum, locust bean gum or galactomannan [14, 16, 17, 28–30, 39, 42]. These formulations are stabilized by strong hydrogen-bond-mediated associations between the particles and the polymer chains, resulting in the formation of extensive 3-dimensional colloidal networks. It is hypothesized that the adsorption of silica particles along the polymer chains leads to the formation of polymer chains with greater rigidity and lower mobility, thus forming a percolated network which possesses significant elasticity. This mechanism of stabilization is shown schematically in Fig. 12.6. The resulting formulations are highly thixotropic, possessing high viscosities and high degrees of elasticity at rest, but demonstrating significant shear-thinning once sufficient shear force is applied to overcome the initial elasticity and rupture these relatively-labile inter-particle bonds. These effects can occur even at relatively modest polymer and particle loadings, especially where the solid particle size is reduced to below 1 μm, and so these systems are attractive as platforms for topical formulation development. It may be deduced from the above explanation of the physico-chemical basis of these formulations that the critical particle attributes are particle size and surface charge.

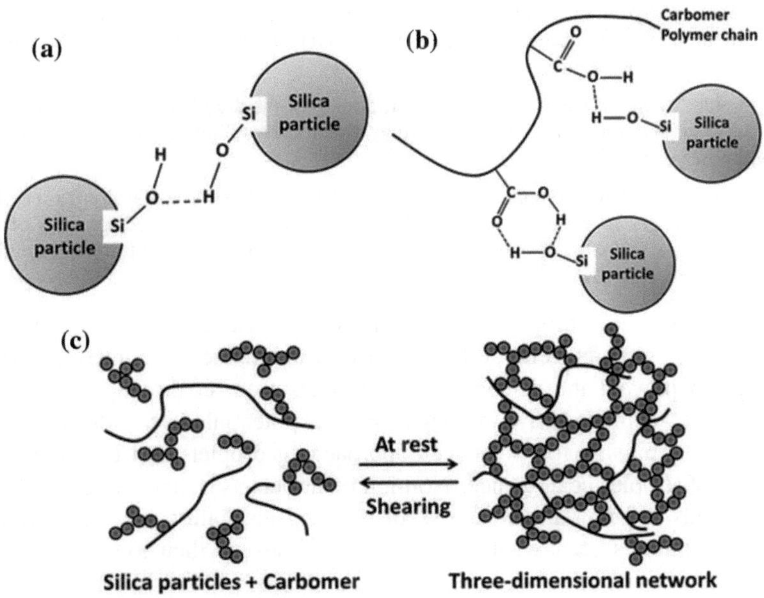

Fig. 12.6 Schematic illustrating likely mechanisms of interaction between silica particles and between silica particles and polymer chains in solid-stabilized gels [17]; copyright Springer; reproduced with permission

12.5 Emulsion Particles

The third broad area of particulate applications in semi-solid products is emulsion particles or droplets, as encountered in cream and some lotion products. Emulsions are colloidal dispersions composed of two immiscible liquids, in which droplets of one of the liquids is dispersed in the other. The external, or continuous, phases of pharmaceutical emulsions are usually aqueous, in which case the internal, or disperse, phases are oleaginous, and these are termed oil-in-water or o/w emulsions; water-in-oil emulsions can also be formulated, but these are much less commonly encountered. The principles of formulation and characterization of pharmaceutical emulsions have been detailed elsewhere [19, 36]; a succinct review is of the most relevant aspects is provided here.

Emulsion products are most commonly formulated for topical therapy in order to leverage the higher solubility of an active substance in the disperse phase of the emulsion as compared to that in the continuous phase. Having determined the need for an emulsion system, it is then necessary to select the appropriate disperse and continuous phases and phase volumes and the appropriate emulsifying agent(s). As outlined earlier, it is necessary for a drug to penetrate the outermost barrier layer of the skin, the stratum corneum, before it can exert a therapeutic effect. The composition of the formulation vehicle can influence rate and extent of absorption of the drug in several ways.

The degree of hydration of the stratum corneum influences the rate of drug permeation through the skin and so occlusive formulations, which retard the drying of the skin, increase the rate of drug permeation;

Drug solubility in the vehicle affects the partitioning of the drug between the formulation and the skin, and thus affects the drug activity which drives diffusion through the permeation barrier of the skin;

Evaporation of volatile components of the formulation following application to the skin results in changes in the concentration of the remaining non-volatile components of the formulation and can also change the state of saturation of the drug, and thus its activity; dissolved drug may also precipitate if supersaturated.

The disperse-phase droplets in an emulsion are stabilized by adsorption of a surfactant (emulsifying agent) at the interface, which reduces the interfacial tension and helps to prevent the disperse phase droplets from coalescing. Charged surfactants increase the positive or negative zeta potential of the droplets, which increases the repulsive force between approaching droplets and thus reduces the likelihood of droplet coalescence. Non-ionic surfactants generally result in lower zeta potentials and thus a greater risk of physical instability, but are preferred in topical formulations because of their lower toxicities and their lower sensitivity to other additives.

One of the principles of emulsion formulation is that the surfactant, or emulsifying agent, should be more soluble in the continuous phase of the emulsion than in the disperse phase. This principle is expressed more quantitatively through use of the HLB (Hydrophilic-Lipophilic Balance) system. The HLB of a surfactant is a measure of the degree to which it is hydrophilic or lipophilic, and is calculated based on the composition of the molecule. Surfactants with HLB values of about 3–6 function as w/o (water in oil) emulsifiers, whereas those with HLBs of about 12–16 function as o/w (oil in water) emulsifiers. It has also been widely demonstrated that combinations of two different emulsifying agents confer greater emulsion stability over the use of a single emulsifier, even though this observation is not explained by the HLB approach. A list of oil-phase constituents and emulsifiers commonly used in semi-solid pharmaceutical products is presented in Table 12.4.

Table 12.4 Excipients commonly used in pharmaceutical semi-solid emulsion formulations

Oil-phase constituents	High-HLB emulsifiers (stabilize o/w emulsions)		Low-HLB emulsifiers (stabilize w/o emulsions)	
Cottonseed oil	Triethalolamine oleate	HLB 12.0	Sorbitan sesquioleate	HLB 3.7
Petrolatum	Oleth-10	HLB 12.4	Glyceryl monostearate	HLB 3.8
Dodecane	Ceteth-10	HLB 12.9	Sorbitan monooleate	HLB 4.3
Mineral oil	PEG-8 laurate	HLB 13.0	Sorbitan monostearate	HLB 4.7
Cyclohexane	Polysorbate 80	HLB 15.0	Oleth-2	HLB 4.9
Dimethyl isosorbide	Oleth-20	HLB 15.3	Steareth-2	HLB 4.9
	Steareth-20	HLB 15.3	Glyceryl laurate	HLB 5.2
	Ceteareth-20	HLB 15.7	Ceteth-2	HLB 5.3
	Ceteth-20	HLB 15.7	PEG-4 dilaurate	HLB 6.0
	Polysorbate 20	HLB 16.7	Lecithin	HLB 4–9

Emulsions are thermodynamically unstable and tend to phase-separate over time so as to minimize their interfacial free energy. There are two elements to the process of emulsion phase separation. Firstly, the emulsion droplets may settle as a result of gravity which means, for the most common case of an oil-in-water emulsion, an upward migration of the emulsion droplets ("creaming"). This settling is in accordance with Stokes' law, and so is dependent on the size of the droplets, the density difference between the internal and external phases, and the viscosity of the system; small droplets and a high external phase viscosity both serve to minimize this phenomenon. In practice, for viscous semi-solid products such as creams, droplet mobility is very low, and so creaming is usually negligible; less-viscous lotion formulations are more prone to creaming, however. The second element of the phase separation process is the coalescence of emulsion droplets brought into close proximity as a result of creaming, product flow, Brownian motion, etc. Rupture of the liquid film between two adjacent droplets leads to irreversible coalescence of the droplets. Where particles are prevented from approaching one another closely, by steric interference between the surfactant molecules oriented at the interfaces or by particle-particle repulsion due to their zeta potentials, coalescence is reduced.

A critical aspect of emulsion formulation is the size distribution of the emulsion droplets. Droplet sizes in semi-solid pharmaceutical products generally fall within the range of 1–100 μm, although it must be recognized that emulsions tend to be polydisperse, meaning that they contain a range of droplet sizes, in the same way as do populations of solid particles. Emulsions can be formulated with particle sizes significantly lower than 1 μm ("microemulsions") although those are usually less viscous systems designed for oral or parenteral administration. Larger droplets may be more prone to creaming and ultimately coalescence, and so it is important to achieve a consistently fine droplet size distribution in manufacturing to assure the stability of the product over its shelf life. In principle, monitoring the droplet size distribution of an emulsion provides an understanding of the stability of the product with respect to phase separation. In practice, however, quantifying the size distribution of emulsion particles or droplets, while critical to understanding the stability of the product, can present significant analytical challenges. Several different analytical techniques may be applied to determine the droplet size of emulsions, depending on the characteristics of the formulation, including optical microscopy, laser diffraction, dynamic light scattering, and nuclear magnetic resonance; these have been reviewed extensively elsewhere [23]. In practice, however, many of these techniques are not well suited to the evaluation of semi-solid pharmaceutical systems. For example, concentrated, viscous emulsions may need to be diluted in order to be measureable by techniques such as laser diffraction and dynamic light scattering, but such manipulations may change the characteristics of the droplet size distribution. Microscopic techniques are often hindered by the presence of other interfering structures in the product, which prevent the unambiguous identification of emulsion droplets. Development and validation of a suitable analytical technique for assuring the phase-stability of a semi-solid emulsion formulation remains a significant challenge in many instances.

A more empirical, though still quantitative, approach is to measure the rate of creaming or sedimentation in an emulsion by subjecting a sample to centrifugation. One such instrument is the LUMiFuge stability analyzer, in which a sample of the product is placed in a transparent tube and centrifuged at a controlled temperature. Under centrifugation, creaming or sedimentation of the internal phase droplets followed by phase separation can result in the evolution of clear zones at the top and/or bottom of the sample tube. The rate of evolution of such clear zones can be quantitated by measuring light transmission at points along the length of the tube over time. The rates of settling derived from this test can predict the physical stability of emulsion formulations, or can at least serve to rank-order formulations based on their stability under these test conditions.

12.6 Manufacture of Particulate-Containing Semi-solid Products

The principles of scale-up and manufacture of semi-solid products have been reviewed elsewhere. Pathak and Thassu [31] reviewed product design, scale-up and commercial manufacture for suspension products, and provided an overview of commercial-scale compounding vessels and mixing equipment. Block [2] reviewed transport phenomena, the principles of laminar and turbulent flow regimes, and detailed a range of dynamic mixing devices. Schroën and Berton-Carabin [36] provided a detailed review of the most widely used emulsification devices.

The single most ubiquitous unit operation in manufacturing particulate semi-solid products is mixing. However, the mixing operations in suspension and emulsion product manufacture are more problematic than the corresponding operations in less-viscous, single-phase systems. The mixing of high-viscosity systems is relatively slow and inefficient. The Reynolds numbers of these systems are relatively low—often below 100. (The Reynolds number is a dimensionless parameter used to characterize fluid flow; it is essentially a ratio of the inertial forces to viscous forces within a fluid; a Reynolds number of about 2000 usually signifies the transition from laminar to turbulent flow.) Conventional mixing tanks and propeller-type impellers are generally inadequate; the inertial forces imparted during mixing dissipate quickly, and eddy formation and diffusion are virtually absent. Also, due to the shear-thinning nature of many semi-solid products, wall effects may arise by virtue of the presence of a thin layer of product adjacent to the wall of the vessel which exhibits a lower viscosity than the bulk of the product and allows slippage at the expense of bulk mixing. Alternate mixing geometries are thus needed to ensure thorough mixing, including paddle and anchor agitators and recirculation systems. Heat transfer is often an essential element in semi-solid product processing, particularly for hydrocarbon-based emulsion products which necessitate a melting step. Such processes require jacketed vessels with heating and cooling capabilities and with counter-rotating mixing systems which include wall-scraping so as to prevent the build-up of solidified product on the vessel wall during cooling.

Another critical step in the manufacture of semi-solid particulate products is that of incorporating and dispersing solid particulates into the product. While trivial at laboratory scale, this step can be slow at manufacturing scale and, if improperly performed, can result in incomplete or variable particle dispersion. The preferred techniques for incorporating large quantities of solid particulates into a viscous liquid are to draw the solid material into a liquid stream by vacuum, or to use an in-line disperser such as a Quadro Ytron disperser, in which the solid material is fed into a liquid stream in a high-speed disperser head.

Schroën and Berton-Carabin [36] reviewed the two principal approaches to emulsification at manufacturing scale, namely high-pressure homogenization and rotor-stator homogenization. The same technologies are applicable to the dispersion, deaggregation and, in some cases, the primary particle size reduction of solid particulate materials. Block [2] reviewed the size reduction of solid particulates in liquid systems in some detail.

The majority of processing operations for semi-solid particulate products are carried out in batch or semi continuous modes rather than continuous modes. The use of Process Analytical Technologies (PAT) for real-time monitoring of process performance has been discussed in detail by Pathak and Thassu [31]. It is often desirable to be able to monitor the particle (or droplet) size of a material during processing; e.g., to determine the end point of a dispersion or emulsification step, or to monitor a dispersed system for particle or droplet growth. Given the complexity of many semi-solid formulations, true in-line monitoring of particle size is difficult to achieve, although it remains an important goal in pharmaceutical manufacturing. The technique which is best able to monitor particle size in process at the present time is Focused Beam Reflectance Measurement (FBRM) which was described previously (see Sect. 12.2.1). This technique suffers from the same limitations as other particle sizing techniques, in that it is not able to differentiate between particles of different materials, and so is of limited use in complex systems containing different particle populations. It is likely, however, that continuing advances in analytical techniques will lead to the development of additional tools for in-process monitoring in the future.

12.7 Concluding Remarks

In summary, it can be seen that there are many diverse applications of semi-solid particulate-containing products. The most significant of these are topical pharmaceutical formulations in which the particulate is the active substance, sunscreen formulations containing titanium dioxide and zinc oxide, and semi-solid formulations in which the particles are composed of emulsion droplets. Various attributes of these particles are critical to their performance, with particle size being the most important of these. Particle size, and the closely-related property of surface area, governs the rate and extent of drug delivery from particulate pharmaceutical formulations, as well as the efficacy of particulate sunscreens and the physical stability of emulsion formulations. A range of techniques exists to measure particle size and other particle attributes; of these, laser diffraction sieve analysis and microscopy are the most widely utilized techniques. It should be evident that a thorough understanding of particle attributes, of the influence of particle attributes on product processing and performance, and of the principle methods of characterizing particles are all essential to ensure robust product design and manufacture.

12.8 Abbreviations and Symbols

A	Surface area of drug particle
c	Actual concentration of drug in liquid
c_s	Saturation concentration of drug in liquid
dw/dt	Rate of increase of amount of drug in solution
D	Particle size (diameter of equivalent sphere)
D_{50}	Median size (extra subscript for size distribution basis, e.g. 0 for number basis and 3 for volume/mass basis)
$\langle D_{4;3} \rangle$	Mean size of volume-based distribution
$\mathbb{D}$	Diffusion coefficient of drug in liquid
Δ	Thickness of diffusion layer around drug particle
HLB	Hydrophilic-lipophilic balance
λ	Wavelength of light
η_d	Viscosity of dispersion
η_0	Viscosity of liquid
o/w	Oil-in-water
SPF	Sun protection factor
Φ	Volume fraction of drug particles in suspension
w/o	Water-in-oil
ζ	Zeta potential

References

1. Barua S, Mitragotri S (2014) Challenges associated with penetration of nanoparticles across cell and tissue barriers: a review of current status and future prospects. Nano Today 9:223–243
2. Block LH (2002) Non parenteral liquids and semisolids. In: Levin M (ed) Pharmaceutical scale up. Marcel Dekker Inc., New York
3. Chen S, Øye G, Sjöblom J (2005) Rheological properties of silica particle suspensions in mineral oil. J Dispers Sci Tech 26:791–798
4. Dinsmore AD, Hsu MF, Nikolaides MG, Marquez M, Bausch AR, Weitz DA (2002) Colloidosomes: selectively permeable capsules composed of colloidal particles. Science 298:1006–1009
5. Einstein A (1911) Berichtigung zu meiner arbeit: eine neue bestimmung der moleküldimensionen. Ann Phys 339:591–592

6. Fairhurst D (2014) An overview of physical (particulate) sunscreens. In: Merkus HG, Meesters GMH (eds) Particulate products: tailoring properties for optimal performance. Springer, Netherlands
7. Farris RJ (1968) Prediction of the viscosity of multimodal suspensions from unimodal viscosity data. Trans Soc Rheol 12:281–301
8. Frelichowska J, Bolzinger M-A, Pelletier J, Valour J-P, Chevalier Y (2009) Topical delivery of lipophilic drugs from o/w Pickering emulsions. Int J Pharm 371:56–63
9. Frelichowska J, Bolzinger M-A, Valour J-P, Mouaziz H, Pelletier J, Chevalier Y (2009) Pickering w/o emulsions: drug release and topical delivery. Int J Pharm 368:7–15
10. Ghate D, Edelhauser HF (2006) Ocular drug delivery. Expert Opin Drug Deliv 3:275–287
11. Goupale DC, Rajkapoor B (2011) Evaluation of physical stability of oleogels containing diclofenac diethylamine. Res J Pharm Biol Chem Sci 2:92–99
12. Iacocca RG, Burcham CL, Hilden LR (2010) Particle engineering: a strategy for establishing drug substance physical property specifications during small molecule development. J Pharm Sci 99:51–75
13. International Conference on Harmonization (2010) Analytical sieving general chapter, Q4B annex 12. http://www.ich.org/fileadmin/Public_Web_Site/ICH_Products/Guidelines/Quality/Q4B_Annex_12/Step4/Q4B_Annex_12_Step_4.pdf
14. Jun C (2001) Improved paste formulations. International Patent Application WO2001060409
15. Kassem MA, Abdel Rahman AA, Ghorab MM, Ahmed MB, Khalil RM (2007) Nanosuspension as an ophthalmic delivery system for certain glucocorticoid drugs. Int J Pharm 340:126–133
16. Kennedy JRM, Kent KE, Brown JR (2015) Rheology of dispersions of xanthan gum, locust bean gum and mixed biopolymer gel with silicon dioxide nanoparticles. Mat Sci Eng C 48:347–353
17. Lu Z, Fassihi R (2015) Influence of colloidal silicon dioxide on gel strength, robustness, and adhesive properties of diclofenac gel formulation for topical application. AAPS PharmSciTech 16:636–644
18. Malloy A, Carr B (2006) Nanoparticle tracking analysis—the Halo™ system. Part Part Syst Charact 23:197–204
19. Manoharan C, Basarkar A, Singh J (2010) Various pharmaceutical disperse systems. In: Kulshreshtha AK, Singh ON, Wall GM (eds) Pharmaceutical suspensions: from formulation development to manufacturing. Springer, New York
20. Marku D, Wahlgren M, Rayner M, Sjöö M, Timgren A (2012) Characterization of starch Pickering emulsions for potential applications in topical formulations. Int J Pharm 428:1–7
21. Mathias NR, Hussain MA (2010) Non-invasive systemic drug delivery: developability considerations for alternate routes of administration. J Pharm Sci 99:1–20
22. Merkus HG (2009) Particle size measurements: fundamentals, practice quality. Springer, Netherlands
23. Merkus HG (2014) Basic information for design of particulate products. In: Merkus HG, Meesters GMH (eds) Particulate products: tailoring properties for optimal performance. Springer, Netherlands
24. Merkus HG, Meesters GMH (eds) (2014) Particulate products: tailoring properties for optimal performance. Springer, Netherlands
25. Michniak-Kohn BB, Wertz PW, Al-Khalili M, Meidan VM (2005) Skin: physiology and penetration pathways. In: Rosen MR (ed) Delivery system handbook for personal care and cosmetic products. William Andrew Publishing, Norwich, NY
26. Murthy SN, Shivakumar HN (2009) Topical and transdermal drug delivery. In: Kulkarni VS (ed) Handbook of non-invasive drug delivery systems: science and technology. William Andrew Publishing, Oxford
27. Newman MD, Stotland M, Ellis JI (2009) The safety of nanosized particles in titanium dioxide and zinc oxide based sunscreens. J Am Acad Dermatol 61:685–692
28. Oh M-H, So J-H, Yang S-M (1999) Rheological evidence for the silica-mediated gelation of xanthan gum. J Coll Interfac Sci 216:320–328

29. Oliveira F, Monteiro SR, Barros-Timmons A, Lopes-da-Silva JA (2010) Weak-gel formation in dispersions of silica particles in a matrix of a non-ionic polysaccharide: Structure and rheological characterization. Carbohyd Polym 82:1219–1227
30. Pashkovski EE, Masters JG, Mehreteab A (2003) Viscoelastic scaling of colloidal gels in polymer solutions. Langmuir 19:3589–3595
31. Pathak Y, Thassu D (2010) Scale up and technology transfer of pharmaceutical suspensions. In: Kulshreshtha AK, Singh ON, Wall GM (eds) Pharmaceutical suspensions: from formulation development to manufacturing. Springer, New York
32. Pickering SU (1907) Emulsions. J Chem Soc 91:2001–2021
33. Ramsden W (1903) Separation of solids in the surface-layers of solutions and suspensions. Proc R Soc London 72:156–164
34. Ranade VV, Cannon JB (2011) Transmucosal and ocular drug delivery. In: Drug delivery systems, 3rd ed. CRC Press, Boca Raton
35. Rawle AF (2010) Analytical tools for suspension characterization. In: Kulshreshtha AK, Singh ON, Wall GM (eds) Pharmaceutical suspensions, from formulation development to manufacturing. Springer, New York
36. Schroën K, Berton-Carabin CC (2016) Emulsification: established and future technologies. In: Merkus HG, Meesters GMH (eds) Production, handling and characterization of particulate materials. Springer, Netherlands
37. Schulman JH, Leja J (1954) Control of contact angles at the oil-water-solid interfaces; emulsions stabilized by solid particles (BaSO4). Trans Faraday Soc 50:598–605
38. TGA OTC Medicines Section (2009) A review of the scientific literature on the safety of nanoparticulate titanium dioxide or zinc oxide in sunscreens. https://www.tga.gov.au/sites/default/files/sunscreens-nanoparticles-2009.pdf. Accessed 27 Feb 2017
39. Trompette JL, Charnay C, Partyka S (1998) Storage stability and rheological behavior of talc-polyacrylylglycinamide gelified suspensions. Langmuir 14:4475–4481
40. van Duynhoven JPM, Maillet B, Schell J, Tronquet M, Goudappel G-JW, Trezza E, Bulbarello A, van Dusschoten D (2007) A rapid benchtop NMR method for determination of droplet size distributions in food emulsions. Eur J Lipid Sci Technol 109:1095–1103
41. Vermeir L, Sabatino P, Balcaen M, Declerck A, Dewettinck K, Martins JC, Van der Meeren P (2016) Effect of molecular exchange on water droplet size analysis in W/O emulsions as determined by diffusion NMR. J Coll Interfac Sci 463:128–136
42. Zhang Q, Archer LA (2004) Optical polarimetry and mechanical rheometry of poly(ethylene oxide)-silica dispersions. Macromol 37:1928–1936

David Harris is a formulation scientist with over 25 years experience in the development of new chemical entities. He has a Bachelor of Pharmacy degree from the University of Bath, and a Ph.D. in Pharmaceutics from the University of Manchester, both in the United Kingdom, and subsequently spent two years as a Post-Doctoral Fellow at the University of Wisconsin—Madison, in the laboratory of the late Prof. Joseph Robinson. Since then he has been employed in various roles in formulation development, first with Schering-Plough and subsequently with Merck & Co., Inc. in the U.S.

Dr. Harris has extensive experience in pharmaceutical product development, spanning oral solid formulation development through oral liquid and topical/dermatological formulation development. He has experience across the development lifecycle, and has made significant contributions to many development programs, including a number of successful commercial products. Dr. Harris has particular interests in semi-solid and liquid formulation development and in pediatric product development.

Chapter 13
The Side Effects of Drugs: Nanopathological Hazards and Risks

Antonietta M. Gatti and Stefano Montanari

Abstract Drugs are aimed at the prevention or treatment of diseases, but their positive effect cannot be guaranteed in any case. The chapter describes a few clinical cases of diseases contracted by the patient from particulate contaminants contained as foreign bodies in the drug either as such or in the form of their precursors. The cases were studied analysing the biopsic pathological samples by means of a scanning electron microscope coupled with an X-ray microprobe of an energy dispersive system. Once the foreign bodies are identified, an anamnestic work is performed to trace them in the patient's environment and lifestyle, including the drugs he/she takes.

13.1 Introduction

"Pharmakon" is an ancient Greek name meaning both a remedy and a poison. And, as a matter of fact, there is no drug free from side effects that, in some cases, can be much more serious than the disease meant to be cured. So, before prescribing or taking any medicament, it is essential to put on the scales the effects that are desired on one side and the negative ones that are to be expected or, at least, that can't be excluded on the other side.

As one progresses up along the zoological scale, things grow increasingly complex. If unicellular organisms can be reasonably expected to react in a predictable way to a chemical stimulus, reactions become more and more uncertain as the organism's differentiation becomes more refined, and no living being is more complicated and, in a way, more refined than Man. From the anatomical and biological point of view we are certainly not more complicated than any ape, but to

A. M. Gatti (✉)
President of Health, Law and Science Association,
Quai du Sujeut 16-18, Geneva, Switzerland
e-mail: gatti@nanodiagnostics.it

S. Montanari
Nanodiagnostics, via E. Fermi 1L, 41051 San Vito, MO, Italy

H. G. Merkus et al. (eds.), *Particles and Nanoparticles in Pharmaceutical Products*,
AAPS Advances in the Pharmaceutical Sciences Series 29,
https://doi.org/10.1007/978-3-319-94174-5_13

anatomy and biology we must add the huge influence that our psyche exerts on our body as any pharmacologist busy working on placebo and paradoxical effects knows.

So, predicting which unwanted side effects are more or less likely to occur with a certain drug administered to a certain subject is very hard indeed. Even more complicated is any prediction when the drugs at stake are more than one and no one can exclude mutual chemical interference. And as to excipients, coloring agents, preservatives, stabilizers and all the many other components added to drugs, their safety can't be taken for granted, if only given the diversity between individuals.

The last complication is that linked to pollutants: i.e. something that is not supposed to be contained anywhere, let alone drugs, but which sometimes is there.

In the vast majority of cases, drugs are prepared following the various stages of chemical processes. and in most instances there are substances useful for technical preparation but harmful to the patients. For that reason, they must be removed but, unfortunately, complete removal is not successful in all circumstances. The obvious consequence is that traces of those substances can remain in the final product (for example mercury in vaccines) (internet reference, Schlegl et al. [6, 10, 14]), with all the acute risks and those of possible accumulations that may occur in those who normally take the medicine. The applicable laws state that the presence of a certain substance may not be declared if it is below a certain threshold, so, as a rule, those presences are not reported and nobody is aware of them.

This chapter describes several case studies where solid, inorganic particles coming from pharmaceutical products (as carrier or pollutant) caused serious or even fatal illness to patients. In these studies, which were possible while being in charge of two European research projects [EU projects 3, 4], Scanning Electron Microscopy (SEM- Quanta 200-Field Emission Gun Environmental Scanning Electron Microscope by FEI Company. The Netherlands) and Energy Dispersive Spectroscopy (EDS by EDAX, USA) were applied for particle detection and identification.

13.2 Case Studies

Back in the 1990s, we discovered that solid, inorganic, non biodegradable particles produced by many different sources, often inadvertently or as a side effects of some industrial processes, can enter the human body without it being able to eliminate them [5, 7–9, 11, 12]. Their dispersion in the body can interest all the organs [1, 2, 13].

Once in the body, they settle in any tissue and organ and trigger the typical foreign-body reaction. The branch of medicine that deals with the diseases induced by micro- and nano-particles is called nanopathology. Number of particles per volume of tissue (though with a considerable difference with classical toxicology), size, shape and chemical composition are the main factors responsible for the pathological event or events that can result, be they on a particular organ or tissue or be they systemic.

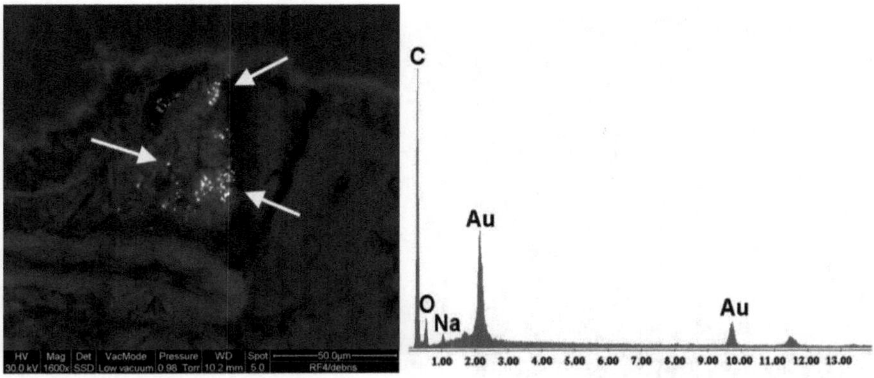

Fig. 13.1 SEM image and EDS spectrum of a liver biopsy showing the presence of gold nanoparticles in the liver affected by "cryptogenic" granulomatosis

13.2.1 Case: Gold in Medicine

The first nanopathological effect linked to drugs we came across at the beginning of this century was not caused by particulate pollutants but by the pure medicament itself.

The Royal Free Hospital of London sent us the documents about the case of a lady who had died of a liver cancer as the evolution of a liver granulomatosis. In a way, a normal pathogenesis and nothing that wasn't already perfectly known. But what was not clear was the origin of that granulomatosis. We checked the biopsic samples (by means of a Scanning Electron Microscope (SEM) and an Energy Dispersive Spectroscopy (EDS)) that had been sent us and discovered that the tissue was literally invaded by nanometric (50 nm) gold particles: the typical colloidal configuration (Fig. 13.1). We asked to consult the medical history of the lady and, doing so, it was easy to understand what had happened.

Years before, the patient had suffered from a serious form of arthritis in her knees and her doctor had injected colloidal gold directly into the articulation. The immediate result was excellent but, unfortunately, our body has no mechanisms to get rid of those particles. So, slowly, they had migrated to the liver, the prince of emunctory organs, where they had settled and become the origin of a foreign-body inflammation and finally, after years, of a cancer.

13.2.2 Case: Steel Contaminant

Another interesting case dating back to not long after the one described above was that of a gentleman who had breathing problems and blackish spots on his skin. The chest X-ray showed radio-opaque zones in his lungs that the radiologist could not

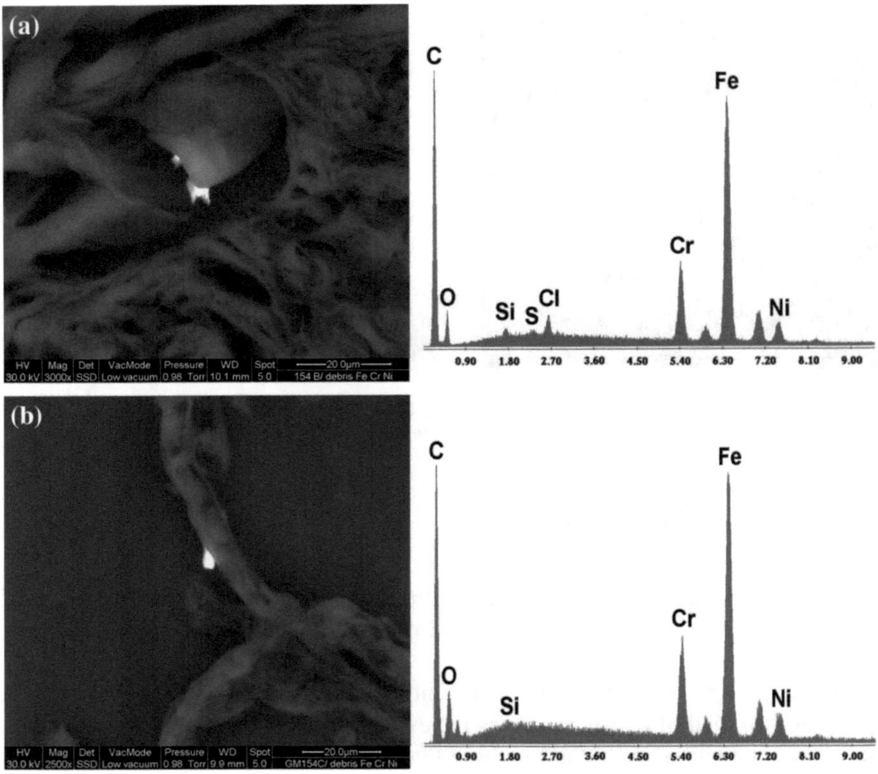

Fig. 13.2 SEM images of biopsies of the derma (**a**) and the lung (**b**). Both contain debris of stainless steel (iron-chromium-nickel)

read. Our SEM examination of a sample of the patient's skin showed the presence of steel particles whose shape was that typical of cold machining (Fig. 13.2a, b). After having questioned at some length the patient, we learned that for many years he had used nasal drop to treat a chronic form of his nasal mucosa edema. Observed under the microscope, that drug showed its content of steel particles as a pollutant (Fig. 13.3). Our hypothesis is that the liquid, administered through the nose, had in part reached the lungs where the particles had stopped and, non biodegradable and impossible to get rid of as they are, accumulated. This gives account of the radio-opaque areas of the lungs. Apparently, some of the particles had somehow managed to escape but could not go beyond the skin. Unlike the former case, what prompted the health problem here was not the medicament itself but its particulate pollutant. Where those particles came from is somewhat harder to say, but the cold working of certain plants that were used as the drug's components could be the origin of those metal particles. Anyhow, whatever their origin, their presence had gone unnoticed.

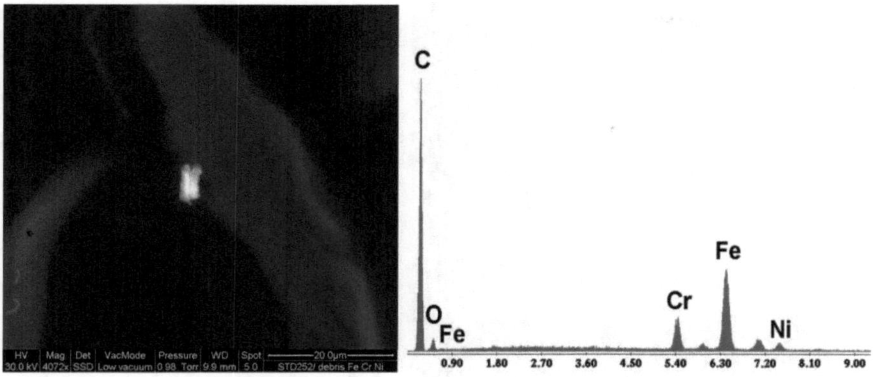

Fig. 13.3 SEM image of a drop of the nasal drug. It contains stainless-steel debris probably coming from the crushing tools used for the herbal drug preparation

13.2.3 Case: Excipients Magnesium Silicate (Talc) and Their Purity

Another interesting example came from the University Hospital of Modena (Italy) where a 26-year-old patient underwent surgery on his colon for a Crohn's disease, a pathology characterized by bowel inflammation, whose origin is debated and, in fact, remains mysterious. The surgical samples of the colon resection showed clearly the presence of granulomas, inside which a great number of silicate particles were detected.

The patient was a student who did not undergo any apparent exposure to particularly important environmental pollution but, since he was six, he had been treated every spring and autumn, for a total of about six months a year, with pills against a chronic form of allergy, and those pills contained the same particles found in his bowel (Fig. 13.4). As a matter of fact, inorganic particles are often used in drugs for oral use, but no one has ever really bothered to check if those particles are effectively eliminated from the body or if, at least in part, they remain inside. In that case, how much and with what consequences.

Figure 13.5 presents particles of talc entrapped in a liver affected by a cryptogenic granulomatosis.

We analyzed other pills and found that all of them contained non-biodegradable excipients such as silicates, or titanium dioxide (Figs. 13.6 and 13.7). So, some side effects after their chronic uptake could be related to the accumulation in filter organs of those inorganic particles.

In one case, studying the commercial talc, we found that the talc used was contaminated with Uranium (Fig. 13.8).

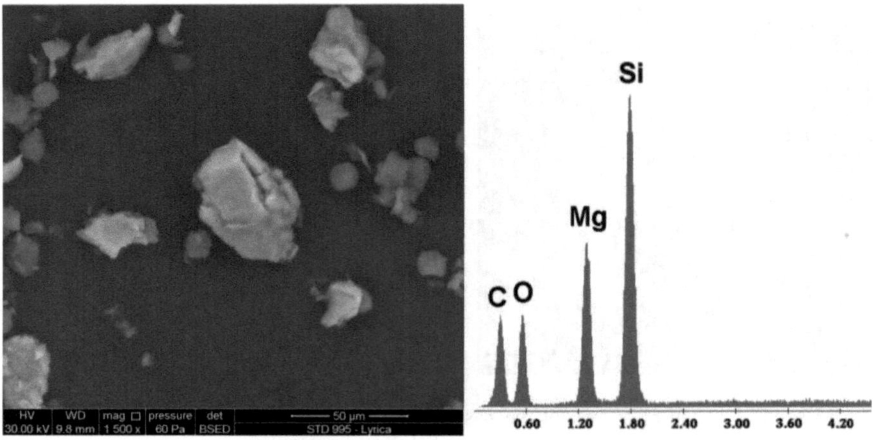

Fig. 13.4 SEM image showing particles of an anti-allergic drug, which EDS analyzed to consist of Mg and Si. The excipient is composed of talc, i.e. magnesium silicate

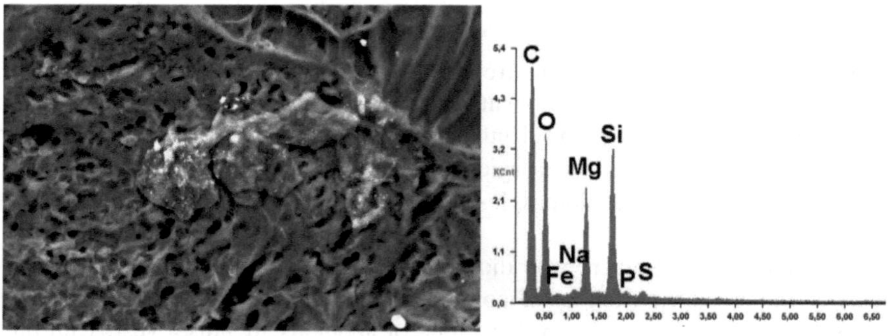

Fig. 13.5 SEM Image of a liver biopsy affected by "cryptogenic" granulomatosis. The foreign bodies were identified by EDS as being composed of silicon and magnesium, i.e. talc

13.2.4 Case: Zinc in Mouthwash

A further case we came across was that of a university teacher who was diagnosed suffering from amyotrophic lateral sclerosis (ALS). He had started a few years before expressing problems with his tongue movements. Hence, difficulties in speaking and swallowing. Then, slowly, his body had paralyzed. Since the problem had started from the tongue, we asked for a biopsy of that organ and in it we found many zinc particles (Fig. 13.9). It was only after having questioned long as is our custom the patient that we came to know that many years ago, after a small dental work, the patient had used for a long time a mouthwash. Though the dentist had prescribed its use just for a few days, the patient had kept using it daily, and many times a day, for at least twenty years.

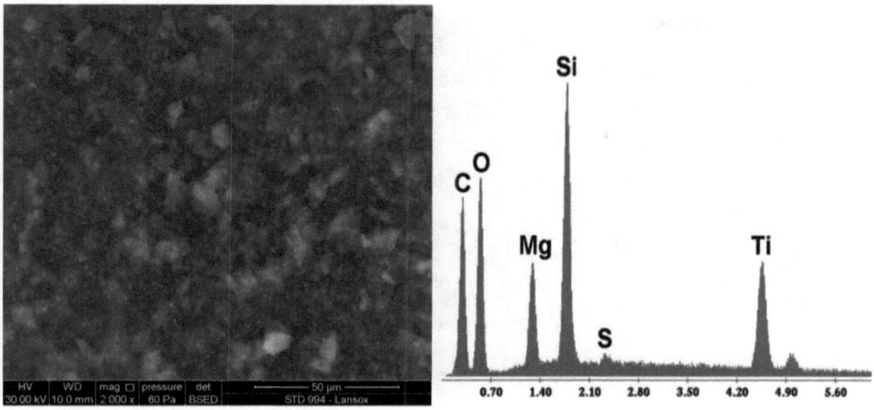

Fig. 13.6 SEM image of the surface of a pill. Besides the active component, it contains magnesium, silicon, titanium (E171) and sulfur as excipients

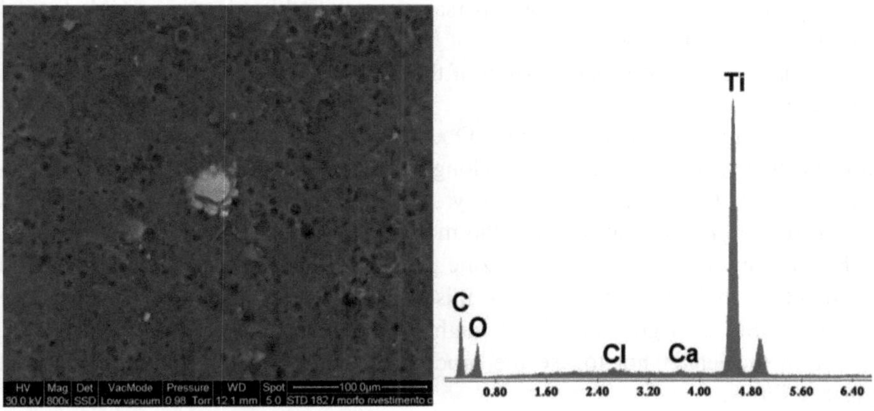

Fig. 13.7 SEM image and EDS spectrum show particles of titanium dioxide (E171) entrapped in the pill

Fig. 13.8 EDS spectrum of an excipient containing silicon-magnesium (talc), sulfur and uranium. Uranium is a possible contaminant of talc

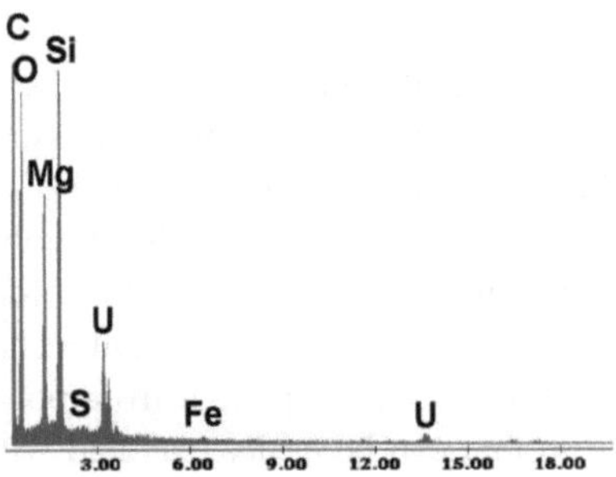

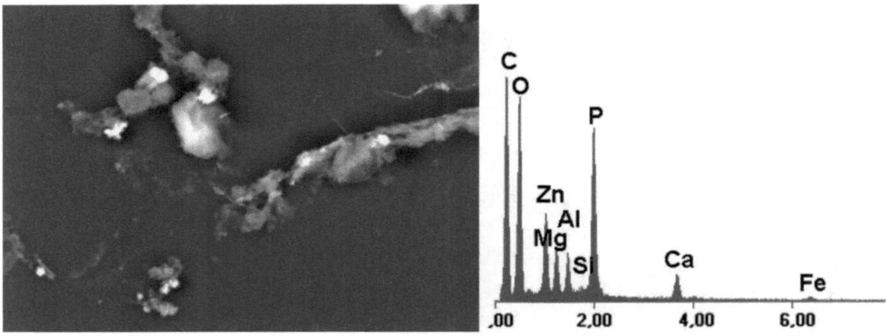

Fig. 13.9 SEM image shows a tongue biopsy where EDS analysis found zinc precipitates

We analyzed that drug and saw that it contained a zinc salt dissolved in an alcohol solution. What had happened was that alcohol, a peripheral vessel dilator, had made the oral cavity more permeable and the zinc salt had entered the tongue, reaching, after a while, a state of supersaturation, resulting in the precipitation of zinc, probably as phosphate.

We can't say if that was the origin of the paralysis and, eventually, the cause of the professor's death.

We can only remind that Günter Oberdörster of the University of Rochester showed that particles can move along the nerves [2, 13] and, just as an undemonstrated hypothesis of ours, they could have entered the peripheral nervous system of the *patient* starting from the mouth.

Besides any other consideration, zinc particles were actually in the tongue and the mouthwash had been obviously misused, going far beyond any reasonable indication. So, as a general rule, it might be prudent to warn patients and write clearly in the leaflet not to use the product beyond certain time, frequency and quantity limits.

13.2.5 Case: Chelating Agents

However, it is not rare that drugs are abused, and anti-cancer medicaments used off-label are one of the many possible examples. Another case of inappropriate use we see occurring more and more often is that of chelating agents. Chelation is a chemical reaction generally involving an organic compound-a ligand, i.e. the chelating agent—and an atom or ion, generally of metallic nature. Two coordinate covalent bonds are formed and a chemical complex is the result. Chelation is much used to detoxify a patient from poisonous metal agents, by converting them into a form that can be easily excreted without further interaction with the body. Salts of ethylenediaminetetraacetic acid (EDTA) are the most popular agents for many forms of heavy-metal poisonings, despite the fact that the U.S. Food and Drugs

Administration approves calcium disodium only for serious cases of lead poisoning and no other use.

The FDA has made this restriction for application because chelating components may induce a severe or even fatal hypocalcemia.

The main problem with the use of chelating agents exceeding a relatively short time limit is that they bind elements (often oligo-elements) that are essential for the correct human metabolism. It must be understood that chelators can bind atoms and ions but not micro- and nanoparticles. In fact, particles are far too big and, in any case, can't behave as a Lewis acid, i.e. any molecule or ion capable of forming a covalent bond, accepting a couple of electrons. It is only obvious that using chelating agents in the hope that they can do what is chemically impossible can only harm, sometimes seriously, the patient.

13.2.6 Case: Combined Phosphate and Calcium Administration

A further case of pharmacological abuse we came across back in 2006 is that involving the administration of sodium phosphate to a dehydrated, old patient (83) with a decreased renal function, an intestinal inflammatory condition, an intestinal sub-occlusion and a colostomy. He was hospitalized for influenza-like illness after a week of diarrhea and he showed a stasis of liquids in his stomach with some episodes of vomiting. 100 ml of sodium phosphate were administered through the colostomy with consequent emission of feces, than he was hydrated via trans-venous. The second day he received another dose of 100 ml of sodium phosphate without emission of feces and so the third day. On the fourth day, the patient began to vomit copiously and was given 300 ml more of sodium phosphate though his colostomy. A sudden, dramatic dehydration occurred and blood tests revealed a marked hypernatremia and hypocalcaemia. So, the saline administration was interrupted and four vials of calcium gluconate were administered. Immediately conditions of anuria and respiratory failure were established, followed by a cardiac arrest which proved fatal. The post-mortem showed a perforation of the intestinal wall.

With our electron-microscopy methods we analyzed tissue samples taken from different organs. The kidney sample showed the presence of needle-shaped crystals composed of carbon, oxygen, calcium and phosphorus localized in the renal tubules and in the pelvis (Fig. 13.10). Usually, the calcified areas show the widespread presence of calcium and phosphorus and contain either just calcium or calcium and phosphorus with a calcium/phosphorus ratio = 1.67. That inorganic compound is apatite and is naturally present, for example, in the bone. In the case in question, the ratio of calcium/phosphorus of the crystals detected was 0.9 and, therefore, did not fall within the normal compounds present in the human body.

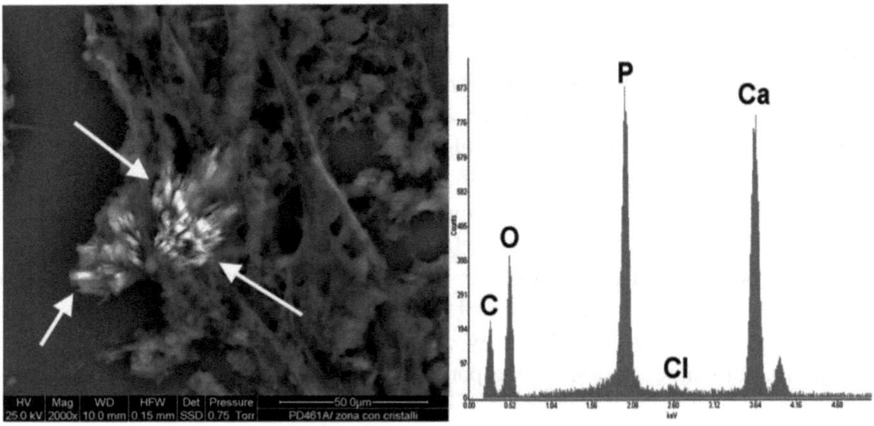

Fig. 13.10 The SEM image and EDS spectrum of a kidney autopsic sample where crystals of calcium phosphate

The needle-like morphology of these crystals definitely indicated a recent formation. In fact, even small movements of the internal organs or the pulsing blood flow would have destroyed the delicate morphology and, thus, compromised the kidney's functionality. Their size and location inside the renal tubules coupled with their fragility suggested that they formed where they were found. In fact, they couldn't have passed through the glomeruli in the conditions in which they were found, and their formation suggests a local excess of ions such as calcium and phosphorus. When the water is reabsorbed in the tubules, if the fluid contains an excess of ions, for example phosphorus and calcium (e.g. from the calcium gluconate administered), they concentrate and, when their concentration has reached the saturation point, their precipitation and the consequent formation of crystals occur. This is the only known chemical mechanism which may explain their presence in the kidney sample. Their growth occluded the tubules at various levels, thus causing the occlusion of the renal tract with the consequent anuria. Their pointed shape could explain the intestinal perforation observed.

It may be interesting to note that similar crystals were visible in post-mortem samples of lungs, liver, heart, brain and cerebellum (Figs. 13.11 and 13.12), where they broke the morphology and organ structures.

13.2.7 Case: Vaccine Purity

The topic in the field of drugs we dealt more with and for a longer time is that of vaccines.

At the time of writing this chapter we have analyzed 31 different vaccines coming from Italy, France and Switzerland, sometimes in just one sample, often in a

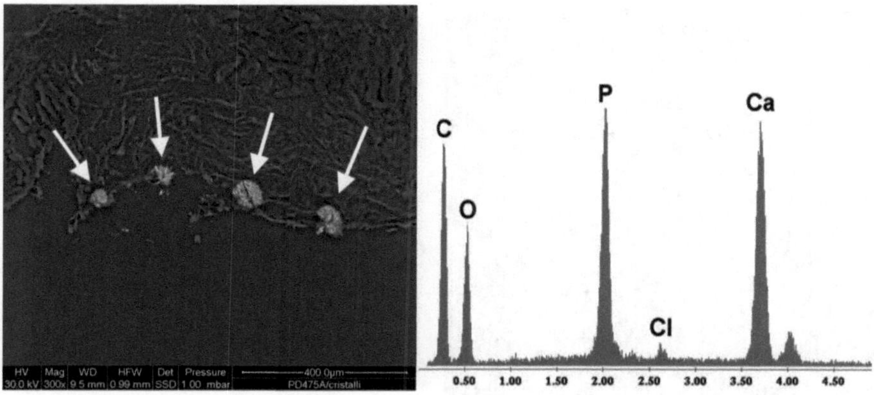

Fig. 13.11 SEM image and EDS spectrum of a heart tissue sample showing calcium phosphate crystals

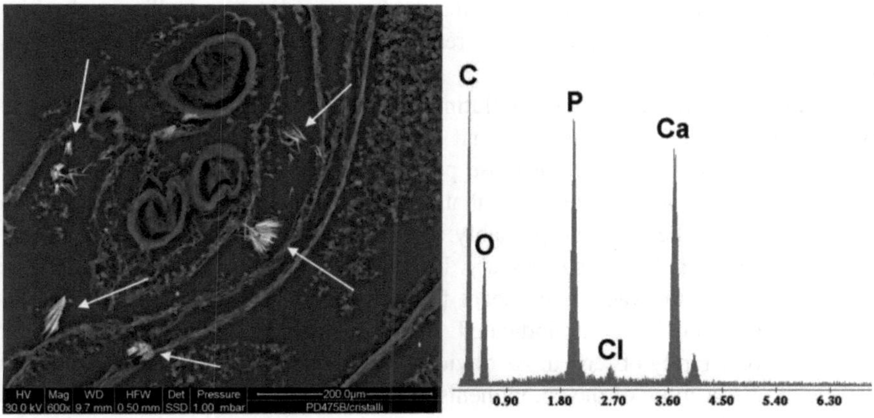

Fig. 13.12 SEM image and EDS spectrum of a brain sample showing the presence of calcium phosphate needle-shaped crystals

number of samples and, for example as far as flu vaccines is concerned, in the course of several years. Our result indicate that all human-use samples are more or less heavily polluted by inorganic micro- and nanoparticles. The only sample that was free from particulate contaminants was an animal-use vaccine (Feligen by Virbac). [6]

It is not our intention to discuss the utility, effectiveness and side effects of these much-discussed drugs. The only things we can say, proof in hand, is that they contain pollutants that are incompatible with the human body and are, therefore, potentially pathogenic. We don't know where those pollutants come from, although it is likely that stainless-steel and tungsten particles could come from dust that could be contained inside the needle, originating from the sharpening process. It is only

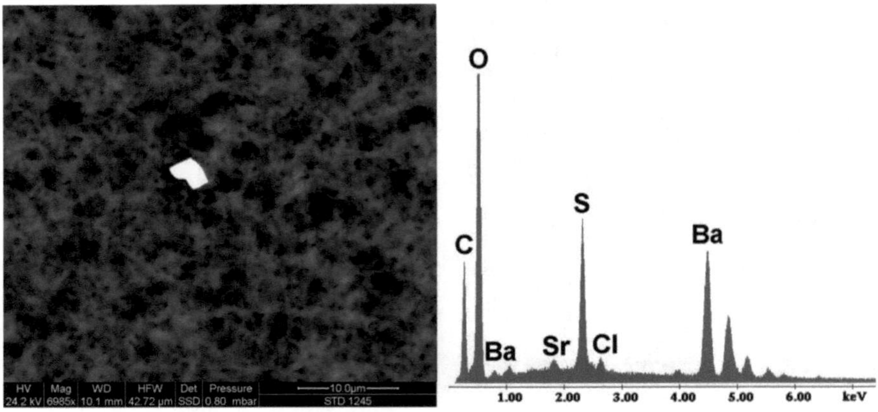

Fig. 13.13 SEM-EDS analysis showing a particle of barium sulphate entrapped in a vaccine drop

evident that the controls performed by the producers and the controlling authorities are not sufficiently careful. Also, their reactions to the complaint of these findings have been disappointing.

Arguing that the amount of polluting particles is small means ignoring the fundamentals of nanopathology. What should be done, instead, is repeat the investigation, trace the source of those particles and produce clean vaccines.

Good medical practice requires that, before administering a vaccine, some checks are done. The first, and probably more obvious among them as it is for any drug, is making sure that it is necessary. If the subject has already suffered from the infectious disease the vaccine is meant for, the administration is not only unnecessary but is thoroughly contraindicated. A simple and indisputable precaution: yet this rule is very rarely observed. As obvious is also checking that the subject is not allergic to one or more of the components of the drug. Again, also that basic rule is never observed. Whatever the reasons, vaccinations are always practiced blindly, and that increases the risk of adverse reactions. Special attention should be given to infants. Apart from any other consideration, their microbiota is quite different from that of adults and interference with the many components of vaccines, including mercury (generally present for production reasons), aluminum (added as an adjuvant) and particulate pollutants (of unknown origin), is something that is the potential source of very serious health problems, especially neurological ones.

Figures 13.13, 13.14 and 13.15 show the presence of particulate-matter contamination in different vaccines: flu, hexavalent: diphtheria, tetanus, pertussis, poliomyelitis, Haemophilus influenzae type B and hepatitis B, and tetravalent: diphtheria, tetanus, pertussis (acellular) and hepatitis B. The chemical composition of this contamination is not fully understood. We think that during the vaccine preparation an involuntary contamination can occur. In many vaccines there is an important quantity of Aluminum hydroxide or phosphate or Aluminum nanoparticles, but there is no scientific study of the fate of this element in the human bosy and it possible interactions with proteins, enzymes or cells.

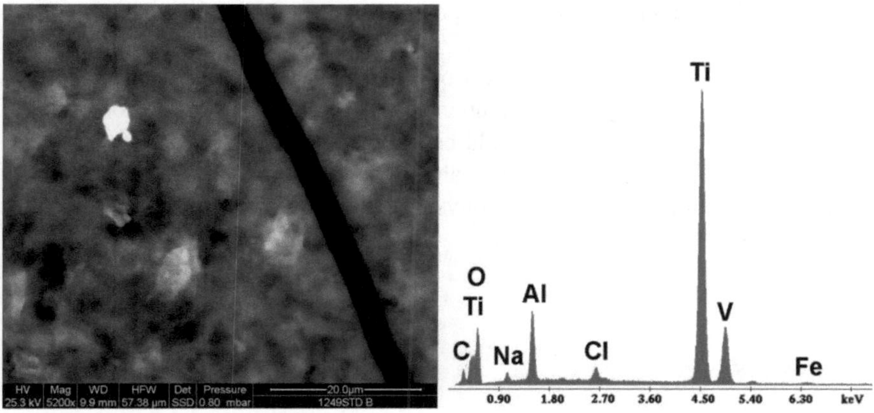

Fig. 13.14 SEM-EDS analysis showing a particle of an alloy of titanium-aluminum-vanadium-iron in a hexavalent vaccine

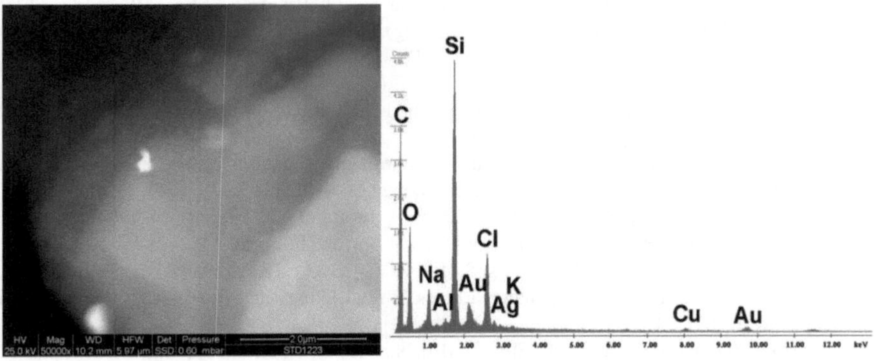

Fig. 13.15 SEM-EDS analysis showing 2 particles in a tetravalent vaccine drop, composed of silicon-sodium-aluminum-gold-chlorine-silver-copper

13.3 Conclusions

It is not our intention to discuss the role and use of curative and preventive drugs. Our aim is to attract attention to some objective problems involving internal-use medications. It is our opinion that

1. The responsible international and governmental agencies should reconsider the real capacity of human organs to cope with components and, more in particular, with potential contaminants of pharmaceutical products for internal use.
2. More attention should be paid by manufacturers to the design, manufacture and indications of pharmaceutical products for internal use, so that safety for the

users can be guaranteed. Especially, potential presence of contaminants deserves attention.

3. Testing of pharmaceutical products on the presence of particulate matter (e.g. stainless steel) and inorganic components that might lead to the formation of harmful particles in patients should be improved.
4. Patients should be better warned about the potential dangers of misuse of pharmaceutical products by their physician and/or pharmacist, as well as in the documentation of the medicament.

References

1. Elder A, Oberdörster G (2006) Translocation and effects of ultrafine particles outside of the lung. Rev Clin Occup Environ Med 5(4):785–796
2. Elder A, Gelein R, Silva V, Feikert T, Opanashuk L, Carter J, Potter R, Maynard A, Ito Y, Finkelstein J, Oberdörster G (2006) Translocation of inhaled ultrafine manganese oxide particles to the central nervous system. Environ Health Perspect 114(8, Aug):1172–1178; Erratum in: Environ Health Perspect 114(8, Aug):1178
3. EU Project:Nanopathology FP4-QOL-2002-147
4. EU Project: DIPNA-FP5-NMP-2006-9-03231
5. Gatti A, Montanari S (2008) Nanopathology: the health impact of nanoparticles, PanStanford Publishing Pte.Ltd, Singapore, pp 1–298
6. Gatti AM, Montanari S (2017) New quality-control investigations on vaccines: micro- and nanocontamination. Int J of Vaccines and Vaccination 4(1): 1–13
7. Gatti A, Quaglino D, Sighinolfi GL (2009) A morphological approach to monitor the nanoparticle-cell interaction. Int J Imaging 2(S09 Spring (Editorial 1)):2–21
8. Gatti A, Bosco P, Rivasi F, Bianca S, Ettore G, Gaetti L, Montanari S, Bartoloni G, Gazzolo D (2011) Heavy metals nanoparticles in fetal kidney and liver tissues. Front Biosci (Elite Ed, E3) 1:221–226
9. Hansen T, Clermont G, Alves A, Eloy R, Brochhausen C, Boutrand JP, Gatti A, Kirkpatrick (2006) Biological tolerance of different materials in bulk and nanoparticulate form in a rat model: sarcoma development by nanoparticles. J R Soc Interface 3:767–775
10. http://www.impfkritik.de/pressespiegel/2017020402.html?cmode=form&creplyto=8910
11. Iannitti T, Capone S, Gatti A, Capitani F, Francesco C, Palmieri B (2010) Intracellular heavy metal nanoparticle storage: progressive accumulation within lymph nodes with transformation from chronic inflammation to malignancy. Int J Nanomed 5:955–960
12. Lucarelli M, Gatti A, Savarino G, Quattroni P, Martinelli L, Monari E, Boraschi D (2004) Innate defence function of macrophages can be biased by nano-sized ceramic and metallic particles. Cytokin Netw 15(4):339–346
13. Oberdörster G, Sharp Z, Atudorei V, Elder A, Gelein R, Kreyling W, Cox C (2004) Translocation of inhaled ultrafine particles to the brain. Inhal Toxicol 16(6–7, Jun):437–445
14. Schlegl R, Weber M, Wruss J, Low D, Queen K, Stilwell S, Lindblad EB, Möhl M (2015) Influence of elemental impurities in aluminum hydroxide adjuvant on the stability of inactivated Japanese Encephalitis vaccine. IXIARO Vaccine 33(44):5989–5996

Antonietta Morena Gatti has been in charge of the Laboratory of Biomaterials at the University of Modena (Italy) for 32 years, was Associate Professor at the National Council of Research of Italy, Institute of Science and Technology of Ceramic Materials and she is President of the Health, Law and Science Association in Switzerland. In 2012 she was awarded and listed among the International Fellows of the Societies of Biomaterials and Engineering. She founded Nanodiagnostics, a research company whose main missions are investigating on nanopathologies and performing forensic investigations on the exposure to nanosized contamination. She is a consultant to the Italian Government's Commissions for the pathologies contracted by the soldiers engaged in war (or peace) missions (Depleted-Uranium diseases). Dr. Gatti is also Visiting Professor at International Clean Water Institute (Washington, USA), an Institute of the US Department of State, and was a consultant to the European Science Foundation. She was in charge of many national and international Projects: EU Project Nanopathology, EU Project DIPNA of nanotoxicity and INESE, a nanoecotoxicological project of the Italian Institute of Nantotechnology. She had many collaborations for research with the Ministry of Defence (Batnan, Venam projects). Dr. Gatti is the discoverer of "nanopathologies", i.e. the pathologies due to engineered or incidental nanoparticles that can cause illnesses like cancer and other diseases whose origin is still largely unknown. Her pioneering studies were supported by the European Commission, by the Italian Institute of Technology and by the Italian Ministry of Defense. She was active as international coordinator of many research projects. For her studies she was invited twice at the House of Lords in London for a hearing. Dr. Gatti is the author of more than 250 articles, 2 books and 7 chapters in other author's books. She belongs to the editorial Board of the Int. J. of Nanomedicine and Research and is a referee to other international journals. She was member of a few European Commissions and NGOs (ANEC, ECOS) and worked at ISO and OECD for the Nanomaterials Standardization.

Stefano Montanari has a Degree in Pharmacology, University of Modena (1972) (Italy). He is the Co-founder and CEO of Nanodiagnostics.srl, a EU start-up. He was scientific Consultant of BOSA S.p.A., Milan for cardiosurgery, cardiology and haemodialysis. Scientific Consultant of Angiocor, Lille (France) on vena cava filters, Scientific Director of the Biosa proiect for the development of a cardiac valve with bovine pericardium and scientific Consultant of ALN—Bormes les Mimosas (France) for the development of a temporary vena cava filter and for the CUPID project (NewYork, USA) for an automatic system to identify cardiac pathologies. Dr. Montanari spent 1-month Stage at Malmo University Hospital, Nephrology and haemodialysis; 1-month Stage in cardiosurgery at St. Joseph's Hospital and Medical Center (Phoenix, Arizona) Prof. E. Dietrich; 2-month Stage in cardiosurgery at the Karolinska Institut (Stockholm, Sweden) Prof. Viking O. Björk. He was Lecturer of Citometry (University Carlo Bo, Urbino, Italy) and Lecturer of nanotechnologies and nanopathologies at IFOA School (Reggio Emilia, Italy). He held seminars about vena cava filters at various universities and hospitals. At present he is Scientific Director of the laboratory of Research Nanodiagnostics (Modena). He wrote 8 books in Italian and was co-author with Dr. Gatti of 2 books (Nanopathology: the health impact of nanoparticles, Pan Stanford Pub. Singapore, 2008; 1–298 and Case studies in nanotoxicology and particle toxicology, Elsevier, USA, 2015;1–260).

Index

H. G. Merkus et al. (eds.), *Particles and Nanoparticles in Pharmaceutical Products*,
AAPS Advances in the Pharmaceutical Sciences Series 29,
https://doi.org/10.1007/978-3-319-94174-5

MIX
Papier aus verantwortungsvollen Quellen
Paper from responsible sources
FSC® C105338

If you have any concerns about our products,
you can contact us on
ProductSafety@springernature.com

In case Publisher is established outside the EU,
the EU authorized representative is:
Springer Nature Customer Service Center GmbH
Europaplatz 3, 69115 Heidelberg, Germany

Printed by Libri Plureos GmbH
in Hamburg, Germany